T0190300

Communications
in Computer and Information Science　　　2151

Rationale

The CCIS series is devoted to the publication of proceedings of computer science conferences. Its aim is to efficiently disseminate original research results in informatics in printed and electronic form. While the focus is on publication of peer-reviewed full papers presenting mature work, inclusion of reviewed short papers reporting on work in progress is welcome, too. Besides globally relevant meetings with internationally representative program committees guaranteeing a strict peer-reviewing and paper selection process, conferences run by societies or of high regional or national relevance are also considered for publication.

Topics

The topical scope of CCIS spans the entire spectrum of informatics ranging from foundational topics in the theory of computing to information and communications science and technology and a broad variety of interdisciplinary application fields.

Information for Volume Editors and Authors

Publication in CCIS is free of charge. No royalties are paid, however, we offer registered conference participants temporary free access to the online version of the conference proceedings on SpringerLink (http://link.springer.com) by means of an http referrer from the conference website and/or a number of complimentary printed copies, as specified in the official acceptance email of the event.

CCIS proceedings can be published in time for distribution at conferences or as post-proceedings, and delivered in the form of printed books and/or electronically as USBs and/or e-content licenses for accessing proceedings at SpringerLink. Furthermore, CCIS proceedings are included in the CCIS electronic book series hosted in the SpringerLink digital library at http://link.springer.com/bookseries/7899. Conferences publishing in CCIS are allowed to use Online Conference Service (OCS) for managing the whole proceedings lifecycle (from submission and reviewing to preparing for publication) free of charge.

Publication process

The language of publication is exclusively English. Authors publishing in CCIS have to sign the Springer CCIS copyright transfer form, however, they are free to use their material published in CCIS for substantially changed, more elaborate subsequent publications elsewhere. For the preparation of the camera-ready papers/files, authors have to strictly adhere to the Springer CCIS Authors' Instructions and are strongly encouraged to use the CCIS LaTeX style files or templates.

Abstracting/Indexing

CCIS is abstracted/indexed in DBLP, Google Scholar, EI-Compendex, Mathematical Reviews, SCImago, Scopus. CCIS volumes are also submitted for the inclusion in ISI Proceedings.

How to start

To start the evaluation of your proposal for inclusion in the CCIS series, please send an e-mail to ccis@springer.com.

Andrew M. Olney · Irene-Angelica Chounta ·
Zitao Liu · Olga C. Santos · Ig Ibert Bittencourt
Editors

Artificial Intelligence in Education

Posters and Late Breaking Results, Workshops and Tutorials, Industry and Innovation Tracks, Practitioners, Doctoral Consortium and Blue Sky

25th International Conference, AIED 2024
Recife, Brazil, July 8–12, 2024
Proceedings, Part II

Springer

Editors
Andrew M. Olney 🆔
University of Memphis
Memphis, TN, USA

Irene-Angelica Chounta 🆔
University of Duisburg-Essen
Duisburg, Germany

Zitao Liu 🆔
Jinan University
Guangzhou, China

Olga C. Santos 🆔
UNED
Madrid, Spain

Ig Ibert Bittencourt 🆔
Universidade Federal de Alagoas
Maceio, Brazil

ISSN 1865-0929 ISSN 1865-0937 (electronic)
Communications in Computer and Information Science
ISBN 978-3-031-64311-8 ISBN 978-3-031-64312-5 (eBook)
https://doi.org/10.1007/978-3-031-64312-5

This Springer imprint is published by the registered company Springer Nature Switzerland AG
The registered company address is: Gewerbestrasse 11, 6330 Cham, Switzerland

If disposing of this product, please recycle the paper.

Preface

The 25th International Conference on Artificial Intelligence in Education (AIED 2024) was hosted by Centro de Estudos e Sistemas Avançados do Recife (CESAR), Brazil from July 8 to July 12, 2024. It was set up in a face-to-face format but included an option for an online audience. AIED 2024 was the next in a longstanding series of annual international conferences for the presentation of high-quality research on intelligent systems and the cognitive sciences for the improvement and advancement of education. Note that AIED is ranked A in CORE (top 16% of all 783 ranked venues), the well-known ranking of computer science conferences. The AIED conferences are organized by the prestigious International Artificial Intelligence in Education Society, a global association of researchers and academics, which has already celebrated its 30th anniversary, and aims to advance the science and engineering of intelligent human-technology ecosystems that support learning by promoting rigorous research and development of interactive and adaptive learning environments for learners of all ages across all domains.

The theme for the AIED 2024 conference was "AIED for a World in Transition". The conference aimed to explore how AI can be used to enhance the learning experiences of students and teachers alike when disruptive technologies are turning education upside down. Rapid advances in Artificial Intelligence (AI) have created opportunities not only for personalized and immersive experiences but also for ad hoc learning by engaging with cutting-edge technology continually, extending classroom borders, from engaging in real-time conversations with large language models (LLMs) to creating expressive artifacts such as digital images with generative AI or physically interacting with the environment for a more embodied learning. As a result, we now need new approaches and measurements to harness this potential and ensure that we can safely and responsibly cope with a world in transition. The conference seeks to stimulate discussion of how AI can shape education for all sectors, how to advance the science and engineering of AI-assisted learning systems, and how to promote broad adoption.

AIED 2024 attracted broad participation. We received 334 submissions for the main program, of which 280 were submitted as full papers, and 54 were submitted as short papers. Of the full paper submissions, 49 were accepted as full papers, and another 27 were accepted as short papers. The acceptance rate for full papers and short papers together was 23%. These accepted contributions are published in the Springer proceedings volumes LNAI 14829 and 14830.

The submissions underwent a rigorous double-masked peer-review process aimed to reduce evaluation bias as much as possible. The first step of the review process was done by the program chairs, who verified that all papers were appropriate for the conference and properly anonymized. Program committee members were asked to declare conflicts of interest. After the initial revision, the program committee members were invited to bid on the anonymized papers that were not in conflict according to their declared conflicts of interest. With this information, the program chairs made the review assignment, which consisted of three regular members to review each paper plus a senior member to provide a meta-review. The management of the review process (i.e., bidding, assignment, discussion, and meta-review) was done with the EasyChair platform, which was configured so that reviewers of the same paper were anonymous to each other. A subset of the program committee members were not included in the initial assignment but were asked to be ready to do reviews that were not submitted on time (i.e., the emergency review period). To avoid a situation where program committee members would be involved in too many submissions, we balanced review assignments and then rebalanced them during the emergency review period.

As a result, each submission was reviewed anonymously by at least three Program Committee (PC) members and then a discussion was led by a Senior Program Committee (SPC) member. PC and SPC members were selected based on their authorship in previous AIED conferences, their experience as reviewers in previous AIED editions, their h-index as calculated by Google Scholar, and their previous positions in organizing and reviewing related conferences. Therefore all members were active researchers in the field, and SPC members were particularly accomplished on these metrics. SPC members served as meta-reviewers whose role was to seek consensus to reach the final decision about acceptance and to provide the corresponding meta-review. They were also asked to check and highlight any possible biases or inappropriate reviews. Decisions to accept/reject were taken by the program chairs. For borderline cases, the contents of the paper were read in detail before reaching the final decision. In summary, we are confident that the review process assured a fair and equal evaluation of the submissions received without any bias, as far as we are aware.

Beyond paper presentations, the conference included a Doctoral Consortium Track, Late-Breaking Results, a Workshops and Tutorials Track, and an Industry, Innovation and Practitioner Track. There was a WideAIED track, which was established in 2023, where opportunities and challenges of AI in education were discussed with a global perspective and with contributions coming also from areas of the world that are currently under-represented in AIED. Additionally, a BlueSky special track was included with contributions that reflect upon the progress of AIED so far and envision what is to come in the future. The submissions for all these tracks underwent a rigorous peer review process. Each submission was reviewed by at least three members of the AIED community, assigned by the corresponding track organizers who then took the final decision about acceptance. The participants of the conference had the opportunity to attend three keynote talks: "Navigating Strategic Challenges in Education in the Post-Pandemic AI Era" by Blaženka Divjak, "Navigating the Evolution: The Rising Tide of Large Language Models for AI and Education" by Peter Clark, and "Artificial Intelligence in Education and Public Policy: A Case from Brazil" by Seiji Isotani. These contributions are published in the Springer proceedings volumes CCIS 2150 and 2151.

The conference also included a Panel with experts in the field and the opportunity for the participants to present a demonstration of their AIED system in a specific session of Interactive Events. A selection of the systems presented is included as showcases on the web page of the IAIED Society[1]. Finally, there was a session with presentations of papers published in the International Journal of Artificial Intelligence in Education[2],

---

[1] https://iaied.org/showcase.
[2] https://link.springer.com/journal/40593.

the journal of the IAIED Society indexed in the main databases, and a session with the best papers from conferences of the International Alliance to Advance Learning in the Digital Era (IAALDE)[3], an alliance of research societies that focus on advances in computer-supported learning, to which the IAIED Society belongs.

For making AIED 2024 possible, we thank the AIED 2024 Organizing Committee, the hundreds of Program Committee members, the Senior Program Committee members, and the AIED proceedings chairs Paraskevi Topali and Rafael D. Araújo. In addition, we would like to thank the Executive Committee of the IAIED Society for their advice during the conference preparation, and specifically two of the working groups, the Conference Steering Committee, and the Diversity and Inclusion working group. They all gave their time and expertise generously and helped with shaping a stimulating AIED 2024 conference. We are extremely grateful to everyone!

July 2024

Andrew M. Olney
Irene-Angelica Chounta
Zitao Liu
Olga C. Santos
Ig Ibert Bittencourt

---

[3] https://alliancelss.com/.

Organization

Conference General Co-chairs

Olga C. Santos — UNED, Spain
Ig Ibert Bittencourt — Universidade Federal de Alagoas, Brazil

Program Co-chairs

Andrew M. Olney — University of Memphis, USA
Irene-Angelica Chounta — University of Duisburg-Essen, Germany
Zitao Liu — Jinan University, China

Doctoral Consortium Co-chairs

Yu Lu — Beijing Normal University, China
Elaine Harada T. Oliveira — Universidade Federal do Amazonas, Brazil
Vanda Luengo — Sorbonne Université, France

Workshop and Tutorials Co-chairs

Cristian Cechinel — Federal University of Santa Catarina, Brazil
Carrie Demmans Epp — University of Alberta, Canada

Interactive Events Co-chairs

Leonardo B. Marques — Federal University of Alagoas, Brazil
Ben Nye — University of Southern California, USA
Rwitajit Majumdar — Kumamoto University, Japan

Industry, Innovation and Practitioner Co-chairs

Diego Dermeval   Federal University of Alagoas, Brazil
Richard Tong   IEEE Artificial Intelligence Standards
                                        Committee, USA
Sreecharan Sankaranarayanan   Amazon, USA
Insa Reichow   German Research Center for Artificial
                                        Intelligence, Germany

Posters and Late Breaking Results Co-chairs

Marie-Luce Bourguet   Queen Mary University of London, UK
Qianru Liang   Jinan University, China
Jingyun Wang   Durham University, UK

Panel Co-chairs

Julita Vassileva   University of Saskatchewan, Canada
Alexandra Cristea   Durham University, UK

Blue Sky Co-chairs

Ryan S. Baker   University of Pennsylvania, USA
Benedict du Boulay   University of Sussex, UK
Mirko Marras   University of Cagliari, Italy

WideAIED Co-chairs

Isabel Hilliger   Pontificia Universidad Católica, Chile
Marco Temperini   Sapienza University of Rome, Italy
Ifeoma Adaji   University of British Columbia, Canada
Maomi Ueno   University of Electro-Communications, Japan

Local Organising Co-chairs

Rafael Ferreira Mello	Universidade Federal Rural de Pernambuco, Brazil
Taciana Pontual	Universidade Federal Rural de Pernambuco, Brazil

AIED Mentoring Fellowship Co-chairs

Amruth N. Kumar	Ramapo College of New Jersey, USA
Vania Dimitrova	University of Leeds, UK

Diversity and Inclusion Co-chairs

Rod Roscoe	Arizona State University, USA
Kaska Porayska-Pomsta	University College London, UK

Virtual Experiences Co-chairs

Guanliang Chen	Monash University, Australia
Teng Guo	Jinan University, China
Eduardo A. Oliveira	University of Melbourne, Australia

Publicity Co-chairs

Son T. H. Pham	Nha Viet Institute, USA
Pham Duc Tho	Hung Vuong University, Vietnam
Miguel Portaz	UNED, Spain

Volunteer Co-chairs

Isabela Gasparini	Santa Catarina State University, Brazil
Lele Sha	Monash University, Australia

Proceedings Co-chairs

Paraskevi Topali   National Education Lab AI, Radboud University,
                                Netherlands
Rafael D. Araújo   Federal University of Uberlândia, Brazil

Awards Co-chairs

Ning Wang   University of Southern California, USA
Beverly Woolf   University of Massachusetts, USA

Sponsorship Chair

Tanci Simões Gomes   CESAR, Brazil

Scholarship Chair

Patrícia Tedesco   Universidade Federal de Pernambuco, Brazil

Steering Committee

Noboru Matsuda   North Carolina State University, USA
Eva Millan   Universidad de Málaga, Spain
Sergey Sosnovsky   Utrecht University, Netherlands
Ido Roll   Israel Institute of Technology, Israel
Maria Mercedes T. Rodrigo   Ateneo de Manila University, Philippines

Blue Sky Track Program Committee Members

Giora Alexandron   Weizmann Institute of Science, Israel
Gautam Biswas   Vanderbilt University, USA
Mingyu Feng   WestEd, USA
Neil Heffernan   Worcester Polytechnic Institute, USA
Sharon Hsiao   Santa Clara University, USA
Kenneth Koedinger   Carnegie Mellon University, USA
Roberto Martinez-Maldonado Monash University, Australia

Gordon McCalla	University of Saskatchewan, Canada
Agathe Merceron	BHT Berlin, Germany
Tanja Mitrovic	University of Canterbury, New Zealand
Luc Paquette	University of Illinois at Urbana-Champaign, USA
Anna Rafferty	Carleton College, USA
Ana Serrano Mamolar	University of Burgos, Spain
Beverly Woolf	University of Massachusetts, USA

WideAIED Track Program Committee Members

Ifeoma Adaji	University of British Columbia, Canada
Gabriel Astudillo	Pontificia Universidad Católica de Chile, Chile
Paolo Fantozzi	LUMSA University, Italy
Kazuma Fuchimoto	University of Electro-Communications, Japan
Shili Ge	Guangdong University of Foreign Studies, China
Isabel Hilliger	Pontificia Universidad Católica de Chile, Chile
Zuzana Kubincová	Comenius University, Slovakia
Carla Limongelli	Roma Tre University, Italy
Ivana Marenzi	L3S Research Center, Germany
Ronald Pérez-Álvarez	Universidad de Costa Rica, Costa Rica
Mar Perez-Sanagustin	Université Paul Sabatier Toulouse III, France
Valerio Rughetti	Università Telematica Internazionale Uninettuno, Italy
Daniele Schicchi	ITD, National Council of Research, Italy
Filippo Sciarrone	Universitas Mercatorum, Italy
Andrea Sterbini	Sapienza University of Rome, Italy
Davide Taibi	ITD, National Council of Research, Italy
Juan Andrés Talamás	Tecnológico de Monterrey, Mexico
Marco Temperini	Sapienza University of Rome, Italy
Maomi Ueno	University of Electro-Communications, Japan
Giacomo Valente	University of L'Aquila, Italy
Ignacio Villagran	Pontificia Universidad Católica de Chile, Chile
Esteban Villalobos	Université Paul Sabatier, France

Late-Breaking Results Track Program Committee Members

Ifeoma Adaji	University of British Columbia, Canada
Mohammad Alshehri	University of Jeddah, Saudi Arabia
Pablo Arnau González	Universidad de Valencia, Spain
Yuya Asano	University of Pittsburgh, USA

Gabriel Astudillo	Pontificia Universidad Católica de Chile, Chile
Francisco Leonardo	.
Aureliano Ferreira	UFERSA, Brazil
Shiva Baghel	Extramarks Education India Pvt. Ltd., India
Rabin Banjade	University of Memphis, USA
Sami Baral	Worcester Polytechnic Institute, USA
Hemilis Joyse Barbosa Rocha	UFPE, Brazil
Luca Benedetto	University of Cambridge, UK
Raul Benites Paradeda	State University of Rio Grande do Norte, Brazil
Jiang Bo	East China Normal University, China
Diana Borza	Babeş-Bolyai University, Romania
Matthieu Branthome	Université de Bretagne Occidentale CREAD, France
Minghao Cai	University of Alberta, Canada
May Kristine Jonson Carlon	RIKEN Center for Brain Science, Japan
Liangyu Chen	East China Normal University, China
Jade Cock	EPFL, Switzerland
Clayton Cohn	Vanderbilt University, USA
Syaamantak Das	Indian Institute of Technology Bombay, India
Deep Dwivedi	Indraprastha Institute of Information Technology, India
Yo Ehara	Tokyo Gakugei University, Japan
Abul Ehtesham	The Davey Tree Expert Company, USA
Lara Ceren Ergenç	King's College London, UK
Luyang Fang	University of Georgia, USA
Zechu Feng	University of Hong Kong, China
Rafael Ferreira Mello	CESAR School, Brazil
Ritik Garg	Extramarks Education Pvt. Ltd., India
Sahin Gokcearslan	Gazi University, Turkey
Sara Guerreiro-Santalla	Universidade da Coruña, Spain
Shivang Gupta	Carnegie Mellon University, USA
Songhee Han	University of Texas at Austin, USA
John Hollander	University of Memphis, USA
Xinying Hou	University of Michigan, USA
Mingliang Hou	Dalian University of Technology, China
Rositsa V. Ivanova	University of St. Gallen, Switzerland
Yuan-Hao Jiang	East China Normal University, China
Stamos Katsigiannis	Durham University, UK
Ekaterina Kochmar	MBZUAI, UAE
Elizabeth Koh	Nanyang Technological University, Singapore
Tatsuhiro Konishi	Shizuoka University, Japan

Marco Kragten	Amsterdam University of Applied Science, Netherlands
Ehsan Latif	University of Georgia, USA
Morgan Lee	Worcester Polytechnic Institute, USA
Junyoung Lee	Nanyang Technological University, Singapore
Liang-Yi Li	National Taiwan Normal University, Taiwan
Mengting Li	University of Hong Kong, China
Jianwei Li	Beijing University of Posts and Telecommunications, China
Jeongki Lim	Parsons School of Design, USA
Yu Lu	Beijing Normal University, China
Pascal Muam Mah	AGH University of Krakow, Poland
Leonardo Brandão Marques	University of São Paulo, Brazil
Paola Mejia Domenzain	EPFL, Switzerland
Fatma Miladi	MIRACL, Tunisia
Yoshimitsu Miyazawa	National Center for University Entrance Examinations, Japan
Yasuhiro Noguchi	Shizuoka University, Japan
Ange Adrienne Nyamen Tato	École de Technologie Supérieure, France
Chaohua Ou	Georgia Institute of Technology, USA
Qianqian Pan	Nanyang Technological University, Singapore
Andrew Petersen	University of Toronto, Canada
Vijay Prakash	Indian Institute of Technology Bombay, India
Ethan Prihar	École polytechnique fédérale de Lausanne, Switzerland
Kun Qian	Columbia University, USA
Emanuel Queiroga	Federal University of Pelotas, Brazil
Krishna Chaitanya Rao Kathala	University of Massachusetts Amherst, USA
Vasile Rus	University of Memphis, USA
Martin Ruskov	University of Milan, Italy
Sabrina Schork	UAS Aschaffenburg, Germany
Sabine Seufert	University of St. Gallen, Switzerland
Mohammed Seyam	Virginia Tech, USA
Atsushi Shimada	Kyushu University, Japan
Kazutaka Shimada	Kyushu Institute of Technology, Japan
Jinnie Shin	University of Florida, USA
Aditi Singh	Cleveland State University, USA
Anuradha Singh	Banaras Hindu University, India
Jasbir Singh	University of Otago, New Zealand
Kelly Smalenberger	Belmont Abbey College, USA
Michael Smalenberger	University of North Carolina at Charlotte, USA

Álvaro Sobrinho	Federal University of Pernambuco, Brazil
Aakash Soni	ECE Paris, France
Victor Sotelo	UNICAMP, Brazil
Zhongtian Sun	University of Cambridge, UK
Rushil Thareja	IIIT Delhi, India
Danielle R. Thomas	Carnegie Mellon University, USA
Liu Tianrui	University of Sydney, Australia
Anne Trumbore	University of Virginia, USA
Maomi Ueno	University of Electro-Communications, Japan
Jim Wagstaff	Noodle Factory Learning, Singapore
Deliang Wang	Beijing Normal University, China
Jiahui Wu	Beijing University of Posts and Telecommunication, China
Yuhang Wu	University of Hong Kong, China
Yu Xiong	Chongqing University of Posts and Telecommunications, China
Mingrui Xu	Beijing University of Posts and Telecommunications, China
R. Yamamoto Ravenor	Ochanomizu University, Japan
Koichi Yamashita	Tokoha University, Japan
Andres Felipe Zambrano	University of Pennsylvania, USA
Kamyar Zeinalipour	University of Siena, Italy
Ivy Chenjia Zhu	University of Hong Kong, China

Industry, Innovation, and Practitioner Track Program Committee Members

Cristian Cechinel	Federal University of Santa Catarina, Brazil
Geiser Challco	Federal University of the Semi-arid Region, Brazil
Rafael Ferreira Mello	Universidade Federal Rural de Pernambuco, Brazil
Elyda Freitas	University of Pernambuco, Brazil
Huy Nguyen	Carnegie Mellon University, USA
Ranilson Paiva	Federal University of Alagoas, Brazil
Cleon Xavier Pereira Júnior	Federal Institute of Goiás, Brazil
Faisal Rachid	German Research Center for Artificial Intelligence, Germany
Luiz Rodrigues	Federal University of Alagoas, Brazil
Shashi Kant Shankar	Amrita University, India
Tiago Thompsen Primo	Federal University of Pelotas, Brazil

International Artificial Intelligence in Education Society

Management Board

President

Olga C. Santos   UNED, Spain

Secretary/Treasurer

Vania Dimitrova   University of Leeds, UK

Journal Editors

Vincent Aleven   Carnegie Mellon University, USA
Judy Kay   University of Sydney, Australia

Finance Chair

Benedict du Boulay   University of Sussex, UK

Membership Chair

Benjamin D. Nye   University of Southern California, USA

IAIED Officers

Yancy Vance Paredes   North Carolina State University, USA
Son T. H. Pham   Nha Viet Institute, USA
Miguel Portaz   UNED, Spain

Executive Committee

Akihiro Kashihara   University of Electro-Communications, Japan
Amruth Kumar   Ramapo College of New Jersey, USA
Christothea Herodotou   Open University, UK
Jeanine A. Defalco   CCDC-STTC, USA
Judith Masthoff   Utrecht University, Netherlands
Maria Mercedes T. Rodrigo   Ateneo de Manila University, Philippines
Ning Wang   University of Southern California, USA

Olga Santos	UNED, Spain
Rawad Hammad	University of East London, UK
Zitao Liu	Jinan University, China
Bruce M. McLaren	Carnegie Mellon University, USA
Cristina Conati	University of British Columbia, Canada
Diego Zapata-Rivera	Educational Testing Service, Princeton, USA
Erin Walker	University of Pittsburgh, USA
Seiji Isotani	University of São Paulo, Brazil
Tanja Mitrovic	University of Canterbury, New Zealand
Noboru Matsuda	North Carolina State University, USA
Min Chi	North Carolina State University, USA
Alexandra I. Cristea	Durham University, UK
Neil Heffernan	Worcester Polytechnic Institute, USA
Andrew M. Olney	University of Memphis, USA
Irene-Angelica Chounta	University of Duisburg-Essen, Germany
Beverly Park Woolf	University of Massachusetts, USA
Ig Ibert Bittencourt	Universidade Federal de Alagoas, Brazil

AIED 2024 Panel

Do We Need AIED If We Have Powerful Generative AI?

Julita Vassileva[1] (iD) and Alexandra I. Cristea [2] (iD)

[1]University of Saskatchewan, Saskatoon, SK, Canada
jiv@cs.usask.ca
[2]University of Durham, Durham, UK
alexandra.i.cristea@durham.ac.uk

Abstract. The advent of advanced AI systems, particularly generative pre-trained transformers (GPTs) and large language models (LLMs) like ChatGPT, Gemini, Claude, has significantly changed the landscape of artificial intelligence and its applications across various domains. GPTs have demonstrated impressive capabilities in generating human-like text and solving complex problems. Educators are finding creative ways to deploy GPTs systems in their practice [1]. Many schools are struggling with the question of whether to allow or ban the use of AI, because of the potential risks (overreliance, academic integrity). The anticipation of change in the education system is palpable. What does the future hold? This raises important questions about the role of artificial intelligence in education (AIED) research. This panel will discuss the reasons why and ways in which AIED research should or should not remain essential in these transformative times for education, as well as ways in which it should change.

Keywords: AI systems · generative AI · LLMs · AIED

1 Panel Description

Generative AI has entered Education. Many teachers are finding creative ways to deploy the myriads of GPT applications in their classes; schools are debating whether to ban or allow the use of generative AI tools. Is there a place for AIED research in this new world? There are a lot of reasons for optimism and many open research directions. Education is a deeply individualized process. AIED focuses not just on delivering information but on understanding and adapting to individual learning styles, needs, and learners' emotional states. While GPTs can provide answers and explanations, AIED systems are designed to engage learners more interactively and adaptively, aiming to mimic the nuanced responses of a human teacher. While the deployment of generative AI in educational settings raises concerns about pedagogical effectiveness, AIED systems developed with specific educational theories as underpinning can be tailored to foster not just cognitive skills but also critical thinking, creativity, and emotional intelligence. On the other hand, several papers in the AIED'2024 conference program demonstrate that the research community already incorporates GPTs to power systems that accomplish these aspirations.

However, the AIED community needs to be careful to avoid raising too high expectations. Many of us remember the "AI winter" (a term coined at the AAAI conference in 1984 [2]), which lasted nearly 20 years. According to some [3], Generative AI has reached the 'Peak of Inflated Expectations' in the Gartner Hype Cycle curve in September 2023, where the initial excitement and hype are at their highest. However, the critical concern is whether the generative AI technologies will descend into the 'Trough of Disillusionment,' as past AI technologies have, potentially leading to another AI winter. Many factors can lead to this, including concerns about privacy, misinformation, job displacement, and environmental concerns, leading to public and regulatory backlash. Specific areas of concern in applying generative AI applications in education are the current limitations of GPTs in understanding context, reasoning, and providing explanations based on correlations rather than causal relationships [4]—also, their computational resource intensity. If access to Generative AI systems is limited to populations in affluent regions, educational disparities, inherent biases, digital divides, and inequities will be amplified, and vulnerable workers in specific sectors will be disproportionately displaced. Moreover, disillusionment with deep models may lead to general disillusionment with AI, as society tends to use these terms almost interchangeably [5]. Preventing a new AI winter depends on new research to minimize the algorithmic bias and the computational resources needed to run GPTs, as well as on the global society to address these challenges through improved access, developing regulatory frameworks, and effective methods for public awareness and education.

From an AI in education perspective, there seems to be a need for new research about how good pedagogical practices [6] and clearer pathways to psychological methodological underpinning, such as specific, measurable ways for embedding, extracting and evaluating motivational theories [20] for educational contexts [7] can be applied within GPTs, or implemented alongside GPTs. Do other classic AIED topics, such as learner models [8, 9, 10], semantic web and concept models [11, 12], personalized game-based learning [13] and gamification [14, 15, 16], affect [17] and engagement [7, 8], equity, inclusion and ethics [14, 18], scrutable AIED [19], and low-cost solutions for developing societies [14] still have a place, and if so, how?

The AIED community will need to address these issues. We hope the panel will bring new insights into how to approach them and raise new questions that may lead to future avenues for the community to explore.

References

1. Johnson, M.: Generative AI in Computer Science Education. Commun. ACM (2024)
2. Russell, S.J., Norvig, P.: Artificial Intelligence: A Modern Approach, 2nd edn. Prentice Hall, Upper Saddle River (2023). ISBN 0-13-790395-2

3. Gartner (2023). https://www.gartner.com/en/articles/what-s-new-in-the-2023-gartner-hype-cycle-for-emerging-technologies

4. Feder, A., et al.: CausaLM: causal model explanation through counterfactual language models. Comput. Linguist. **47**(2), 333–386 (2021). https://doi.org/10.1162/coli_a_00404

5. IBM Data and AI team, AI vs. Machine Learning vs. Deep Learning vs. Neural Networks: What's the difference? (2023). https://www.ibm.com/blog/ai-vs-machine-learning-vs-deep-learning-vs-neural-networks/

6. Al-Khresheh, M.H.: Bridging technology and pedagogy from a global lens: Teachers' perspectives on integrating ChatGPT in English language teaching. Comput. Educ. Artif. Intell. **6**, 100218 (2024). https://doi.org/10.1016/j.caeai.2024.100218

7. Cristea, A.I., et al.: The engage taxonomy: SDT-based measurable engagement indicators for MOOCs and their evaluation. User Model. User-Adap. Inter. (2023). https://doi.org/10.1007/s11257-023-09374-x

8. Orji, F. A., and Vassileva, J.: Modeling the impact of motivation factors on students' study strategies and performance using machine learning. J. Educ. Technol. Syst. **52**(2), 274–296 (2023). https://doi.org/10.1177/00472395231191139

9. Li, Z., Shi, L., et al.: Sim-GAIL: a generative adversarial imitation learning approach of student modelling for intelligent tutoring systems. Neural Comput. Appl. **35**, 24369–24388 (2023). https://doi.org/10.1007/s00521-023-08989-w

10. Alrajhi, L., et al.: Solving the imbalanced data issue: automatic urgency detection for instructor assistance in MOOC discussion forums. User Model. User-Adap. Inter. (2023). https://doi.org/10.1007/s11257-023-09381-y

11. Gkiokas, A., Cristea A.I.: Cognitive agents and machine learning by example: representation with conceptual graphs. Comput. Intell. **34**, 603–634. (2018). https://doi.org/10.1111/coin.12167

12. Aljohani, T., Cristea, A.I.: Learners demographics classification on MOOCs during the COVID-19: author profiling via deep learning based on semantic and syntactic representations. Front. Res. Metrics Anal. **6**, 673928–673928 (2021)

13. Kiron, N., Omar, M.T., Vassileva, J.: Evaluating the impact of serious games on study skills and habits. In: European Conference on Games Based Learning, vol. 17, issue number 1, pp. 326–335 (2023)

14. Toda, A., et al.: Gamification through the looking glass - perceived biases and ethical concerns of brazilian teachers. In: Rodrigo, M.M., Matsuda, N., Cristea, A.I., Dimitrova, V. (eds.) AIED 2022. LNCS, vol. 13356, pp. 259–262. Springer, Cham (2022). https://doi.org/10.1007/978-3-031-11647-6_47

15. Toda, A.M., et al.: Analysing gamification elements in educational environments using an existing gamification taxonomy. Smart Learn. Environ. **6**, 16 (2019). https://doi.org/10.1186/s40561-019-0106-1

16. Toda, A., et al.: Gamification Design for Educational Contexts: Theoretical and Practical Contributions. Springer, Cham (2023). https://link.springer.com/book/10.1007/978-3-031-31949-5

17. Sümer, Ö., et al.: Multimodal engagement analysis from facial videos in the classroom. IEEE Trans. Affect. Comput. **14**(2), 1012–1027 (2023). https://doi.org/10.1109/TAFFC.2021.3127692

18. Jafari, E., Vassileva, J.: Ethical issues in explanations of personalized recommender systems. In: ACM Conference on User Modeling Adaptation and Personalization, UMAP 2023 (2023). https://dl.acm.org/doi/abs/10.1145/3563359.3597383. ISBN 978-1-4503-9891-6

19. Kay, J., et al.: Scrutable AIED. In: Handbook of Artificial Intelligence in Education, pp. 101–125. Edward Elgar Publishing (2023)

20. Orji, F.A., Gutierrez, F.J., Vassileva, J.: Exploring the influence of persuasive strategies on student motivation: self-determination theory perspective. In: Baghaei, N., Ali, R., Win, K., Oyibo, K. (eds.) PERSUASIVE 2024. LNCS, vol. 14636, pp. 222–236. Springer, Cham (2024). https://doi.org/10.1007/978-3-031-58226-4_17

AIED 2024 Keynotes

Navigating Strategic Challenges in Education in the Post-pandemic AI Era

Blaženka Divjak🆔

Faculty of Organization and Informatics, University of Zagreb, Croatia
blazenka.divjak@foi.unizg.hr

Abstract. Navigating strategic challenges in education in the post-pandemic AI era has the characteristics of a *complex* decision-making context. The current challenges and possible approaches, supporting decisions about future-oriented education, are discussed.

Keywords: strategic decision-making · education · pandemic · artificial intelligence

1 Introduction

Strategic planning in education needs to consider the contemporary challenges and the type of a decision-making context. Today's education has been significantly impacted by two major unprecedented events which have had a strong transformative effect: the COVID-19 pandemic and the rapid spreading of AI.

To better understand the decision-making in such contexts, it is useful to look at them through the lens of the Cynefin framework. As explained by Snowden [1], "Cynefin creates four open spaces or domains of knowledge all of which have validity within different contexts". In Cynefin, contexts are "defined by the nature of the relationship between cause and effect" [2]. In the four contexts, simple, complicated, complex, and chaotic, leaders need to recognize the situation and act appropriately.

The COVID-19 context was a chaotic one, in which it was not possible to identify the relationships between cause and effect and there were no patterns, so leaders had to act immediately and work to shift from *chaos* to *complexity*. While in chaos *trust* in leaders is necessary, it also presents a danger for democratic processes if it is blind.

Once the world shifted to a *complex* context, another challenge emerged: the mainstreaming of AI. Both challenges have a significant impact on education today, on the global, national, institutional, and on the level of individual learners and educators.

Strategic decision-making today is therefore double-burdened. On the one hand, it has to consider the possibility of new major global threats, like a pandemic or a war, and their possible tremendous effect on education. On the other hand, careful thought should be given to AI, which can be a powerful assistant to deal with threats like this, but also a hidden enemy to ethical and meaningful teaching and learning.

2 Decision Making in the *Chaotic* Domain

Research on educational decision-making in the chaotic domain, characterized by the imperative to master unfamiliar threats, has been scarce, at least until the unprecedented global challenge posed by the pandemic. The pandemic presented us with a rich research environment, enabling us to investigate and draw conclusions which can be generalized and applied to other chaotic contexts.

Educational decision-making during the pandemic was done at different levels: global, (multi)national, institution, individual (teacher, student). So, when analysing the decisions and decision-making processes, it makes sense to consider the said levels.

At the onset of the pandemic, I was the chair of the Council of the European Union (EU), gathering ministers in charge of education of the 27 EU Member States. So, looking at the multinational perspective, I personally witnessed an increase in the inclination of educational ministers to work together closely in making decisions in this *chaotic* context. Without delay, new formal and informal instruments for fast data collection, mutual support, learning and exchange of important decisions and processes were developed, as well as the sharing of good and bad practices. We collected primarily qualitative data, based on the responses of the Member States' educational authorities, that were aimed to support the understanding of the situation and finding/recognizing common denominators (patterns). There was no time to collect structured quantitative data, as decision-making had to be instant, making sure that lives and health were safeguarded, and educational processes continued in alternative formats. The public was often not aware of the efforts done at decision-making levels, as in this *chaotic* context [2], it was essential to trust the leaders and their dedication to ensuring the wellbeing of those related to educational institutions. Similar processes striving for fast and open collaboration were also recognized on institutional [3] and global levels (e.g., [4]).

Regardless of the global and regional efforts and collaboration, there were still differences in the success of educational systems' responses to the pandemic. Importantly, previous research has shown that the agility and fitness-for-change of educational systems and their main actors is essential in responding to and coping with major challenges [5].

3 Decision Making in the *Complex* and *Complicated* Domain

One of the most important goals in the *chaotic* domain is to lower the level of chaos and push towards the *complex* context, where we have a clearer picture of what we do and do not know, and have a toolbox of decision-making theories and instruments that can be used to identify at least some patterns for decision-making. In this context, we still need to be agile in decision-making, exploiting the benefits of concepts and tools already at hand, as well as the emerging technologies and approaches.

Currently, we are working in a *complex* context, significantly challenged by the rapid development and increasing accessibility of AI, intertwined with the consequences of the pandemic. To respond to the contemporary requirements, educators should be continuously strengthened and motivated to harness the potential of learning design [6], as a possible universal language, and innovative pedagogies. Meaningful learning design and implementation of future- and learner-oriented teaching and learning practices can be enhanced by the ethical and creative use of AI. Sensible use of AI and minimising the black-box effect require changes in curricula, as well as initial education and continuous professional development of educators [7].

Besides being used as a teaching and learning tool, AI can provide important assistance in decision-making, particularly as an indispensable source of technology, methods and tools for learning analytics. We strive for evidence-based decision-making supported by explainable AI, but this opens up a range of questions. How can we make and implement decisions about learners? How do regulatory frameworks streamline the process of decision-making supported by AI? For example, on the EU level, the new overarching AI regulation places AI systems supporting some kinds of decisions about learners among high-risk AI systems. Such frameworks require adjustments at both national and institutional level, to enable lawful, ethical and meaningful use of AI and learning analytics.

An important area of learning analytics, strongly supported by AI, refers to predictive learning analytics, relying on the development of predictive models. Although predictive modelling has been increasingly used and gaining significance, there are concerns regarding its adequacy in decision-making without human supervision [8]. A need for stronger governance has been identified, as algorithms no longer serve only for informing, but also provide and steer decision-making. As such, it is essential that algorithms are explainable [9], and learning analytics trustworthy. To achieve this trustworthiness, we should take a look through the human-centred lens and consider the perspectives of potential beneficiaries, primarily learners [10]. But above all, we should not trust machines to autonomously make decisions about humans: human-related decisions should be human-made.

Besides being evidence-based and oriented towards hard goals, decision-making should be balanced and fair. This calls for participatory decision-making, which is easily achievable in *complex* and *complicated* domains, because of well-established group decision-making approaches. However, we should consider different perspectives, especially at the higher decision-making (managerial) levels. How can we answer the common situation in education in which men are more involved in managerial decision-making, while a majority of the teaching workforce are women [11]?

Finally, it is crucial that decision-makers are ready to adequately respond to the given *complex* context, being agile, collaborative and relying on relevant expertise and evidence provided by learning analytics [2]. Both educators and decision-makers need future-literacy to be able to imagine possible futures, as well as practical skills to streamline education towards the desired future.

4 Conclusion

The recent developments in AI, with the lessons learnt from the pandemic, and the EdTech legacy, leave us at a new beginning. Besides challenges, we have a valuable opportunity to rethink and streamline education to be more future-looking. It is crucial to innovate decision-making processes as well. We should be aware of the level of complexity of the context and mindful that opening up and widening participation contributes to the relevance of our decisions and the quality of education. Finally, we should make sure that human-related decisions are made by humans, and not machines.

Acknowledgments. This work was partially supported by the *Trustworthy Learning Analytics and Artificial Intelligence for Sound Learning Design* (TRUELA) research project financed by the Croatian Science Foundation (IP-2022-10-2854).

References

1. Snowden, D.: Complex acts of knowing: paradox and descriptive self-awareness. J. Knowl. Manag. **6**(2), 100–111 (2002). https://doi.org/10.1108/13673270210424639
2. Snowden, D.J., Boone, M.E.: A leader's framework for decision making. Harv. Bus. Rev. 85**(11), 68–76 (2007)**
3. Beauvais, A., Kazer, M., Rebeschi, L.M., Baker, R., Lupinacci, J.H.: Educating nursing students through the pandemic: the essentials of collaboration. SAGE Open Nurs. **7**, 237796082110626 (2021). https://doi.org/10.1177/23779608211062678
4. OECD. The state of higher education: One year into the COVID-19 pandemic, June 2021. Accessed 21 Sept. 2021. https://www.oecd-ilibrary.org/docserver/83c41957-en.pdf?expires=1632215384&id=id&accname=guest&checksum=CA1E82861929056554205870BB8DDCEB
5. Svetec, B., Divjak, B.: Emergency responses to the COVID-19 crisis in education: a shift from Chaos to complexity. In: EDEN Conference Proceedings, no. 1, pp. 513–523, September 2021. https://doi.org/10.38069/edenconf-2021-ac0051
6. Grabar, D., Svetec, B., Vondra, P., Divjak, B.: Balanced learning design planning. J. Inf. Organ. Sci. **46**(2), 361–375 (2022). https://doi.org/10.31341/jios.46.2.6
7. Rienties, B., et al.: Online professional development across institutions and borders. Int. J. Educ. Technol. High. Educ. **20**(1), 30 (2023). https://doi.org/10.1186/s41239-023-00399-1
8. Khosravi, H., et al.: Intelligent learning analytics dashboards: automated drill-down recommendations to support teacher data exploration. J. Learn. Anal. **8**(3), 133–154 (2021). https://doi.org/10.18608/jla.2021.7279
9. McConvey, K., Guha, S., Kuzminykh, A.: A human-centered review of algorithms in decision-making in higher education. In: Proceedings of the 2023 CHI Conference on Human Factors in Computing Systems, pp. 1–15, April 2023. https://doi.org/10.1145/3544548.3580658
10. Divjak, B., Svetec, B., Horvat, D.: Learning analytics dashboards: what do students actually ask for? In: LAK23: 13th International Learning Analytics and Knowledge Conference, pp. 44–56, March 2023. https://doi.org/10.1145/3576050.3576141
11. Grinshtain, Y., Addi-Raccah, A.: Domains of decision-making and forms of capital among men and women teachers. Int. J. Educ. Manag. **34**(6), 1021–1034 (2020). https://doi.org/10.1108/IJEM-03-2019-0108
12. Divjak, B., Svetec, B., Horvat, D., Kadoić, N.: Assessment validity and learning analytics as prerequisites for ensuring student-centred learning design. Br. J. Educ. Technol. **54**(1), 313–334 (2023). https://doi.org/10.1111/bjet.13290

Navigating the Evolution: The Rising Tide of Large Language Models for AI and Education

Peter Clark

Allen Institute for AI, Seattle, WA, USA
petec@allenai.org

1 Introduction

AI has labored for 60 years trying to build systems with some degree of "understanding" of the world, with the promise of many potential benefits including in the world of Education and Intelligent Tutoring Systems. After all, a system that can answer questions and explain its answers should surely be a boon for education? Now, with the emergence of language models (LMs), arguably such systems have suddenly appeared seemingly from nowhere. Yet we don't really understand how they work, they make mistakes, and can be inconsistent. What should we make of this new world? To answer this, I'll briefly retrace my journey through this rapidly changing space, and then look to the future.

2 The Rapidly Changing World

2.1 Project Halo

The late Paul Allen, co-founder of Microsoft, had a long-standing dream to build a "Digital Aristotle" - an AI system that contained and understood much of the world's knowledge, so as to help humans in their endeavors, with particular emphasis on science and education. Project Halo, which ran from 2003-2013, was an embodiment of this vision, aiming to build a knowledgeable AI system with a deep understanding of a specific subject - college-level biology - that could be used for educational purposes [4]. The project constructed a large, formal knowledge-base that could answer novel biology questions and explain them by drawing on its internal rules, allowing users to understand the underlying biology behind its answers. This capability was deployed in an educational setting, as an interactive eTextbook called Inquire, where students could not only browse and explore the book but also ask it questions and get reasoned answers back [1]. The application was exciting, and extensive evaluations showed substantial educational benefit. However, the cost of manually building the knowledge base was prohibitive, and ended up as a show-stopper for the work.

2.2 Project Aristo

In 2014, this quest for a knowledgeable AI system was restarted at the then-new Allen Institute for AI (AI2), with machine learning and natural language processing (NLP) at the fore. Specifically, rather than manually build a knowledge-base, the new system - called Aristo - acquired its knowledge through reading and processing of text. With the addition of early language models, specifically BERT and RoBERTa, Aristo's ability accelerated, and in 2019 was able to score over 90% on the non-diagram, multiple choice part of the Regents 8th Grade Science Exams, reflecting the breakthrough that language models (LMs) was bringing to AI [2, 6].

However, illustrating the flip side of LMs, this success came at a cost, namely (unlike in Project Halo) Aristo was unable to systematically explain its answers, a problem which still plagues LM technology today. In particular, if the target application is in an educational setting, offering systematic, reliable explanations for answers is a key requirement, and one that Aristo did not meet. Much of our work since then has been devoted to identifying and extracting the "rational" domain models that underly a LM's normally typically opaque answers.

2.3 Human and Machine Mental Models

Part of teaching involves conveying an accurate, *mental model* of the world from the teacher to a student, allowing the student to understand observations and make good predictions. Do LMs even have "mental models"? Arguably, early LMs did not – their answers were highly correlated with superficial linguistic patterns rather than reflecting a deep understanding of a topic [7]. However, modern LMs perform amazingly well on a variety of tasks, including answering complex, novel questions, suggesting that they form *some* kind of internal world model when answering [5]. If we can expose these internal models that LMs have, many opportunities open up for learning and understanding.

Of course, we humans do not directly communicate our neural, mental models directly either (indeed, we are not fully aware of what all the neurons in our brain are doing). Rather, we express our thoughts in a symbolic formalism – natural language - that *can* be communicated easily, reflecting our understandings of the world that we have been taught, and that we have formed ourselves. Expressing our knowledge in natural language can be viewed as a *constructive* process, in which we generate (statements of) facts and rules that reflect our beliefs, and that chain together coherently to produce useful new knowledge. Indeed, a good textbook will convey concepts, facts, and rules about the world that have been carefully designed to reflect reality and allow new knowledge to be inferred.

2.4 Entailer

Can we evoke this process in LMs, namely have them articulate their "mental models" (coherent set of beliefs) also, and hence have them "teach" a user what they (the LMs) know? It turns out that we can, through a combination of (a) asking a model to generate systematic explanations for its answers and (b) checking the model actually believes its generations (i.e., it "believes what it says"). In a system called Entailer, we trained a LM to explain its answers in a *systematic* way, by producing a chain of reasoning (facts and rules expressed in NL) that objectively concludes the answer (rather than just an informal textual paragraph) [8]. This opens a window into the LM's model of the world, and - to the extent that window is accurate - provides a mechanism for conveying its underlying knowledge to a user.

2.5 TeachMe

Of course, LMs are not perfect, and have erroneous "beliefs" about the world. Systems like Entailer, above, can help users identify those machine misconceptions. Specifically, if the LM produces a wrong answer justified by a model-believed chain of reasoning, the user can then isolate the source of the LM's mistake by tracing down that chain to find the problem. For example, in simple reasoning about physics, Entailer incorrectly concluded that *a penny is magnetic*. By examining the LM's explanation, the user could trace the error to the (bad) LM belief in the reasoning chain that *metals are magnetic* (while in reality, copper is not magnetic). Furthermore, if the user provides the corrected knowledge to the LM (*copper is not magnetic*), they can correct the system's reasoning. This gave rise to a "flipped" educational tool that we developed, called TeachMe, in which the student learns by "teaching" the AI system [3]. Given an AI system that is making mistakes on a test, the student is tasked with identifying the system's mistakes and correct them (including using auxiliary material, if necessary), and in so doing, develops and reinforces their own understanding of the topic. This follows Feynman's maxim that: "If you want to master something, teach it."

2.6 A Never-Ending Research Assistant

Even in the last year, LMs have continued to rapidly advance. Today's LMs not only chat and answer questions, but can search document repositories, read technical documents, generate and execute software code, run (software) experiments, and even generate research hypotheses that might be worthy of exploration. Our current work seeks to integrate these capabilities together into a smart research assistant, able to assist scientists in their work in a variety of ways. The AI assistant can help manage the vast information overload that scientists experience today, discuss specific technical papers with the scientist, and semi-autonomously execute complex workflows (experiments, dataset analyses). Similarly, the human scientist can provide top-level research directions and hypotheses to expore, and guide and direct the AI assistant. In this way, both the human scientist and AI Assistant can learn from each others' strengths. And, following the earlier theme, a key task for the AI assistant is maintaining a rational understanding ("mental model") of the research topic: namely a graph of the claims, hypotheses, evidence, experiments, and the inferential relationships between them all, and ensuring that this understanding remains consistent. And, unlike the earlier systems, there is one key difference: this understanding is dynamic, constantly changing as new papers are published and new research results obtained - and, with a bit of luck, perhaps help the scientist discover breakthroughs that would otherwise have been missed.

3 Conclusion

The rapidly evolving capabilities of LMs is perhaps one of the most surprising results that the field of AI has had, and is one that is likely to touch all aspects of life, including in education. I've traced the rapid evolution of my group's quest to build knowledgeable systems, from hand-built knowledge bases to a modern, LM-based research assistant agent. While the possibilities are endless, let me also offer three concluding thoughts: embrace the technology; be cautious (the technology is still highly fallible); and, to the extent possible, don't only look at the LM's answers but also at the knowledge and reasoning used arrive at those answers – I've outlined some of the techniques we have been using to expose that information, and in the end it is that which will help us have confidence in the information that LMs are providing, and help us learn from and interact with these new systems appropriately.

References

1. Chaudhri, V.K., et al.: Inquire biology: a textbook that answers questions. AI Mag. **34**, 55–72 (2013). https://api.semanticscholar.org/CorpusID:9127231
2. Clark, P., et al.: From 'f' to 'a' on the n.y. regents science exams: an overview of the aristo project. AI Mag. **41**, 39–53 (2019). https://api.semanticscholar.org/CorpusID: 202539605
3. Dalvi, B., Tafjord, O., Clark, P.: Towards teachable reasoning systems: Using a dynamic memory of user feedback for continual system improvement. In: EMNLP (2022)
4. Gunning, D., et al.: Project halo update – progress toward digital aristotle. AI Mag. **31** (2010)
5. Hinton, G.: Two paths to intelligence. ACL (2023). https://doi.org/10.48448/hf7y-x909. (Keynote Talk)
6. Metz, C.: A breakthrough for AI technology: passing an 8th-grade science test. NY Times (2019). https://www.nytimes.com/2019/09/04/technology/artificial-intell igence-aristo-passed-test.html
7. Ribeiro, M.T., Guestrin, C., Singh, S.: Are red roses red? Evaluating consistency of question-answering models. In: ACL (2019)
8. Tafjord, O., Dalvi, B., Clark, P.: Entailer: Answering questions with faithful and truthful chains of reasoning. In: EMNLP (2022)

Contents – Part II

Doctoral Consortium

Workshops and Tutorials

Contents – Part I

WideAIED

Late-Breaking Results

Late-Breaking Results

A Schema-Based Approach to the Linkage of Multimodal Learning Sources with Generative AI

Christine Kwon[1]([✉]) [iD], James King[2], John Carney[2] [iD], and John Stamper[1] [iD]

[1] Carnegie Mellon University, Pittsburgh, PA 15213, USA
ckwon2@andrew.cmu.edu, jstamper@cmu.edu
[2] MARi LLC, Alexandria, VA 22314, USA
{james.king,john.carney}@mari.com

Abstract. Learning how to execute a complex, hands-on task in a domain such as auto maintenance, cooking, or guitar playing while relying exclusively on text instruction from a manual is often frustrating and ineffective. Despite the need for multimedia instruction to enable the learning of complex, manual tasks, learners often rely exclusively on text instruction. However, through widespread usage of user-generated content platforms, such as YouTube and TikTok, learners are no longer limited to standard text and are able to watch videos from easily accessible platforms to learn such procedural tasks. As YouTube consists of a large corpus of diverse instructional videos, the accuracy of videos on sensitive and complex tasks has yet to be validated in comparison to "golden standard" manuals. Our work provides a unique LLM-based multimodal pipeline to interpret and verify task-related key steps in a video within organized knowledge schemas, in which demonstrated video steps are automatically extracted, systematized, and validated in comparison to a text manual of official steps. Applied to a dataset of twenty-four videos on the task of flat tire replacement on a car, the LLM-based pipeline achieved high performance on our metrics, identifying an average of 98% of key task steps, with 86% precision and 92% recall across all videos.

Keywords: Generative AI · Large Language Models (LLMs) · Multimodal Learning · Video Training

1 Introduction

The execution of complex tasks from steps in written form from a manual can be difficult for novice learners. Pedagogical models of demonstration and apprenticeship have long served as the foundation for teaching complex hands-on skills, reflecting the nuanced interplay between observation, practice, and feedback. This pedagogical approach, deeply rooted in the cognitive apprenticeship theory, emphasizes learning in context [5], where the learner engages in authentic tasks with the guidance of a more knowledgeable expert. The effectiveness of these models in complex skill acquisition is largely attributed to the scaffolded support they offer, allowing learners to gradually develop competence through iterative engagement and personalized feedback [13].

© The Author(s), under exclusive license to Springer Nature Switzerland AG 2024
A. M. Olney et al. (Eds.): AIED 2024 Workshops, CCIS 2151, pp. 3–10, 2024.
https://doi.org/10.1007/978-3-031-64312-5_1

In recent times, the surge in digital technologies has paved the way for innovations to these traditional learning models. Platforms such as YouTube and TikTok are now widely used for the demonstration and teaching of complex tasks in areas such as auto maintenance, sports, or learning to play a musical instrument. While these videos can provide better observation than a written manual, students may still need to follow and understand the steps in a manual to achieve mastery. In this research, we recognize the multimodal aspects of this kind of learning and are building the foundation of a technical infrastructure to allow learners to seamlessly move between multiple modes of learning by organizing complex tasks into written steps that can then be instantly attached to video or audio content at the correct time point.

To accomplish this, we have proposed a structural ontology around the concept of a knowledge object (KO) [7]. We define a KO around a specific task that is composed of one or more steps and the metadata associated with each step. There are several reasons why the idea of defining KOs is important for task-based learning. The primary reason is to easily organize tasks for learning purposes. A secondary and equally important reason is the ability to search for tasks and examples on demand. Scenarios where KOs excel are mechanical tasks like automotive repair (changing oil, spark plugs, tire, etc.) where finding the task currently is quite easy, but a specific instance for a particular vehicle may be much harder to find and may have details specific to that instance. Having the right metadata in the right structure is critical, especially where the metadata links multimodal datastreams for a task. It is possible to define a hierarchical structure to a set of KOs where parent-child relationships exist connecting tasks at multiple levels, although for the purpose of broad search we are most interested in the higher level KOs.

A noteworthy development in KOs is the integration of Large Language Models (LLMs), such as GPT-4, and frameworks like LangChain, which have shown potential to revolutionize the way we approach the organization and interpretation of multimodal data in educational contexts [14]. These advanced LLMs, with both generative capabilities and nuanced understanding of natural language, offer unprecedented support in structuring learning experiences around complex tasks. They not only facilitate the organization of diverse data types—ranging from textual instructions and procedural guidelines to visual demonstrations and interactive simulations—but also enhance the interactive learning experience by providing real-time, contextually relevant support and feedback.

The incorporation of LLMs to link demonstration and apprenticeship models in KOs enables a more dynamic, adaptive, and personalized learning environment. By leveraging the power of LangChain schemas to interlink and contextualize multimodal information, educators and learners can now navigate complex hands-on tasks with greater ease and efficiency. This paper aims to explore the implications of this integration, highlighting the potential of LLMs to augment learning support and reshape the pedagogical approaches to complex task training.

2 Related Work

There has been a large body of prior work in the area of video-based learning for procedural tasks, as well as comparisons of video-based learning to text-based learning. We are also interested in previous work that investigates various methods of extraction of learning materials, including LLM-based methods, from instructional videos.

2.1 Comparing Video-Based and Text-Based Learning

Using YouTube-like videos has become a core source of learning for many students, yet there is still an insufficient understanding of how video-based learning juxtaposes with text-based learning, especially for complex procedural tasks. A prime example of a highly sensitive task that benefits from video-based learning is surgical procedures [3, 11]. In particular, prior research compared video and text-based learning material to investigate which modality of learning promotes higher student learning in basic laparoscopic suturing and clinical procedures and skills [3, 11]. Sonnenfeld et al. investigated training procedures in electronic and distance learning approaches in the context of flight crew training, in which they found that video-based learning was effective for procedure-based training in providing flight crew trainees with interactive opportunities and formal training content [12]. While video-based learning has the potential to teach and deliver information that is difficult to convey from standard text material [10], it is especially important to understand how to conduct effective video-based learning for high-risk tasks that are dependent on a correct ordering of steps.

2.2 Video-Based Step-By-Step Feature Extraction

Instructional videos are extremely diverse in content, especially in the hierarchy of steps and key moments in learning how to conduct a certain task. Hence, there has been a noticeable growing interest in improving the searchability and indexing of these instructional videos by extracting key informational features from these videos [16]. The detection of key steps within instructional videos is difficult, which rely on both the chronological and temporal ordering of steps in accomplishing a task [18]. Recent studies employ a joint method that evaluates text-based extractions in conjunction with visual feature detectors [1, 8]. Additional methods require extracted visual actions and objects to have a high degree of semantic relatedness with the textual information attached to a procedural video [9]. These extraction methods require an examination of the relevance of these features to an empirical measure of comparison. Our work aims to automate this comparative process by using an LLM-based pipeline to accurately extract task step-related features and systematize the hierarchy of steps verified by an empirical measure of comparison to support more complex task-based learning.

2.3 LLM-Based Step-By-Step Information Extraction.

Currently, Large Language Models (LLMs) are showing their immense potential in efficiently carrying out diverse NLP tasks [2, 19]. However, we have yet to completely rely on LLM-based text generation as little is known about the ability of LLMs to verify their content output [17]. To improve the ability of complex reasoning and content generation of LLMs, Wei et al. introduced the "chain of thought" (CoT) prompting method, which induces these language models to deconstruct a problem into multiple in-depth reasoning steps [15]. However, verification methods using prompt chaining may not be completely efficient for extracting and verifying multiple ordered steps of a task, especially if the task data is unstructured. On the other hand, another prior study used an LLM-assistance pipeline to extract organized annotations and informational features from unstructured clinical data [6]. While these prior studies separately investigated the validation of LLM-generated content and organized extraction of information from unstructured data, we have yet to see work that combines these processes, especially to verify task-based video content.

3   Methodology

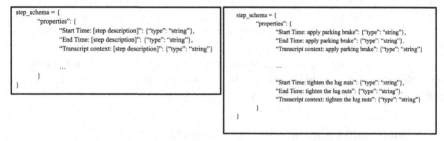

Fig. 1. The left figure is the outline of the schema input, which consists of three features: the start time and end times of the demonstrated step and the transcript context used to determine the time interval of each step occurrence. The right figure is the schema input used to extract key task steps for each video on replacing a flat tire on a car.

We employed LangChain, a language model integration framework powered by Large Language Models (LLMs) to develop and extricate organized schemas from any text source [4]. We specifically used LangChain in conjunction with GPT 4.0 to create and extract schemas that align the retrieved steps with affirmed key steps from any official manual.

Using the LangChain framework, we created an input schema, labeled "step_schema," shown in Fig. 1, that lists fundamental key steps on a procedural task, in this case, car tire replacement, from an official manual within an organized knowledge metastructure. Within the schema, we listed the desired extracted features for each step, which include the starting timestamp of when a step is first mentioned, the ending timestamp when a step is no longer mentioned, and the transcript context used to extract each step occurrence from a video narration. The extracted features for each step of a task are also explained in Table 1. Figure 1 also presents the specific schema input used in our approach to extract the key steps listed in Table 2 from videos on car flat tire replacement.

Table 1. This table presents the unique features that are listed in the properties for each task step of the input schema. Each feature is listed along with its type and definition.

Feature Name	Feature Type	Feature Definition
Start Time: [step description]	string	Starting timestamp on first narrated mention of step
End Time: [step description]	string	Ending timestamp on last narrated mention of step
Transcript context: [step description]	string	The portion of text on the full occurrence of step

Table 2. This table lists 12 official manual steps of flat tire replacement on a car used to list and verify the retrieved steps within the extracted schema.

Manual Steps (1–6)	Manual Step	Manual Steps (7–12)	Manual Step
1	Apply parking brake	7	Remove lug nuts
2	Remove the spare tire from the car	8	Remove flat tire
3	Use wheel chocks to block the wheels opposite of the wheel you're changing	9	Place the spare tire
4	Loosen the lug nuts from the tire	10	Screw on the lug nuts
5	Loosen the jack	11	Use the jack to lower the car
6	Use the jack to lift up the car	12	Tighten the lug nuts

In providing an input schema with key task steps, this LLM-based pipeline can structure an output schema isomorphic to the input schema. By systematizing task steps and their features within an organized schema, crucial step information is easily accessible and directly aligned with authenticated task steps from a verified source.

4   Preliminary Results and Discussion

We collected 24 videos from YouTube on replacing a flat tire on a car. YouTube-generated transcripts from each video, which included toggled timestamps, were collected as input data for our LLM-based pipeline. This pipeline extracted an isomorphic schema to the input schema shown in Fig. 1 for each video. To evaluate the performance of the pipeline, two human evaluators determined a correctly identified extracted task step of replacing a flat car tire if its starting and ending timestamps overlap with the time interval of the same step in each video. The LLM-based pipeline achieved high-performance metrics, identifying an average of 98% of key task steps with 86% precision and 92% recall across all videos. The pipeline was highly successful in identifying missing steps for most videos. However, we also encountered some drawbacks with this LLM-based approach. Though the LLM-based pipeline retrieved and identified 100% of key task steps for approximately 92% of videos in the dataset, at least some steps were incorrectly identified for 73% of the videos. For instance, while the pipeline was successful in recognizing most task steps in each video, it also was prone to misidentify task steps not demonstrated in videos. Additionally, there were instances where the pipeline incorrectly identified the sections of the video in which a task step was demonstrated. We intend to address these limitations by evaluating and improving the LLM-based pipeline performance on diverse video datasets.

The ability to auto-identify the start and stop time of each required step in a "how-to" video lays the groundwork for instructional videos to be systematically checked for accuracy at scale. The benefits of automatically extracting accurate content are most important in mission-critical spaces like defense, aviation, and healthcare. When a novice is learning to execute a lower-stakes procedure such as car dent removal, for example, perhaps the wisdom of the crowd conveyed via the most-liked videos on YouTube is a sufficient filter to ensure useful content reaches learners. However, when learning a high-stakes procedure like aircraft repair, execution of each step in the required order is critical; the learner must be able to trust the accuracy of a single video.

The schema-driven approach that links related instructional content to a golden standard official procedure can also streamline updates to instructional content. For example, consider the official procedures and documentation for AI/ML tools that change frequently. If the plethora of YouTube tutorial videos describing how to use said AI/ML tools was linked with "gold standard" documentation, relevant moments of the videos could be automatically flagged as deprecated when appropriate. Importantly, the current implementation described above compares only a video's time-stamped transcript against the official procedure. Therefore, the step of the procedure must be described or referred to aloud in the video for the algorithm to identify its presence. The below Future Work section describes the extensibility of the transcript-procedure comparison approach as well as plans to parse objects and actions detected in videos.

5   Conclusion and Future Work

Automatic alignment of crowdsourced instructional videos, like those on YouTube, and official "gold standard" procedure steps, that are found in written manuals, will enable learners of tasks in diverse domains to quickly jump from a step in a manual to an instructional video moment detailing a task step of interest. Mission-critical and machine repair contexts such as aviation and medicine often involve reliance on official step-by-step procedure documents that lack the detail needed to convey a task's proper execution. Linkage of an official procedure step and related multimedia segments could enable learners to toggle between a text and video depiction of the same task, empowering the learner to access detailed instruction from a trusted source.

Further, the automatic tagging of multimedia content segments with their associated official procedure steps will enable the recommendation of video moments to individual learners or groups based on observed task and/or procedure proficiency. For example, a content recommendation system could leverage tags to serve a video moment that depicts "How to replace Part X on a given aircraft" to a learner with relevant task proficiency below a threshold. In this way, the automatic alignment of official procedure steps and video moments will enable targeted content pushes that preempt a learner's proactive query. The ability to proactively push tutorial content segments to learners could yield performance benefits during cases of "unconscious incompetence", in which the learner does not know that they lack the ability to effectively execute a task.

A next step of our current approach, which compares a video transcript against an official procedure, is integrating automated video intelligence. Detecting on-screen objects and actions will enable the content-procedure alignment algorithm to identify a step included in a video that is not mentioned aloud. Development of object and action recognition video intelligence methods is a high priority going forward and will involve fine-tuning existing large models with data relevant to each use case (e.g., car tire replacement, de-icing aircraft). As the use of mixed reality learning content to address skill gaps in hands-on domains grows, the same content-procedure alignment method will also be applied to mixed reality (MR) content.

Acknowledgments. This work was supported by US Navy STTR #N68335-21-C-0438.

References

1. Alayrac, J.B., Bojanowski, P., Agrawal, N., Sivic, J., Laptev, I., Lacoste-Julien, S.: Unsupervised learning from narrated instruction videos. In: Proceedings of the IEEE Conference on Computer Vision and Pattern Recognition, pp. 4575–4583 (2016)
2. Ampel, B.M., Yang, C.H., Hu, J., Chen, H.: Large language models for conducting advanced text analytics information systems research (2023). arXiv preprint arXiv:2312.17278
3. Buch, S.V., Treschow, F.P., Svendsen, J.B., Worm, B.S.: Video-or text-based e-learning when teaching clinical procedures? A randomized controlled trial. Adv. Med. Educ. Pract. 257–262 (2014)
4. Chase, H.: LangChain. https://langchain.com/. Accessed on 1 Aug 2023
5. Dennen, V.P., Burner, K.J.: The cognitive apprenticeship model in educational practice. In: Handbook of research on educational communications and technology, pp. 425–439. Routledge (2008)
6. Goel, A., et al.: LLMS accelerate annotation for medical information extraction. In: Machine Learning for Health (ML4H), pp. 82–100. PMLR (2023)
7. Kwon, C., Stamper, J., King, J., Lam, J., Carney, J.: Multimodal data support in knowledge objects for real-time knowledge sharing. In: Proceedings of CROSSMMLA Workshop at the 13th International Conference on Learning Analytics & Knowledge (2023)
8. Malmaud, J., Huang, J., Rathod, V., Johnston, N., Rabinovich, A., Murphy, K.: What's cookin'? interpreting cooking videos using text, speech and vision (2015). arXiv preprint arXiv:1503.01558
9. Manju, A., Valarmathie, P.: Organizing multimedia big data using semantic based video content extraction technique. In: 2015 International Conference on Soft-Computing and Networks Security (ICSNS), pp. 1–4. IEEE (2015)
10. Navarrete, E., Nehring, A., Schanze, S., Ewerth, R., Hoppe, A.: A closer look into recent video-based learning research: a comprehensive review of video characteristics, tools, technologies, and learning effectiveness (2023). arXiv preprint arXiv:2301.13617

11. Routh, D., Rao, P.P., Sharma, A., Arunjeet, K.: To compare the effectiveness of traditional textbook-based learning with video-based teaching for basic laparoscopic suturing skills training-a randomized controlled trial. Medical Journal of Dr. DY Patil University (2023)

12. Sonnenfeld, N., Nguyen, B., Boesser, C.T., Jentsch, F.: Modern practices for flightcrew training of procedural knowledge. In: 84th International Symposium on Aviation Psychology, p. 303 (2021)

13. Stamper, J., Barnes, T., Croy, M.: Enhancing the automatic generation of hints with expert seeding. In: Intelligent Tutoring Systems: 10th International Conference, ITS 2010, Pittsburgh, PA, USA, June 14–18, 2010, Proceedings, Part II 10, pp. 31–40. Springer (2010)

14. Topsakal, O., Akinci, T.C.: Creating large language model applications utilizing langchain: a primer on developing LLM apps fast. In: International Conference on Applied Engineering and Natural Sciences, vol. 1, pp. 1050–1056 (2023)

15. Wei, J., et al.: Chain-of-thought prompting elicits reasoning in large language models. Adv. Neural. Inf. Process. Syst. **35**, 24824–24837 (2022)

16. Zala, A., et al.: Hierarchical video-moment retrieval and step-captioning. In: Proceedings of the IEEE/CVF Conference on Computer Vision and Pattern Recognition, pp. 23056–23065 (2023)

17. Zhang, X., Gao, W.: Towards LLM-based fact verification on news claims with a hierarchical step-by-step prompting method (2023). arXiv preprint arXiv:2310.00305

18. Zhong, Y., Yu, L., Bai, Y., Li, S., Yan, X., Li, Y.: Learning procedure-aware video representation from instructional videos and their narrations. In: Proceedings of the IEEE/CVF Conference on Computer Vision and Pattern Recognition, pp. 14825–14835 (2023)

19. Zhu, Y., et al.: Large language models for information retrieval: A survey (2023). arXiv preprint arXiv:2308.07107

Towards a Human-in-the-Loop LLM Approach to Collaborative Discourse Analysis

Clayton Cohn[1(✉)], Caitlin Snyder[1(✉)], Justin Montenegro[2(✉)], and Gautam Biswas[1(✉)]

[1] Vanderbilt University, Nashville, TN 37240, USA
{clayton.a.cohn,caitlin.r.snyder,gautam.biswas}@vanderbilt.edu
[2] Martin Luther King, Jr. Academic Magnet High School, Nashville, TN 37203, USA
justin.montenegro@mnps.org

Abstract. LLMs have demonstrated proficiency in contextualizing their outputs using human input, often matching or beating human-level performance on a variety of tasks. However, LLMs have not yet been used to characterize synergistic learning in students' collaborative discourse. In this exploratory work, we take a first step towards adopting a human-in-the-loop prompt engineering approach with GPT-4-Turbo to summarize and categorize students' synergistic learning during collaborative discourse. Our preliminary findings suggest GPT-4-Turbo may be able to characterize students' synergistic learning in a manner comparable to humans and that our approach warrants further investigation.

Keywords: LLM · Collaborative Learning · Human-in-the-Loop · Discourse Analysis · K12 STEM

1 Introduction

Computational modeling of scientific processes has been shown to effectively foster students' Science, Technology, Engineering, Mathematics, and Computing (STEM+C) learning [5], but task success necessitates *synergistic learning* (i.e., the simultaneous development and application of science and computing knowledge to address modeling tasks), which can lead to student difficulties [1]. Research has shown that problem-solving environments promoting synergistic learning in domains such as physics and computing often facilitate a better understanding of physics and computing concepts and practices when compared to students taught via a traditional curriculum [5]. Analyzing students' collaborative discourse offers valuable insights into their application of both domains' concepts as they construct computational models [8]. Unfortunately, manually analyzing students' discourse to identify their synergistic processes is time-consuming, and programmatic approaches are needed.

© The Author(s), under exclusive license to Springer Nature Switzerland AG 2024
A. M. Olney et al. (Eds.): AIED 2024 Workshops, CCIS 2151, pp. 11–19, 2024.
https://doi.org/10.1007/978-3-031-64312-5_2

In this paper, we take an exploratory first step towards adopting a *human-in-the-loop* LLM approach from previous work called *Chain-of-Thought Prompting + Active Learning* [3] (detailed in Sect. 3) to characterize the synergistic content in students' collaborative discourse. We use a large language model (LLM) to summarize conversation segments in terms of how physics and computing concepts are interwoven to support students' model building and debugging tasks. We evaluate our approach by comparing the LLM's summaries to human-produced ones (using an expert human evaluator to rank them) and by qualitatively analyzing the summaries to discern the LLM's strengths and weaknesses alongside a physics and computer science teacher (the Educator) with experience teaching the C2STEM curriculum (see Sect. 3.1). Within this framework, we analyze data from high school students working in pairs to build kinematics models and answer the following research questions: **RQ1**) How does the quality of human- and LLM-generated summaries and synergistic learning characterizations of collaborative student discourse compare?, and **RQ2**) What are the LLM's strengths, and where does it struggle, in summarizing and characterizing synergistic learning in physics and computing?

As this work is exploratory, due to the small sample size, we aim not to present generalizable findings but hope that our results will inform subsequent research as we work towards forging a human-AI partnership by providing teachers with actionable, LLM-generated feedback and recommendations to help them guide students in their synergistic learning.

2   Background

Roschelle and Teasley [7] define collaboration as *"a coordinated, synchronous activity that is a result of a continuous attempt to construct and maintain a shared conception of a problem"*. This development of a shared conceptual understanding necessitates multi-faceted collaborative discourse across multiple dimensions: social (e.g., navigating the social intricacies of forming a consensus [12]), cognitive (e.g., the development of context-specific knowledge [8]), and metacognitive (e.g., socially shared regulation [4]). Researchers have developed and leveraged frameworks situated within learning theory to classify and analyze collaborative problem solving (CPS) both broadly (i.e., across dimensions [6]) and narrowly (i.e., by focusing on one CPS aspect to gain in-depth insight, e.g., argumentative knowledge construction [12]). In this paper, we focus on one dimension of CPS that is particularly important to the context of STEM+C learning: students' cognitive integration of synergistic domains.

Leveraging CPS frameworks to classify student discourse has traditionally been done through hand-coding utterances. However, this is time-consuming and laborious, leading researchers to leverage automated classification methods such as rule-based approaches, supervised machine learning methods, and (more recently) LLMs [10]. Utilizing LLMs can help extend previous work on classifying synergistic learning discourse, which has primarily relied on the frequency counts of domain-specific concept codes [5,8]. In particular, the use of LLMs can help address the following difficulties encountered while employing traditional methods: (1) concept codes are difficult to identify programmatically, as rule-based approaches like regular expressions (regex) have difficulties with misspellings and homonyms; (2) the presence or absence of concept codes is not analyzed in a conversational context; and (3) the presence of cross-domain concept codes is not necessarily indicative of synergistic learning, as synergistic learning requires students to form connections between concepts in both domains.

Recent advances in LLM performance capabilities have allowed researchers to find new and creative ways to apply these powerful models to education using *in-context learning* (ICL) [2] (i.e., providing the LLM with labeled instances during inference) in lieu of traditional training that requires expensive parameter updates. One prominent extension of ICL is *chain-of-thought reasoning* (CoT) [11], which augments the labeled instances with "reasoning chains" that explain the rationale behind the correct answer and help guide the LLM towards the correct solution. Recent work has found success in leveraging CoT towards scoring and explaining students' formative assessment responses in the Earth Science domain [3]. In this work, we investigate this approach as a means to summarize and characterize synergistic learning in students' collaborative discourse.

3   Methods

This paper extends the previous work of 1) Snyder et al. on log-segmented discourse summarization defined by students' model building segments extracted from their activity logs [9], and 2) Cohn et al. on a human-in-the-loop prompt engineering approach called *Chain-of-Thought Prompting + Active Learning* [3] (the Method) for scoring and explaining students' science formative assessment responses. The original Method is a three-step process: 1) *Response Scoring*, where two human reviewers manually label a sample of students' formative assessment responses and identify disagreements (i.e., sticking points) the LLM may similarly struggle with; 2) *Prompt Development*, which employs few-shot CoT prompting to address the sticking points and help align the LLM with the humans' scoring consensus; and 3) *Active Learning*, where a knowledgeable human (e.g., a domain expert, researcher, or instructor) acts as an "oracle" and identifies the LLM's reasoning errors on a validation set, then appends additional few-shot instances that the LLM struggled with to the prompt and uses CoT reasoning to help correct the LLM's misconceptions. We illustrate the Method in Fig. 1. For a complete description of the Method, please see [3].

In this work, we combine log-based discourse segmentation [9] and CoT prompting [3] to generate more contextualized summaries of students' discourse segments to study students' synergistic learning processes by linking their model construction and debugging activities with their conversations during each problem-solving segment. We provide Supplementary Materials[1] that include 1) additional information about the learning environment, 2) method application details (including our final prompt and few-shot example selection methodology), 3) a

---

[1] https://github.com/oele-isis-vanderbilt/AIED24_LBR.

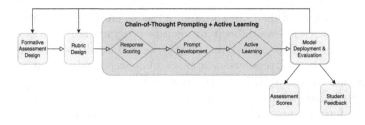

Fig. 1. *Chain-of-Thought Prompting + Active Learning*, identified by the green box, where each blue diamond is a step in the Method. Yellow boxes represent the process's application to the classroom detailed in prior work [3].

more in depth look at our conversation with the Educator, and 4) a more detailed analysis of the LLM's strengths and weaknesses while applying the Method.

3.1   STEM+C Learning Environment, Curriculum, and Data

Our work in this paper centers on the C2STEM learning environment [5], where students learn kinematics by building computational models of the 1- and 2-D motion of objects. C2STEM combines block-based programming with domain-specific modeling blocks to support the development and integration of science and computing knowledge as students create partial or complete models that simulate behaviors governed by scientific principles. This paper focuses on the 1-D *Truck Task*, where students use their knowledge of kinematic equations to model the motion of a truck that starts from rest, accelerates to a speed limit, cruises at that speed, then decelerates to come to a stop at a stop sign.

Our study, approved by our university Institutional Review Board, included 26 consented high school students (aged 14–15) who completed the C2STEM kinematics curriculum. Students' demographic information was not collected as part of this study (we began collecting it in later studies). Data collection included logged actions in the C2STEM environment, saved project files, and video and audio data (collected using laptop webcams and OBS software). Our data analysis included 9 dyads (one group had a student who did not consent to data collection, so we did not analyze that group; and we had technical issues with audio data from other groups). The dataset includes 9 h of discourse transcripts and over 2,000 logged actions collected during one day of the study. Student discourse was transcribed using Otter.ai and edited for accuracy.

3.2   Approach

We extend the Method, previously used for formative assessment scoring and feedback, to prompt GPT-4-Turbo to summarize segments of students' discourse and identify the *Discourse Category* (defined momentarily) by answering the following question: "Given a discourse segment, and its environment task context and actions, is the students' conversation best characterized as *physics-focused* (i.e., the conversation is primarily focused on the physics domain), *computing-focused* (i.e., the conversation is primarily focused on the computing domain), *physics-and-computing-synergistic* (i.e., students discuss concepts from both domains, interleaving them throughout the conversation, and making connections between them), or *physics-and-computing-separate* (i.e., students discuss both domains but do so separately without interleaving)?" We use the recently released GPT-4-Turbo LLM (gpt-4-0125-preview) because it provides an extended context window (128,000 tokens).

We selected 10 training instances and 12 testing instances (10 additional segments were used as a validation set to perform Active Learning) prior to Response Scoring, using stratified sampling to approximate a uniform distribution across Discourse Categories for both the train and test sets. Note that the student discourse was segmented based on which element of the model the students were working on (identified automatically via log data). During Response Scoring, the first two authors of this paper (Reviewers R1 and R2, respectively) independently evaluated the training set segments, classifying each segment as belonging to one of the four Discourse Categories. For each segment the Reviewers disagreed on, the reason for disagreement was noted as a sticking point, and the segment was discussed until a consensus was reached on the specific Discourse Category for that segment. R1 and R2 initially struggled to agree on segments' Discourse Categories (Cohen's $k = 0.315$). This is because segments often contained concepts from both domains that may or may not have been interwoven, so it was not always clear which Discourse Category a segment belonged to. Because of this, the Reviewers ultimately opted to label all segments via consensus coding.

During *Prompt Development*, we provided the LLM with explicit task instructions, curricular and environment context, and general guidelines (e.g., instructing the LLM to cite evidence directly from the students' discourse to support its summary decisions and Discourse Category choice). We supplemented the prompt with extensive contextual information not found in previous work [9], including the Discourse Categories, C2STEM variables and their values, physics and computing concepts and their definitions, and students' actions in the learning environment (derived from environment logs). Four labeled instances were initially appended to the prompt as few-shot examples (one per Discourse Category). Active Learning was performed for a total of two rounds over 10 validation set instances, at the end of which one additional few-shot instance was added.

Before testing, R1 wrote summaries (and labeled Discourse Categories) for the 12 test instances. R2 then compared the human-generated summaries to two LLMs' summaries: GPT-4-Turbo and GPT-4. We compare GPT-4 to GPT-4-Turbo to see which LLM is most promising for use in future work. To evaluate RQ1, R2 used "ranked choice" to rank the three summaries from best to worst for each test set instance without knowledge of whether the summaries were generated by a human, GPT-4-Turbo, or GPT-4 (the Competitors). Three rankings were used for the scoring: (1) *Wins* (the number of times each Competitor was ranked higher than another Competitor across all instances, i.e., the best Competitor for an individual segment receives two "wins" for outranking the other two Competitors for that segment); (2) *Best* (the number of instances each Competitor was selected as the best choice); and (3) *Worst* (the number of instances each Competitor was selected as the worst choice). To answer RQ1, we used the Wilcoxon signed-rank test to determine if the difference in rankings between the Human and GPT-4-Turbo's summaries was statistically significant. We also qualitatively compared the differences between the summaries. To answer RQ2, we performed qualitative analysis using the constant comparative method and interviewed the Educator to derive GPT-4-Turbo's strengths and weaknesses using the Method for our task.

4   Findings

To answer **RQ1**, we first present the test results for the *Wins, Best,* and *Worst* rankings for the 12 test set instances for all three Competitors in Table 1. For all three metrics, the human outperformed GPT-4-Turbo and GPT-4-Turbo outperformed GPT-4, as evaluated by R2. While ranking, R2 remarked on several occasions that GPT-4-Turbo's responses stood out as being the most detailed and informative regarding the students' problem-solving processes. GPT-4-Turbo also correctly identified a segment's Discourse Category and explained why it did not belong to another category even though the distinction was nuanced. Conversely, GPT-4 suffered from hallucinations in a number of instances and failed to produce a *Best* summary. For example, GPT-4 included a physics concept in its summary that was not part of the discourse and cited irrelevant evidence (e.g., it cited a student who said "I got it" as evidence of a computing concept). R2 also remarked GPT-4 was prone to generating summaries that lacked depth and detail. There were no discernible trends in the LLMs' abilities to classify segments across different Discourse Categories.

Table 1. Competitors' rankings across all test set instances. The best-performing Competitor for each segment is in **bold**.

	R1	GPT-4-Turbo	GPT-4
Wins	**17**	12	7
Best	**8**	4	0
Worst	3	4	5

To quantify our answer for **RQ1**, we tested the differences between the three Competitors' raw rankings via Wilcoxon signed-rank tests, which yielded $p = 0.519$ and $p = 0.266$ comparing the Human to GPT-4-Turbo and GPT-4-Turbo to GPT-4, respectively; and $p = 0.052$ comparing the Human to GPT-4, implying the ranking differences were not significant at the $p = 0.05$ level for all three comparisons. Future work with a larger sample size is necessary to determine if humans outperform GPT-4-Turbo or GPT-4, as our study cannot rule this out (especially given the low p-value comparing the Human to GPT-4). Contrary to the quantitative findings, our qualitative analysis revealed that GPT-4-Turbo exhibited several strengths relative to the Human, such as providing greater detail and explaining why a Discourse Category was not appropriate for a given segment. This is especially useful in supporting classroom instructors and generating automated, adaptive scaffolding for students. To answer **RQ1**, the Method enables GPT-4-Turbo to perform similarly to humans for this task and dataset but both Competitors exhibit nuances that warrant further investigation.

We answer **RQ2** by analyzing GPT-4-Turbo's test set generations via the constant comparative method to discern its strengths and weaknesses and by integrating insights from our conversation with the Educator. The LLM consistently followed prompt instructions, cited relevant discourse pieces similarly to humans (often citing the exact same discourse pieces as R1), adhered to the CoT reasoning chains outlined in the few-shot examples, and selectively extracted relevant information from the segments for summarization. These results corroborate previous findings [3]. Notably, the LLM seamlessly integrated the additional context (see Sect. 3.2) and exhibited accurate coreference resolution, correctly identifying entities like physics and computing concepts when ambiguous pronouns such as "it" or "that" were used. The LLM effectively recognized ambiguous segments, and explained when multiple Discourse Categories may be applicable and why the less relevant one was not chosen. The model also showcased adept zero-shot identification of concepts defined in the prompt but not used in the reasoning chains.

The Educator highlighted the LLM's ability to pinpoint specific items the human may have missed. In one instance, the Educator was shown an LLM-generated segment summary that he initially believed to be *physics-and-computing-synergistic*, but he later agreed with the LLM that the segment was best categorized as *physics-and-computing-separate*, as the students were merely discussing the domains sequentially and not interleaving and forming connections between the cross-domain concepts. The Educator also valued the LLM's ability to highlight when students may be in need of teacher assistance and provided several ideas for enhancing the human-AI partnership by using the LLM's summaries to generate actionable insight to support students' STEM+C learning (e.g., using the LLM's summaries to create a graphical timeline to capture students' conceptual understanding).

Despite GPT-4-Turbo's capabilities, there are notable areas for improvement. The autoregressive nature of LLMs introduces challenges related to reliance on keywords and phrases. As the LLM considers every token generated previously in subsequent iterations, any hallucinations (or misinterpretations of the prompt or its own generation) can propagate forward and compromise the overall integrity of its response. An instance of this occurred when the LLM fixated on two physics concepts during summarization when the segment was almost entirely focused on the computing domain. GPT-4-Turbo's initial focus on the physics concepts caused it to label the segment as *physics-and-computing-synergistic*, even though both Reviewers and GPT-4 all considered the segment to be *computing-focused*. Additionally, the LLM's ability to integrate environment actions in its summaries was limited, often addressing them superficially without connecting them to the broader discourse context. The Educator also suggested the LLM should consider the temporality of segments by incorporating "pause" identification and duration from prosodic audio via timestamps and suggested highlighting instances where students expressed uncertainty by saying things like "Um..." or "I'm still a little stumped", as both may help teachers identify students' difficulties.

5 Conclusion, Limitations, and Future Work

The primary limitation of this exploratory study is its small test sample size (12 segments). Additionally, only one researcher ranked the three sets of test set summaries (one human-generated and two LLM-generated). The constraints we faced were the time-cost of manually labeling, summarizing, analyzing, and evaluating the individual segments and summaries (up to two hours and ≈256 tokens per segment summary). While these results cannot be generalized, we have demonstrated the Method's potential for summarizing and characterizing students' synergistic discourse in a manner that can deepen educators' insights into students' cross-domain conceptual understandings. In future work, we will conduct an extensive evaluation to test the generalizability of our approach, including evaluating our Method's performance across various learning tasks.

Acknowledgments. This work is supported under National Science Foundation awards DRL-2112635 and IIS-2327708.

Disclosure of Interests. The authors have no competing interests to declare.

References

1. Basu, S., et al.: Identifying middle school students' challenges in computational thinking-based science learning. Res. Pract. Technol. Enhanc. Learn. **11**(1), 13 (2016)
2. Brown, T.B., et al.: Language models are few-shot learners. arXiv e-prints arXiv:2005.14165 (2020)
3. Cohn, C., Hutchins, N., Le, T., Biswas, G.: A chain-of-thought prompting approach with LLMs for evaluating students' formative assessment responses in science. In: Proceedings of the AAAI Conference on Artificial Intelligence, vol. 38, no. 21 (2024)
4. Hadwin, A., et al.: Self-regulated, co-regulated, and socially shared regulation of learning. Handb. Self-regulation Learn. Perform. **30**, 65–84 (2011)
5. Hutchins, N., et al.: C2STEM: a system for synergistic learning of physics and computational thinking. J. Sci. Educ. Technol. **29** (2020)
6. Meier, A., Spada, H., Rummel, N.: A rating scheme for assessing the quality of computer-supported collaboration processes. IJCSCL **2**, 63–86 (2007)
7. Roschelle, J., Teasley, S.D.: The construction of shared knowledge in collaborative problem solving. In: O'Malley, C. (ed.) Computer Supported Collaborative Learning. NATO ASI Series, vol. 128, pp. 69–97. Springer, Heidelberg (1995). https://doi.org/10.1007/978-3-642-85098-1_5
8. Snyder, C., et al.: Analyzing students' synergistic learning processes in physics and CT by collaborative discourse analysis. In: CSCL (2019)
9. Snyder, C., Hutchins, N.M., Cohn, C., Fonteles, J.H., Biswas, G.: Analyzing students collaborative problem-solving behaviors in synergistic STEM+C learning. In: Proceedings of the 14th Learning Analytics and Knowledge Conference (2024)

10. Suraworachet, W., Seon, J., Cukurova, M.: Predicting challenge moments from students' discourse: a comparison of GPT-4 to two traditional natural language processing approaches. arXiv preprint arXiv:2401.01692 (2024)
11. Wei, J., et al.: Chain-of-thought prompting elicits reasoning in large language models. arXiv e-prints arXiv:2201.11903 (2022)
12. Weinberger, A., Fischer, F.: A framework to analyze argumentative knowledge construction in computer-supported collaborative learning. Comput. Educ. **46**(1), 71–95 (2006)

G-Learn: A Graph Machine Learning Content Recommendation System for Virtual Learning Environments

Hugo Firmino Damasceno[1](✉), Leonardo Sampaio Rocha[1](✉), and Antonio de Barros Serra[2](✉)

[1] Graphs and Computational Intelligence Lab - LAGIC, Postgraduate Program in Computer Science - PPGCC, Center for Science and Technology - CCT, State University of Ceará - UECE, Fortaleza, Ceará, Brazil
`hugo.damasceno@aluno.uece.br, leonardo.sampaio@uece.br`
[2] Institute of Technological Networks and Energies - IREDE, Fortaleza, Ceará, Brazil
`prof.serra@gmail.com`

Abstract. Recommender Systems in Virtual Learning Environments (VLEs) provide personalized suggestions to users based on preferences, interaction history, and behavior. They enhance learning by offering personalized content, increasing engagement, and improving teaching effectiveness. Challenges in VLEs include the cold start problem, data sparsity, and limited coverage. To address these, we propose G-Learn, a recommendation system operating in both supervised and unsupervised models. It utilizes graph machine learning, keyword mining, and similarity techniques to recommend educational materials tailored to each student's performance. We demonstrate G-Learn's effectiveness in a real scenario using data from Homero, a VLE for computer science education developed for the brazilian federal government. Validation shows an average f1-score of 0.64 in unsupervised model and 0.95 in supervised model.

Keywords: Recommendation systems · Virtual learning environments · Graph machine learning

1 Introduction

Recommendation systems are information systems that provide personalized suggestions or recommendations to users based on their preferences, interaction history, and behavior patterns. They can recommend products, services, digital content, or any other type of item relevant to the user [8].

A relevant domain for the use of recommendation systems is Virtual Learning Environments (VLEs), remote learning platforms where students can access educational materials, activities, assessments, etc. These systems have been increasingly used because they allow for greater scope and flexibility in education, as well as cost reduction.

A. M. Olney et al. (Eds.): AIED 2024 Workshops, CCIS 2151, pp. 20–28, 2024.
https://doi.org/10.1007/978-3-031-64312-5_3

In [10], the authors identify key challenges in educational recommendation systems, namely the cold start problem and data sparsity. The cold start problem refers to difficulties in providing accurate recommendations for new users or items with limited history. Data sparsity arises from a lack of interactions between users and items. Additionally, limited coverage poses challenges in recommending all available items in the system [8].

To address these challenges, we propose and validate G-Learn, a content recommendation system for VLEs that uses machine learning on graphs to recommend educational materials based on students' knowledge levels observed in assessments.

G-Learn allows for the recommendation of educational materials tailored to each student's performance, assessed based on their responses to questions within the platform. The system can operate in both unsupervised and supervised models, depending on the availability of a training dataset.

Besides the practical interest of G-Learn, the main research questions that guide this work are: Can a robust graph-based machine learning recommendation system be developed for questions and materials within a virtual learning environment, effectively addressing the challenges of cold start, data sparsity and limited coverage? Furthermore, can it achieve high F1-scores in comparison to recommendations provided by experts?

The work is structured as follows: Sect. 2 references similar proposals and provides scientific background. Section 3 presents the concepts and methodologies used in the proposal's development. Section 4 showcases the results obtained and compares them with results from related work, and finally, Sect. 5 outlines future work intentions.

2   Related Works

The study in [5] presents an algorithm that combines knowledge graph and collaborative filtering to recommend courses on an online learning platform. It embeds semantic information into a low-dimensional space using knowledge graph representation learning, calculates semantic similarity between items, and integrates this into a collaborative filtering recommendation algorithm. Results suggest the approach improves recommendation effectiveness over methods solely using collaborative filtering.

In [13] the authors propose a knowledge graph approach for the recommendation of personalized learning resources. Their supervised algorithm is based on the interest similarity and knowledge connection degree of items. They validate the recommendation algorithm by experiments that show the correctness and effectiveness of their approach.

In [3] the authors propose a model built to facilitate reaching the appropriate learning path for each learner among a large group of lessons. The supervised model is based on knowledge graphs, where at the first step they group users by their profile and then connect each group of users to its appropriate learning path.

In [12] they present MKR, a multi-task feature learning approach for Knowledge graph enhanced Recommendation. They prove that MKR is a generalized framework over several representative methods of recommender systems and multi-task learning. Through experiments on real-world datasets, they show that MKR achieves substantial gains in movie, book, music, and news recommendation, over state-of-the-art methods.

Following the proposal of MKR, in [2] it is proposed MKCR, that allows to enhancement of online course source recommendations when the interaction between students and courses is extremely sparse. MKCR is an end-to-end framework that utilizes a knowledge graph embedding task to assist recommendation tasks. The experiment data partially come from the MOOC platform of Chinese universities. The experiments show that MKCR has better performance than other methods.

In [7], a method for personalized exercise recommendation is introduced. It utilizes a knowledge graph depicting study areas and prerequisite relationships within a discipline. Each student receives a personalized graph with weights assigned to vertices based on their competence in each subject, predicted using data from a training set. The algorithm suggests exercises in areas where the student struggles, i.e., where their competence is below average, resulting in improved student scores compared to baseline methods.

Table 1 shows that G-Learn's main differentials from previous works are: being able to function in supervised and unsupervised model, which is important in the absence of training data; being able to recommend both materials and questions; and focusing on content features, which is an important differential that allows the system to avoid cold start, data sparsity, and limited coverage problems.

Table 1. Comparison of related works

	Recommendation				RS Technique			ML Technique	
	Ma	LP	Cs	Qu	CF	Ct	KG	Su	Un
KHALIL [5]			✓		✓	✓	✓	✓	
WEI [13]	✓						✓	✓	
CHETOUI [3]		✓					✓	✓	
WANG [12]					✓		✓		
CHEN [2]		✓			✓		✓	✓	
LV [7]				✓			✓	✓	
G-LEARN	✓			✓		✓	✓	✓	✓

Ma - Materials, **Qu** - Questions, **KG** - Knowledge Graph, **LP** - Learning Path, **CF** -Collaborative Filtering, **Cs** - Courses, **Ct** - Content, **Su** - Supervised, **Un** - Unsupervised

3   Our Proposal

G-Learn is an end-to-end framework that takes as input a course in a VLE, composed of a set of Learning Modules (LMs), each of which is composed of Learning Objects (LOs), that can be either questions or materials. The architecture of G-Learn is described in Fig. 1, and we detail the steps below.

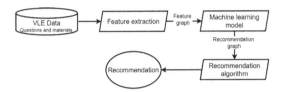

Fig. 1. The main steps of the G-learn framework

Feature Extraction and the Feature Graph. The objective of this step is to generate a graph $G_F = (V, E)$ whose vertices $V = Q \cup M$ are formed from the set of materials M and questions Q, and $E = \{qm \in Q \times M\}$, i.e., a complete bipartite graph. Associated with each vertex and edge of the graph will be a set of features that will enable machine learning models to recommend appropriate materials for each question.

To obtain the graph's features, the first step involves transcribing the learning objectives files from the VLE, commonly in text and video formats such as PDF and MP4, into strings. For this step, several tools are available, such as the *Tika* and *SpeechRecognition* libraries in Python, which were used in this work.

From the transcribed data of materials and questions, keyword mining techniques are applied to the set of transcripted learning objects in each module. In this work, we considered techniques such as Rake [9], Yake [1], and TF-IDF [11], although other techniques can be used.

The extracted keywords undergo a filtering process based on the titles associated with the modules and learning objectives of the course on the platform so that irrelevant keywords are not considered in the context. Although we have not presented this comparison due to space limitations, in our experiments, this filtering proved to be of great importance for the performance of recommendations.

After the keyword mining and filtering process, we obtain the set $K = k_1, \ldots, k_k$, consisting of all k mined keywords. For each node $x \in Q \cup M$ of the graph, a *keyword vector*, $key(v) = (p_1, \ldots, p_k)$, is associated, where position i corresponds to keyword $k_i \in K$ and contains the frequency of that keyword in node v, see Fig. 2.

For each edge $mq \in E(G_F)$, the feature $sim(mq) \in \mathbb{R}$, $0 \leq sim(mq) \leq 1$, will measure the similarity between the material $m \in M$ and the question $q \in Q$. To obtain the similarity values of edge $mq \in E(G_F)$, the keyword vectors $key(m)$ and $key(q)$ will be applied to the cosine similarity measure [6], see (1).

$$cos(key(m), key(q)) = \frac{key(m) \cdot key(q)}{\|key(m)\| \cdot \|key(q)\|} \qquad (1)$$

Finally, other features will be added to the vertices, such as weighted degree, which is the sum of the similarities of the edges incident on a vertex; the number of characters in the learning object; and the material type, 0 for text and 1 for video, applicable only to vertices corresponding to materials.

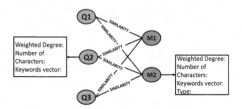

Fig. 2. Feature graph

Machine Learning Model and the Recommendation Graph

With the feature graph in hand, it becomes possible to use a graph machine learning model to generate the recommendation graph. The recommendation graph $G_R = (V, E)$ has vertices $V(G_R) = M \cup Q$, and its set of edges $E(G_R)$ consists only of pairs m, q for which the machine learning model has predicted the edge mq as positive, i.e., the material m is sufficient for solving question q.

To predict these edges, it is possible to opt for either a supervised or unsupervised algorithm, depending on the availability of a training dataset for the considered VLE.

As the unsupervised approach, for this work, we considered a minimum similarity threshold for an edge to be predicted. Thus, $mq \in E(G_R)$ if $sim(m, q) \geq l$, where $0 < l \leq 1$ is the threshold.

As the supervised approach, we used the *Random Forest* model [4], which is commonly used in machine learning research.

The choice of other machine learning models is at the discretion of the framework user, without the need for significant adaptations.

Recomendation Algorithm. Once the recommendation graph has been created, it is used by the recommendation algorithm for generating the recommendations. Given that a student has encountered difficulties with a set of questions Q', a simple approach is to select a set of materials M' in a greedy manner, choosing in each iteration a material that covers the maximum number of questions from Q', until there are no more questions $q \in Q'$ that can be covered by a material $m \notin M'$.

A more sophisticated approach, which takes into account other aspects such as minimizing the number of recommended materials, the size of the materials, and the type of media, among others, can be obtained through a multi-objective optimization metaheuristic, such as evolutionary algorithms.

4   Experiments and Results

In this section, we describe the experiments and results of applying the proposed approach to a dataset obtained from the Homero platform, related to the Data Structures course.

4.1   Dataset

The validation set data was collected by three undergraduate Computer Science students who completed the Data Structures course. They associated each question in the course with the materials relevant to solving it. The validation set was constructed based on associations indicated by the majority of students. It was used to evaluate recommendation graphs generated in unsupervised and supervised models, considering metrics like Accuracy, F1-Score, Precision, and Recall. The dataset was formed by applying feature extraction on learning modules and course objects related to the Data Structures course. Each sample in the dataset represents a triple (q_{Mod}, m_{Mod}, Mod), where Mod is a learning module, and q_{Mod} and m_{Mod} are a question and material from the same module, with associated features. Each q_{Mod}, m_{Mod} pair is labeled as 0 or 1 based on majority student identification of their relationship. The Data Structures course used comprises 131 materials and 257 questions divided into 20 learning objectives.

4.2   Supervised Model

As a supervised model, in our experiments we considered the Random Forest algorithm and the keyword extraction method RAKE, evaluating its performance through k-fold cross-validation with $k = 5$.

Table 2 presents the results obtained. With the model achieving an average accuracy of 0.95, a f1-score of 0.95, a precision of 0.94, and a recall of 0.96.

Table 2. Supervised Results

	RAKE			
	Accuracy	F1-Score	Precision	Recall
Average	0.95	0.95	0.94	0.96
Median	0.97	0.95	0.94	0.96
Standard Deviation	0.03	0.04	0.06	0.04

The model was able to achieve good results in all learning modules, as shown Fig. 3, displays the results of the F1-Score metric for each learning module, exhibiting a similar pattern to precision regarding the behavior across the learning modules.

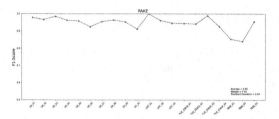

Fig. 3. Attained F1-score for each Learning Module using the supervised model.

4.3   Unsupervised Model

In our experiments, we employed an unsupervised model to generate the recommendation graph based on similarity thresholds, as described in Sect. 3. An important aspect of using this model is to find a threshold value that allows us to achieve the highest effectiveness of the method in the recommendation process.

To find the best threshold value, we utilized the ROC Curve, which enables us to assess the performance of a classifier as the classification threshold varies. The Area Under the Curve (AUC) was 0.63 for RAKE, 0.60 for YAKE, and 0.53 for TFIDF, which made RAKE the choice for keyword extraction in our model. We chose the threshold as 0.4, as it provides a good balance between the true positive rate and the false positive rate.

We then generated the recommendation graph with RAKE and the threshold of 0.4 and verified the results based on the validation set. The model achieved an average accuracy of 0.59, and a f1-score of 0.64. The highlight was the recall metric, with an average of 0.84, in contrast to a 0.55 precision. Improving the precision is an important challenge, as avoiding false positives is crucial in our recommendation task. This leaves place for improvements in the feature graph and in the unsupervised model.

4.4   Comparison with Related Works

Among the surveyed works that are based on supervised models, the one that achieved the best F1-score result was [12] with a score of 0.69, along with the highest recall and precision at 0.36 and 0.52, respectively. Meanwhile, the study [7] reported the highest accuracy, with a value of 0.88. However, G-learn with our supervised model proposal surpassed these results, attaining an F1 score of 0.95, with precision at 0.94 and recall at 0.96.

On the other hand, G-learn with our unsupervised model proposal achieved an F1-score of 0.64, with a high recall of 0.84 but a low precision of 0.55. While the F1-score is reasonable for an unsupervised model, there is a clear need for improvements to enhance precision, a crucial metric for our task. It's worth noting that the related works did not utilize unsupervised models, hence the lack of comparison in this regard.

5   Conclusion

G-learn achieved a high f1-score compared to expert recommendations, demonstrating the robustness of our approach in handling the "cold start" problem, as all necessary data for recommendations were available in the dataset. We also overcame challenges of data scarcity and limited coverage by considering all relationships between materials and questions. For future work, we aim to test the model's performance across different courses, collect new user data to enhance recommendation effectiveness, and conduct user experiments in Virtual Learning Environments (VLEs) to evaluate their perception of the recommendations.

Acknowledgement. This project is supported by the Ministry of Science, Tech- nology, and Innovation, with resources from Law No. 8,248, of October 23, 1991, within the scope of PPI-Softex, coordinated by Softex and published in Project Residency in TIC 10.

References

1. Campos, R., Mangaravite, V., Pasquali, A., Jorge, A., Nunes, C., Jatowt, A.: YAKE! Keyword extraction from single documents using multiple local features. Inf. Sci. **509**, 257–289 (2020)
2. Chen, X., Sun, Y., Zhou, T., Wen, Y., Zhang, F., Zeng, Q.: Recommending online course resources based on knowledge graph. In: Zhao, X., Yang, S., Wang, X., Li, J. (eds.) WISA 2022. LNCS, vol. 13579, pp. 581–588. Springer, Cham (2022). https://doi.org/10.1007/978-3-031-20309-1_51
3. Chetoui, I., El Bachari, E., El Adnani, M.: Course recommendation model based on knowledge graph embedding. In: 2022 16th International Conference on Signal-Image Technology & Internet-Based Systems (SITIS), pp. 510–514. IEEE (2022)
4. Cutler, A., Cutler, D.R., Stevens, J.R.: Random forests. In: Ensemble Machine Learning: Methods and Applications, pp. 157–175 (2012)
5. Khalil, A.M., Xu, G., Jia, G., Shi, L., Zhang, Z.: Personalized course recommendation system fusing with knowledge graph and collaborative filtering. Comput. Intell. Neurosci. **2021**, 9590502 (2021)
6. Li, X., Li, F., Zhang, J.: A comprehensive evaluation of text similarity measures. J. Inf. Sci. **46**(6), 957–976 (2020)
7. Lv, P., Wang, X., Xu, J., Wang, J.: Intelligent personalised exercise recommendation: a weighted knowledge graph-based approach. Comput. Appl. Eng. Educ. **29**(5), 1403–1419 (2021)
8. Ricci, F., Rokach, L., Shapira, B.: Recommender systems: introduction and challenges. In: Recommender Systems Handbook, pp. 1–34 (2015)
9. Rose, S., Engel, D., Cramer, N., Cowley, W.: Automatic keyword extraction from individual documents. Text Min.: Appl. Theory 1–20 (2010)
10. da Silva, F.L., Slodkowski, B.K., da Silva, K.K.A., Cazella, S.C.: A systematic literature review on educational recommender systems for teaching and learning: research trends, limitations and opportunities. Educ. Inf. Technol. **28**, 3289–3328 (2023)
11. Sparck Jones, K.: A statistical interpretation of term specificity and its application in retrieval, pp. 132–142. Taylor Graham Publishing, GBR (1988)

12. Wang, H., Zhang, F., Zhao, M., Li, W., Xie, X., Guo, M.: Multi-task feature learning for knowledge graph enhanced recommendation. In: The World Wide Web Conference, WWW 2019, pp. 2000–2010. Association for Computing Machinery, New York (2019)
13. Wei, Q., Yao, X.: Personalized recommendation of learning resources based on knowledge graph. In: 2022 11th International Conference on Educational and Information Technology (ICEIT), pp. 46–50 (2022)

Exploring the Role, Implementation, and Educational Implications of AI in Early Childhood Education

Iro Voulgari[1]([⊠]) [iD], Konstantinos Lavidas[2] [iD], Spyridon Aravantinos[2] [iD], Soultana Sypsa[3], and Maria Sfyroera[1]

[1] National and Kapodistrian University of Athens, 10676 Athens, Greece
{voulgari,msfyroera}@ecd.uoa.gr
[2] University of Patras, 26504 Rion, Achaia, Greece
{lavidas,aravantinos_spyridon}@upatras.gr
[3] University of Piraeus, 18534 Piraeus, Greece

Abstract. In this paper we are examining the evolving landscape of the implementation of Artificial Intelligence (AI) in early childhood education. We are reviewing journal papers published over the past decade (2012–2023) (N = 29) and analyse them descriptively and through thematic analysis. We focus on the learning context and educational content of the AI applications in preschool settings. Research in the field has focused on areas such as the design and development of AI-powered tools and intelligent systems, the interaction of children with AI technologies and their impact on children's skills and development, while more recently AI is being also viewed as a learning objective in the context of AI education and literacy. Our findings suggest that AI can be leveraged to foster a range of cognitive, social, and emotional skills through appropriate pedagogical frameworks and carefully designed pedagogical practices. The trends identified in the paper suggest the need for multidisciplinary approaches where technology, pedagogy, education, psychology, and policymaking intersect.

Keywords: Artificial Intelligence · AI literacy · AI education · preschool education · early childhood education

1 Introduction

This paper examines the emerging landscape of Artificial Intelligence (AI) in early childhood education (ECE), reviewing studies from the last decade to analyse trends and pedagogical implications. While AI education tools and practices are gradually being explored mainly in higher and secondary education [1], their inclusion in preschool settings is still limited. This study, builds upon previous reviews on AI in ECE [2–4] and adopts a pedagogical lens to focus on the educational approaches involved in the design of tools and implementation [5].

Given the increasing presence of AI environments mediating children's entertainment, learning, and aspects of their everyday life, such as video and music recommendations, image and voice recognition, smart toys and devices, there is a significant need

A. M. Olney et al. (Eds.): AIED 2024 Workshops, CCIS 2151, pp. 29–37, 2024.
https://doi.org/10.1007/978-3-031-64312-5_4

to examine how these technologies are introduced, their content, and ensure the children's safety and privacy [6–8]. Our study aims to contribute to this area by identifying gaps and appropriate pedagogical frameworks that consider the unique needs of young learners as well as the whole learning ecosystem (e.g., role of educators) [9]. Previous reviews have indicated limitations of targeted educational strategies for this age group, highlighting the importance of our focus on pedagogically sound AI applications [6, 10]. The research questions guiding our study are: a) what is the role of the teachers in AI education in preschool? What skills do they need to have? b) what is the role of the children? c) what is the AI-related educational content and AI tools used? and d) what are the pedagogical foundations and frameworks implemented?

2 Methods

2.1 Data Collection and Selection

We selected journal papers published between 2012 and 2023, with a search conducted on 18 August 2023. Keywords included the terms: preschool* OR "early childhood" OR kindergarten OR "pre-primary" OR "early years" AND "Artificial Intelligence" OR "Machine Learning" OR "AI-assisted" AND education OR learning, in the title, abstract, or keywords of the papers.

After removing duplicates, we screened titles and abstracts to exclude studies not addressing ECE, reviews, metareviews, and non-peer-reviewed papers. Our inclusion criteria centered on studies discussing AI's role in supporting learning and education at the preschool level. The final sample comprised 29 papers; for a detailed breakdown of these studies, see https://bit.ly/reviewedpapers and Table 1.

Table 1. Sample selection process.

Step	Total/Databases
Documents identified in Scopus, Education Research Complete, and Web of Science	960
Records after duplicates removed	733
Records excluded after title/abstract screening	631
Records removed after full text examination	73
Documents included in qualitative analysis	29 (Education Research Complete 9; Scopus 6; Web of Science 14)

2.2 Data Analysis

A coding framework which included axes from the framework proposed in [8] (e.g., Learning Content, AI tool used, Learning Outcomes), for comparison of findings, and also axes based on our research questions (e.g., Teacher role, Teacher skills) was designed. It included axes such as research methods, ethical issues, and future research, but the axes we are focusing on here are: Research Goal, Type of Study, Participants, Learning Content, Learning Activity, Learning Outcomes, AI tool used, Pedagogy, Teacher role, and Teacher skills. Based on these axes, 3 of the authors analysed the papers thematically, through focused coding [11] to identify themes and categories grounded on the data rather than existing theory [12]. After coding, the categories and themes that emerged were discussed and negotiated among the 3 coders until a consensus was reached [13]. These themes are discussed in the Results section.

3 Results

3.1 Types of Papers and Research Goals

Since 2019, a rising trend in publications was observed, with 15 papers in 2022 and 6 in 2023, aligning with reviews of AI in K-12 and ECE by [5] and [14] respectively. The journals reveal diverse domains exploring the field: journals typically focusing on technological aspects, journals on child development, special education, and the intersection of education and technology. Most papers were published on journals on the technology implementation in education, and engineering and technology (Table 2). There was a strong trend of studies on the development and implementation of AI-related technologies and tools, while others examined perceptions, and discussed guidelines for a safe and effective use of AI (Table 3).

Table 2. Domain of published papers reviewed.

Domain	Journals	Number of papers
Technology in Education	Computers and Education: Artificial Intelligence; International Journal of Artificial Intelligence in Education; Interactive Learning Environments; Education and Information Technologies; Informatics in Education; Journal of Research on Technology in Education; International Journal of Child-Computer Interaction	10

(continued)

Table 2. (*continued*)

Domain	Journals	Number of papers
Engineering and Technology	AI EDAM-Artificial Intelligence for Engineering Design Analysis and Manufacturing; Applied Artificial Intelligence; Electronics; Displays; Future Internet; Sensors	6
Interdisciplinary and Applied Research	AI Magazine; Computational and Mathematical Methods in Medicine; Computational Intelligence and Neuroscience; Mathematical Problems in Engineering; Mobile Information Systems	5
Child Development and Psychology	Child Development Perspectives; Developmental Psychology; European Early Childhood Education Research Journal	3
Special Education and Inclusive Education	International Journal of Early Childhood Special Education; Occupational Therapy International	3
General Education and Policy	London Review of Education	1

Table 3. Research focus of papers

Category	Description	Number of Papers
Technology Design and Development	Studies focusing on the design and development of new technologies or tools	9
Technology Implementation and Impact	Studies assessing the implementation of technologies in various learning settings (e.g. experimental, classrooms) and their impact on learning and child development	9
Policies and Guidelines	Studies discussing policies and guidelines for the safe, ethical, and effective use of AI and technology in preschool education	6
Perceptions and Attitudes	Studies exploring the perceptions and attitudes of various stakeholders (i.e., teachers, parents, policymakers, public)	5

3.2 Learning Content and Activities

Fifteen of the studies involved children aged 2–9, mainly involved in studies on technology "Design and Development" or "Implementation and Impact", assessing the effectiveness of technologies.

Activities. Examining the activities the children engaged in, the following broader categories emerged: a) *Children as tutors* where they used platforms such as Google's Teachable Machine to build models, programmed a game to play Rock-Paper-Scissors, or taught the computer to recognize emotions. In these cases the children learnt by teaching the system to perform tasks; b) *Creative expression* where the children explored the learning content (e.g., learnt about generative AI) by creating music remixes, drawing, making concept maps, creating personal stories, and used various tools such as Quick Draw and PictoBox; c) *Learning concepts* such as numbers, vocabulary, colours, shapes, animals; d) *Engaging in inquiry, exploration, and experimentation*, where they conducted chemistry experiments, observed colour changes, or reflected on their gameplay with smart toys (Fig. 1).

Learning Outcomes. Through the analysis of the learning objectives, outcomes, or skills addressed in the studies, the following themes emerged: a) *Social and Emotional Skills*, such as collaboration, communication and cooperation skills, moral qualities, good habits, creative collaboration, empathy, and emotional enquiry were the most frequently cited skills (16 instances); b) the second most frequently cited theme was *Cognitive Development Skills* including memory, observation, critical thinking, problem-solving, imagination, creative inquiry, cognitive flexibility, and externalising ideas; c) *Literacy and Language Skills* referring to reading literacy, pronunciation, language learning, vocabulary, alphabet, and idioms; d) *Numeracy and Mathematics* referring to math, natural numbers, and number sense; e) *Physical and Motor Skills* such as fine motor skills, spatial skills, and drawing shapes; f) *AI and Technological Literacy* refereeing to understanding concepts and implications of AI such as algorithmic bias, and e) *Specific Content Knowledge* such as animal names, colours, shapes (Fig. 2).

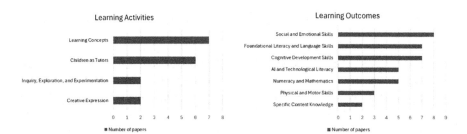

Fig. 1. Learning activities of the children.   **Fig. 2.** Learning outcomes reported.

A horizontal learning goal or outcome, underlying a wide range of studies, was leveraging the *motivation and engagement* of the children, by attracting their attention, supporting their efforts, and increasing their concentration. The *learning domains* addressed in the studies spanned a wide range of fields, from STEAM, chemistry, mathematics, to language, music, and the Arts.

3.3 Pedagogy of Activities or AI Tools

The learning frameworks cited reflect a shift towards more interactive and learner-centred approaches supporting the development of higher-order thinking skills, collaborative learning, and problem-solving. However, in many cases, no learning frameworks were reported.

Specifically, 5 broader themes of cited learning approaches emerged (Fig. 3): a) *constructivist and sociocultural approaches* with references to Papert's Constructionism, Vygotsky's Sociocultural Theory and Zone of Proximal Development, Vygotsky's Mediation Theory, socio-culturally constructed understanding, and constructivist learning; b) *personalised and adaptive learning and teaching*, where the learning systems adapt to the needs, learning style, requirements of the learners (e.g., smart learning environments, personalised books) or the teachers adapt the learning material to learning goals and needs of the students using AI applications; c) *collaboration and social learning*, with mentions to collaborative learning, participatory learning, and socialisation; d) *active and experiential learning*, with references to immersive and interactive experiences, problem-solving, game-based learning, project-based approaches, and learning-by-teaching (children as tutors of the computer system). And finally, e) *multimodal pedagogy* and learning through audio, images etc.

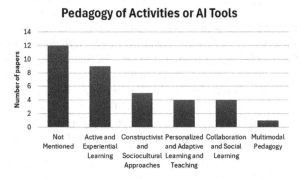

Fig. 3. Pedagogies of activities or tools reported.

The AI tools and platforms used to directly support the activities of the children (i.e., excluding references to image and voice recognition systems, databases, etc.) mainly involved *digital educational platforms and tools* (11) such as Scratch, App Inventor, Snap!, Google's Quick Draw!, and AI for Oceans, with Google's Teachable Machine being the most referenced (3), or *tangible objects* such as robots and smart toys (8).

3.4 The Multiple Roles of the Teachers and Teacher Skills

Educators participated in 10 studies, primarily involved in the design, implementation, and testing of technologies. Their perceptions and attitudes were examined in 3 studies, and 1 study involved them as experts on a panel proposing assessment tools and approaches.

The analysis revealed 3 main roles of educators: a) *facilitators and guides*, helping to scaffold learning experiences, engage children, and support their skills, and emotional and social development; b) *mediators* ensuring the safety and equitable access to technology and addressing ethical challenges; and c) *providers of expert knowledge* in designing and evaluating educational technologies, understanding children's needs, and collaborating with researchers.

Two key skill areas emerged: a) *pedagogical expertise*, including content knowledge and supporting children's curiosity and social development (3 cases), and b) *technological and AI literacy*, crucial for integrating and evaluating AI systems and ensuring safe and ethical technology use (4 cases). Both skill sets were required in 5 cases. Two studies underscored the importance of continuous professional development. This range of skills indicates the demanding nature of the teachers' role and the necessity for continuous training and teachers' support.

4   Discussion and Conclusions

This review builds upon existing research to highlight emerging trends and critical areas in AI applications in ECE, aiming to foster a more robust discourse to guide future studies and inform practice and policymaking. Although our small sample size, as in [14], reflects the nascent state of AI education research at lower educational levels [5], expanding our sample to include conference papers could help uncover early trends [14].

Most studies in our sample primarily focus on technology design, development, and implementation, suggesting a possible technological bias. While these are valuable, there's a scarcity of research on the long-term effects of AI on children's learning, emotional and social outcomes. However, the interdisciplinary application of AI, involving domains such as STEAM, underlines the need for multidisciplinary approaches where technology, pedagogy, education, psychology, and policymaking intersect.

Educators were actively involved, mainly in technology implementation, yet there is insufficient information on their AI-related pedagogical training and ongoing support, which are vital for sustainable integration [9, 14]. A broad spectrum of technological and pedagogical skills is essential. More active engagement of stakeholders, especially children and educators, through approaches such as participatory design, could be suggested, to address their needs and requirements.

Digital tools were mostly reported for supporting learning, yet the rising use of tangible AI tools like robots suggests a shift towards more hands-on educational approaches [14]. Although many studies do not focus on a specific pedagogical framework, those reported promote active, constructivist, and collaborative learning, positioning children in active roles that foster creative expression and inquiry, moving beyond behaviouristic approaches which aim at the passive acquisition of information and content knowledge.

Future research should explore diverse cultural, social, and educational contexts to tackle the digital divide and ensure inclusivity and equitable AI access [14]. Addressing overlooked issues such as data privacy, AI bias, and ethical concerns is crucial. Empirical studies in naturalistic settings are recommended to understand AI's practical impacts better [14]. Interdisciplinary collaborations are necessary to develop a cohesive framework for ethically integrating AI into early childhood curricula.

Certainly, since this was a qualitative approach we cannot generalise the findings, but the dimensions emerged could be a basis for further, in-depth examination of aspects such as the features and quality of the relevant educational content, the effectiveness of the learning approaches integrated in the material and the activities, and the training needs and requirements of teachers and children.

Disclosure of Interests.. The authors have no competing interests to declare that are relevant to the content of this article.

References

1. Tenório, K., Olari, V., Chikobava, M., Romeike, R.: Artificial Intelligence Literacy Research field: a bibliometric analysis from 1989 to 2021. In: Proceedings of the 54th ACM Technical Symposium on Computer Science Education V. 1, pp. 1083–1089. Association for Computing Machinery, New York, NY, USA (2023). https://doi.org/10.1145/3545945.3569874
2. Crescenzi-Lanna, L.: Literature review of the reciprocal value of artificial and human intelligence in early childhood education. J. Res. Technol. Educ. **55**, 21–33 (2023). https://doi.org/10.1080/15391523.2022.2128480
3. Antonenko, P., Abramowitz, B.: In-service teachers' (MIS)conceptions of artificial intelligence in K-12 science education. J. Res. Technol. Educ. **55**, 64–78 (2023)
4. de Haas, M., et al.: Engagement in longitudinal child-robot language learning interactions: disentangling robot and task engagement. Int. J. Child-Comput. Inter. **33**, 100501 (2022). https://doi.org/10.1016/j.ijcci.2022.100501
5. Zafari, M., Bazargani, J.S., Sadeghi-Niaraki, A., Choi, S.-M.: Artificial intelligence applications in K-12 education: a systematic literature review. IEEE Access **10**, 61905–61921 (2022). https://doi.org/10.1109/ACCESS.2022.3179356
6. UNICEF Office of Global Insight & Policy: AI for children. https://www.unicef.org/global insight/featured-projects/ai-children. Accessed 06 Feb 2024
7. Ng, D.T.K., Leung, J.K.L., Chu, S.K.W., Qiao, M.S.: Conceptualizing AI literacy: an exploratory review. Comput. Educ.: Artif. Intell. **2**, 100041 (2021). https://doi.org/10.1016/j.caeai.2021.100041
8. Su, J., Ng, D.T.K., Chu, S.K.W.: Artificial intelligence (AI) literacy in early childhood education: the challenges and opportunities. Comput. Educ.: Artif. Intell. **4**, 100124 (2023). https://doi.org/10.1016/j.caeai.2023.100124
9. Voulgari, I., Stouraitis, E., Camilleri, V., Karpouzis, K.: Artificial intelligence and machine learning education and literacy: teacher training for primary and secondary education teachers. In: Handbook of Research on Integrating ICTs in STEAM Education, pp. 1–21. IGI Global (2022). https://doi.org/10.4018/978-1-6684-3861-9.ch001
10. Williams, R., Park, H.W., Breazeal, C.: A is for artificial intelligence: the impact of artificial intelligence activities on young children's perceptions of robots. In: Proceedings of the 2019 CHI Conference on Human Factors in Computing Systems - CHI 2019, pp. 1–11. ACM Press, Glasgow, Scotland UK (2019). https://doi.org/10.1145/3290605.3300677
11. Saldaña, J.: The Coding Manual for Qualitative Researchers. Sage, London (2009)
12. Braun, V., Clarke, V.: Using thematic analysis in psychology. Qual. Res. Psychol. **3**, 77–101 (2006). https://doi.org/10.1191/1478088706qp063oa

13. Wiltshire, G., Ronkainen, N.: A realist approach to thematic analysis: making sense of qualitative data through experiential, inferential and dispositional themes. J. Crit. Realism **20**, 159–180 (2021). https://doi.org/10.1080/14767430.2021.1894909
14. Su, J., Yang, W.: Artificial intelligence in early childhood education: a scoping review. Comput. Educ.: Artif. Intell. **3**, 100049 (2022). https://doi.org/10.1016/j.caeai.2022.100049

Eighth Graders and a Math Intelligent Tutoring System: A Deep Neural Network Analysis

Kelun Lu[1]([✉]), Lingxin Hao[2], I-Jeng Wang[3], and Anqi Liu[4]

[1] School of Education, Johns Hopkins University, Baltimore, MD, USA
Kelun.Lu@jhu.edu
[2] Department of Sociology, Johns Hopkins University, Baltimore, MD, USA
hao@jhu.edu
[3] Applied Physics Laboratory, Johns Hopkins University, Baltimore, MD, USA
I-Jeng.Wang@jhuapl.edu
[4] Department of Computer Science, Johns Hopkins University, Baltimore, MD, USA
aliu@cs.jhu.edu

Abstract. Intelligent tutoring systems (ITS) typically employ scaffolding to provide learning opportunities for needed students. To assess the ITS effectiveness, most past research relies on the end-of-year exam outcome. However, it is important to assess ITS' immediate effectiveness after the scaffolding. To address this gap, this paper draws principles from theory on scaffolding and zone of proximal development to guide the analysis of micro-level, time-stamped student action data from the ASSISTments. Using a deep neural network to analyze the data, we reveal that the appropriate use of scaffolds in the ITS fosters modest growth of math problem-solving skills. Excessive use, however, is counterproductive when students are overly reliant on ITS. Our findings suggest that scaffolding in ITS or AI tutoring should encourage student active learning and prevent over-reliance on intelligent systems.

Keywords: Intelligent tutoring system · Scaffolding · Deep neural network · Student action Data

1 Introduction

Intelligent tutoring systems (ITS), an educational technology to assist student learning is a significant supplement to K-12 formal education. ITS, e.g. the ASSISTments for 8th grade math tutoring [6], have employed scaffolding in assisting student learning. Much research on the effectiveness of ITS focuses end-of-year outcomes such as exam scores or college enrollment [3,10]. On the other hand, the immediate outcome of micro scaffolding sequences for student users has rarely been examined. This paper attempts to fill in this gap by utilizing detailed student action data of the 2004–2005 and 2005–2006 ASSISTment, which is the only publicly available source of micro scaffolding sequences.

A. M. Olney et al. (Eds.): AIED 2024 Workshops, CCIS 2151, pp. 38–46, 2024.
https://doi.org/10.1007/978-3-031-64312-5_5

Despite further developments in ITS, we believe our investigation into the micro process embedded in ASSISTment can provide useful insights and advance our understanding of the functioning and effectiveness of scaffolding in ITS and future AI-powered tutoring that employ scaffolding types and strategies similar to ASSISTment. In particular, we seek answers to two questions: 1) To what extent does scaffolding increase student user's problem-solving skills? 2) What are the features that help or hinder such gain of student users?

2 Related Work

Theory on scaffolding: The metaphor of scaffolding is introduced by the work of Wood and colleagues [17]. Scaffolding refers to tutoring or assistance to enable a student to solve a problem that would be beyond the student's unassisted effort. Scaffolding is closely related to the concept of the zone of proximal development (ZPD), which refers to the difference between what a student can do independently and what s/he can do with the help of a more knowledgeable other [16]. Linking the two concepts, we specify scaffolding as the support provided within a student's ZPD. Scaffolds have evolved with technological advancements in tools, strategies, or guides to gain higher orders of understanding [13] [14].

Scaffolding types: The classification of scaffolding functions was proposed [5,7] and adopted widely including the ASSISTments. Scaffolding functions include conceptual (what to consider directly), metacognitive (what to consider differently), procedural (what is available to use), and strategic (how to approach). The effectiveness of scaffolding types in improving students' academic learning has been generally supported by empirical studies. A meta-analysis of research on digital game-based learning (DGBL) suggests that the use of scaffolding in any form can be effective in improving students' learning outcomes [2]. These authors also find that scaffolding is more effective for math and science than for politics and business. Computer-based scaffolding significantly impacts cognitive outcomes in problem-based learning in STEM education and varies slightly to moderately across contexts [8]. Similar support for the effectiveness of scaffolding is also found in early educational stages [15].

The ASSISTments: This study complements the literature by examining the process of scaffolding at the micro level and assessing whether the student's ability to solve the next problem in the same areas improves as a result of the scaffolds received, utilizing the student action data of the ASSISments. The ASSISTments, one of the most used ITS [1], is a Massachusetts-based online math tutoring system for 8th graders, incorporates the scaffolding support at the core of its algorithm with teacher-designed scaffolding and hint messaging sub-steps [6,11]. Figure 1 shows how students go through the scaffolding steps and hint sub-steps. The student action log data at the ASSISTments permit us to perform our analysis. This micro-level time-stamped data, however, is only available for the 2004–2005 and 2005–2006 cohorts of 8th graders. The test items on the ASSISTments were drawn from the Massachusetts Comprehensive Assessment System (MCAS). The schools requested students to use the ASSISTments.

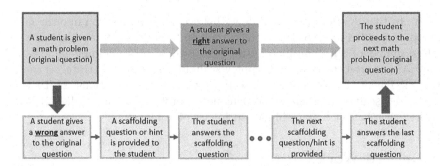

Fig. 1. Scaffolding in the ASSISTments. The process starts with presenting students with the test item. If a student's answer is wrong, scaffolds are provided with hints in between. Students must complete the scaffolding before proceeding to the next question.

On average, 4,000 students use ASSISTments each weekday, about half of them use it for nightly homework, while the rest use it in school [6]. The action data records fine-grained details about how students go through scaffolds and ask for hints.

3   Data from the ASSISTments

The dataset used in this study comprises 321,617 records of students' actions from the publicly available ASSISTments dataset. The dataset includes a total of 1,595 students from the cohorts 2004–2005 and 2005–2006, the only public-available student-action sequence data. During the interactions between students and the platform, a total of 1,183 original questions and 35,326 scaffolding sequences were presented to the students. On average, each scaffolding sequence consisted of 3.23 scaffolding questions and 1.64 hints. Since students were allowed to retry the scaffolding questions, the average number of action records per scaffolding sequence is 9.10. These students' action data contains information about whether a given question is an original question or scaffolding question, whether a student's response was correct, whether a student used a scaffolding/hint, and the time interval that a student used to respond to each question.

Compared with the later released 2009–10 and 2012–13 data, the 2004–06 data contains necessary student-platform interactions through steps and sub-steps, which are not available in later released data. In particular, the 2004–06 data records the timestamps of students' every action, while the 2009–10 and 2012–13 data aggregate students' actions at the question-level. Therefore, the 2004–06 student action data is the only suitable dataset for our task of examining students' scaffolding learning process.

4   Research Design

We design three steps in our investigation. First, we measure students' math proficiency and test item difficulty using the psychometric Rasch model, capturing the trade-off between student abilities and test item difficulties. Second, we employ the deep neural network to predict whether students can correctly answer the next test item in the same topic area. Third, we utilize the feature importance analysis to identify the important underlying factors in the scaffolding process that affect student learning.

The objective of our deep neural networks (DNNs) is to predict whether a scaffolding process enables a student to correctly answer the next test item in the topic area. The distribution of this target shows 37.0% in the positive class (correct answer) and 63.0% in the negative class (wrong answer). Because scaffolds are offered in a sequence, we use Recurrent Neural Networks (RNNs) which is appropriate for modeling a sequence of inputs in deep learning [12]. Further, to mitigate the issue of vanishing gradients that may arise in the backpropagation process when dealing with lengthy sequences up to 12 actions, we use Long Short-Term Memory networks (LSTM). The LSTM enables the analysis of long sequences without vanishing gradients [4].

The input features used in this study encompass a wide range of aspects related to students' learning, such as the difficulty levels of both current and next original questions (test items), knowledge areas and question types of the current original question and the next original question, the number of hints used by learners, the correctness of students' answers to the scaffolding questions, students' response time to each scaffolding question, if students use hints and other sequence-related information. After normalizing the input features to the same scale, we proceed to feed them into our neural network models.

From the 1,594 students in our dataset used for the neural network analysis, we randomly selected 70% of the student data to constitute the training set and the rest 30% for the test set. We randomly sample students rather than actions because students are the units of analysis in our study.

In the target variable, the number of students in the negative class is about twice that in the positive class. This class imbalance issue can lead to the non-optimal performance of the machine learning classifier. To tackle the class imbalance challenge, we employed the weighted binary cross entropy as the loss function. This weighted loss function allocates higher weights to the positive class and lower weights to the negative class when evaluating how well our model fits the data. In particular, we weight the examples using the inverse propensity scores of the class such that the model is optimized for a balanced class distribution in the test domain.

We employ the Shapley Additive Explanations (SHAP) method to assess the importance of each feature in our neural network models [9]. Different from the feature importance analysis based on mutual information scores, the SHAP is a post-hoc analysis that measures the association between the input features and the predicted outcome. In other words, the goal of SHAP is to interpret the neural network's prediction by computing the contribution made by each input feature. The equation for computing the SHAP value is as follows:

$$g(z') = \phi_0 + \sum_{j=1}^{K} \phi_j z'_j \tag{1}$$

In Eq. (1), each j corresponds to an input feature and ϕ_j is the Shapley value measuring the feature j's contribution to the model prediction. Using the concept of "coalition" in the game theory, SHAP employs a coalition vector $z' \in \{0,1\}^K$ to designate feature j as "present" or "absent" and K is the maximum coalition size. To be specific, if z'_j is set to "1", then it means the feature j is "present" and switching it to "0" means the feature j is absent. By simulating only some features as "present" while others are "absent", we can get an estimate of each feature's "marginal contribution" to each prediction output. To estimate a feature's global importance across all data points, we can compute the average of absolute Shapley values across the entire dataset using Eq. 2 below. A larger SHAP value I_j indicates that the feature is more important.

$$I_j = \frac{1}{n} \sum_{i=1}^{n} \left| \phi_j^{(i)} \right| \tag{2}$$

5   Results

5.1   Predicting Correct Answer to the Next Test Item

In this section, we use deep neural networks to predict if the student can answer the next test item in the same knowledge area correctly. This prediction is based on the student's interaction patterns with the preceding scaffolding sequence. For instance, if the scaffolding process is effective, it will boost students' ability to tackle the next question on the same area of knowledge, while an ineffective scaffolding process may undermine students' ability to give correct answers by distracting, disengaging or confusing them.

To achieve the best prediction performance, we tested several different RNN and LSTM models, including those with or without dropout, with or without weighted binary cross-entropy. After training our machine learning models on the training set, we proceed to evaluate the performance of our different models on unseen test data in terms of their accuracy, recall, specificity, precision, F1 scores and AUC scores. These performance metrics offer a comprehensive evaluation of how well the models predict positive and negative instances in the testing data. A strong model performance suggests that the model effectively captures the relevant information regarding students' learning from the data. With this assurance, we can then proceed to employ SHAP analysis to extract the factors that play important roles in influencing students' learning.

Table 1 shows that, apart from the initial RNN model, all the subsequent models exhibit decent capabilities for the prediction task. Among them, the LSTM with dropout layer and weighted bidirectional performs slightly better than other models given that it has the highest accuracy, precision score, specificity score, F1 score and AUC scores. Therefore, we used the LSTM with dropout and weighted binary cross entropy as our final model.

Table 1. Model performance comparison

	Accuracy	Recall	Specificity	Precision	F1 Score	AUC score
RNN	0.38	1.00	0.00	0.38	0.55	0.50
RNN-Weighted	0.67	0.65	0.69	0.56	0.60	0.74
RNN-Dropout	0.67	0.62	0.70	0.56	0.59	0.73
RNN-Dropout+Weighted	0.68	0.63	0.70	0.56	0.59	0.74
LSTM	0.68	0.68	0.68	0.56	0.61	0.74
LSTM-Weighted	0.68	0.67	0.68	0.56	0.61	0.74
LSTM-Dropout	0.68	0.66	0.69	0.56	0.61	0.74
LSTM-Dropout+Weighted	0.69	0.67	0.70	0.58	0.62	0.75
Bidirectional LSTM-Dropout+Weighted	0.69	0.67	0.70	0.57	0.62	0.75

5.2   Identified Important Features in the Scaffolding Process

After predicting the correctness of students' response using the LSTM with dropout and weighted binary cross entropy, this study uses the SHAP Beeswarm plot and the bar plot (Fig. 2) to visualize the importance of different features in the prediction process of our final model.

In Fig. 2, the SHAP bar plot on the left shows the ranked top ten features that contribute most to our final classifier's prediction. These features are: (1) the next test item's difficulty; (2) the number of actions; (3) the current test item's difficulty; (4) a scaffolding question; (5) the number of correct responses so far; (6) correct student response; (7) the total number of first responses attempted by the student so far; (8) the first response being a help request; (9) mistakes repeatedly made on one question; and (10) using ASSISTments while in school.

On the right of Fig. 2, the SHAP Beeswarm plot further demonstrates the direction of the associations between the top ten important features and the predicted outcome. Each dot in the SHAP Beeswarm plot corresponds to an observation in the data and the dot's color shows if the value of the input feature is higher (toward the red end) or lower (toward the blue end) for that observation. The horizontal location of a dot represents the sign and magnitude of its impact, measured by the SHAP value, on the model prediction.

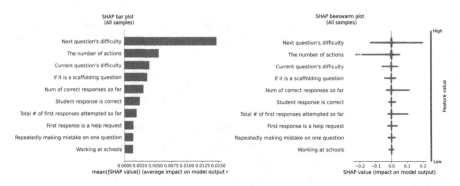

Fig. 2. Left: SHAP bar plot and beeswarm plot visualizes the overall importance of input features, represented by the length of the bars. Right: The SHAP beeswarm plot represents the direction of the association between input features and prediction.

For instance, if a large cluster of red dots (i.e., the value of the input feature is higher) on the left side of the vertical line at 0, it suggests that the input feature negatively impacts the outcome. Following this interpretation, we find that the top two most influential factors, which are the next question's difficulty and the total number of actions, are negatively associated with the predicted outcome. Furthermore, it is also worth noting from Fig. 2 that the difficulty of the current question and whether the scaffolding question is provided are positively associated with whether the student can correctly answer the next test item within the same knowledge area, albeit to a modest extent.

6 Conclusions and Future Research Directions

This study contributes to the literature by examining the micro-level, time-stamped student action log data of the ASSISTments and provides answers to the inquiries regarding the effectiveness of the scaffolding process using the next test item correctness. We focus on students' problem-solving ability immediately after a scaffolding sequence complements the usual end-of-year academic outcomes examined in the literature. Some of the limitations of this study should be noted. First, the only public-available student action log data of ASSISTments is relatively old (2004–05 and 2005–06) while ITS technology keeps improving. This data limitation misses the opportunity to observe students' interaction with the more developed ASSISTments. At the same time, our study focuses on the evaluation of scaffolding not provided by teachers or other more knowledgeable persons based on students' behaviors with outcomes and conclusions that still have substantial relevance today. Second, our study looks at scaffolding sequences as if they are independent of one another, missing the opportunity to understand longitudinal patterns of scaffolding learning for students. Nonetheless, the understanding of a single scaffolding sequence lays out the basis for analyzing multiple processes.

With these caveats, our study offers two major findings. First, the scaffolding process built in the ASSISTments is moderately effective in assisting students to increase their ability of problem-solving. Second, we find that frequent help-seeking behaviors can weaken or even nullify the benefits of scaffolding support. In other words, when students actively learn from scaffolds they gain knowledge, whereas when they overly rely on the intelligent system, they lose the opportunity.

Future research directions can use our study as a launching pad to evaluate the effectiveness of scaffolding in terms of the growth of student problem-solving skills, capitalizing on longitudinal data of student behaviors and single test item outcomes. Moreover, future studies are needed to expand the scope to include a broader range of subjects and encompass diverse student populations. More broadly, our finding about the impeding over-reliance on ITS by students reminds the growing research on AI for education of incorporating a key component of AI agents to encourage human learners to engage in active learning.

References

1. Baker, R.S.: Stupid tutoring systems, intelligent humans. Int. J. Artif. Intell. Educ. **26**, 600–614 (2016)
2. Cai, Z., Mao, P., Wang, D., He, J., Chen, X., Fan, X.: Effects of scaffolding in digital game-based learning on student's achievement: a three-level meta-analysis. Educ. Psychol. Rev. **34**(2), 537–574 (2022)
3. Feng, M., Heffernan, N., Koedinger, K.: Addressing the assessment challenge with an online system that tutors as it assesses. User Model. User-Adap. Inter. **19**, 243–266 (2009)
4. Gers, F.A., Schmidhuber, J., Cummins, F.: Learning to forget: continual prediction with LSTM. Neural Comput. **12**(10), 2451–2471 (2000)
5. Hannafin, M., Land, S.M., Oliver, K.: Open learning environments: foundations, methods, and models. In: Reigeluth, C. (ed.) Instructional Design Theories and Models, pp. 115–140. Lawrence Erlbaum Associates, Mahwah (1999)
6. Heffernan, N.T., Heffernan, C.L.: The ASSISTments ecosystem: building a platform that brings scientists and teachers together for minimally invasive research on human learning and teaching. Int. J. Artif. Intell. Educ. **24**, 470–497 (2014)
7. Hill, J.R., Hannafin, M.J.: Teaching and learning in digital environments: the resurgence of resource-based learning. Educ. Tech. Res. Dev. **49**, 1042–1629 (2001)
8. Kim, N.J., Belland, B.R., Walker, A.E.: Effectiveness of computer-based scaffolding in the context of problem-based learning for stem education: Bayesian meta-analysis. Educ. Psychol. Rev. **30**, 397–429 (2018)
9. Lundberg, S.M., Lee, S.I.: A unified approach to interpreting model predictions. In: Advances in Neural Information Processing Systems, vol. 30 (2017)
10. Makhlouf, J., Mine, T.: Analysis of click-stream data to predict stem careers from student usage of an intelligent tutoring system. J. Educ. Data Min. **12**(2), 1–18 (2020)
11. Pardos, Z.A., Baker, R.S., San Pedro, M.O., Gowda, S.M., Gowda, S.M.: Affective states and state tests: investigating how affect and engagement during the school year predict end-of-year learning outcomes. J. Learn. Anal. **1**(1), 107–128 (2014)

12. Pascanu, R., Mikolov, T., Bengio, Y.: On the difficulty of training recurrent neural networks. In: International Conference on Machine Learning, pp. 1310–1318. PMLR (2013)

13. Saye, J.W., Brush, T.: Scaffolding critical reasoning about history and social issues in multimedia-supported learning environments. Educ. Tech. Research Dev. **50**(3), 77–96 (2002)

14. Simons, K.D., Klein, J.D.: The impact of scaffolding and student achievement levels in a problem-based learning environment. Instr. Sci. **35**, 41–72 (2007)

15. Sun, L., Kangas, M., Ruokamo, H., Siklander, S.: A systematic literature review of teacher scaffolding in game-based learning in primary education. Educ. Res. Rev. 100546 (2023)

16. Vygotsky, L.S., Cole, M.: Mind in Society: Development of Higher Psychological Processes. Harvard University Press (1978). Google-Books-ID: RxjjUefze_oC

17. Wood, D., Bruner, J.S., Ross, G.: The role of tutoring in problem solving. J. Child Psychol. Psychiatry **17**(2), 89–100 (1976)

Heuristic Technique to Find Optimal Learning Rate of LSTM for Predicting Student Dropout Rate

Anuradha Kumari Singh$^{(\boxtimes)}$ (ID) and S. Karthikeyan

Department of Computer Science, Banaras Hindu University, Varanasi 200105, Uttar
Pradesh, India
{anuradha,karthik}@bhu.ac.in

Abstract. Predictive analytics is being increasingly recognized as being important for evaluating university students' academic achievement. Utilizing big data analytics particularly students' demographic information, offers valuable insights to bolster academic success and enhance completion rates. For instance, learning analytics is a vital element of big data within university settings, offering strategic decision-makers the chance to conduct time series analyses on learning activities. We have used semesters first and second records of Polytechnic Institute of Portalegre students. Advanced deep learning methods, such as the Long short-term memory (LSTM) model are employed to analyze students at risk of retention issues. Typically, the best design for a deep neural network model was found by trial and error, which is a laborious and exponential combinatorial challenge. Hence we proposed the heuristic technique to configure the parameters of the neural network. The parameter taken into account in this work is the learning rate of the Adam optimizer. In this study, we have presented the ant colony optimization (ACO) technique to determine the ideal learning rate for model training. Experimental results obtained with the predictive model indicated that prediction of student retention is possible with a high level of accuracy using ACO-LSTM approach.

Keywords: Ant colony optimization · Deep learning · Educational Data Mining · Long short-term memory · Dropout prediction

1 Introduction

Academic underachievement in Higher Education (HE) can be categorized into four primary types [1]; inconsistency in credit/exam completion, prolonged environment status (referred to as out-of-school education), non-linear career paths(eg., course transfers), and ultimately, the abandonment of the academic journey resulting in a departure from the university system without obtaining a degree. The occurrence of students leaving university prematurely brings about various adverse impacts beyond its repercussions. Bean proposed the "Student

© The Author(s), under exclusive license to Springer Nature Switzerland AG 2024
A. M. Olney et al. (Eds.): AIED 2024 Workshops, CCIS 2151, pp. 47–54, 2024.
https://doi.org/10.1007/978-3-031-64312-5_6

attrition" model, based on the attitude-behavior of the student, which evaluates individual and institutional factors and their interactions to predict University dropout [2]. Pascasella's model proposed [3] which emphasizes the pivotal significance of students' academic success is the emphasis on cultivating informal connections with teachers. This underscores the essential role that casual interactions and relationships between students and teachers play in fostering a positive learning environment and contributing to overall achievement.

Previous studies have indicated that a student's voluntary withdrawal from their course of study can be impacted by factors related to both personal and institutional aspects [4]. The present study examined the relationship between the first-semester and the second-semester performance of students on their retention rate. Along with semester performance students' demographic data and their prior qualification is also examined. Furthermore, it offers a distinctive practical contribution to strategic planning by enhancing our comprehension of the correlation between students' decisions to stay enrolled in their course of study and the primary factors influencing their decision-making process.

In practice, empirical approaches are typically used to manually select an optimization function and set learning rate parameters through trials and errors. This paper proposed the idea of using a heuristic algorithm Ant colony optimization to find the optimum learning rate for the training of LSTM model. Learning rate is an important hyperparameter to tune for effective training of deep neural networks. even for the baseline of a constant learning rate, it is non-trivial to choose a good constant value for training a deep neural network. The main aim of this study is to find the optimum learning rate for an LSTM model that can be used in other universities, defining and predicting dropout students' characteristics. We propose the following research questions(RQ):

> **RQ1:** *By using LSTM it will be possible to predict student dropout for an entire cohort of students using demographic and academic characteristics.*
> **RQ2:** *By using Ant-colony optimization(ACO) it will be possible to obtain the best learning rate to train the model for getting a high level of accuracy.*

The rest of the article is structured as follows: Sect. 2 contains the relevant existing literature and provides a more detailed explanation of the core concept presented in this paper. Moving on, Sect. 3 gives an introduction to the proposed methodology, while Sect. 4 showcases the results and verifies their reliability. Finally, Sect. 5 wraps up with the conclusion.

2   Related Literature

In this section, we first explore prior research that delved into university dropout using Educational Data Mining techniques (EDM) [5] and later the dropout rate at massive open online courses (MOOCs) platform. The literature analysis of university dropout reveals that the Decision Tree (DT)algorithm is frequently employed for constructing predictive models aimed at identifying dropout [6]. An Indian study assessed models developed by the DT algorithm using metrics such as accuracy, precision, recall, and F1 measure [7]. Another research project utilized the DT algorithm by incorporating socio-economic, academic, and institutional data [8]. In a separate study at the Budapest University of Technology and Economics, six types of algorithms were utilized to identify students at risk of dropout, employing data from 15,285 university students [9]. To explore students at risk of retention, predictive deep learning techniques, specifically the bidirectional long short-term memory (BLSTM) model were employed using a two-year retrospective analysis of student learning data from the University of Ha'il [10].

The optimum architecture of the deep neural network model was generally determined through the trial and error process which is a tedious task. Many studies are using the heuristic approach to obtain the optimum neural network configuration. This study [11] used a genetic algorithm for constructing the optimal architecture of a multilayer perceptron for solving classification has been developed. Another study [12] in which a genetic algorithm-determined Deep Feedforward neural network architecture is proposed for both day-ahead hourly and week-ahead daily electricity consumption of a real-world campus building in the United Kingdom. We find a research gap in that there is no such approach for configuring optimal neural network architecture for predicting student dropout rate, hence we have proposed LSTM baseline deep neural model with ACO approach.

3 Methodology

Enhancements to retention rates can be initiated by establishing a student demographic dataset as an indicator of undergraduate student performance at the university. The dataset includes demographic data, socioeconomic and macroeconomic data, data at the time of student enrollment, and data at the end of the first and second semesters. These include data from 17 undergraduate degrees from different fields of knowledge. This paper examines the effect of ACO optimization on LSTM model performance.

The proposed method is illustrated in Fig. 1.

In the Algorithm 1 where min_lr refers to minimum learning rate, max_lr refers to maximum learning rate, n_ants refers to the number of ant counts, n_iterations refers to the number of iterations, and n_best refers to the number best solution. In our experiment, min_lr is 0.0001, max_lr is 0.5, n_ants is 5, n_iteration is 10, and n_best is 2. Here decay is the rate at which the pheromone levels are reduced or decayed at each iteration, in our experiment we set the decay value as 0.5. Alpha and Beta are the parameters that control the relative importance of the heuristic information (accuracy of learning rates) versus the pheromone levels in the decision-making process of the ants, in our experiment alpha, and beta are assigned to 1.

Data Collection. We have used a public dataset [13] of the Polytechnic Institute of Portalegre. The dataset size is 743KB and it contains 4424 records with 36 attributes. The data refer to records of students enrolled between the academic

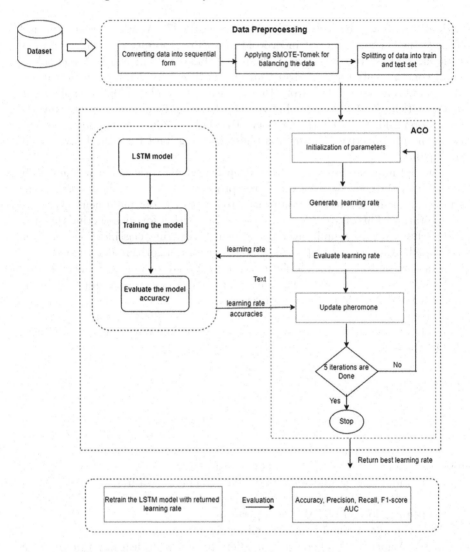

Fig. 1. Graphical representation of the proposed framework

years 2008/2009 to 2018/2019. These consist of data from 17 undergraduate degrees from different fields of knowledge, such as agronomy, design, education, nursing, journalism, management, social service, and technologies. The dataset includes demographic data, socioeconomic and macroeconomic data, and admission data and consists of semester first and semester second internal and external exam data from the institution.

Data Preprocessing. The dataset had 3 targets class dropout, graduate, and enrolled, but for our work, we only used 2 classes: dropout and graduate. Hence after removing enrolled records, we are left with 3630 records. Transforming data

Algorithm 1. ACO-LSTM approach to obtain optimal learning rate for Adam optimizer

---

Input : Parameters min_lr, max_lr, n_ants, n_best, n_iterations, decays, alpha, and beta

Output: best_learning_rate , best_score

1: For each iteration in n_iterations:
2:     Generate learning rates for each ant uniform between min_lr and max_lr
3:     Evaluate the performance for each learning rate
4:         For each ant:
5:             Train an LSTM with the given learning rate
6:             Evaluate the accuracy of the trained model
7:             Store the accuracy
8:         Update pheromone levels based on the accuracy of each learning rate:
9:             Sort the accuracies in descending order to determine ranks
10:            Decay pheromone levels
11:            Update pheromone levels
12:            Update pheromone levels for the best n_best learning rates
13: Return the learning rate with the highest accuracy along with its accuracy.

---

by encoding categorical variables, scaling features, and creating new features. Splitting the data into training and testing sets with an 80 : 20 ratio, 80% for the training set and 20% for the testing set. Balanced the dataset using the Synthetic Minority Over-sampling Technique(SMOTE).

Data Transformation. During data transformation, we convert the preprocessed data into sequential format with window size 1. Then the splitting of the dataset into training and testing sets. After splitting of data we standardized the training features, each feature is transformed to have a mean of 0 and a standard deviation of 1. This means the data is centered around 0 and has a spread of 1.

Ant colony Optimization. Ant colony optimization(ACO) draws inspiration from the foraging behavior of certain ant species. In this approach, ants deposit pheromones on the ground to mark favorable paths guiding other members of the colony to follow these identified routes [14].

Long Short-Term Memory (LSTM). LSTM is a recurrent neural network (RNN) widely used in classifications and predictions based on time series data. It can be used to process entire sequences and retain information in a compressed form using sequence labeling methods. LSTMs are used in applications such as named entity recognition, part-of-speech tagging, sentiment analysis, and many more. This advantage inspired us to formulate the prediction problem from a sequence perspective. The previous step tries to characterize students with their features, and then, we need to train the LSTM method over the student features.

4   Experimental Results and Discussion

All the experiments have been performed using the Python programming language (version 3.7), the scikit-learn framework (version 0.22.1) for data preprocessing tasks, and the tensorflow which provides access to the implementation of LSTM model. Training and testing run on Google Colab environment with T4 GPU. The principal objective was to investigate the number of students at risk of discontinuing their study as identified by dropout risk and completion within the time risk. In our dataset, three classes were presented that were enrolled, dropout and graduated. We removed the enrolled record and worked only on the dropout and graduated label data. The preprocessed dataset was divided into two groups 80% of which were used as training datasets and 20% of which were used as testing datasets. 2904 samples were used for training and 858 samples were used for testing, according to the dataset's sample count. LSTM was used as a deep learning model for predictions, the parameters and values of which are displayed in Table 1. ACO optimization is applied to obtain the best learning rate to train features using the LSTM model. We compiled the model with a binary cross entropy loss function equation (1).

$$\mathbf{L(y, y') = (y \ log(y') + (1 - y) \ log(1 - y'))} \tag{1}$$

Table 1. Parameters used for LSTM model

Parameters	Descriptions
Number of layers	2
Number of neurons in input unit	50
Batch size	16
Learning rate	0.001
Optimizer	Adam
Epochs	200
Timestep	1
Activation function	Sigmoid
Computation mode	T4 GPU

Three evaluation metrics were used to examine the performance of our prediction model: Precision, Recall, and F1-score. We also compared LSTM and LSTM-ACO performance using ROC. ROC(Receiver Operating Characteristics) is a graphical representation of a classifier's performance across different threshold settings. The ROC is created by plotting the true positive rate(sensitivity) against the false positive rate(1-specificity) at various threshold settings. The area under the ROC curve (AUC-ROC) is often used as an evaluation metric for the performance of a classification model. Table 2 shows the comparison of the LSTM-ACO with LSTM model on the metrics of precision, recall, f1-score, accuracy, and AUC. From the results, we can see that there is an improvement in LSTM performance by using ACO.

Table 2. Comparision of performance evaluation.

Models	Precision	Recall	F1-score	Accuracy	AUC
LSTM	0.8534	0.8298	0.8414	0.8377	0.84
LSTM-ACO ant counts=5	0.8581	0.9029	0.88	0.8715	0.88
LSTM-ACO ant counts=10	**0.9082**	**0.8713**	**0.8826**	**0.89**	0.88

From the results in Table 2, we can conclude: (1) our proposed LSTM-ACO model is effective in solving dropout prediction problems in universities with large amounts of data and achieves better results than LSTM. (2) In the optimization stage, ant counts of different scales have a certain impact on obtaining the learning rate.

5   Conclusion

The persistence of students in their chosen academic disciplines and the overall graduation rates over a defined timeframe are matters of importance for higher education institutions. University educators emphasize the need for continuous efforts to enhance student retention rates. This paper has a two-fold emphasis. Firstly, it delves into the application of data preprocessing and ACO optimization to train the model to obtain the best learning rate of the LSTM model without the trial and error approach. Second, on the training of train features using LSTM to determine the optimal weight for classification of the train features. Our work was started with two research questions. First, "By using LSTM it will be possible to predict student dropout for an entire cohort of students using demographic and academic characteristics". To answer the first question, yes it will be possible to predict student dropout using LSTM. The second question was "By using ACO it will be possible to obtain the best learning rate to train the model for getting a high level of accuracy", yes ACO will be able to obtain the best learning rate. In our future work, we will explore more heuristic optimization techniques with deep neural networks to predict student dropout rates in universities.

Conflict of Interest. The Authors declare that they have no conflict of interest.

References

1. Tanucci, G., Fasanella, A.: Orientamento e carriera universitaria. Ingressi ed abbandoni in 5 Facoltà dell'Università di Roma "La Sapienza" nel nuovo assetto didattico (2006)
2. Bentler, P.M., Speckart, G.: Models of attitude-behavior relations. Psychol. Rev. **86**(5), 452 (1979)
3. Pascarella, E.T., Terenzini, P.T.: Predicting freshman persistence and voluntary dropout decisions from a theoretical model. J. High. Educ. **51**(1), 60–75 (1980)

4. Raju, D., Schumacker, R.: Exploring student characteristics of retention that lead to graduation in higher education using data mining models. J. Coll. Student Retention: Res. Theory Pract. **16**(4), 563–591 (2015)
5. Bala, M., Ojha, D.B.: Study of applications of data mining techniques in education. Int. J. Res. Sci. Technol. **1**(4), 1–10 (2012)
6. Alban, M., Mauricio, D.: Neural networks to predict dropout at the universities. Int. J. Mach. Learn. Comput. **9**(2), 149–153 (2019)
7. Sivakumar, S., Venkataraman, S., Selvaraj, R.: Predictive modeling of student dropout indicators in educational data mining using improved decision tree. Indian J. Sci. Technol. (2016)
8. Pereira, R.T., Romero, A.C., Toledo, J.J.: Extraction student dropout patterns with data mining techniques in undergraduate programs. In: International Conference on Knowledge Discovery and Information Retrieval, vol. 2. SCITEPRESS (2013)
9. Nagy, M., Molontay, R.: Predicting dropout in higher education based on secondary school performance. In: 2018 IEEE 22nd International Conference on Intelligent Engineering Systems (INES). IEEE (2018)
10. Uliyan, D., et al.: Deep learning model to predict students retention using BLSTM and CRF. IEEE Access **9**, 135550–135558 (2021)
11. Domashova, J.V., et al.: Selecting an optimal architecture of neural network using genetic algorithm. Procedia Comput. Sci. **190**, 263–273 (2021)
12. Luo, X.J., et al.: Genetic algorithm-determined deep feedforward neural network architecture for predicting electricity consumption in real buildings. Energy AI **2**, 100015 (2020)
13. Realinho, V., et al.: Predicting student dropout and academic success. Data **7**(11), 146 (2022)
14. Dorigo, M., Birattari, M., Stutzle, T.: Ant colony optimization. IEEE Comput. Intell. Mag. **1**(4), 28–39 (2006)

Modeling Learner Memory Based on LSTM Autoencoder and Collaborative Filtering

Tengju Li[1], Cunling Bian[1], Ning Wang[2], Yangbin Xie[3], Kaiquan Chen[1],
and Weigang Lu[1,2(✉)]

[1] Department of Educational Technology, Ocean University of China, Qingdao
266100, China
luweigang@ouc.edu.cn
[2] College of Computer Science and Technology, Ocean University of China,
Qingdao 266100, China
[3] Research Center for Data Hub and Security, Zhejiang Lab,
Hangzhou 311100, China

Abstract. Memory modeling, aimed at predicting the memory states of learners throughout their learning process, has become an integral component in online learning systems. This is particularly crucial for applications such as spaced repetition scheduling and knowledge tracking. However, existing machine learning-based memory modeling methods encounter challenges with weak supervision signals and sparse interaction data. To tackle these issues, we propose a novel approach named LSTM Autoencoder Collaborative Filtering (LACF) for modeling learner memory. Our model utilizes an LSTM autoencoder to extract temporal features from user-item interaction sequences. Additionally, a collaborative filtering module, based on the deepFM architecture, is employed to effectively learn interactions between different features, facilitating comprehensive low-order and high-order feature combinations. Experimental results demonstrate that LACF significantly outperforms existing state-of-the-art memory modeling methods, offering enhanced accuracy and precision in predictions.

Keywords: Memory modeling · Autoencoder · Collaborative filtering · Time series prediction

1 Introduction

In the process of learning, combating forgetting is crucial, and repeated memorization plays a pivotal role. Cognitive psychology research highlights the spacing effect, revealing that spaced repetition enhances memory retention more effectively than massed repetition [9]. This principle guides the development of learning systems like SuperMemo[1] and Anki[2], which help learners schedule reviews to

---

[1] https://www.supermemo.com.
[2] https://apps.ankiweb.net.

A. M. Olney et al. (Eds.): AIED 2024 Workshops, CCIS 2151, pp. 55–62, 2024.
https://doi.org/10.1007/978-3-031-64312-5_7

prevent forgetting. Notably, predicting learners' current memory state emerges as a vital task within spaced repetition [4], enabling timely reminders for review sessions before memory decay occurs.

Traditionally, research on memory has heavily relied on controlled psychological experiments. However, the proliferation of online learning platforms in recent years has yielded abundant data on learner memory behaviors, facilitating more fine-grained memory modeling approaches [6]. Concurrently, machine learning has emerged as a promising tool in this field. Despite numerous memory prediction methods proposed, two persistent challenges remain:

– Weak supervision signals: Directly observing learners' memory states without controlled psychological experiments is challenging [4]. Understanding memory states often relies on subtle memory cues. Machine learning models must infer learners' current memory states based on whether they successfully recall information during a review session. However, binary recall results serving as supervisory signals for machine learning models are inherently weak and heavily laden with noise. While some researchers have proposed that hand-crafted memory features might mitigate this challenge [6,10,11], such an approach necessitates domain expertise and intricate feature engineering.
– Sparse interaction data: In online learning systems, users have limited interactions with items compared to the vast number of available items [8]. Consequently, the scarcity of interaction data exacerbates the challenge of training models to learn robust user and item feature representations. Although employing collaborative filtering methods can partially address this issue of data sparsity [3], identifying the most efficacious model from a variety of collaborative filtering approaches requires additional exploration.

To tackle these challenges, we present LSTM Autoencoder Collaborative Filtering (LACF). It uses a Long Short-Term Memory (LSTM) autoencoder to extract temporal features from sequences of user-item interactions efficiently. Then, it combines these features with sparse features like user IDs and item IDs using a collaborative filtering module based on the deepFM architecture. By integrating these models, LACF benefits from the LSTM autoencoder's capability for feature extraction without the need for manual engineering, along with deepFM's ability to handle sparse data. Experimental results demonstrate that LACF achieves more accurate predictions of learner memory.

2   Related Work

Predicting learner memory states from behavioral data is a key goal in memory modeling research. Current approaches can be broadly categorized into two categories, depending on the formulation of the problem:

– Regression Approaches: Some studies treat memory prediction as regression problems, mapping behavioral features to continuous values representing forgetting or recall probability. Settles and Moerland [6] used a linear regression model to predict memory half-life. Su et al. [7] employed Recurrent Neural Networks (RNNs) for this task. Regression methods address challenges in weak supervision signals but assume uniformity in memory states, possibly overlooking individual differences. However, those method assumes homogeneity among learners or items, which may not accurately reflect individual differences in memory processes.

- Binary Classification Approaches: Alternatively, memory prediction tasks can be framed as binary classification problems, predicting recall or not in next interactions. Reddy et al. [5] proposed an exponential forgetting curve model considering interaction counts, results, and item difficulty. Mozer and Lindsey [4] combined collaborative filtering with cognitive models for prediction, while Choffin et al. [1] improved this model with Factorization Machine techniques. Binary classification allows for personalized memory modeling but heavily relies on feature engineering, potentially limiting generalizability.

In summary, while binary classification offers more personalized predictions, exploring RNN efficacy within it and refining collaborative filtering models remain areas of interest. To bridge these gaps, we propose LACF, leveraging LSTM autoencoders and collaborative filtering for improved memory state predictions.

3   Methods

3.1   Problem Formulation

Consider a dataset with users U and items V. Here, u represents a user, and v signifies an item. A memory event is characterized as a tuple (u, v, t, r), which indicates that a user u engaged with an item v at a specific time t, resulting in either recall (represented as $r = 1$) or forgetting (denoted by $r = 0$) of the item. The goal of memory modeling is as follows: for a user u and a particular item v, given the historical memory interaction sequence $x_{1:i}$ (comprising time interval sequence $\Delta t_{1:i}$ and recall sequence $r_{1:i}$), predict the outcome of the next interaction, x_{i+1} (i.e., forecasting the value of interaction result, r_{i+1}, after a time interval Δt_{i+1} has passed).

3.2   Memory Modeling Method

The architecture of LACF, as depicted in Fig. 1, consists of two primary components: firstly, the interaction sequence autoencoder module, which is built upon the LSTM architecture for efficient sequence processing, and secondly, the feature combination module, using the deepFM architecture for enhanced feature integration.

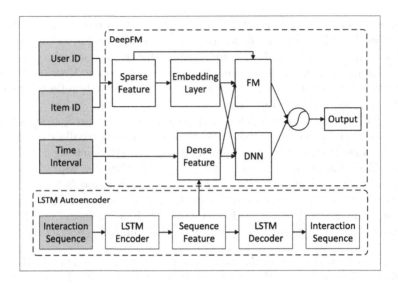

Fig. 1. Model architecture of LACF.

LSTM Autoencoder. To avoid feature engineering, we aim to utilize neural networks for autonomously extracting sequence features from interaction sequences. Ye's research highlights the effectiveness of RNNs in modeling memory interactions [10]. However, in our study, we have chosen a different approach. Our experimental results, as outlined in Sect. 4.2, reveal the limitations of RNNs due to the lack of recall probability as a explicit supervision signal. To address this issue, we adopted a self-supervised method using an autoencoder to learn directly from raw data, which effectively tackles limitations posed by weak supervision signals.

The LSTM autoencoder comprises LSTM encoder and decoder units with distinct parameters. Initially, the historical memory interaction sequence $x_{1:i}$ is inputted into the LSTM encoder, which condenses it into a feature vector h. Then, the feature vector h is used as the input for the LSTM decoder. The decoder attempts to reconstruct the original sequence. The degree of reconstruction accuracy is quantified by the reconstruction error E_r, as defined in Eq. 1:

$$E_r = \sum_{i=1}^{L} \|x_i - \hat{x}_i\|_2^2,\tag{1}$$

where L represents the sequence length, x_i denotes the i-th element of the input sequence, and $\hat{x}_i$ represents the i-th element of the reconstructed sequence. By minimizing the reconstruction error E_r, the LSTM encoder learns the feature representation of the original sequence.

DeepFM. To accurately predict memory behavior and calculate recall probability, we integrates a variety of features from user-item interactions. Drawing inspiration from Choffin's research [1], we implement a collaborative filtering approach based on the DeepFM model [2]. This model is adept at capturing both low-order and high-order feature combinations. The DeepFM model comprises two key components: the FM component and the Deep Neural Network (DNN) component, both sharing the same input data. Sparse features such as user IDs and item IDs are first transformed into dense representations via an embedding layer, then inputted into the network. Concurrently, dense features such as sequence features and time intervals, are directly inputted into the network.

The FM component learns first and second-order feature combinations, which are calculated as follows:

$$y_{FM} = <w, x> + \sum_{i=1}^{d} \sum_{j=i+1}^{d} <V_i, V_j> x_i \cdot x_j, \tag{2}$$

where $x = [x_{user}, x_{item}, x_{time}, x_{sequence}]$ represents a d-dimensional vector. Here, x_{user}, x_{item}, x_{time}, and $x_{sequence}$ correspond to vector representations of various features. The DNN component captures intricate interactions via a multi-layer fully connected neural network. In our model, the DNN comprises two hidden layers. The output of the DNN is computed as:

$$y_{DNN} = \mathbf{W}^{(2)} ReLU\left(\mathbf{W}^{(1)} x + b^{(1)}\right) + b^{(2)}, \tag{3}$$

where ReLU represents the Rectified Linear Unit activation function. The final recall probability is calculated using:

$$\hat{y} = \text{sigmoid}\left(y_{FM} + y_{DNN}\right). \tag{4}$$

For model training, Binary Cross Entropy Loss (BCELoss) is employed:

$$\text{loss}(y, \hat{y}) = -\frac{1}{N} \sum_{i=1}^{N} \left(y_i \log\left(\hat{y}_i\right) + (1 - y_i) \log\left(1 - \hat{y}_i\right)\right). \tag{5}$$

4   Experiments

4.1   Experimental Settings

Datasets and Baselines. Our experiments use the maimemo dataset [10], sourced from a popular online language learning platform. This real-world dataset provides rich interaction records between learners and words. Given the large size of the original dataset, we randomly sampled 300 users' interactions with words, yielding 326,000 records for our experimental dataset. To assess our method's performance, we compare it with several existing approaches:

- Mean-based method [10]: This approach categorizes words by difficulty and computes recall probability based on the proportion of individuals who correctly recall them.

- HLR [6]: HLR employs feature engineering and a linear regression model to estimate memory half-life. For binary classification, the loss function is modified to BCELoss.
- N-HLR [11]: An enhanced version of HLR, N-HLR incorporates a neural network to improve the accuracy of memory half-life modeling.
- GRU-HLR [7]: This approach utilizes a GRU model to directly predict memory half-life from interaction sequences, which is particularly effective for regression problems.
- DAS3H [1]: DAS3H employs feature engineering and an FM model to directly predict recall probability, demonstrating excellent performance on binary classification problems.

Evaluation Metrics and Training. We employ standard binary classification metrics, including Accuracy, Precision, F1 score, and AUC (Area Under the Curve). For evaluation purposes, we adopt standard k-fold cross-validation (with k = 5) for all models. In this setup, each fold involves allocating 10% of the records as the test set, 10% as the validation set, and 80% as the training set. For each fold, we use the validation set to perform early stopping and tune the parameters for every method.

4.2   Performance Comparison

The performance comparison results are detailed in Table 1, yielding several key observations:

Table 1. Experimental results on the maimemo dataset.

	Accuracy	Precision	F1	AUC
Mean-based method	0.6346	0.6196	**0.7283**	0.6214
HLR	0.6246	0.6624	0.6606	0.6441
N-HLR	0.6588	0.6591	0.7214	0.6876
GRU-HLR	0.6333	0.6227	0.7222	0.6412
DAS3H	0.6590	0.6643	0.7168	0.6901
LACF	**0.6710**	**0.6782**	0.7229	**0.7211**

Firstly, our LACF model demonstrates superior performance, achieving the highest scores in Accuracy, Precision, and AUC, significantly outperforming state-of-the-art baselines. This highlights the effectiveness of combining LSTM autoencoder and collaborative filtering, enhancing accurate memory predictions. Secondly, the mean-based method shows relatively lower performance, particularly in Accuracy and AUC, suggesting limitations due to fundamental assumptions in regression problems, ultimately hindering personalized memory predictions. Thirdly, despite its advanced architecture, the GRU-HLR model does not perform as well as HLR and N-HLR, especially in Precision and AUC. This indicates challenges in effectively capturing sequence features in binary classification tasks without specific input features like recall probability, emphasizing the importance of our self-supervised component, the LSTM autoencoder. Lastly, models employing collaborative filtering methods, such as the DAS3H model and ours, outperform others, highlighting the efficacy of collaborative filtering methods in memory predictions.

To investigate the influence of sparsity on the model, we further examine the performance of the LACF model among users with limited interaction records. Figure 2 presents the model's performance across various levels of sparsity. The LACF model consistently outperforms others, demonstrating its robustness and its capability to maintain accuracy even with sparse data. This highlights the pivotal role of the autoencoder and collaborative filtering components in enriching the information extracted from limited user interactions.

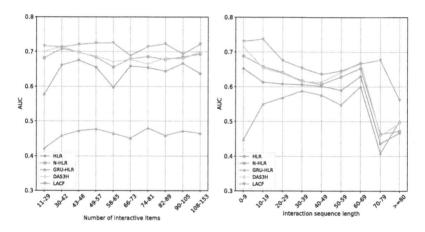

Fig. 2. Performance comparison across diverse user group sparsity levels. **Left:** Grouping of users based on the number of interactive items. **Right:** Grouping of users based on the length of interaction sequences.

5   Conclusion

In this paper, we introduce LACF, a novel model designed to tackle two critical challenges in learner memory modeling. LACF leverages a self-supervised LSTM autoencoder to extract temporal features from user-item interaction sequences, effectively overcoming the issue of weak supervision signals. Moreover, the model incorporates a collaborative filtering module based on the deepFM architecture, adeptly handling both low-order and high-order feature combinations, thus addressing the problem of sparse interaction data. Compared to existing models,

LACF demonstrates superior adaptability in memory prediction tasks by eliminating the need for manual feature engineering. The experimental results on real-world data consistently shows that the proposed LACF outperforms state-of-the-art memory models, especially in scenarios involving cold-start users with limited interactions. Future work will focus on integrating additional structural information, such as knowledge graphs, to further improve prediction accuracy.

Acknowledgements. This work has been supported by National Natural Science Foundation of China (No. 62277045) and Natural Science Foundation of Shandong Province (No. ZR 2021MF011).

References

1. Choffin, B., Popineau, F., Bourda, Y., Vie, J.: DAS3H: modeling student learning and forgetting for optimally scheduling distributed practice of skills. In: Proceedings of the 12th International Conference on Educational Data Mining, pp. 29–38 (2019)
2. Guo, H., Tang, R., Ye, Y., Li, Z., He, X.: DeepFM: a factorization-machine based neural network for CTR prediction. In: Proceedings of the Twenty-Sixth International Joint Conference on Artificial Intelligence, pp. 1725–1731 (2017)
3. Lindsey, R.V., Shroyer, J.D., Pashler, H., Mozer, M.C.: Improving students' long-term knowledge retention through personalized review. Psychol. Sci. **25**(3), 639–647 (2014)
4. Mozer, M.C., Lindsey, R.V.: Predicting and improving memory retention: psychological theory matters in the big data era. In: Big Data in Cognitive Science, pp. 34–64 (2017)
5. Reddy, S., Labutov, I., Banerjee, S., Joachims, T.: Unbounded human learning: optimal scheduling for spaced repetition. In: Proceedings of the 22nd ACM SIGKDD International Conference on Knowledge Discovery and Data Mining, pp. 1815–1824 (2016)
6. Settles, B., Meeder, B.: A trainable spaced repetition model for language learning. In: Proceedings of the 54th Annual Meeting of the Association for Computational Linguistics (Volume 1: Long Papers), pp. 1848–1858 (2016)
7. Su, J., Ye, J., Nie, L., Cao, Y., Chen, Y.: Optimizing spaced repetition schedule by capturing the dynamics of memory. IEEE Trans. Knowl. Data Eng. **35**(10), 10085–10097 (2023)
8. Vie, J.J., Kashima, H.: Knowledge tracing machines: factorization machines for knowledge tracing. In: Proceedings of the AAAI Conference on Artificial Intelligence, vol. 33, no. 01, pp. 750–757 (2019)
9. Yang, Z., Shen, J., Liu, Y., Yang, Y., Zhang, W., Yu, Y.: TADS: learning time-aware scheduling policy with dyna-style planning for spaced repetition. In: Proceedings of the 43rd International ACM SIGIR Conference on Research and Development in Information Retrieval, pp. 1917–1920 (2020)
10. Ye, J., Su, J., Cao, Y.: A stochastic shortest path algorithm for optimizing spaced repetition scheduling. In: Proceedings of the 28th ACM SIGKDD Conference on Knowledge Discovery and Data Mining, pp. 4381–4390 (2022)
11. Zaidi, A., Caines, A., Moore, R., Buttery, P., Rice, A.: Adaptive forgetting curves for spaced repetition language learning. In: Bittencourt, I.I., Cukurova, M., Muldner, K., Luckin, R., Millán, E. (eds.) AIED 2020. LNCS (LNAI), vol. 12164, pp. 358–363. Springer, Cham (2020). https://doi.org/10.1007/978-3-030-52240-7_65

Personalized Pedagogy Through a LLM-Based Recommender System

Nasrin Dehbozorgi$^{(\boxtimes)}$ (ID), Mourya Teja Kunuku (ID), and Seyedamin Pouriyeh (ID)

College of Computing and Software Engineering, Kennesaw State University, Georgia, USA

{dnasrin,spouriye}@kennesaw.edu, mkunuku@students.kennesaw.edu

Abstract. The educational landscape is evolving with the integration of AI, large language models (LLMs), and generative AI, requiring educators to adopt state-of-the-art technologies and strategies in their pedagogical practices. Pedagogical Design Patterns (PDPs) have garnered attention for disseminating best practices and bridging the gap between research and practice. However, their widespread adoption is hindered by limited publicly available resources and fragmented publishing platforms. To address this, we propose leveraging LLMs to recommend pedagogical practices, drawing from existing PDPs. Our model utilizes a local knowledge base and the Retrieval Augmented Generation (RAG) framework to create query contexts for LLM prompts. Initial findings show promise, with an accuracy score of 0.83 and high relevance of recommendations to input queries. This study presents early results of our ongoing project, supporting further development of the model. The proposed system aims to empower novice educators by providing expert wisdom to enrich their teaching methodologies.

Keywords: Large Language Models · Pedagogical Design Patterns · Educational Recommender System · Retrieval Augmented Generation

1 Introduction

Recent advancements in AI, LLMs, and generative AI are reshaping the educational landscape. The increasing application of AI technologies is providing customized learning tools, establishing scaffolding, and enhancing the overall educational experience for both students and educators [1]. In this era, educators are not only tasked with imparting knowledge but are also expected to identify and apply the most effective technologies and strategies in their teaching practices. Consequently, the teaching profession has transformed into a design science, demanding that educators develop innovative and evidence-based approaches to enhance their practices [2]. Despite numerous advancements in educational technologies and methodologies the valuable insights of educational researchers and innovative pedagogical practices often remain local [2]. This underscores the need for streamlined channels and frameworks to facilitate the exchange of pedagogical design ideas. Pedagogical Design Patterns (PDPs) act as a formalized

A. M. Olney et al. (Eds.): AIED 2024 Workshops, CCIS 2151, pp. 63–70, 2024.
https://doi.org/10.1007/978-3-031-64312-5_8

framework for the exchange of pedagogical interventions and tactics, serving as a link between research and practice [3]. PDPs offer structured, effective strategies and methodologies, drawn from the collective wisdom and practices of educators across various disciplines. The concept of PDPs is similar to design patterns in software engineering, offering well-established solutions to recurring problems faced by system developers. Similarly, PDPs empower educators to access solutions to pedagogical challenges based on learning theories or empirical rationale. These solutions are not prescriptive and just serve as a universal guide, allowing for creativity in their implementation based on individual discretion. They provide a valuable resource equally for both experienced educators and those new to teaching, fostering a culture of continuous improvement in the educational domain [3]. However, despite their potential to transform educational practices, PDPs are dispersed across numerous sources and platforms, making it difficult to adopt them into teaching practices [4]. The scattered nature of these resources poses a significant hurdle for instructors seeking innovative solutions to classroom challenges such as engaging students more effectively, integrating technology into classes, or designing inclusive learning activities [2]. The time and effort required to search through disjointed repositories can be prohibitive, often leaving educators to rely on tried and tested methods, potentially at the expense of more innovative or effective approaches. To address this challenge we propose the development of an accessible framework that can assist instructors in identifying research-based solutions and apply them to their unique classroom challenges. For this purpose, we adopt advanced AI methods to develop a PDP Recommender System that serves both as a platform for hosting existing peer-reviewed PDPs, and a personalized LLM-based approach to recommend pedagogical practices to educators and identify evidence-based solutions for their pedagogical challenges. In the subsequent sections, we will present the related work, followed by our proposed methodology. Lastly, we present the early result of our model assessment, and plan for future development and refinement of the system.

2   Related Work

Recommender Systems (RS) play a vital role in different domains by offering users tailored recommendations amidst the overwhelming surge in information availability on the web [5]. Various types of RSs exist, including Content-Based Filtering [7], Collaborative Filtering [8], and Context-Aware systems [9], each facing distinct challenges such as over-specialization, scalability issues, sparsity, and cold start problems. Deep Learning-Based models, represent a significant evolution, effectively handling complex relationships between users and items by leveraging various data layers [10]. Conversational recommender systems introduce interactive dialogue approaches, mitigating cold start issues and offering flexibility in recommendations [11]. However, they may suffer from reduced processing speed and inefficiencies in the absence of advanced Natural Language Processing (NLP) resources [12]. After the emergence of LLMs as powerful tools in NLP they have gained significant attention in the field of RS [6,13]. Trained on extensive datasets, these models hold the potential to enhance various aspects of RS through techniques like fine-tuning and prompt tuning. An important aspect of integrating LLMs into RS is their ability to leverage extensive external knowledge, enabling the creation of high-quality textual representations and establishing correlations between users and items [13]. Current research on utilizing LLMs in recommender systems (RS) categorizes them into two main paradigms: a) Discriminative LLM for Recommendation (DLLM-RS) and b) Generative LLM for Recommendation (GLLM-RS) [13]. Additionally, recent reviews of the application of LLMs in RS indicate two distinct approaches. The first approach involves directly employing LLMs as recommendation systems, such as in movie/book RS [14] and job recommendation systems [15]. The second approach integrates LLM tokens and embeddings within existing RS frameworks, as demonstrated in news recommendation systems [16,17]. LLM-based models surpass traditional recommendation systems by capturing contextual information and comprehending user queries and item descriptions [18]. This understanding enhances recommendation accuracy and relevance, increasing user satisfaction. Additionally, LLMs mitigate data sparsity issues by introducing novel capabilities like zero or few-shot recommendations, leveraging pre-training to provide reasonable suggestions even for unseen items or users [19]. The integration of generative models into RS can lead to more innovation, and practical applications and enhance the recommendation interpretability. Particularly, LLM-based models by using their language generation capabilities can provide explanations about the recommended items, aiding users in understanding recommendation factors [13]. The effectiveness of these models in overcoming data sparsity and efficiently tackling challenges encountered by previous RS systems makes their adoption in the academic domain a promising avenue. Particularly, GLLM-RS systems, by offering more comprehensive explanations of recommended features, can significantly enhance the provision of educational information and feedback to learners within the educational sector. In this study, we propose the adoption of the GLLM-RS approach, by employing LLMs to recommend pedagogical practices to educators through prompt tuning and the RAG framework. The subsequent section presents our methodology.

3   Methodology

The motivation behind our research initiative is the need to simplify the complex and time-consuming procedure that educators face when searching for PDPs that are tailored to address their specific challenges. The goal of our project is twofold: 1) to create a knowledge base that stores a wide range of PDPs from the existing resources, and 2) Integration of LLM into the RS using the RAG framework to provide accurate and contextually relevant solutions in response to specific user queries. The initial phase of the project involved collecting, unifying, and preprocessing a comprehensive set of published PDPs from academic literature, journals, educational websites, and conference proceedings. These PDPs cover diverse pedagogical issues such as team formation, assessment strategies, content delivery methods, and student engagement techniques. Emphasis was placed on sourcing evidence-based, peer-reviewed PDPs that reflect current educational research and practices. The dataset underwent rigorous curation, categorizing PDPs based on the addressed problems. This process led to the collection of 300 PDPs in 11 categories of Assessment, Assignment, Diversity, Feedback, Learning, Lectures, Passive Learning, Preparation, Problem Solving, Procrastination and Teamwork. To ensure consistency in our knowledge base, we employed an object-based model [3] focusing on key elements of 'problem' and 'solution' in published PDPs. Each PDP was annotated with metadata, including instructional objectives and relevant references. This enhanced searchability and retrieval of the PDPs. We developed a repository of the structured PDPs, each comprising a quadruple: label (category), problem (academic challenges), solution (empirical evidence and solution), and reference (source). Transformer-based embedding models were utilized to generate meaningful embeddings of the PDPs, stored in a dedicated vector database for subsequent tasks. The next step in the model development involves integrating LLM into the RS by leveraging the RAG framework. As pedagogical practices evolve and undergo changes, so do the PDPs. The dynamic nature of PDPs, coupled with their growing number over time, presents challenges in maintaining a current and comprehensive knowledge base. To overcome this challenge, we integrate LLM into the RS to dynamically extract up-to-date pedagogical practices from online sources thus surpassing the limitations of a static knowledge base. While integrating the LLM into the recommendation process holds promise for accommodating novel or unfamiliar situations that were not addressed in the local knowledge base, it also introduces certain limitations. One inherited pitfall in LLMs is that due to the vast amount of trained data their outputs often display substantial heterogeneity in the resources from which the response is generated. Sometimes these sources are not validated or are out of date. Additionally, lack of LLM access to the local knowledge base may potentially result in the loss of contextual information and specificity in the generated response. Thus, the generated answers merely based on the LLM may surpass the confines of the local knowledge base, leading to responses that vary in terms of abstraction and relevance to the given context. To address this challenge we adopt the RAG framework with LLM to enhance the recommendation result. RAG combines the strengths of information retrieval and generative language models. This framework is designed to augment LLMs by integrating them with an external knowledge retrieval process, thereby extending their utility and relevance. The process of model development involves three key stages: 1)Indexing: building an index of external documents (PDP knowledge base). In this phase, a comprehensive database of PDPs is collected and stored in the local knowledge base that can be efficiently queried, 2)Retrieval: querying this index based on specific educational challenges (in the form of user input queries)to fetch the most relevant documents (PDPs), and 3) Generation: inputting these retrieved documents into an LLM, which then generates or suggests actionable pedagogical strategies grounded in the retrieved information. Indexing involves converting PDPs into numerical vectors, a process that simplifies the comparison of documents and allows for the efficient identification of relevant content (a key step for RAG framework). This conversion is facilitated by embedding methods, which compress text into dense, fixed-length vector representations. These vectors, capturing the semantic meaning of the text, are then stored in a vector database.

The indexing process involves three steps 1) Segmentation: which segments documents (PDPs) into smaller portions to meet the limitations of the embedding models, 2) Vector Embedding: each segment of the document is embedded into a vector using embedding models, ensuring a rich semantic representation, and 3) Vector Store Creation: The resulting vectors are indexed in a vector store, linking them back to their original text segments for easy retrieval. Following the indexing stage, the retrieval component of our framework plays a key role in pinpointing PDPs from indexed repositories that are most semantically relevant to the user query. The retrieval process includes two steps: 1) creating semantic embeddings of user queries by transforming them into numerical vectors, similar to the process of indexing documents. This enables comparison in a high-dimensional space and facilitates quantitative assessment of semantic similarity, 2) Conducting semantic search which involves performing a similarity search within a vector space. By projecting both documents and queries into the same semantic space, the framework identifies documents closely aligned with the query. This is achieved through neighborhood search techniques, where the system identifies documents (or segments thereof) that are nearest, in terms of semantic similarity, to the query vector.

In the final step, LLM generates contextually relevant answers by integrating retrieved documents into its context window. This process, called generation, produces pedagogically valuable insights for user queries. First, retrieved documents (PDPs) are segmented and embedded into the LLM's context window to ensure access to relevant information. Next, prompts are constructed as structured placeholders by combining context (retrieved PDPs) and user queries. Finally, LLM generates responses by interpreting user queries and retrieved document contexts, resulting in comprehensive answers. The model's high-level architecture is depicted in Fig. 1, with the three main stages of indexing, retrieval, and generation-color-coded in blue, green, and orange, respectively, for clarity. As shown in the figure, upon receiving the user input query, the system employs the 'similarity index processor' module to determine the most relevant PDPs by assessing the semantic closeness of vector pairs. A similarity score above 0.5 is considered to indicate semantic similarity. The retriever module then extracts the top 5 PDPs from the knowledge base based on these similarity scores and formulates a set of prompts to be forwarded to the LLM, taking into account the user input query (query context). The prompt chains, constructed using the LangChain framework, are sent to a Hugging Face API for input into the LLM. The selected LLM for this study is Llama 2-13B-Chat-H, a pre-trained and fine-tuned open-source model with 13 billion parameters developed by Meta. The resulting LLM response is stored in a cache data store for subsequent use in similar queries, optimizing response generation efficiency over time in terms of both time and computing resources. The generated response from the LLM is also sent back to the Hugging Face API. In this phase, the response undergoes processing in the 'output parser' module for final formatting. The formatted response is then transmitted to the Dispatcher API, which sends it to the UI for presentation to the user.

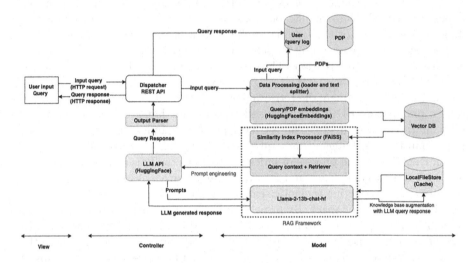

Fig. 1. Architecture of the proposed system

We evaluated the effectiveness of our model through a mixed-method approach, prioritizing the precision and factual integrity of the answers provided by both the RAG and LLM. The focus was to pinpoint instances of erroneous information or 'hallucinations' by leveraging cross-encoders that utilize transformer technology to differentiate between accurate information and incorrect outputs. Our findings indicate an acceptable level of accuracy, with an F-score of 0.83. We also integrated the Giskard platform to test the RAG model, powered by the LLaMA 2 13B LLM, for bias and hallucinations, assessing the model's overall efficacy. We queried the model with a set of 22 PDPs to assess bias, accuracy, and identify areas for performance improvement and reliability enhancement. By utilizing Giskard, we conducted thorough checks for biases, such as undue sycophancy, and ensured response plausibility. Among the 22 queries, three instances of sycophancy were noted. Given the assessment was conducted on a limited set of PDPs, the early findings indicate a relatively sound reliability, with no irrational responses detected. A qualitative study was also conducted to assess the quality and relevance of the model's responses. Nine diverse queries on educational challenges were selected for a survey distributed to 16 academic individuals. Participants rated each LLM response's relevance on a Likert scale and individually evaluated solutions for each query. Feedback revealed that 87% of the model's solutions were highly or moderately relevant, while 13% were deemed not pertinent. While a bigger sample size is necessary for robust conclusions on the reliability and performance of the proposed model, the initial findings are promising and showcase the model's efficiency in resolving user queries effectively.

4   Conclusion and Future Work

This study presents a novel approach to recommend research-based pedagogical practices by integrating the LLama2 large language model into the core of the RS. The proposed conversational RS leverages an extensive set of existing PDPs from academic publications as its local knowledge base. When a user input query is received, the RAG framework is employed to extract contextually relevant PDPs from the knowledge base. These, along with the user input query, form the query context to be passed to the LLM. The novelty of this model lies in the generated query context from the well-curated knowledge base of PDPs. This unique feature ensures outputs are aligned with existing published PDPs, enhancing relevancy. The open-source architecture minimizes operational costs, eliminates token-based transaction limitations, and supports community-driven development for continuous improvement. To evaluate the model's accuracy, Giskard's framework was employed, resulting in an accuracy score of 83% for query responses. In addition, a qualitative analysis was conducted with 16 academic professionals rating the relevancy of responses to nine sample queries. The data revealed an overall 87% of the responses were perceived as highly to moderately relevant, with only 13% as irrelevant. These early results are promising, motivating further optimization and enhancement of the model. Future work includes expanding the knowledge base with more PDPs and conducting broader evaluations to refine the model. We will further address the tendency of LLM to generate 'hallucinated' content, and plan to incorporate a robust verification method for resource reliability. Moreover, a user feedback mechanism will be introduced, enabling highly relevant query responses to be added to the local knowledge base. Will further explore the approaches for implementing contextual recommendations based on the user query log to provide more personalized recommendations to the user. The ultimate goal is to integrate this model into the LMS, to provide a secure, and cost-effective tool for enriching educational practices. This system and the developed knowledge-based will be publicly available to the educational community and will have a great potential to function as a repository of collective wisdom among educators, serving both experts and novice educators in exchanging the best pedagogical practices.

References

1. Ryoo, J., Winkelmann, K.: Innovative Learning Environments in STEM Higher Education: Opportunities, Challenges, and Looking Forward (2021)
2. Laurillard, D.: Teaching as a design science: Building pedagogical patterns for learning and technology. Routledge (2013)

3. Dehbozorgi, N., MacNeil, S., Maher, M. & Dorodchi, M.: A comparison of lecture-based and active learning design patterns in CS education. In: 2018 IEEE Frontiers In Education Conference (FIE), pp. 1-8 (2018)

4. Inventado, P.S., Scupelli, P.: Towards a community-centric pattern repository. In Proceedings of the 22nd European Conference on Pattern Languages of Programs, pp. 1-4 (July 2017)

5. Dehbozorgi, N., Norkham, A.: An architecture model of recommender system for pedagogical design patterns,. In: 2021 IEEE Frontiers in Education Conference (FIE) (Manuscript Accepted for Publication). IEEE (2021)

6. Di Palma, D.: Retrieval-augmented recommender system: enhancing recommender systems with large language models. In: Proceedings of the 17th ACM Conference on Recommender Systems, pp. 1369-1373 (September 2023)

7. Lops, P., De Gemmis, M., Semeraro, G.: Content-based recommender systems: State of the art and trends. Recommender systems handbook, pp. 73–105 (2011)

8. Cui, L. Z., Guo, F. L., Liang, Y. J.: Research overview of educational recommender systems. In: Proceedings of the 2nd International Conference on Computer Science and Application Engineering, pp. 1-7 (October 2018)

9. He, C., Parra, D., Verbert, K.: Interactive recommender systems: a survey of the state of the art and future research challenges and opportunities. Expert Syst. Appl. **56**, 9–27 (2016)

10. Sun, F., et al.: BERT4Rec: sequential recommendation with bidirectional encoder representations from transformer. In: CIKM, pp. 1441–1450. ACM (2019)

11. Barria-Pineda, J.: Exploring the need for transparency in educational recommender systems. In: Proceedings of the 28th ACM Conference on User Modeling, Adaptation and Personalization, pp. 376- 379 (July 2020)

12. Yanes, N., Mostafa, A.M., Ezz, M., Almuayqil, S.N.: A machine learning-based recommender system for improving students learning experiences. IEEE Access **8**, 201218–201235 (2020)

13. Wu, L., et al.: A survey on large language models for recommendation. arXiv preprint arXiv:2305.19860 (2023)

14. Bao, K., Zhang, J., Zhang, Y., Wang, W., Feng, F., He, X.: Tallrec: an effective and efficient tuning framework to align large language model with recommendation. In: Proceedings of the 17th ACM Conference on Recommender Systems, pp. 1007-1014 (September 2023)

15. Wu, L., Qiu, Z., Zheng, Z., Zhu, H., Chen, E.: Exploring large language model for graph data understanding in online job recommendations. arXiv preprint arXiv:2307.05722 (2023)

16. Liu, Q., Chen, N., Sakai, T., Wu, X.M.: A first look at llm-powered generative news recommendation. arXiv preprint arXiv:2305.06566 (2023)

17. Wu, C., Wu, F., Qi, T., Huang, Y.: Empowering news recommendation with pre-trained language models. In: Proceedings of the 44th international ACM SIGIR Conference on Research and Development in Information Retrieval, pp. 1652-1656 (July 2021)

18. Geng, S., Liu, S., Fu, Z., Ge, Y., Zhang, Y.: Recommendation as language processing (RLP): a unified pretrain, personalized prompt & predict paradigm (P5). In: Proceedings of the 16th ACM Conference on Recommender Systems, pp. 299-315 (September 2022)

19. Sileo, D., Vossen, W., Raymaekers, R.: Zero-shot recommendation as language modeling. In: Hagen, M., et al. (eds.) ECIR 2022. LNCS, vol. 13186, pp. 223–230. Springer, Cham (2022). https://doi.org/10.1007/978-3-030-99739-7_26

Role of Ethical and Responsible AI in Education for Next Generation Inventors

Krishna Chaitanya Rao Kathala[1]([✉]), Ivon Arroyo[1]([✉]),
and Nishith Reddy Mannuru[2]([✉])

[1] University of Massachusetts Amherst, Amherst, USA
{kkathala,ivon}@umass.edu
[2] University of North Texas, Denton, USA
NishithReddyMannuru@my.unt.edu

Abstract. The integration of Artificial Intelligence (AI) in education holds transformative potential and requires careful ethical navigation. This paper explores the crucial role of ethical and responsible AI in developing the next generation of inventors poised to lead technological advancements. We critically analyze existing ethical frameworks and address specific considerations essential for utilizing AI to foster innovation. Drawing from interdisciplinary insights, we propose a comprehensive framework that emphasizes collaboration, ethical literacy, inclusive design, governance structures, and continuous evaluation. We highlight strategies to embed AI ethics within STEM curricula, thereby cultivating crucial decision-making skills among students. Additionally, we provide policy recommendations to support the responsible adoption of AI in educational settings. Ultimately, instilling robust ethical principles in future inventors is essential for promoting responsible technological progress that prioritizes societal well-being.

Keywords: Artificial Intelligence · Ethics in Education · Responsible Innovation · STEM Education · Next Generation Inventors

1 Introduction

The education sector is no exception to the rapid penetration of artificial intelligence (AI) across various fields. Artificial Intelligence (AI) has the potential to completely transform teaching and learning environments in educational settings. AI promises to improve educational outcomes by offering educators and students more power through individualized learning pathways, intelligent tutoring systems, automated grading, and assessment tools (Zawacki-Richter et al., 2019). But even as we welcome AI's transformative potential in education, it is crucial to navigate its implementation responsibly and ethically.

This paper aims to explore the pivotal role of ethical and responsible AI in education, with a specific focus on its implications for nurturing the next

A. M. Olney et al. (Eds.): AIED 2024 Workshops, CCIS 2151, pp. 71–78, 2024.
https://doi.org/10.1007/978-3-031-64312-5_9

generation of inventors. We seek to create a strong framework that directs the creation and application of AI technologies in educational settings by looking at the areas where ethics, AI, and education converge. Three goals are the focus of this study: First, in order to find any gaps, difficulties, or areas that could need improvement, a critical analysis of the current ethical frameworks and guidelines pertaining to the use of AI in education is necessary. Secondly, to look into the particular ethical issues-such as bias, privacy, and transparency-that come up when using AI to support the development of future inventors. Finally, to suggest a thorough ethical, and responsible AI framework designed for educational environments, with a focus on inspiring the next generation of inventors.

The results of this endeavor are important because they acknowledge that AI, while immensely powerful, is not inherently neutral. The values, principles, and ethical underpinnings ingrained in its creation and execution influence its effects (Floridi & Cowls, 2019). We can harness the transformative potential of AI in education while reducing potential risks and unintended consequences by proactively addressing ethical concerns and establishing responsible practices.

This study also emphasizes how critical it is to instill ethical principles in the next generation of inventors from their inception. These future innovators need to have the moral fortitude and critical thinking abilities to propel responsible and sustainable progress as they traverse an increasingly complex technological landscape (Schomberg, 2013). We can raise a new generation of ethically grounded and technically skilled inventors who can prioritize societal well-being and create solutions that are consistent with human values by incorporating ethical and responsible AI principles into educational curricula and practices, Nguyen et al. (2023).

In this pursuit, the paper draws upon a diverse range of interdisciplinary perspectives, bridging the fields of education, ethics, AI, and innovation. We seek to thoroughly grasp the possibilities as well as the challenges that exist at the nexus of AI, ethics, and education for the next generation of inventors by combining knowledge from theoretical frameworks, empirical research, and real-world case studies.

2   Landscape of AI in Education

The integration of AI in educational environments has opened up game-changing possibilities for improving teaching and learning experiences for students. Through intelligent tutoring systems that adapt instructional content, provide real-time feedback, and offer tailored support, AI-driven personalized learning responds to each student's unique needs, styles, and paces (Mousavinasab et al., 2021; Kulik & Fletcher, 2016). Furthermore, through analysis of massive amounts of data from student interactions and assessments, AI-powered learning analytics enable data-informed decisions by tracking student progress, identifying struggles, and recommending personalized interventions (Viberg et al., 2020). This data-driven approach holds the potential to revolutionize instructional design and delivery as per Liedtka (2018).

To properly utilize AI's full potential, ethical issues must be resolved. Data security and privacy are critical, and as AI systems gather and process student data, strict safeguards must be put in place to prevent misuse or illegal access. Another serious problem is algorithmic bias, whereby AI models trained on data reflecting past biases may produce biased predictions and recommendations that continue discrimination (Mehrabi et al., 2021). To tackle this, debiasing techniques must be used, diverse datasets must be purchased, and AI systems must be audited for fairness. An important obstacle to ensuring that AI-enabled educational technologies are adopted and accessible to all is the digital divide, Vassilakopoulou and Hustad (2021). Inequalities already in place may be made worse by socioeconomic gaps, inadequate infrastructure, and low levels of digital literacy (Mousavinasab et al., 2021). To close this gap, comprehensive plans focusing on digital inclusion, digital skill development, and technology access are required.

AI has enormous potential to change pedagogy, tailor learning experiences, and promote more inclusive and equitable educational outcomes in the classroom, even in the face of these obstacles, Drigas et al. (2017). Through proactive resolution of ethical issues, advocacy of conscientious behaviors, and emphasis on diversity, equity, and inclusion, the incorporation of AI can facilitate the delivery of specialized, high-quality education that equips every student to thrive in an increasingly complex world.

3   Ethical Considerations in AI-Enhanced Education

With AI becoming more and more prevalent in education, it is imperative that important ethical considerations be made in order to avoid sustaining systemic biases, violating privacy, and undermining principles of equity and inclusion. Recent developments, in AI-driven systems, have raised significant privacy concerns and fairness issues. The risk is that AI systems trained on biased data will magnify current disparities through skewed assessments and evaluations (Kizilcec & Lee, 2022). AI grading algorithms and personalized learning tools risk disadvantaging marginalized groups if not designed inclusively. Building trust and accountability requires openness and interpretability. Many AI systems function as mysterious "black boxes", making it difficult to understand how they make decisions and making it more difficult to spot and correct biases or mistakes (Arrieta et al., 2020). Transparency can be promoted by methods such as explainable AI.

To prevent breaches or misuse of students' sensitive personal data collected by AI ed-tech tools, robust data governance and strict security protocols are essential (Alam, A., 2023). Having transparent policies and oversight procedures is essential to preserving stakeholder trust. Moreover, the integration of AI mustn't compromise fundamental educational principles such as fostering creativity, critical thinking, and socio-emotional development (Popenici & Kerr, 2017). Artificial intelligence (AI) should enhance learning but shouldn't completely replace human teachers or minimize interpersonal pedagogy.

To address these ethical concerns, it will take a multi-stakeholder effort including ethicists, educators, policymakers, and technologists to jointly develop best practices and guidelines. The OECD AI Principles (OECD, 2019) and IEEE's Ethically Aligned Design (K. Shahriari & M. Shahriari, 2017) are two comprehensive frameworks that offer guidance on integrating values of accountability, fairness, privacy, and transparency into the design and implementation of AI systems. Fostering a culture of ethical mindfulness and responsible innovation requires developing ethical literacy through awareness-raising initiatives.

4   Framework for Ethical and Responsible AI in Education

The integration of AI in educational contexts necessitates a comprehensive and carefully designed framework that prioritizes ethical and responsible practices. This proposed framework seeks to ensure adherence to moral and ethical principles, promote cooperation among important stakeholders, and provide a solid foundation for the development and application of AI technologies in education.

The idea of cross-disciplinary collaboration is integral to this framework. This approach acknowledges the complex nature of artificial intelligence (AI) in education and promotes the convergence of various viewpoints from educators, technologists, ethicists, and students. The framework fosters a co-creative process that makes use of the diverse but complementary perspectives of all stakeholders by bringing them together. The first pillar of this framework is to enhance the ethical literacy of all participants. Developing a thorough understanding of ethical frameworks, concepts, and principles that are pertinent to AI in education is a necessary component of ethical literacy. Stakeholders can acquire the critical thinking abilities needed to resolve difficult ethical challenges and reach wise decisions by means of targeted training courses, seminars, and instructional materials (Peters D, et al., 2020). By enabling proactive steps to mitigate potential risks, biases, and unintended consequences associated with AI technologies, ethical literacy empowers individuals. Developing on this ethical literacy base, the framework highlights the significance of inclusive and collaborative design processes. To ensure that AI solutions in education are equitable, accessible, and reflective of the needs and values of all learners, it is imperative to incorporate diverse voices and perspectives from the outset (McBride, et al., 2021). The framework fosters responsible and socially responsible AI implementations by actively involving students, educators, and community members in the design and development phases. This fosters a sense of ownership, trust, and accountability. The framework's establishment of explicit accountability and governance mechanisms is another essential element. Principles, rules, and procedures for the ethical and responsible application of AI in education should be outlined in effective governance frameworks (Khosravi, H. et al., 2022). These frameworks should cover topics like algorithmic transparency, data privacy, and bias mitigation. They should also include procedures for oversight, monitoring, and redress in the event that any ethical violations or unexpected consequences occur. The framework also highlights the significance of iterative improvement and ongoing assessment. It is crucial to continually evaluate the effects and ramifications of AI solutions in education because these technologies are developing quickly and because social settings are changing. Stakeholders can better align AI implementations with ethical principles and educational objectives, identify areas for improvement, and respond to emerging ethical challenges by putting rigorous monitoring and evaluation processes in place (Holmes, W et al., 2022).

This framework is based on a dedication to educating the next generation of innovators and inventors about ethical and responsible AI literacy. The framework seeks to provide future leaders with the knowledge, abilities, and mindsets required for fostering responsible and sustainable technological advancement by incorporating ethical considerations and responsible AI principles into educational curricula and learning experiences (Aler Tubella, A. et al., 2024). By placing a strong focus on ethical AI literacy, we can make sure that the next generation of innovators is equipped to make ethical decisions that put society's welfare and human-centric values first. For this framework to be implemented effectively, an ecosystem that encourages cooperation, knowledge exchange, and ongoing education is needed. In addition to fostering partnerships between academic institutions, businesses, and civil society organizations, this ecosystem should encourage the exchange of best practices, case studies, and cutting-edge research on ethical and responsible AI in education Vahrenhold and Parameswaran (2021). Through the adoption of this comprehensive framework, educational establishments and interested parties can take advantage of artificial intelligence's transformative potential while maintaining the highest ethical standards and making sure that AI solutions put equity, inclusivity, and the welfare of all students first. Through interdisciplinary collaboration, ethical literacy, inclusive design, robust governance, and continuous improvement, this framework paves the way for a future where AI enhances educational experiences responsibly and ethically, nurturing the next generation of ethically grounded innovators and inventors.

5 Preparing the Next Generation of Inventors, Policy Recommendations and Future Directions

To effectively prepare the next generation of inventors, educators must prioritize the integration of AI ethics into STEM (Science, Technology, Engineering, and Mathematics) curricula. This interdisciplinary approach fosters a holistic understanding of AI's societal impacts and cultivates the essential skills required for ethical decision-making and responsible innovation. One effective strategy involves incorporating case studies and real-world examples that highlight the ethical implications of AI technologies. By analyzing the successes, failures, and unintended consequences of AI applications, students can develop critical thinking and problem-solving abilities, enabling them to identify potential risks and devise ethical solutions. Additionally, interactive workshops and simulations can provide hands-on experiences, allowing students to grapple with ethical dilemmas and practice navigating complex scenarios. Furthermore, educational institutions should promote cross-disciplinary collaboration, encouraging students to engage with experts from diverse fields, such as ethics, law, and social sciences (Chan, C. K., 2023). This exposure to varied perspectives and disciplines equips students with a comprehensive understanding of AI's multifaceted impact, fostering empathy, ethical reasoning, and the ability to consider diverse stakeholder interests. Ultimately, the goal is to instill in the next generation of inventors a deep-rooted commitment to ethical and responsible innovation. By cultivating these values and skills from an early stage, we can empower future innovators to drive technological progress while prioritizing societal well-being, equity, and human-centric values.

5.1   Policy Recommendations

Together, policymakers, educators, and technologists must develop strategies to support ethical AI in education and foster responsible inventors. Key measures include establishing national frameworks for AI ethics covering data privacy, algorithmic transparency, bias mitigation, and governance (Chan, C. K., 2023); investing in AI ethics education for ethical decision-making skills (Chan, C. K., 2023); promoting interdisciplinary collaboration (Xu, W., & Ouyang, F., 2022); advancing AI research focused on fairness and accountability (Arrieta et al., 2020); and implementing robust governance with oversight (Kamalov, F. et al., 2023). Continuous adaptation of ethical norms and practices is essential, requiring ongoing vigilance and a commitment to improvement to align AI with educational standards and societal values.

5.2   Future Directions

As artificial intelligence continues to evolve within educational settings, ongoing dialogue among stakeholders-students, educators, technologists, policymakers, and the community-is essential. To harness AI's full potential responsibly, educational policies must evolve to include ethical AI training and support interdisciplinary teams that prioritize ethical considerations. Additionally, sustained investment in AI research is crucial for developing pedagogical models that reflect current ethical standards. These efforts will ensure that AI technology not only enhances educational outcomes but also aligns with societal values, preparing a new generation of inventors to lead with innovation and integrity.

6   Conclusion

In the rapidly evolving realm of AI in education, it is crucial to adopt a principled approach that upholds ethical and responsible practices. The way we educate the next generation of innovators, who will drive technological advancements, is fundamentally influenced by the impact of ethical AI. A thorough examination of ethical issues in current AI applications within educational settings underscores the importance of frameworks that support inclusive design, robust governance, ethical literacy, interdisciplinary collaboration, and continual improvement. These elements ensure that AI's integration into education respects human-centered learning principles, fostering innovation, critical thinking, equity, accountability, and transparency.

Empowering future innovators to ethically navigate AI's complexities involves embedding responsible AI practices into STEM education, and adjusting to evolving ethical, technological, and societal landscapes. Policymakers, stakeholders, and educators must engage in ongoing dialogue and embrace lifelong learning. By following the guidelines discussed in this paper and remaining adaptable, we can ethically maximize AI's educational potential, nurturing a generation committed to ethical innovation and positively impacting the global community.

References

Floridi, L., Cowls, J.: A unified framework of five principles for AI in society. Harvard Data Sci. Rev. **1**(1) (2019). https://doi.org/10.1162/99608f92.8cd550d1

Liedtka, J.: Why design thinking works. Harv. Bus. Rev. **96**(5), 72–79 (2018)

Schomberg, R.V.: A vision of responsible research and innovation. In: Responsible Innovation: Managing the Responsible Emergence of Science and Innovation in Society, pp. 51–74. Wiley (2013)

Zawacki-Richter, O., Marìn, V.I., Bond, M., Gouverneur, F.: Systematic review of research on artificial intelligence applications in higher education - where are the educators? Int. J. Educ. Technol. High. Educ. **16**(1), 1–27 (2019). https://doi.org/10.1186/s41239-019-0171-0

Nguyen, A., Ngo, H.N., Hong, Y., et al.: Ethical principles for artificial intelligence in education. Educ. Inf. Technol. **28**, 4221–4241 (2023). https://doi.org/10.1007/s10639-022-11316-w

Drigas, A.S., Argyri, K.: Accessibility for learners with disabilities in tertiary education. Int. J. Educ. Inf. Technol. **11**, 96–101 (2017)

Seale, J.: E-learning and Disability in Higher Education: Accessibility Research and Practice. Routledge (2013)

Kulik, J.A., Fletcher, J.D.: Effectiveness of intelligent tutoring systems: a meta-analytic review. Rev. Educ. Res. **86**(1), 42–78 (2016). https://doi.org/10.3102/0034654315581420

Mehrabi, N., Morstatter, F., Saxena, N., Lerman, K., Galstyan, A.: A survey on bias and fairness in machine learning. ACM Comput. Surv. **54**(6), 1–35 (2021). https://doi.org/10.1145/3457607

Mousavinasab, E., Zarifsanayei, N., Niazi, S.R., Faghri, F., Roshani, N., Nazemi, K.: Intelligent tutoring systems: a systematic review of characteristics, applications, and evaluation methods. Interact. Learn. Environ. **1–22**,(2021). https://doi.org/10.1080/10494820.2021.1932241

Vassilakopoulou, P., Hustad, E.: Bridging digital divides (2021): a literature review and research agenda for information systems research. Inf. Syst. Front. **25**(3), 955–969 (2023). https://doi.org/10.1007/s10796-020-10096-3. Epub. PMID: 33424421; PMCID: PMC7786312

Viberg, Khalil, M., Baars, M.: Self-regulated learning and learning analytics in online learning environments: a review of empirical research. In: Proceedings LAK 2020: 10th International Conference on Learning Analytics and Knowledge, pp. 524–533 (2020)

McBride, E., Peters, E.S., Judd, S.: Design considerations for inclusive AI curriculum materials. In: Proceedings of the 52nd ACM Technical Symposium on Computer Science Education, pp. 1370–1370 (2021)

Peters, D., Vold, K., Robinson, D., Calvo, R.A.: Responsible AI-two frameworks for ethical design practice. IEEE Trans. Technol. Soc. **1**(1), 34–47 (2020)

Aler Tubella, A., Mora-Cantallops, M., Nieves, J.C.: How to teach responsible AI in higher education: challenges and opportunities. Ethics Inf. Technol. **26** (2024). https://doi.org/10.1007/s10676-023-09733-7

Khosravi, H., et al.: Explainable artificial intelligence in education. Comput. Educ. Artif. Intell. **3**, 100074 (2022)

Vahrenhold, J., Parameswaran, N.: Ethical AI in education: ethical and responsible use of AI in education. In ICERI 2021: Proceedings of the 14th annual International Conference of Education, Research and Innovation, pp. 1882–1892. IATED Academy (2021)

Holmes, W., et al.: Ethics of AI in education: towards a community-wide framework. Int. J. Artif. Intell. Educ. 1–23 (2022)

Chan, C.K.: A comprehensive AI policy education framework for university teaching and learning. Int. J. Educ. Technol. High. Educ. **20**(1), 1–25 (2023). https://doi.org/10.1186/s41239-023-00408-3

Kamalov, F., Calong, D.S., Gurrib, I.: New Era of Artificial Intelligence in Education: Towards a Sustainable Multifaceted Revolution. arXiv abs/2305.18303 (2023)

Xu, W., Ouyang, F.: The application of AI technologies in STEM education: a systematic review from 2011 to 2021. Int. J. STEM Educ. **9**(1), 1–20 (2022). https://doi.org/10.1186/s40594-022-00377-5

Alam, A.: Developing a curriculum for ethical and responsible AI: a university course on safety, fairness, privacy, and ethics to prepare next generation of AI professionals. In: Rajakumar, G., Du, K.L., Rocha, A. (eds.) ICICV 2023. LNDECT, vol. 171, pp. 879–894. Springer, Singapore (2023). https://doi.org/10.1007/978-981-99-1767-9_64

Kizilcec, R.F., Lee, H.: Algorithmic Fairness in Education. Routledge, London (2022). https://doi.org/10.4324/9780429329067-10

Popenici, S.A., Kerr, S.: Exploring the impact of artificial intelligence on teaching and learning in higher education. Res. Pract. Technol. Enhanc. Learn. **12**(1), 1–13 (2017). https://doi.org/10.1186/s41039-017-0062-8

OECD. Recommendation of the Council on Artificial Intelligence (2019). https://legalinstruments.oecd.org/en/instruments/OECD-LEGAL-0449

Shahriari, K., Shahriari, M.: IEEE standard review - ethically aligned design: a vision for prioritizing human wellbeing with artificial intelligence and autonomous systems. In: 2017 IEEE Canada International Humanitarian Technology Conference (IHTC), Toronto, ON, Canada, pp. 197–201 (2017). https://doi.org/10.1109/IHTC.2017.8058187

Arrieta, A.B., et al.: Explainable Artificial Intelligence (XAI): concepts, taxonomies, opportunities and challenges toward responsible AI. Inf. Fusion **58**, 82–115 (2020). https://doi.org/10.1016/j.inffus.2019.12.012

Collaborative Essay Evaluation with Human and Neural Graders Using Item Response Theory Under a Nonequivalent Groups Design

Kota Aramaki$^{(\boxtimes)}$ and Masaki Uto$^{(\boxtimes)}$ (ORCID)

The University of Electro-Communications, Tokyo, Japan
{aramaki,uto}@ai.lab.uec.ac.jp

Abstract. In the assessment of essay writing, reliably measuring examinee ability can be difficult owing to bias effects arising from rater characteristics. To address this, item response theory (IRT) models that incorporate rater characteristic parameters have been proposed. These models estimate the ability of examinees from scores assigned by multiple raters while considering their scoring characteristics, thereby achieving more accurate measurement of ability compared with a simple average of scores. However, issues arise when different groups of examinees are assessed by distinct sets of raters. In such cases, test linking is required to standardize the scale of ability estimates among multiple examinee groups. Traditional test linking methods require administrators to design groups in which either examinees or raters are partially shared—a requirement that is often impractical in real-world assessment settings. To overcome this problem, we introduce a novel linking method that does not rely on common examinees and raters by utilizing a recent automated essay scoring (AES) method. Our method not only facilitates test linking but also enables effective collaboration between human raters and AES, which enhances the accuracy of ability measurement.

Keywords: item response theory · automated essay scoring · test linking · essay writing assessments

1 Introduction

An increasing need to evaluate higher-order abilities, such as logical reasoning and expressive skills, has led essay writing assessments to become more prominent [1]. These assessments are designed to measure the abilities of examinees to write essays in response to specific prompts, based on scores assigned by human raters. However, the human grading process often involves difficulties in reliably measuring examinee ability due to bias effects arising from individual rater characteristics, such as severity and consistency. To address this issue, item response theory (IRT) models that incorporate parameters representing rater characteristics have been proposed [9,13]. While one well-known model is the many facet Rasch model (MFRM) [5], recent studies have explored extensions of the MFRM aimed at relaxing some of its strong assumptions [8,9,13]. A representative example is the generalized MFRM (GMFRM) [13], which allows for a flexible representation of rater characteristics, thus providing more accurate ability measurement compared with traditional MFRM.

However, challenges arise when essays from distinct groups of examinees are evaluated by different sets of raters—a scenario often encountered in real-world testing. For example, in academic settings such as university admissions, individual departments might employ separate pools of raters to evaluate essays from specific applicant groups. Similarly, in large-scale standardized tests, different sets of raters might be assigned to assess essays across various test dates or locations. When applying IRT models like the GMFRM to such a scenario, examinee abilities are not directly comparable among different groups. Specifically, assume that one group consists of high-ability examinees evaluated by severe raters and another comprises low-ability examinees assessed by lenient raters. Such situations are frequently referred to as a *nonequivalent groups design* [2,14]. In such designs, IRT gives the examinees' abilities on different scales for each group. To ensure comparability of examinees' abilities across these nonequivalent groups, *test linking*, a procedure designed to unify the measurement scales across multiple groups [10,14], becomes essential. Conventional test linking methods generally require some overlap of examinees or raters across the groups being linked [4,7,11]. However, the design of such overlapping groups can often be impractical in real-world testing scenarios.

To address these problems, we introduce a novel test linking method that utilizes a recent deep neural automated essay scoring (AES) technique [12] capable of predicting IRT-based examinee abilities from their essays. The key idea of our method is to train the AES model using data for a reference group, whose ability scale is targeted for standardization. We then use the trained model to predict the abilities of examinees in focal groups, whose ability scale we aim to align with that of the reference group. Since the AES model is trained to predict abilities on the scale of the reference group, it can predict the abilities of examinees in focal groups by aligning these predictions with the scale of the reference group. Consequently, employing AES-predicted abilities as prior information in IRT-based ability estimation for the focal group aids in aligning the obtained scale with that of the reference group, enabling comparison of examinee abilities across different nonequivalent groups. Our contributions are as follows:

1. Our method estimates the abilities of examinees in focal groups by integrating evaluations from both human raters and AES technology. This approach not only facilitates test linking but also enables effective collaboration between human raters and AES under nonequivalent groups designs. It helps to achieve high accuracy in ability measurement and ensures comparability of ability scales among nonequivalent groups.
2. We provide a completely original solution for the test linking problem, a crucial challenge in educational measurement. This has the potential to offer a new perspective on foundational test theory in educational measurement.

2 Setting

In this study, two groups of examinees responded to the same writing question, and their written essays were assessed by two distinct sets of raters following the same scoring rubric. One group was designated as the *reference group*, serving as the benchmark for the scale, while the other was designated as the *focal group*, whose scale we aim to align with that of the reference group. The score data obtained from each group are defined as

$$U^{ref} = \left\{ u_{jr} \in \{\mathcal{K} \cup -1\} | j \in \mathcal{J}^{ref}, r \in \mathcal{R}^{ref} \right\}, \tag{1}$$

$$U^{foc} = \left\{ u_{j'r'} \in \{\mathcal{K} \cup -1\} | j' \in \mathcal{J}^{foc}, r' \in \mathcal{R}^{foc} \right\}, \tag{2}$$

where $u_{jr} \in \mathcal{K} = \{1, \ldots, K\}$ represents the score given by rater r to the essay of examinee j, and $u_{jr} = -1$ denotes missing data. K represents the number of score categories. Additionally, $\mathcal{J}^{ref}$ and $\mathcal{R}^{ref}$ denote the sets of examinees and raters in the reference group, respectively, while $\mathcal{J}^{foc}$ and $\mathcal{R}^{foc}$ represent the sets of examinees and raters in the focal group.

The purpose of this study was to measure IRT-based abilities of examinees from two sets of data on the same scale and in conditions without common examinees and raters, $\mathcal{J}^{ref} \cap \mathcal{J}^{foc} = \emptyset$ and $\mathcal{R}^{ref} \cap \mathcal{R}^{foc} = \emptyset$.

3 Item Response Theory

IRT, a test theory grounded in mathematical models, has been applied in various educational measurement situations. However, traditional IRT models are not directly applicable to essay-writing test data, where examinees' responses to test items are assessed by multiple human raters. Thus, extended IRT models with rater parameters have been proposed to address this issue [5,15]. In the present study, we employ GMFRM [13] as one of the latest IRT models with rater parameters.

When a single writing question is offered, GMFRM defines the probability P_{jrk} that the essay of examinee j for the writing question will receive a score of k from rater r as

$$P_{jrk} = \frac{\exp \sum_{m=1}^{k} [D\alpha_r (\theta_j - \beta_r - d_{rm})]}{\sum_{l=1}^{K} \exp \sum_{m=1}^{l} [D\alpha_r (\theta_j - \beta_r - d_{rm})]}. \tag{3}$$

where θ_j is the latent ability of examinee j, α_r indicates the consistency of rater r, β_r is the severity of rater r, and d_{rk} is a parameter representing the severity of rater r for score k. $D = 1.7$ is a scaling constant that minimizes the difference between the normal and logistic distribution functions. For parameter identification, the restrictions $d_{r1} = 0$, and $\sum_{k=2}^{K} d_{rk} = 0$, and a normal prior distribution for the ability θ_j, are assumed.

Since we assume a scenario in which the same writing question is assigned to both examinee groups, we use this formulation of the GMFRM as the IRT model for estimating the ability of examinees.

4   Test Linking

The goal of this study is to estimate the abilities of examinees and the rater parameters in the GMFRM from U^{ref} and U^{foc} while ensuring the comparability of these estimates between groups. However, when the GMFRM is simply applied to datasets from the two different groups, the scales of the estimated parameters typically vary between them because IRT allows for arbitrary scaling of parameters within each dataset. An exception to this occurs when equivalency in the distributions of examinee abilities and rater parameters between groups can be assumed [6,11]. However, in real-world testing scenarios, this assumption may not always hold, meaning that these distributions are nonequivalent between groups. *Test linking* becomes essential for making parameter estimates between nonequivalent groups comparable.

Test linking is a statistical procedure for standardizing the scale of the IRT parameters derived from different groups' data. One widely used approach for test linking is *linear linking*. In the context of the essay-writing test considered in this study, implementing linear linking necessitates some overlap in examinees (and/or raters) between the two groups. However, arranging multiple groups with partially overlapping examinees (and/or raters) can often be impractical in real-world testing environments. To address this limitation, we aim to facilitate test linking without common examinees and raters by leveraging AES technology.

5   Using AES to Predict IRT-Based Examinee Ability

An AES model using deep neural networks that can predict examinee ability θ_j in the GMFRM has recently been proposed [12]. This model uses BERT (bidirectional encoder representations from transformers) [3], a pre-trained transformer model, as the basis model. This AES model receives essay texts with an additional special [CLS] token as input and produces an output vector corresponding to the [CLS] token as a distributed representation vector for a given essay, a fixed-dimensional real-valued vector that aggregates the information of the essay. Then, using the distributed representation vector $\boldsymbol{h}$, the model predicts the ability of the examinee who wrote the essay, as $\hat{\theta} = 6 \times sigmoid(\boldsymbol{W}\boldsymbol{h} + b) - 3$, where $sigmoid(\cdot)$ is the sigmoid function, and $\boldsymbol{W}$ and b are the weight vector and bias parameters. Note that the reason for multiplying by 6 and then subtracting 3 is that while the output of the sigmoid function ranges from 0 to 1, the ability values in the GMFRM typically range from -3 to 3 because of the assumption of a standard normal distribution for the prior distribution of the ability.

This model is trained using a dataset consisting of a collection of essays and scores given by multiple raters, like U^{ref} and U^{foc}, as follows.

1. Estimate the abilities of examinees based on scores assigned to their essays by multiple raters. We denote $\boldsymbol{\theta}$ as the vector of estimated abilities, with θ_j representing its j-th element, which corresponds to the estimated ability of the j-th examinee.

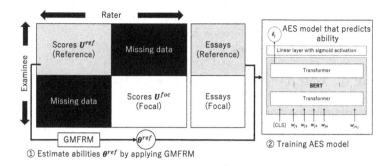

Fig. 1. Procedures 1 and 2 of our proposed method.

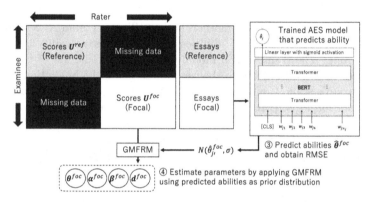

Fig. 2. Procedures 3 and 4 of our proposed method.

2. Train the AES model using the ability estimates as the objective function by minimizing the mean squared error (MSE) loss function $\frac{1}{J}\sum_{j=1}^{J}(\theta_j - \hat{\theta}_j)^2$, where $\hat{\theta}_j$ represents the AES-predicted ability of examinee j, and J is the number of training samples.

This method was proposed to enable robust training of AES models by mitigating the impact of rater bias in the training data. However, our primary interest lies in leveraging its secondary feature, which allows for the prediction of examinee ability from essay texts, aligned with the scale of the training data.

6   Proposed Method

The specific procedures of our method are outlined in detail below and are visually represented in Figs. 1 and 2.

1. For the reference group examinees $\mathcal{J}^{ref}$, estimate their abilities $\boldsymbol{\theta}^{ref}$ using the GMFRM from the corresponding data $\boldsymbol{U}^{ref}$.

2. Train the AES model introduced above using the estimated abilities and the essays of examinees in the reference group.

3. For the focal group examinees $\mathcal{J}^{foc}$, predict their abilities $\hat{\boldsymbol{\theta}}^{foc}$ using their essays and the trained AES model.

4. Estimate the GMFRM parameters by incorporating the predicted abilities $\hat{\boldsymbol{\theta}}^{foc}$ as the prior distribution for each examinee's ability. Specifically, the prior distribution for examinee $j' \in \mathcal{J}^{foc}$ is $N(\hat{\theta}_{j'}^{foc}, \sigma)$, where $\hat{\theta}_{j'}^{foc}$ represents the j'-th element of $\hat{\boldsymbol{\theta}}^{foc}$ and σ is a constant standard deviation. In this study σ is calculated as the root mean square error (RMSE), derived as the square root of the MSE loss function introduced in Sect. 5. The parameter estimation is made through Bayesian estimation using a Markov chain Monte-Carlo algorithm [13].

The key concept behind this method is the utilization of an AES model that is trained to predict examinees' abilities from their essays using data from a reference group. The trained AES model enables the prediction of the abilities of examinees in focal groups from their essays, aligning the predictions with the reference group's scale. Consequently, employing the AES-predicted abilities as prior information in GMFRM parameter estimation for the focal group aids in aligning the obtained ability scale with that of the reference group. This enables us to compare examinee abilities across different groups.

Moreover, our method estimates the abilities of examinees by using both scores from human raters and abilities predicted by an AES model. This collaborative process helps to achieve high accuracy in ability measurement while enabling comparability of ability scales among nonequivalent examinee groups.

7   Experiments

In this section, we demonstrate the effectiveness of our method through experiments using actual data.

In our experiments, we used the dataset created by Ref. [12] based on the automated student assessment prize (ASAP) dataset[1], a widely used benchmark dataset for AES studies. It consists of essays written in English by 1805 students from grades 7 to 10, along with scores from 38 raters for these essays. The raters were native English speakers recruited from Amazon Mechanical Turk, a popular crowdsourcing platform. The raters were asked to grade the essays using a holistic rubric with five rating categories. For detailed dataset information, see Ref. [12].

We first conducted the following experiment to evaluate the linking accuracy of our proposed method using the actual dataset.

1. We divided the dataset into a reference group and a focal group, ensuring there was no overlap of examinees and raters between them. During this division we arranged for the distributions of examinee abilities and rater severities

---

[1] https://kaggle.com/competitions/asap-aes
This is a publicly available dataset following the privacy policy of the Hewlett Foundation.

Table 1. Data separation conditions.

	Groups	Examinee ability θ_j		Rater severity β_r	
		Mean	Variance	Mean	Variance
Condition 1	Reference	High	Wide	Low	Wide
	Focal	Low	Wide	High	Wide
Condition 2	Reference	High	Wide	High	Wide
	Focal	Low	Narrow	Low	Wide

to be different in each group. We examined two separation conditions, each of which is detailed in Table 1.

2. Using the data corresponding to the reference and focal groups, we applied our proposed method to obtain the GMFRM parameters for the focal group on the parameter scale of the corresponding reference group.
3. We computed the RMSE between the obtained parameters and their gold-standard values, which were estimated from the complete dataset. Note that the parameter scale of the gold-standard values was adjusted to align with that of the reference group to facilitate an appropriate comparison with the parameter estimates derived from our proposed method.
4. We repeated the above procedures 10 times for each data separation condition and calculated the average RMSE for each condition.

The above experiments were also conducted without using our method for comparison purposes; the GMFRM parameters for the focal group were estimated exclusively from the data of the focal group.

The experimental results are presented in Table 2. The *Prop.* column displays the results when our proposed method was applied, while the *Base.* column shows the results obtained without using our method. As shown in Table 2, applying our method leads to a significant reduction in the RMSE for all parameters compared to the baseline. This suggests that the proposed method is effective at facilitating test linking.

Table 2. Results of the linking accuracy evaluation.

	θ_j			α_r		β_r		d_{rm}	
	Base.	Prop.	AES	Base.	Prop.	Base.	Prop.	Base.	Prop.
Condition 1	0.78	**0.41**	0.61	0.44	**0.38**	0.77	**0.36**	0.19	**0.18**
Condition 2	0.82	**0.38**	0.51	0.50	**0.43**	0.75	**0.27**	0.33	**0.27**

Furthermore, another feature of the proposed method is its capability for collaborative evaluation between human raters and neural AES. To assess the effectiveness of this collaborative evaluation approach, we calculated the RMSEs between the ability values predicted by the AES model and the gold-standard ability values for the focal group. The results of this analysis are presented in the *AES* column of Table 2. They indicate that the RMSEs for the abilities predicted by the AES model are not sufficiently accurate compared to the results obtained using our method. This finding suggests that the integration of the AES model with human raters significantly enhances ability measurement accuracy.

8 Conclusion

In this study, we introduced a novel IRT-based linking method for essay writing assessments in a nonequivalent groups design without common examinees and raters, leveraging recent advancements in AES technology. Our experimental results demonstrated that our method effectively accomplishes test linking without the need for overlapping examinees or raters across groups. Additionally, our findings demonstrate that integrating the AES model with human raters significantly improves the accuracy of ability measurement. In this research, we assumed that a single essay writing question was administered across multiple groups. We plan to explore the extension of our method to tests with multiple shared questions in future research.

References

1. Abosalem, Y.: Assessment techniques and students' higher-order thinking skills. Int. J. Secondary Educ. **4**, 1–11 (2016)
2. Bock, R.D., Zimowski, M.F.: Handbook of Modern Item Response Theory. Springer, New York (1997)
3. Devlin, J., Chang, M.W., Lee, K., Toutanova, K.: BERT: pre-training of deep bidirectional transformers for language understanding. In: Proceedings of the Annual Conference of the North American Chapter of the Association for Computational Linguistics: Human Language Technologies, pp. 4171–4186 (2019)
4. Engelhard, G.: Constructing rater and task banks for performance assessments. J. Outcome Meas. **1**, 19–33 (1997)
5. Linacre, J.M.: Many-Faceted Rasch Measurement. MESA Press (1989)
6. Linacre, J.M.: A user's guide to FACETS: Rasch-model computer programs (2014)
7. Loyd, B.H., Hoover, H.D.: Vertical equating using the Rasch model. J. Educ. Meas. **17**(3), 179–193 (1980)
8. Patz, R.J., Junker, B.W.: Applications and extensions of MCMC in IRT: multiple item types, missing data, and rated responses. J. Educ. Behav. Stat. **24**(4), 342–366 (1999)
9. Patz, R.J., Junker, B.W., Johnson, M.S., Mariano, L.T.: The hierarchical rater model for rated test items and its application to large-scale educational assessment data. J. Educ. Behav. Stat. **27**(4), 341–384 (2002)
10. Sinharay, S., Holland, P.W.: A new approach to comparing several equating methods in the context of the NEAT design. J. Educ. Meas. **47**, 261–285 (2010)
11. Uto, M.: Accuracy of performance-test linking based on a many-facet Rasch model. Behav. Res. Methods **53**(4), 1440–1454 (2021)

12. Uto, M., Okano, M.: Learning automated essay scoring models using item-response-theory-based scores to decrease effects of rater biases. IEEE Trans. Learn. Technol. **14**(6), 763–776 (2021)
13. Uto, M., Ueno, M.: A generalized many-facet Rasch model and its Bayesian estimation using Hamiltonian Monte Carlo. Behaviormetrika **47**(2), 469–496 (2020)
14. Wiberg, M., Branberg, K.: Kernel equating under the non-equivalent groups with covariates design. Appl. Psychol. Measur. **39** (2015)
15. Wilson, M., Hoskens, M.: The rater bundle model. J. Educ. Behav. Stat. **26**(3), 283–306 (2001)

Exploring the Impact of Gender Stereotypes on Motivation, Flow State, and Learning Performance in a Gamified Tutoring System

Kelly Silva[1], Jário Santos[1], Geiser Chalco[2], Marcelo Reis[1], Leonardo Marques[1], Alan Silva[1], Álvaro Sobrinho[3]([✉]), Diego Dermeval[1], Rafael Mello[4,5], Ig Ibert Bittencourt[1,7], and Seiji Isotani[6,7]

[1] Federal University of Alagoas, Maceió, Brazil
{kbas,alanpedro,ig.ibert}@ic.ufal.br, leonardo.marques@cedu.ufal.br,
diego.matos@famed.ufal.br
[2] Federal Rural University of the Semiarid, Mossoró, Brazil
geiser@alumni.usp.br
[3] Federal University of the Agreste of Pernambuco, Garanhuns, Brazil
alvaro.alvares@ufape.edu.br
[4] Federal Rural University of Pernambuco, Recife, Brazil
rafael.mello@ufrpe.br
[5] CESAR School, Recife, Brazil
[6] University of São Paulo, São Paulo, Brazil
sisotani@icmc.usp.br
[7] Harvard Graduate School of Education, Cambridge, USA

Abstract. Since the COVID-19 pandemic, the demand for online education has intensified, evidencing the need for educational technologies such as tutoring systems based on artificial intelligence to support student learning. As a strategy for improving educational technologies, gamification can enhance learning effectiveness by engaging students in enjoyable learning tasks. However, existing literature emphasizes the importance of tailoring gamification elements, considering factors such as the student's gender. Neglecting such factors may lead to adverse effects stemming from stereotypes and diminishing the learning experience. Thus, we conducted a 2 × 3 factorial experimental study with 122 students focusing on a gamified intelligent tutoring system for teaching logic in Brazilian higher education. Our findings revealed that gender stereotypes significantly motivated men to reject and counteract these stereotypes when perceived as a threat. In a gamified intelligent tutoring system with female stereotypes, males were motivated to alter their positions in the rankings, underscoring the impact of stereotype threat on their perception of relevance. Our results also evidenced that gamification did not impact the flow state of students and, across all scenarios, the learning performance of males consistently exceeded that of females.

Keywords: stereotype threat · stereotyped gamified environment · stereotype boost · stereotype lift · motivation

A. M. Olney et al. (Eds.): AIED 2024 Workshops, CCIS 2151, pp. 88–96, 2024.
https://doi.org/10.1007/978-3-031-64312-5_11

1   Introduction

There is evidence that gamification can positively affect students' cognitive, motivational, and behavioral elements [10]. These effects contribute to improving learning performance and achieving pedagogical objectives. Gamification aims to make learning effective and efficient by encouraging students to engage in learning tasks and activities with pleasure and a love for learning. Thus, the goal of gamification is to promote the flow experience, which relates to the mental state of complete immersion, focus, and concentration, where the participant is fully engaged in the activity for the sheer pleasure of it [3]. Achieving a flow state is the desired outcome for any gamified educational technology.

However, designing game elements without considering the gender or sexual identity of participants may reinforce harmful stereotypes. For instance, in Science, Technology, Engineering, and Mathematics (STEM) courses, introducing game elements that appeal primarily to men may reinforce gender stereotypes and create an environment where girls may not feel welcome, as they may perceive that they do not belong to the social group. Indeed, stereotypes lead to a decline in performance and other adverse effects related to various affective, cognitive, and motivational mechanisms [8].

We conducted an experimental study with 122 students to assess the effects that gender stereotypes can have on a gamified logic intelligent tutoring system concerning motivation, flow experience, and learning performance. Our system relies on a knowledge base and artificial intelligence techniques to support logic students. The students participating in the study are from diverse regions across Brazil and attended a higher education institution. The experimental design employs a 2 × 3 factorial structure, considering both students' gender (male and female) and the versions of our artificial intelligence-based gamified logic tutoring system (stereotyped male, stereotyped female, and non-stereotyped).

2   Method

2.1   Participants

The study involved adult participants, comprising 58 women and 64 men, aged 18 to 54 years, with a mean age of 24.02. Among the participants, 117 had graduated, while 5 were still undergraduate students.

Regarding ethnicity, 62 participants identified as Caucasians, 42 as pardos (mixed), 13 as black, and 5 chose not to declare their ethnicity. Regarding social status, 101 participants were single, 12 were married, four were in a stable relationship, two were divorced, and three chose not to disclose their social status. Finally, regarding their residences, 107 participants were from the northeast of Brazil, 11 were from the southeast of Brazil, and three preferred not to declare.

2.2   Materials and Procedure

We conducted the experimental study using a 2 × 3 factorial design, considering the students' gender (male or female) and the presence of stereotypes in the gamified intelligent tutoring system. The three conditions included: a gender-stereotyped environment (stFemale); a male-stereotyped environment (stMale); and a control group without stereotypes. Therefore, the study also assessed three conditions induced by stereotypes: inBoost, inThreat, and control.

Table 1 outlines the null hypotheses established for our experimental study. We developed three versions of the gamified intelligent tutoring system to test these hypotheses. We adapted these versions from the specifications proposed by Albuquerque et al. [1]. Specifically, we modified the scoring and ranking system, awarding a point for each correct answer without penalizing errors.

Table 1. Null hypotheses formulated for our experimental study.

ID	Description
H1	There is no significant difference in motivation between stereotype conditions inBoost, inThreat and control
H2	There is no significant difference in the motivation of the participants according to their gender and the stereotype in the gamified tutor system
H3	There is no significant difference in the flow experience between the stereotype conditions inBoost, inThreat and control
H4	There is no significant difference in the flow experience of the participants according to their gender and the stereotype in the gamified tutor system
H5	There is no significant difference in learning performance between the stereotype conditions inBoost, inThreat and control
H6	There is no significant difference in participants' learning performance according to their gender and stereotype in the gamified tutor system

We conducted the data collection process online, with voluntary participation from higher education students reached through platform links shared within higher education communities. We disseminated the study through email and instant messaging, targeting students from STEM courses at universities.

The gamified intelligent tutoring system randomly assigned each participant to one of the three versions of the system. This random distribution ensured that the study adhered to a single-anonymized design, where participants were unaware of the existence of other environments. The execution of the experiment involved six steps: (1) introduced a consent form through the technology; (2) participants encountered a Dispositional Flow Scale (DFS) questionnaire; (3) the system presented one of the three versions of educational technology; (4) after solving the logical reasoning problems, participants answered the Instructional Materials Motivation Survey (IMMS) questionnaire; (5) participants were presented with the short validated version (nine items) of the Flow State Scale-2 (FSS-2) questionnaire; and (6) we administered a demographic and socioeconomic questionnaire to gather information about the participants.

3   Results and Discussion

3.1   Effects on Motivation (H1 and H2)

Table 2 displays the descriptive statistics for participants' motivation, as measured by the IMMS questionnaire, across the stereotyped conditions (inBoost, inThreat, and control) in which they participated during the study. The results of ANOVA tests indicated no statistically significant differences based on the stereotyped condition, leading to the non-rejection of null hypothesis H1.

Table 2. Descriptive statistics of motivation according to the stereotyped condition.

	N	Attention		Confidence		Relevance		Satisfaction		Motivation	
		M	SE	M	SE	M	SE	M	SE	M	SE
control	40	3.79	0.11	3.48	0.09	3.26	0.09	3.95	0.10	3.62	0.08
inBoost	50	3.55	0.10	3.56	0.09	3.14	0.11	3.66	0.12	3.48	0.09
inThreat	32	3.87	0.11	3.81	0.11	3.42	0.14	3.87	0.15	3.74	0.11

Table 3 presents the descriptive statistics for motivation, as measured by the IMMS questionnaire, categorized by their gender and the gender-stereotyped environments (stMale, stFemale, and default) to which participants were assigned. The results of ANOVA tests revealed that null hypothesis H2 was rejected. There were statistically significant differences observed for attention ($F(2,116) = 3.08$, $p = 0.05$, and $ges = 0.05$), relevance ($F(2,116) = 5.072$, $p = 0.008$, and $ges = 0.08$), confidence ($F(1,116) = 6.05$, $p = 0.015$, and $ges = 0.05$), and motivation ($F(2,116) = 3.08$, $p = 0.05$, and $ges = 0.05$).

Table 3. Descriptive statistics of motivation according to the environment and gender.

	N	Attention		Relevance		Confidence		Satisfaction		Motivation	
		M	SE	M	SE	M	SE	M	SE	M	SE
default:male	21	3.77	0.12	3.27	0.11	3.50	0.10	3.81	0.12	3.59	0.09
default:female	19	3.81	0.19	3.25	0.16	3.45	0.15	4.10	0.16	3.71	0.14
stFemale:male	18	4.03	0.11	3.62	0.18	4.00	0.10	4.01	0.17	3.96	0.11
stFemale:female	25	3.68	0.16	3.38	0.15	3.45	0.15	3.79	0.18	3.68	0.15
stMale:male	25	3.43	0.10	2.90	0.16	3.67	0.08	3.53	0.16	3.31	0.10
stMale:female	14	3.67	0.20	3.17	0.22	3.56	0.19	3.70	0.28	3.43	0.20

Through pairwise comparisons with the Estimated Marginal Means (EMMs) and p-values adjusted with the Bonferroni method, we identified that the mean motivation for male participants in a gamified intelligent tutoring system with female stereotypes ($adjM = 3.908$ and $SD = 0.45$) was significantly higher than the mean in a gamified intelligent tutoring system with male stereotypes ($adjM = 3.39$ and $SD = 0.491$) with $p.adj = 0.006$.

This finding suggests that male participants were motivated to reject the female stereotypes. Similar to observations in the study by Christy and Fox [2], where females performed well under a stereotype threat caused by a ranking, our study revealed an increase in the motivation of male participants due to the stereotype threat. We can infer that this result reflects, in some way, a form of stereotype lift [11]. However, instead of being a performance boost caused by a negative stereotype [5,6], in our case, we observed a motivation boost.

The attention of male participants in the system with female stereotypes ($adjM$ = 4.02 and SD = 0.467) was significantly higher than in a system with male stereotypes ($adjM$ = 3.449 and SD = 0.439) with $p.adj$ = 0.008. We observed a form of stereotype lift where the stereotype threat led to an unexpected effect. The attention of male participants was lower in a system with male stereotypes. A possible explanation is that participants may have expended cognitive resources processing information from game elements and the environment rather than focusing on the logic quiz. This aligns with findings in other studies highlighting similar effects of stereotypes, referred to as cognitive interference [5]. However, this effect was observed under stereotype threats.

For male participants, the mean confidence in a gamified intelligent tutoring system with female stereotypes ($adjM$ = 3.98 and SD = 0.338) was significantly higher than in a system without stereotypes ($adjM$ = 3.503 and SD = 0.394) with $p.adj$ = 0.014. This result further supports the notion of a stereotype lift, indicating that the confidence of male participants increased when they were under a condition of stereotype threat.

This was also evident when examining the impact of female stereotypes in a gamified intelligent tutoring system. In this case, the confidence of male participants ($adjM$ = 3.98 and SD = 0.338) was significantly higher than that of female participants ($adjM$ = 3.467 and SD = 0.698) with $p.adj$ = 0.002. Both results contrast observations in group activities, where girls under stereotype threats exhibited lower confidence levels [4].

Irrespective of the stereotype condition, the study of Robinson et al. [9] found that male participants showed higher confidence, attributing it to their previous experience. Given that most participants in our study had completed their graduation, it is plausible to argue that their prior experience with logic quizzes could explain why the confidence of male participants increased under the condition of stereotype threat.

The relevance for male participants in an environment with female stereotypes ($adjM$ = 3.609 and SD = 0.692) was significantly higher than the mean in an environment with male stereotypes ($adjM$ = 2.91 and SD = 0.71) with $p.adj$ = 0.003. When male participants were under the condition of stereotype threat, they demonstrated a heightened sense that solving the logic quiz was important, useful, and meaningful to them.

This suggests that male participants assigned greater relevance to contradicting what the stereotype implied. We can infer that, in a female-stereotyped gamified intelligent tutoring system, male participants were motivated to change their positions in the rankings, highlighting the influence of stereotype threat on their perception of relevance.

3.2 Effects on Flow State (H3 and H4)

Table 4 presents the descriptive statistics for disposition to the flow state, as measured by the DFS-2 questionnaire before interacting with the intelligent tutoring systems. Additionally, the table includes the flow state measured after participants interacted with the systems and their adjusted values using EMMs.

We conducted ANCOVA tests following the control for linearity in the disposition to the flow state (DFS-2). The results revealed a statistically significant difference between the disposition to the flow state (DFS-2) and the flow state (FSS-2) with $F(1, 118) = 45.091$, $p < 0.001$, and $ges = 0.276$ (effect size). The null hypothesis H3 is not rejected. While the stereotyped conditions did induce positive changes in flow states, these effects were not significantly different.

Table 4. Descriptive statistics of flow state according to the stereotyped condition.

	N	Before (DFS-2)		After (FSS-2)		Adjusted	
		M	SE	M	SE	M	SE
control	40	3.567	0.062	3.844	0.098	3.795	0.079
inBoost	50	3.445	0.076	3.810	0.078	3.838	0.071
inThreat	32	3.461	0.093	3.950	0.105	3.968	0.088

Table 5 displays the descriptive statistics obtained from the DFS-2 and FSS-2 questionnaires, organized based on two factors: participants' gender and the environments in which they participated. Adjusted values of the flow state using EMMs are also present in the table.

Following the control for the linearity of disposition to the flow state (DFS-2), ANCOVA tests considering these two factors indicated that there was only a statistically significant effect between the DFS-2 and FSS-2 with $F(1, 110) = 33.557$, $p < 0.001$, and $ges = 0.234$ (effect size). This implies that gender stereotypes induce changes in flow state, but these changes are not different based on gender and environment, leading to the non-rejection of null hypotheses H4.

Table 5. Descriptive statistics of flow state according to the environment and gender.

	N	Before (DFS-2)		After (FSS-2)		Adjusted	
		M	SE	M	SE	M	SE
default:male	21	3.677	0.074	3.926	0.122	3.824	0.108
default:female	19	3.477	0.099	3.760	0.154	3.777	0.112
stFemale:male	17	3.695	0.084	4.212	0.079	4.100	0.120
stFemale:female	23	3.211	0.103	3.748	0.125	3.923	0.106
stMale:male	23	3.718	0.097	3.884	0.088	3.758	0.104
stMale:female	14	3.192	0.167	3.659	0.188	3.844	0.135

Therefore, these results suggest that gamification does influence the flow state, aligning with numerous previous studies [7]. However, these effects are not contingent on gender stereotypes.

3.3 Effects on Learning Performance (H5 and H6)

Table 6 provides the descriptive statistics for activity points obtained by participants, categorized by the stereotyped condition (inBoost, inThreat, and control) in which they participated during the study. Activity points are a proxy measure of learning performance, indicating the number of correct answers when participants interacted with the intelligent tutoring system. Based on the ANOVA tests on this data, we can observe no statistically significant differences, leading to the non-rejection of null hypothesis H5.

Table 6. Descriptive statistics of hits according to the stereotyped condition.

	N	M	SE
control	36	15.303	0.383
inBoost	42	15.644	0.321
inThreat	28	16.214	0.346

Table 7 displays the descriptive statistics for activity points obtained by participants based on their genders and the stereotyped environments in which they participated. The ANOVA test results indicate the rejection of null hypothesis H6, suggesting a significant effect dependent on participants' gender with $F(1, 108) = 23.661$, $p < 0.001$, and $ges = 0.18$.

Table 7. Descriptive statistics of hits according to the environment and gender.

	default			stFemale			stMale		
	N	M	SE	N	M	SE	N	M	SE
male	21	16.048	0.480	18	16.603	0.424	19	17.211	0.347
female	19	14.274	0.620	23	15.133	0.408	14	14.643	0.529

In a gamified intelligent tutoring system without stereotypes, the mean activity points obtained by male participants ($adjM = 16.048$ and $SD = 2.202$) were significantly higher than the mean for female participants ($adjM = 14.274$ and $SD = 2.704$) with $p.adj = 0.008$. When female stereotypes existed, the activity points of males ($adjM = 16.603$ and $SD = 1.8$) were significantly higher than those for females ($adjM = 15.133$ and $SD = 1.955$) with $p.adj = 0.026$. Similarly, when this system contained male stereotypes, the activity points of males ($adjM = 17.211$ and $SD = 1.512$) were significantly higher than the mean for female participants ($adjM = 14.643$ and $SD = 1.98$) with $p.adj < 0.001$.

Across all environments (without stereotypes and with male or female stereotypes), the learning performance of male participants exceeded that of female participants. These results align with previous studies [5,6], in which the authors used stereotypes to motivate female students to reject them.

4 Conclusions

We conducted an experiment involving 58 women and 64 men to evaluate the impact of gender stereotypes on the motivation, flow state, and learning performance of participants in a gamified intelligent tutoring system for logic. The results revealed that under the threat of gender stereotypes, men were significantly motivated to reject such stereotypes. Employing the ARCS model of motivation, our study observed positive effects on male participants' attention, confidence, and relevance when exposed to a gamified intelligent tutoring system with a female stereotype. Furthermore, across all environments, the learning performance of men was significantly higher than that of women's.

When participants experience stereotype threat but recognize errors, their motivation to reject it may enhance their efforts and learning performance. A similar phenomenon may occur in our study but with male participants. Nevertheless, further research is needed to evaluate this assumption thoroughly.

Our results may appear paradoxical, but they suggest that striving to create neutral environments, such as gamified intelligent tutoring systems without stereotypes, may not be ideal. While neutral systems have the potential to promote gender equality, they might miss opportunities to motivate students to achieve better learning outcomes.

Disclosure of Interests. The authors have no competing interests to declare.

References

1. Albuquerque, J., Bittencourt, I.I., Coelho, J.A., Silva, A.P.: Does gender stereotype threat in gamified educational environments cause anxiety? An experimental study. Comput. Educ. **115**, 161–170 (2017)
2. Christy, K.R., Fox, J.: Leaderboards in a virtual classroom: a test of stereotype threat and social comparison explanations for women's math performance. Comput. Educ. **78**, 66–77 (2014). https://doi.org/10.1016/j.compedu.2014.05.005. https://www.sciencedirect.com/science/article/pii/S0360131514001195
3. Csikszentmihalyi, M.: Flow: The Psychology of Happiness. Random House (2013)
4. Dasgupta, N., Scircle, M.M., Hunsinger, M.: Female peers in small work groups enhance women's motivation, verbal participation, and career aspirations in engineering. Proc. Natl. Acad. Sci. **112**(16), 4988–4993 (2015)
5. Jamieson, J.P., Harkins, S.G.: Mere effort and stereotype threat performance effects. J. Pers. Soc. Psychol. **93**(4), 544 (2007)
6. Jamieson, J.P., Harkins, S.G.: Distinguishing between the effects of stereotype priming and stereotype threat on math performance. Group Processes Intergroup Relat. **15**(3), 291–304 (2012)
7. Oliveira, W., et al.: Does gamification affect flow experience? A systematic literature review. arXiv preprint arXiv:2106.09942 (2021)
8. Pennington, C.R., Heim, D., Levy, A.R., Larkin, D.T.: Twenty years of stereotype threat research: a review of psychological mediators. PLoS ONE **11**(1), e0146487 (2016)

9. Robinson, J., Nieswandt, M., McEneaney, E.: Motivation and gender dynamics in high school engineering groups. In: 2018 CoNECD-The Collaborative Network for Engineering and Computing Diversity Conference (2018)
10. Sailer, M., Homner, L.: The gamification of learning: a meta-analysis. Educ. Psychol. Rev. **32**(1), 77–112 (2020)
11. Walton, G.M., Cohen, G.L.: Stereotype lift. J. Exp. Soc. Psychol. **39**(5), 456–467 (2003)

Design Based Research of Multimodal Robotic Learning Companions

Hae Seon Yun[1]([envelope]), Heiko Hübert[2], Niels Pinkwart[1], and Verena V. Hafner[1]

[1] Humbolt University Berlin, Berlin, Germany
{yunhaese,hafner}@informatik.hu-berlin.de, pinkwart@hu-berlin.de
[2] HTW Berlin, Berlin, Germany
heiko.huebert@htw-berlin.de

Abstract. This paper presents some initial findings from a design-based research study on multimodal robotic language learning companions. The study involved designing four robots with varying physical modalities, including two humanoid (Pepper and Nao) and two non-humanoid (Cozmo and MyKeepOn), to interact empathetically with participants during language learning scenarios. One German teacher participated in the development of these interactive scenarios. Pre- and post-tests were conducted to measure learning gains, and a survey was administered to gather insights from participants. The results showed that all participants exhibited improvements in targeted behaviors between pre- and post-tests, specifically by using longer and more grammatically correct sentences in their responses. However, unlike the strong preference for NAO, the survey revealed that Cozmo had the most characteristics of a learning companion.

Keywords: Design Based Research · HRI · Learning Companion

1 Introduction

The use of companionship during learning has been shown to enhance cognitive processes by providing learners with opportunities for social interaction, which allows them to face challenges and resolve them through elaboration and discussion [6]. To facilitate social interaction in intelligent tutoring systems (ITS), the pedagogical agent known as the learning companion has been employed. Research on virtual/screen-based agents in ITS has spanned decades, and with recent advancements in robotics, there is growing interest in designing robots as learning companions due to their physical embodiment [5].

Embodied agents, such as robots that physically interact with learners, have several advantages over virtual agents in terms of social interaction and engagement [1]. Robots provide a higher level of immediacy through non-verbal communication, which enhances natural social interactions [16]. Additionally, learners perceive embodied agents as more human-like than virtual agents, leading to positive learning outcomes [2,11,15].

A. M. Olney et al. (Eds.): AIED 2024 Workshops, CCIS 2151, pp. 97–104, 2024.
https://doi.org/10.1007/978-3-031-64312-5_12

The use of physical robotic learning companions offers unique visual and physical attributes compared to screen-based learning companions, which may require different design considerations. Additionally, the impact of physical embodiment on learning outcomes and social interactions in this context may be more complex than with virtual agents. To investigate these factors, a study was conducted to explore the design of robotic learning companions with various modalities and features, as well as their effects on learning companion likeness, cognitive gain, and emotion detection through multimodal sensor data analysis in a language learning context. Preliminary results from the study are reported in this paper, including the design process, learning companion likeness, and cognitive gain.

2   Prior Studies and Aim of This Paper

In this work, we aimed to create multiple robots with varying physical modalities based on three previous studies. Firstly, a screen-based teacher agent in the ITS system, Betty's Brain [3], was replaced with a physical humanoid robot, Pepper in [21]. Secondly, Cozmo, a toy-like robot, was explored as a peer robot in a second language learning context in [17,20]. Lastly, the design considerations of learning companions were derived [22,23] and utilized in designing learning companions with various modalities [24].

As a robotic agent should be designed as similar as possible to a human [18,19], when designing humanoid robots such as Pepper and NAO, the human-based contextual gestures that produced human-like gestures were aimed at. For non-humanoid robots such as Cozmo and MyKeepOn, we considered the importance of designing an interactive second language learning scenario and a suitable robot's expressiveness. Specifically, to integrate Cozmo as a learning companion, the lesson learned from the two prior studies [17,20] on utilizing Cozmo as a learning companion for language learning scenarios was taken into account. To maintain an equilibrium in the ratio of humanoid and non-humanoid robotic agents, another robot, MyKeepOn [14] was included, as the study [9] shows that toy-like robots may be more suitable for learning companionship.

To maintain consistency in design across all four robotic agents, we utilized previously published design considerations for learning companions [22,23] and employed the method of design-based research (DBR) [8] throughout the study to ensure that the design process was iterative and user-centered.

The present paper reports some preliminary results from a comprehensive study that examined the relationship between multimodal data, such as learners' facial expressions and physiological sensor information, and the design of multimodal learning companions. Although these aspects are briefly mentioned in this paper, the primary emphasis is on describing the design process for developing robotic learning companions and presenting the results of the interactive human-robot interaction (HRI) experiment conducted during the study.

3   Method

3.1   Interactive HRI Scenario Development

The author worked closely with a German teacher at a vocational school in Cologne for eight months to develop an interactive second language learning scenario that involved human-robot interaction. The final outcome was six equal-difficulty tasks, each requiring students to interact with four different robots (in addition to pre- and post-tests). These tasks required students to answer the following seven questions posed by each robot in German using the information provided on a handout, as shown in Fig. 2(b).

1. How big is your apartment?
2. How many rooms do you have in your apartment?
3. How many bedrooms do you have in your apartment?
4. What kind of furniture do you have in your room?
5. Do you have a balcony or garden where you can go outside?
6. On which floor do you live?
7. With whom do you live together?

The developed task anticipates that the students will employ the following key phrases in their response while engaging with the robots during the learning scenario:

- I have *"Ich habe .."* or My apartment has *"Meine Wohnung hat .."*
- Yes *"Ja.."* or No *"Nein"*
- I live in *"Ich wohne im ..."*
- I live with *"Ich wohne mit ..."*

The students' spoken replies were converted into text through the use of VOSK, the speech recognition software, and compared with the anticipated answers. Depending on their responses, different levels of feedback were given to each student (e.g., complete sentence structure, incorporating key phrases, grammatical error, using more adjectives).

3.2   Design of Multiple Robots with Different Physical Modality

Four multimodal learning companions were developed as part of the task-four humanoid robots (Pepper and NAO) and two non-humanoid robots (Cozmo and MyKeepOn), utilizing a unified framework, Sobotify [10]. All robots were designed to provide motivational feedback by responding to learners' emotional state. The learner's emotional state was detected through the facial expression recognition system (e.g. DeepFace) employed in Sobotify [10], and physiological sensor data was collected using a wearable wristband to relate the emotions to the FER value after the experiment. Emotions were classified as negative, neutral or happy, allowing robots to provide learners with appropriate motivational messages. Additionally, all robots maintained eye contact with learners through their embedded camera or webcam.

Considering different physical features of the robots, the humanoid robots, Pepper and NAO, were designed to gesture similar to a human using its arms while talking to the learner. Nao was standing on the desk and Pepper was placed on the ground and the participant faced each robot sitting down. For Cozmo, it was placed on the battery charger at the beginning and when the task began, the robot came out toward the participant and maintained eye contact with the participant. Cozmo expressed emotions using facial expression, its lift and also by moving around. In contrast to the other three robots that were programmed, MyKeepOn was designed as a toy and did not have an embedded camera. Therefore, an Arduino board was attached to MyKeepOn for control, while an external speaker and webcam were used for communication and maintaining eye contact with the participant. The specific movements of MyKeepOn such as PON, TILT, and SIDE were investigated and applied during the task to express emotional states and move while talking.

3.3   Experiment Setup and Participants

As shown in the Fig. 1, each student interacts with 4 robots after the pre-test and the experiment was completed with the post-test and a short survey.

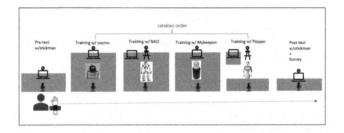

Fig. 1. HRI Experiment Setup with Multiple Robots as Learning Companions

The students arrived one by one to the room and were informed of the experiment's purpose and procedure. Before starting, participants were instructed to wear a wristband with sensors that would collect their physiological data during the experiment for post-hoc analysis of their emotional state. Once the signal acquisition was confirmed, they were given a handout (shown in Fig. 2(b)) and the pre-test task commenced on the computer screen displaying a virtual robot (Stickman) as shown in Fig. 2(a).

Stickman was utilized for pre- and post-test as it allows the communication flow and the pattern consistent with the experiment setup.

After completing the pre-test, the participants were given the option to select any robot they desired for their next task. Based on their choice (or randomly assigned), the subsequent interactive learning companion was chosen. For each task, a new handout on a living situation was provided and similar to the pre-test, each robot initiated interacting with the students by asking questions.

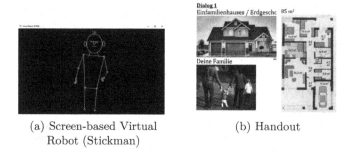

(a) Screen-based Virtual   (b) Handout
Robot (Stickman)

Fig. 2. Screen-based Virtual Robot (Stickman) (a) and Hand-out (b)

During the interaction with each robot, the student was asked to check their answer when they were incorrect and given various hints to correct their answers. With respect to emotion, happy, neutral and negative were detected and based on the detected emotion, the designed verbal feedback was given during the interaction.

After interacting with four robots, the students performed the post-test with the Stickman and once they completed the whole cycle (e.g. pre-test, interacting with 4 robots, post-test), they were given a survey to participate. The survey questionnaire consisted of 16 items, 12 items from the pedagogical agents as learning companions (PAL) [12] in 5 Likert scale for each robot, and four multiple choice questions asking participants to choose one robot that best represents a statement or question aimed at identifying which robot was most enjoyable and suitable as a learning companion.

4   Results

All 6 participants successfully interacted with all 4 robots and an average of 73 min (max: 93 min., min: 53 min.) was taken by each student to complete the experiment (including pre-test, post-test and survey). Out of seven questions, only the last four were considered for their performance improvement because all participants were unfamiliar with the interaction pattern during the first three questions in the pre-test and needed to get used to the turn-taking pattern of interaction. Due to the technical problem occurred during the post test, P5's data was omitted for analysis. The word count (WC), detected by the speech detection software VOSK [4], was used the method to measure the improvement in learning between the pre- and post-tests (Table 1).

In general, there was an increase in WC between pre- and post-test. More importantly, it was observed that after interacting with robotic learning companions, students were using the key phrases shown in Sect. 3.1 more intentionally. For instance, during the pre-test, P1 responded with short answers such as *"three furniture"*, *"no"*, *"third floor"*, and *"my parents and sibling"* without using full sentences. However, after interacting with robotic learning companions, P1 responded in full sentences and also used key phrases like *"I have (=ich habe)"* or *"I live with (=Ich wohne mit)"*. Similarly, P2 showed an improvement in using formal language from *"No I have not"* to *"No, unfortunately, I do not have a balcony."*. Unlike P1 and P2, P3 and P4 utilized many words during the pre-test, so the increase in WC was not notably visible. However, their usage of language improved significantly. For example, P3 used the preposition *"with (mit)"* during the post-test, which made the sentence more grammatically sound. P4 also demonstrated learned behavior by using *"yes (ja)"* or *"no (nein)"* in the post-test. P6 presented the overall differences between pre- and post-test, specifically regarding length, usage of key phrase, usage of adjective and article, and attainment of vocabulary as shown in the Table 1.

Table 1. Difference in words spoken between Pre- to Post-Test of P6 using VOSK Note: * the adjective and articles in German language requires changes according to the language specific rules, which can not simply be translated to English.

Q.	Pre-Test	Post-Test	Diff.
4		**ich habe (I have)** eine welt ein kleine (small) sessel einen (a) schreibtisch	+9
5	ich kenne keine balkon	**ich habe (I have)** keine balkon (balcony)	+4
6	drei starke	**ich wohne (I live)** in act stock stockwerk	+4
7	die familie	**ich wohne mit (I live with)** meiner (my) eltern	+3

According to the survey responses, 4 out of 6 (66%) of the participants chose a humanoid robot (either Pepper or NAO) as the most enjoyable robot to interact with 66% chose a non-humanoid robot (either Cozmo or MyKeepOn) as the least enjoyable robot (e.g. boring) to interact with. In terms of learning companions, 66% of the participants chose NAO robot as their preferred learning companion, while half of them would also choose NAO as their robotic interactive partner for their future task. Interestingly, while participants preferred NAO robot as their learning companion, the dependent variables measuring learners' perceptions of learning companion functionality and self-efficacy presented different results. Analyzing the pedagogical agent as learning companions (PAL) items, all participants evaluated Cozmo as the robot with the most characteristics as a learning companion. Although they considered NAO also as a supportive agent that provided them confidence in task performance, Cozmo was rated as the most effective learning companion due to its informative, useful, motivating, supportive, fun, helpful in gaining attention and concentration, providing relevant information and feedback, and engaging learners into deeper knowledge. MyKeepOn was not rated highly in any of PAL items, while Pepper was selected as a credible agent that encouraged reflection on what they were learning.

5 Discussion

In this study, multiple robots were designed as learning companions for a second language scenario, and preliminary results showed that learners exhibited some improvement in their responses. Despite the preference of participants for the humanoid robot NAO as their learning companion, survey data indicated that Cozmo had the most characteristics of a learning companion. This may be due to several factors: firstly, Cozmo's ability to present emotions through facial expressions unlike other robots may have increased learners' perception of it as a "facilitating" and "engaging" learning companion [13]. Secondly, the dynamic ability of Cozmo to move towards participants and retrieve back may have affected learners' perceptions, making them rate Cozmo as a more suitable learning companion compared to other non-moving robots that resulted in poor interaction [7].

The results of this study are specific to a particular context and due to the low sample size, these findings cannot be generalized to a larger context. However, the significance of this research lies in incorporating DBR into 1) creating an engaging HRI task by involving teachers, 2) designing multiple robots with different modalities as learning companions, and 3) employing various evaluation methods to gain insights into learners' interactions, as well as the design of multimodal learning companions.

Acknowledgements. This research was funded by the Deutsche Forschungsgemeinschaft (DFG, German Research Foundation) under Germany's Excellence Strategy - EXC 2002/1 "Science of Intelligence" - project number 390523135. We would like to express our special thanks to Ellen Donder and Detlef Steppuhn at Erich-Gutenberg-Berufskolleg (EGB) in Cologne for their invaluable inputs and active participation.

References

1. Belpaeme, T., Kennedy, J., Ramachandran, A., Scassellati, B., Tanaka, F.: Social robots for education: a review. Sci. Robot. **3**(21), eaat5954 (2018)
2. Van den Berghe, R., et al.: A toy or a friend? Children's anthropomorphic beliefs about robots and how these relate to second-language word learning. J. Comput. Assist. Learn. **37**(2), 396–410 (2021)
3. Biswas, G., Segedy, J.R., Bunchongchit, K.: From design to implementation to practice a learning by teaching system: betty's brain. Int. J. Artif. Intell. Educ. **26**, 350–364 (2016)
4. Cephei, A.: VOSK offline speech recognition API. https://alphacephei.com/vosk/. Accessed 01 Mar 2022
5. Chan, T.W., Baskin, A.B.: Studying with the prince: the computer as a learning companion. In: Proceedings of the International Conference on Intelligent Tutoring Systems, vol. 194200 (1988)
6. Chan, T.W., Baskin, A.B.: Learning companion systems. Intell. Tutoring Syst. Crossroads Artif. Intell. Educ. **1**, 6–33 (1990)

7. Chu, J., Zhao, G., Li, Y., Fu, Z., Zhu, W., Song, L.: Design and implementation of education companion robot for primary education. In: 2019 IEEE 5th International Conference on Computer and Communications (ICCC), pp. 1327–1331. IEEE (2019)

8. Easterday, M.W., Rees Lewis, D.G., Gerber, E.M.: The logic of design research. Learn. Res. Pract. **4**(2), 131–160 (2018)

9. Hegel, F., Lohse, M., Wrede, B.: Effects of visual appearance on the attribution of applications in social robotics. In: RO-MAN 2009-The 18th IEEE International Symposium on Robot and Human Interactive Communication, pp. 64–71. IEEE (2009)

10. Hübert, H., Yun, H.S.: Sobotify: a framework for turning robots into social robots. In: Proceedings of the 2024 ACM/IEEE International Conference on Human-Robot Interaction (HRI 2024), Boulder, CO, USA, 11–14 March 2024 (2024)

11. Kiesler, S., Powers, A., Fussell, S.R., Torrey, C.: Anthropomorphic interactions with a robot and robot-like agent. Soc. Cogn. **26**(2), 169–181 (2008)

12. Kim, Y.: Desirable characteristics of learning companions. Int. J. Artif. Intell. Educ. **17**(4), 371–388 (2007)

13. Kim, Y., Baylor, A.L., Shen, E.: Pedagogical agents as learning companions: the impact of agent emotion and gender. J. Comput. Assist. Learn. **23**(3), 220–234 (2007)

14. Kozima, H., Michalowski, M.P., Nakagawa, C.: Keepon: a playful robot for research, therapy, and entertainment. Int. J. Soc. Robot. **1**, 3–18 (2009)

15. Lee, H., Lee, J.H.: The effects of robot-assisted language learning: a meta-analysis. Educ. Res. Rev. **35**, 100425 (2022)

16. Liu, W.: Does teacher immediacy affect students? A systematic review of the association between teacher verbal and non-verbal immediacy and student motivation. Front. Psychol. **12**, 713978 (2021)

17. Miller, O.: Design and development of Cozmo as an empathetic languate trainer (2023). https://github.com/milleroski/cozmo_teaching

18. Stenzel, A., Chinellato, E., Bou, M.A.T., Del Pobil, Á.P., Lappe, M., Liepelt, R.: When humanoid robots become human-like interaction partners: corepresentation of robotic actions. J. Exp. Psychol. Hum. Percept. Perform. **38**(5), 1073 (2012)

19. de Wit, J., et al.: The effect of a robot's gestures and adaptive tutoring on children's acquisition of second language vocabularies. In: Proceedings of the 2018 ACM/IEEE International Conference on Human-Robot Interaction, pp. 50–58 (2018)

20. Yun, H.S., Karl, M., Fortenbacher, A.: Designing an interactive second language learning scenario: a case study of Cozmo. In: HCIK 2020: Proceedings of HCI Korea, pp. 384–387 (2020)

21. Yun, H.S., et al.: Challenges in designing teacher robots with motivation based gestures. arXiv preprint arXiv:2302.03942 (2023)

22. Yun, H., Fortenbacher, A., Helbig, R., Pinkwart, N.: Design considerations for a mobile sensor-based learning companion. In: Herzog, M.A., Kubincová, Z., Han, P., Temperini, M. (eds.) ICWL 2019. LNCS, vol. 11841, pp. 344–347. Springer, Cham (2019). https://doi.org/10.1007/978-3-030-35758-0_35

23. Yun, H., Fortenbacher, A., Pinkwart, N.: Improving a mobile learning companion for self-regulated learning using sensors. In: CSEDU (1), pp. 531–536 (2017)

24. Yun, H., Sardogan, A.: Design based research of a sensor based mobile learning companion. In: ICERI2022 Proceedings, pp. 8117–8124. IATED (2022)

StyloAI: Distinguishing AI-Generated Content with Stylometric Analysis

Chidimma Opara[(✉)]

School of Computing, Engineering and Digital Technologies, Teesside University,
Middlesbrough, UK
c.opara@tees.ac.uk

Abstract. The emergence of large language models (LLMs) capable of generating realistic texts and images has sparked ethical concerns across various sectors. In response, researchers in academia and industry are actively exploring methods to distinguish AI-generated content from human-authored material. However, a crucial question remains: What are the unique characteristics of AI-generated text? Addressing this gap, this study proposes *StyloAI*, a data-driven model that uses 31 stylometric features to identify AI-generated texts by applying a Random Forest classifier on two multi-domain datasets. *StyloAI* achieves accuracy rates of 81% and 98% on the test set of the AuTextification dataset and the Education dataset, respectively. This approach surpasses the performance of existing state-of-the-art models and provides valuable insights into the differences between AI-generated and human-authored texts.

Keywords: Stylometric Features · AI in Education · Natural Language Processing · ChatGPT

1 Introduction

The use of AI-generated output has sparked a discussion on the ethical implications of AI in various fields, particularly in Education, where traditional assessment methods like quizzes and essays are foundational to student evaluation. However, studies have shown that AI-powered text generation tools can produce essays, reports, or other written assignments with minimal effort [2,4]. This raises a pivotal ethical question: Is using AI-generated content in academic settings acceptable?

Recent studies have explored the application of AI in enhancing educational experiences [7]. Alongside this, there has been significant research into distinguishing AI-generated content from human-authored texts. The predominant method for such differentiation utilises deep learning techniques for automatic text classification, involving training models such as CNNs, RNNs, LSTM, and Transformer-based models like Roberta and GPT to categorise text into predefined classes [8]. However, despite their effectiveness, these deep learning approaches are often criticised for their "black box" nature, which conceals the

A. M. Olney et al. (Eds.): AIED 2024 Workshops, CCIS 2151, pp. 105–114, 2024.
https://doi.org/10.1007/978-3-031-64312-5_13

specific linguistic cues indicative of AI authorship. Moreover, the focus of these current state-of-the-art models on specific domains raises concerns about their ability to generalise across different domains.

To address the challenges mentioned above, this study's objectives are twofold: first, to identify stylometric features capable of distinguishing AI-generated content from human-written content, and second, to develop a machine learning model that leverages these identified features to accurately classify texts across various domains as either AI-generated or human-written.

This study makes the following contributions:

1. *Identification of stylometric features to Differentiate AI-generated Texts from Human-Authored Content*: This research empirically identified 31 stylometric features across multi-domain datasets. Among these features, 12 represent new contributions to the field which are instrumental in distinguishing between AI-generated and human-authored texts.
2. *StyloAI*: A data-driven model that uses stylometric analysis to attribute authorship of texts within multi-domain datasets: By applying the identified 31 stylometric features on a Random Forest classifier, *StyloAI*, achieves accuracy rates of 81% and 98% on two multi-domain annotated dataset, outperforming existing state-of-the-art models. The study further experimentally validates the significance of *StyloAI*'s features.

The rest of the paper is structured as follows: the next section provides an overview of related works, discussing various techniques proposed for detecting AI-generated texts. Section 3 delves into a detailed description of the proposed model. Following that, Sect. 4 presents a thorough evaluation of the *StyloAI* model. Section 5 concludes the paper.

2   Related Works

Research into distinguishing AI-generated text from human-written content has taken various paths, reflecting a growing interest in developing robust detection tools. One approach involves leveraging industry-developed AI detection tools for text classification. The study by Akram et al. [1] assessed the effectiveness of these tools across diverse datasets, highlighting variations in performance and prompting the need for further innovation in detection methodologies.

Another approach to detecting AI-generated texts is the use of automatic text classification. These studies [3,9] and [10] applied deep learning models to differentiate between AI-generated and human-authored texts. This method has gained traction for its potential to provide high accuracy rates. However, these are "black box" models and the reliance on domain-specific datasets for training raises questions about their applicability across varied contexts.

A promising yet underexplored approach in distinguishing AI-generated content involves integrating handcrafted linguistic features with shallow machine learning classifiers. Recent studies, such as those by [5] and [6], have introduced a range of features aimed at differentiating between human-authored and AI-generated texts. Despite the initial promise of these methodologies, researchers continue to strive for a balance between accuracy, interpretability, and scalability in their proposed solutions.

3   Methodology

This study aims to design and develop a technique to identify AI-generated texts. Stylometric features are extracted from texts to make predictions about their authorship. This section first discusses the feature set used in *StyloAI*, followed by the machine learning algorithms considered. It concludes with the datasets employed to evaluate the proposed model. Figure 1 illustrates the process within the *StyloAI* model.

3.1   *StyloAI* Feature Set

The feature set within the *StyloAI* model comprises 31 stylometric features, 12 of which are new to the detection of AI-generated texts. These features are categorised into six main groups for clarity and ease of understanding. Table 1 summarises these features.

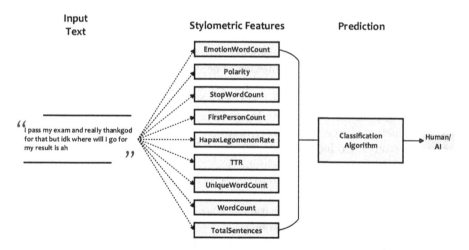

Fig. 1. The *StyloAI* Process: First, textual characteristics are extracted with a feature-engineering approach. Then, a classification algorithm is used to build the model.

6 features categorised under **Lexical Diversity** focus on the range and distribution of words used within a text, highlighting differences in vocabulary richness between AI-generated and human-written content. This study introduces the *Type-Token Ratio* (TTR) and *Hapax Legomenon Rate* (HLR) to evaluate lexical diversity in detecting AI-generated texts. Additionally, basic metrics

obtained from previous studies [5,6] such as *WordCount, UniqueWordCount,* and *CharacterCount,* assess the total number of words, unique words, and characters present in a text, providing further insights into its lexical complexity.

Syntactic Complexity with 12 features, assesses the structural aspects of language, such as sentence length and complexity, to identify patterns unique to human or AI authors. Some features included in this category are the *ComplexVerbsCount* and *ContractionCount.* The hypothesis behind these features is that capturing verbs not in the first most common verbs, using multiple commas, and conjunctions like "and" or "but" indicate more structurally intricate sentences characteristic of human written texts.

The 4 features within the **Sentiment and Subjectivity** category analyse the emotional tone and subjective expressions within the text, providing insights into how emotions and opinions are conveyed. The *Polarity* and *Subjectivity* were obtained from previous studies [5,6] while the *VADERCompound* and the count of words (*EmotionWordCount*) that reflect emotions, including "fear", "joy", and "sadness" were introduced in this study. The hypothesis behind these features is that AI-generated texts might not match the emotional depth of human-written texts. Also, anomalies in emotional word usage, such as overuse, underuse, or misuse in context, could hint at non-human authorship.

The **Readability** comprising 2 features, evaluates how easy it is to understand the text by considering the intended audience's complexity level and educational background. Scores that have been used in other studies [6], such as the *FleschReadingEase* and the metric introduced in this study, the *GunningFogIndex,* are based on factors such as sentence length and word complexity. These metrics provide quantifiable assessments of text accessibility for readers with diverse levels of education and literacy. The differences in the intended audience or content sophistication can be identified by comparing readability scores between human and AI-generated texts. The **Named Entities** category with 4 features focuses on the presence and usage of specific names, places, and organisations, which can vary significantly between AI-generated and human-written texts. This study introduced *DateEntities* and *DirectAddressCount* features. The hypothesis is that incorrect, anachronistic, or overly vague references to dates can be a red flag. AI may either fail to correctly anchor events in time or use dates in ways that don't align with known historical or contextual facts. Also, human-written text should have many instances where the text directly addresses the reader or another hypothetical listener.

Finally, the 3 features grouped under the **Uniqueness and Variety** category focus on how original and diverse the content is, thereby distinguishing between repetitive AI-generated patterns and the dynamic variability seen in human writing. This study introduced the uniqueness of the bigrams/trigrams feature, quantifying how frequently new or varied word pairings appear in the text. Texts with a higher bigram/trigram uniqueness ratio may be considered more creative or novel, as they employ a more comprehensive range of lexical combinations, as expected in human-written texts.

3.2 Model Selection and Implementation

The task of detecting AI-generated text was approached as a binary classification problem, where known AI-generated texts were treated as negative samples and human-written texts as positive samples. Several popular binary classification techniques in machine learning, focusing on these popular options, were explored: Random Forest, SVM, Logistics Regression, Decision Tree, KNN classification and Gradient Boosting.

All experiments for *StyloAI* were conducted on the Google Colab platform, using machine learning classifiers with default settings from the Scikit-Learn library[1]. For reliable outcomes, a 5-fold cross-validation method was employed, dividing our dataset in each iteration into 80% for training, 10% for validation to fine-tune hyperparameters, and 10% for testing with unseen texts.

3.3 Datasets

This study employed two datasets to analyse and evaluate the proposed model. The first, termed the AuTexTification dataset, is from the study by Sarvazyan et al.'s [8], while the second dataset is drawn from Mindner et al.'s study [6].

The Subtask 1 AuTexTification dataset comprises texts from humans and large language models across five domains: tweets, reviews, how-to articles, news, and legal documents sourced from publicly available datasets like MultiEURLEX, XSUM, and Amazon Reviews, among others. This dataset, detailed further in the published article and its official repository, comprises 27,989 AI-generated and 27,688 human-written texts, totalling **55,677** entries, making it one of the largest and most varied corpora used for distinguishing between AI-generated and human-authored content.

The second dataset utilised for assessing *StyloAI* was a smaller, multidomain set curated by Mindner et al. [6], chosen for its relevance to the education sector. This dataset aims to distinguish between human-crafted and AI-generated texts in an educational context. Covering a diverse array of subjects such as biology, chemistry, geography, history, IT, music, politics, religion, sports, and visual arts, it consists of 100 AI-generated texts collected from ChatGPT 3.5[2] interactions and 100 human-written texts collected from Wikipedia[3], resulting in a well-balanced dataset of **200** entries. This dataset makes it particularly useful for evaluating *StyloAI*'s performance in educational settings. Further details on this dataset can be found in the original publication and its official repository.

---

[1] https://scikit-learn.org/stable/.
[2] https://chat.openai.com/chat.
[3] https://en.wikipedia.org/wiki/Main_Page.

Table 1. Comprehensive Features of *StyloAI* Across Six Categories

Category	Features	Description	Reference
Lexical	WordCount	The total number of words in the text	[6]
	UniqueWordCount	The number of unique words used in the text	[6]
	CharCount	The total number of characters in the text, including spaces and punctuation	[6]
	AvgWordLength	Calculated as $\frac{CharacterCount}{WordCount}$	[6]
	TTR	Measures lexical diversity, calculated as $TTR = \frac{UniqueWordCount}{WordCount}$	new
	HapaxLegomenonRate	The proportion of words that appear only once in the text, $\frac{Number\ of\ Words\ Appearing\ Once}{Total\ Words}$	new
Syntactic	SentenceCount	The total number of sentences in the text	
	AvgSentenceLength	Calculated as $\frac{WordCount}{SentenceCount}$	[6]
	PunctuationCount	The total number of punctuation marks in the text	[6]
	StopWordCount	The total number of commonly used words	[6]
	AbstractNounCount	The number of nouns representing intangible concepts or ideas	new
	ComplexVerbCount	The number of verbs not in the most common 5000 words	new
	SophisticatedAdjectiveCount	The number of adjectives with complex suffixes like "ive", "ous", "ic"	new
	AdverbCount	The total number of adverbs in the text	[6]
	ComplexSentenceCount	The number of sentences with more than one clause, indicating complex sentence structures	[5]
	QuestionCount	The total number of questions, as indicated by question marks in the text	[5,6]
	ExclamationCount	The total number of exclamations, as indicated by exclamation marks in the text	[5,6]
	ContractionCount	The total number of contractions in the text, such as "don't" and "can't"	new
Sentiment	EmotionWordCount	The total number of words associated with emotions in the text	new
	Polarity	Measures the text's sentiment orientation (positive, negative, or neutral)	[5,6]
	Subjectivity	Measures the amount of personal opinion and factual information in the text	[5,6]
	VaderCompound	A sentiment analysis score that combines the positive, negative, and neutral scores to give a single compound sentiment score	new
Readability	FleschReadingEase	Calculated as $206.835 - 1.015\left(\frac{Total\ Words}{Total\ Sentences}\right) - 84.6\left(\frac{Total\ Syllables}{Total\ Words}\right)$	[6]
	GunningFog	Estimates the years of formal education needed to understand a text on the first reading, calculated as $0.4\left(\frac{Word\ Count}{Sentence\ Count} + 100\left(\frac{Complex\ Words\ Count}{Word\ Count}\right)\right)$	new
Entity	FirstPersonCount	The number of first-person pronouns	[5,6]
	DirectAddressCount	The number of instances where the text directly addresses the reader or another hypothetical listener	new
	PersonEntities	The count of named individuals mentioned in the text	[5]
	DateEntities	The count of date references within the text	new
Uniqueness	Bigram/trigramUniqueness	These measures calculate the uniqueness of two-word and three-word combinations, indicating the originality and creative combinations of words in the text	new
	SyntaxVariety	The count of all the POS tags in a text	[5]

4   Results

In line with previous studies, The performance of *StyloAI* model was accessed using key metrics, including accuracy, recall, precision, and F1-score.

Tables 2 presents the performance of the extracted features on different shallow learning models on the Subtask 1 AuTexTification dataset [8]. From Table 2, it is evident that the Random Forest classifier consistently outperforms other shallow machine learning algorithms, achieving an 81% value across all metrics. Conversely, the Logistic Regression model demonstrates the lowest performance, scoring 71% across all metrics.

Note: From the analysis above, the extracted stylometric features showed optimal performance with the Random Forest classifier. For the rest of this paper, whenever the *StyloAI* model is used, it indicates the model trained with the Random Forest Classifier.

Table 2. Performance of Stylometric Features on Selected Shallow Machine Learning Classifiers Using Sarvazyan et al.'s [8] AuTexTification Dataset

Model	Precision	Recall	F1-Score	Accuracy	AUC Score
Random Forest	**0.81**	**0.81**	**0.81**	**0.81**	**0.88**
SVM	0.79	0.79	0.79	0.79	0.87
Logistic Regression	0.71	0.71	0.71	0.71	0.77
Decision Tree	0.72	0.72	0.72	0.71	0.85
KNN Classification	0.73	0.73	0.73	0.73	0.80
Gradient Boosting	0.78	0.77	0.77	0.85	0.71

Table 3 presents a comparative analysis of *StyloAI* against state-of-the-art using the Subtask 1 AuTexTification dataset. The comparison includes the TALN-UPF model, which integrates probabilistic token-level features from various GPT-2 models, linguistic features like word frequencies and grammar errors, alongside text representations from pre-trained encoders. Also evaluated are approaches such as random baselines, zero-shot (SBZS) and few-shot (SB-FS) based on text and label embedding similarities, bag-of-words with logistic regression (BOW+LR), Low Dimensional Semantic Embeddings (LDSE), and finely tuned language-specific transformers. A deeper exploration of these baseline models can be found in Sarvazyan et al.'s [8] study.

Among the compared models, TALNUP performed with an F1-score of 0.80. Notably, TALNUP relies on deep learning and word vectors. However, as shown in Table 3, *StyloAI* surpasses it with an F1-score of 0.81. Moreover, StyloAI significantly outperforms the least effective baseline model-the zero-shot model-by 48% in the F1 score, highlighting its superior ability to distinguish AI-generated texts from human-written content.

In the evaluation of *StyloAI* on the second dataset created by Mindner et al., Table 4 illustrates that *StyloAI* performs comparably to the model proposed by Mindner et al. [6], despite Mindner et al.'s reliance on AI-generated features. Achieving a 98% on accuracy demonstrates the robustness of *StyloAI* when applied to text generated from the education sector. Furthermore, in their study, Mindner et al. compared their model's performance with an industry-developed AI detector, revealing that their model outperforms GPTZero by 28.9% relatively in accuracy and 24.2% relatively in F1-score. The hypothesis is that *StyloAI* will have similar performance when compared with GPTZero since its performance is comparable with the result of Mindner et al. on the base dataset.

Table 3. Comparison of *StyloAI* with Sarvazyan et al.'s [8] Models on the AuTexTification Dataset

Model	F1-Score
StyloAI	**0.81**
TALN-UPF	0.80
BOW+LR	0.74
LDSE	0.60
SB-FS	0.59
Transformer	0.57
Random	0.50
SB-ZS	0.33

Table 4. Comparison of *StyloAI* with Mindner et al.'s [6] Models on the Education Dataset

Model	F1-Score	Accuracy
StyloAI	**0.97**	**0.98**
Mindner et al.'s Features + XGBoost	0.90	0.90
Mindner et al.'s Features + Random Forest	0.98	0.98
Mindner et al.'s Features + MLP	0.87	0.87

Feature Importance. From Fig. 2, it is observed that the 4 most significant features highlighted by the Random Forest algorithm are the *UniqueWordCount, StopWordCount, TTR*, and *HapaxLegomenonRate*.

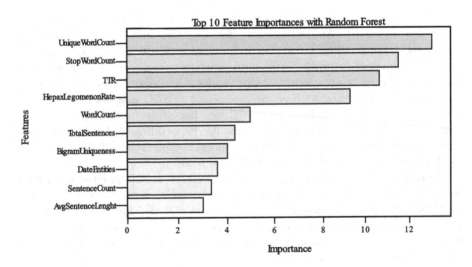

Fig. 2. Ranking of Features and How they Influence the Performance of the Random Forest Classifier

These top 4 features are pivotal in distinguishing AI-generated texts from human-authored ones. The *UniqueWordCount*, which counts non-repetitive words, can reveal AI's tendency to use rare words excessively, aiming for sophistication but resulting in an unnatural count. In contrast, humans write sentences with more stop words to maintain a natural flow. While AI models strive to replicate this pattern, they may deviate due to a lack of human speech rhythm or misuse of stop words.

TTR, which measures lexical diversity, shows that human texts achieve a balance in vocabulary that suits the text's length and complexity. In contrast, AI-generated texts often miss this balance by excessively expanding or restricting their vocabulary range. Additionally, the *HapaxLegomenonRate*, indicating the use of words appearing only once, signifies rich and detailed vocabulary in human writing-a trait that AI texts may not consistently emulate, thereby impacting their authenticity.

5 Conclusion

This paper explored stylometric features to classify text as AI-generated or human-written. Specifically, 31 features categorised under Lexical Diversity, Sentiment and Subjectivity, Readability, Named Entities and Uniqueness and Variety were extracted to determine what characterises AI-generated texts. The proposed model, *StyloAI*, comprising the 31 features trained on a random forest classifier, provided an accuracy of 81% and 98% on two multi-domain datasets. This study also demonstrates that a balanced use of non-repetitive words and a vocabulary that suits the text length and complexity are the top features that differentiate AI-generated texts from human-authored ones. A limitation of the study is that it uses the AuTexTification dataset, generated using the Bloom model rather than the more commonly used GPT models (GPT-3.5 and GPT-4). Future work will focus on investigating features from datasets generated by GPT models.

References

1. Akram, A.: An empirical study of AI generated text detection tools. arXiv preprint arXiv:2310.01423 (2023)
2. Birenbaum, M.: The chatbots' challenge to education: disruption or destruction? Educ. Sci. **13**(7), 711 (2023)
3. Fagni, T., Falchi, F., Gambini, M., Martella, A., Tesconi, M.: Tweepfake: about detecting deepfake tweets. PLoS ONE **16**(5), e0251415 (2021)
4. Khalil, M., Er, E.: Will ChatGPT get you caught? Rethinking of plagiarism detection. In: Zaphiris, P., Ioannou, A. (eds.) HCII 2023. LNCS, vol. 14040, pp. 475–487. Springer, Cham (2023). https://doi.org/10.1007/978-3-031-34411-4_32
5. Kumarage, T., Garland, J., Bhattacharjee, A., Trapeznikov, K., Ruston, S., Liu, H.: Stylometric detection of AI-generated text in Twitter timelines. arXiv preprint arXiv:2303.03697 (2023)

6. Mindner, L., Schlippe, T., Schaaff, K.: Classification of human- and AI-generated texts: investigating features for ChatGPT. In: Schlippe, T., Cheng, E.C.K., Wang, T. (eds.) AIET 2023. LNDECT, vol. 190, pp. 152–170. Springer, Singapore (2023). https://doi.org/10.1007/978-981-99-7947-9_12
7. Ngo, T.T.A.: The perception by university students of the use of ChatGPT in education. Int. J. Emerg. Technol. Learn. (Online) **18**(17), 4 (2023)
8. Sarvazyan, A.M., González, J.Á., Franco-Salvador, M., Rangel, F., Chulvi, B., Rosso, P.: Overview of autextification at IberLEF 2023: detection and attribution of machine-generated text in multiple domains. arXiv preprint arXiv:2309.11285 (2023)
9. Uchendu, A., Le, T., Shu, K., Lee, D.: Authorship attribution for neural text generation. In: Proceedings of the 2020 Conference on Empirical Methods in Natural Language Processing (EMNLP), pp. 8384–8395 (2020)
10. Zellers, R., et al.: Defending against neural fake news. In: Advances in Neural Information Processing Systems, vol. 32 (2019)

Interpretable Methods for Early Prediction of Student Performance in Programming Courses

Ziwei Wang[1](✉), Irena Koprinska[1](✉), and Bryn Jeffries[1,2](✉)

[1] School of Computer Science, The University of Sydney,
Sydney, NSW 2006, Australia
zwan0684@uni.sydney.edu.au, irena.koprinska@sydney.edu.au
[2] Grok Academy, PO Box 144, Broadway, Sydney, NSW 2007, Australia
bryn.jeffries@grokacademy.org

Abstract. Early prediction of student grades is important for identifying students at-risk of failing or students not achieving their goals, and providing timely interventions. There is a need to develop methods that are both accurate and interpretable, to help teachers understand student behaviour and take appropriate actions. In this paper, we present a decision tree based approach for predicting students' final performance in online computer programming courses for high school students, based on log data from the first part of the course. We define and extract suitable features characterising student behaviour and show that it is possible to build compact and accurate decision trees, achieving an overall accuracy of 76% and 82–83% at one-third and mid-course respectively. We provide insights about the important factors affecting student performance and how the rules can be used to improve the student learning outcomes.

1 Introduction

Large-scale online programming courses provide students with interactive content and automatic feedback and teachers with a wealth of valuable data about student activities. However, student data is large and complex, which creates challenges for extracting useful information. There is a need for developing explainable, lightweight and accurate methods for early prediction of student performance, to provide teachers with actionable insights. This would allow them to intervene at key points, improve the course design and prevent dropout.

Furthermore, existing research mainly concentrates on university-level computer courses, despite students' career inclinations forming earlier. Providing more support in programming education during K-12 can boost students' confidence and encourage more of them to pursue computing studies in the future.

In this paper, we propose a novel approach for early prediction of final student performance of high-school students based on student activity data. We consider large-scale programming courses at two levels: beginners and intermediate. We define and extract suitable features pertaining to student activities from content

A. M. Olney et al. (Eds.): AIED 2024 Workshops, CCIS 2151, pp. 115–123, 2024.
https://doi.org/10.1007/978-3-031-64312-5_14

slides, programming problems and modules. Our approach utilises feature selection methods, two types of decision tree classifiers and final tree selection. We chose decision trees as they are intrinsically interpretable, in contrast to post-hoc methods which aim to extract explanations from trained black-box models. Our method has the following characteristics:

1. It is able to produce compact, interpretable and easy to visualise rules, providing actionable insights for teachers, e.g. by revealing key problems and modules students should complete in order to pass.
2. It uses data from student activities only, that is automatically collected and readily available in many courses. The features we define are applicable to other courses, not only programming courses, and the decision tree approach is easy to implement and adapt.

We aim to answer the following questions: (i) How accurately can we predict the final grade at different stages of the course? (ii) Which are the important features for predicting student performance and how compact and interpretable the generated decision trees are?

2   Related Work

Early prediction of final exam grades and at-risk students have been previously investigated, mainly for university courses. In [3,4] information from an autograding system, discussion board and assessment marks was used to predict the exam performance mid-semester. Similarly, in [7] the final student grades were predicted based upon assessment data, activity on the discussion board and time spent on these tasks. A method for predicting student dropout using autograding data from the first two weeks of semester was proposed in [6]. Dsilva et al. [2] applied boosted trees on behavioural data from online courses, to predict student dropout and failure. Asif et al. [1] used machine learning techniques to predict the graduation performance in a four-year university degree from pre-university marks and marks from the first and second year at the university.

Relevant work related to programming courses for high school students includes [5,8]. McBroom et al. [5] proposed progress networks for providing a visualisation which summarises the progression of a student population through a learning task. Zhang et al. [8] developed a method for predicting performance in the last problem of a course module based on the student interactions with content in the same module.

In this paper, we extend previous work by focusing on large-scale programming courses for high school students. We define and extract suitable features and develop explainable and lightweight models to predict student performance early in the course, using data from course activities.

3   Data

We used data collected from the 2019 Australian National Computer Science School Challenge, a 5-week long online Python programming course for high-school students, provided by Grok Academy[1]. We limited our analysis to the Beginners and Intermediate streams, aligned to the years 7–8 and 9–10 of the Australian Digital Technologies curriculum, respectively.

Each course is composed of 10 *modules*. Concepts are introduced in *content* slides within the module. These slides include textual information, explanatory videos, code examples that can be executed in the browser, and interactive activities that encourage students to engage with the material by, for instance, making specific changes to the example code. The number of slides varies depending upon the topic, but there are about 10 content slides per module. This material is supported with *problem* slides that contain programming tasks for students to solve and submit for automated marking and feedback. The student can correct their code based on the feedback and resubmit multiple times until they pass all tests. There are 2–4 problems per module.

The user activity data includes timestamped records of visited content slides and problems. For content slides, a record is also made when the student completes all interactive activities on the slide. For problems, a record is made when the student runs code or submits code for testing, with the outcome of the tests.

In total there were 17,890 students enrolled: 10,558 in Beginners and 7,332 in Intermediate. We excluded the students who did not make any submission to the problem slides in the first half of the course. This resulted in 13,896 active students used in the data analysis: 8,399 in Beginners and 5,497 in Intermediate.

Students are awarded marks for passing the test cases. Each problem is worth 10 marks; 1 mark is deducted after every 5 incorrect submissions, to a minimum mark of 5. The students' total mark for the course was discretized into three intervals: Poor [0%, 25%], Good (25%, 75%], and Excellent (75%, 100%]. The mark distribution (Poor, Good, Excellent) was relatively balanced: 2807, 2914 and 2618 for Beginners and 1968, 1962 and 1567 for Intermediate.

The data was provided in anonymised format, with all student details removed and student identifiers substituted with randomly generated numbers to prevent re-identification.

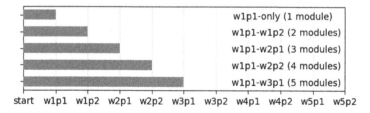

Fig. 1. Predicting the final score (after module 10) based on data at five stages - all data up to module 1, 2, 3, 4 and 5

---

[1] https://grokacademy.org/challenge.

Table 1. Extracted features from content slides, problems and modules. Notation: w*ipj* - module week i part j, q*l* - problem l, s*k* - slide k

Feature Name	Description
Content Slides	
w*ipj*-s*k*	Completion of slide (completed/not completed)
Problems	
w*ipj*-q*l*-runs	Count of times student run their code for the problem
w*ipj*-q*l*-subs	Count of times student submitted code for marking
w*ipj*-q*l*-sessions-outer	Count of sessions between completion of previous and current problem
w*ipj*-q*l*-sessions-inner	Count of sessions from first attempt to completion of current problem
w*ipj*-q*l*-active-time-outer	Total session time between completion of previous and current problem
w*ipj*-q*l*-active-time-inner	Total session time from first attempt to completion of current problem
w*ipj*-q*l*-overall-time-outer	Spanned time between completion of previous and current problem
w*ipj*-q*l*-overall-time-inner	Spanned time from first attempt to completion of current problem
w*ipj*-q*l*-first-attempt	Days from release until the student's first attempt
Modules	
w*ipj*-runs	Count of terminal runs within this module
w*ipj*-subs	Count of submits within this module
w*ipj*-slides-completed-pc	Percentage of content slides completed
w*ipj*-sessions	Count of sessions within this module
w*ipj*-active-time	Total active time within module
w*ipj*-overall-time	Overall time within module (span of all sessions)
w*ipj*-longest-session	Duration of the longest session
w*ipj*-mean-session-time	Average duration of the sessions
w*ipj*-time-efficiency	Total active time/total time

4   Methodology

We predict students' final grades based on their performance at five stages of the each stream. For each stage, we generate prediction models using data from all modules up to that point as shown in Fig. 1. Since the structure of each module varies, particularly in the number of content slides, a different prediction model must be constructed for each case. As a one-third point in the course we consider *w2p1* (3 modules) and as a mid-point – *w3p1* (5 modules).

4.1   Feature Extraction

We extracted features pertaining to activity within content slides, problems and modules, as summarised in Table 1 and described in more detail below.

Content Slides. We only use the slide completion status. Given the large number of slides in each module, we chose to keep the number of slide features low to avoid excessive dimensionality of the feature set.

Problems. We identified a richer set of per-problem features as predictors of course outcome. For instance, the number of times code was run or submitted for testing would likely be higher when students found problems challenging, as well as suggesting higher levels of persistence by the student. We also include the time a problem was available before the student started it, as an indicator, since we expect people to do better when they start problems early.

To get a measure of time on task, we aggregated all event records for a student on a problem that occur within 60 s of a neighbouring event as continuous *sessions*. The number of sessions can, in some sense, indicate how focused students are on their learning or whether they have encountered particularly challenging problems. We considered the total number of sessions whilst attempting the problem ($\texttt{wip}j\texttt{-q}l\texttt{-sessions-inner}$), to cover the period while the student engaged with that problem. Additionally, we also considered all the sessions since the previous problem was completed, up until the end of the current problem, to include the time spend on the slides introducing the assessed topic ($\texttt{wip}j\texttt{-q}l\texttt{-sessions-outer}$).

As well as considering the number of sessions, we also included measures for the sum of the session durations (the *active time*) and the time elapsed from the start of the first session to the end of the last session (the *overall time*).

Modules. The per-module features are similar to the per-problem and per-slide features but are aggregated at a module level.

4.2 Feature Selection

The number of extracted features for each course depends on the number of content slides and problems. To select a smaller subset of informative features, we applied two feature selection algorithms: correlation-based (CFS) and information gain (IG). Appropriate feature selection improves performance and interpretability by reducing overfitting and typically resulting in a smaller decision tree.

After the application of the feature selection algorithms, the number of selected features was reduced to 12–25. The most frequently selected features were: *first-attempt* for problem slides, *active-time*, *overall-time*, *sessions*, *runs* and *subs* for both problems and modules. 89% of selected features were for problems.

4.3 Prediction

We investigated the performance of two types of decision tree classifiers. Decision trees are intrinsically interpretable classifiers by design, in contrast to recently proposed post-hoc methods such as SHAP which analyse the predictions of black-box classifies and assign importance values to the features. They trees are highly appropriate for our context as they generate a set of rules, providing an explanation to teachers why the students were assigned to a particular class (Poor/Good/Excellent), clearly identifying the important features affecting students' performance. They are also fast to train and make predictions.

To generate the decision trees we employed two algorithms: the standard C4.5 and REPTree - a modification of C4.5 which uses reduced error pruning.

For each of the 10 cases (see Table 2) we generated 4 decision trees (2 feature selection algorithms x 2 decision tree algorithms), and then selected the best one based on overall accuracy. If the trees had similar accuracy, within a small tolerance threshold t, the smallest tree was selected. We used $t = 1.5$.

Different combinations of feature selection and decision tree algorithms produced the best trees: CFS+REPTree (4/10 cases), IG+C4.5 (3/10 cases), IG+REPTree (3/10 cases) and CFS+C4.5 (0/10 cases).

The performance was evaluated using 10-fold cross-validation with stratification. The results were compared with a baseline predicting the majority class.

5   Results and Discussion

Table 2 presents the overall accuracy for all classes, the per-class accuracy and the tree size (total number of nodes).

5.1   How Accurately Can We Predict the Final Grade at Different Stages of the Course?

The results show that it is possible to accurately predict the final student score in the middle of the courses, with an overall accuracy of 82–83%. This accuracy is considerably higher than the majority-class baselines of 35–36%. As more student data becomes available, the accuracy gradually increases - from 55–58% after 1 module to 76% after 3 modules and 82–83% after 5 modules. The increase rate is higher at the beginning, until 3 modules are completed, and lower after that.

We found that the class 'Poor' was the easiest to predict with accuracy above 94% at the middle of the course, and these are the students who need help and early intervention most compared to the other groups. Class 'Good' was the most difficult to predict, especially in the early stages. However, as the course progresses after the one-third point, the accuracy for all classes significantly improved, reaching 67–76% for Good and 82–88% for Excellent at mid-point.

5.2   Which Are the Important Features for Predicting Student Performance and How Compact and Interpretable the Trees Are?

Table 3 presents the generated decision trees at two stages: one-third and mid-point of the course. The trees are very compact and interpretable—they contain a small number of rules and only 4 types of features associated with specific problems and modules: *first-attempt*, *subs*, *overall-time-inner* and *overall-time-outer*. Not all features appear in the trees, only the ones with the highest information gain - decision trees can be seen as performing a secondary feature selection.

The *first-attempt* feature is associated with specific problems for each course (e.g., with w2p1-q4 for Beginners at one-third point – Problem 4 from module Week 2 Part 1) and corresponds to the number of days from the release of

Table 2. Accuracy [%] and tree size results for all courses. Per-class accuracies are ordered for 'Poor'/'Good'/'Excellent' classes. One-third point is 3 modules (30%), mid-point is 5 modules (50%). B is predicting the majority class baseline.

		B	1-mod.(10%)	2-mod.(20%)	3-mod.(30%)	4-mod.(40%)	5-mod.(50%)
Beginners	Accuracy	35.4	54.8	60.2	76.3	79.1	83.0
		0	60.8	67.9	95.1	95.1	95.2
	Per-class acc.	100	43.3	49.5	61.7	74.3	76.3
		0	61.4	64.3	73.0	67.3	87.5
	Tree size	–	7	11	5	5	7
Intermediate	Accuracy	35.8	57.6	68.6	76.2	79.3	81.8
		100	58.4	80.0	93.4	96.0	96.0
	Per-class acc.	0	63.3	67.2	67.8	75.1	67.4
		0	49.4	56.0	76.2	63.5	82.0
	Tree size	–	11	7	11	9	7

Table 3. Generated decision trees at one-third and mid-point of the courses. Leaf nodes show (occurrences/misclassifications) counts using whole dataset.

One-third point - 3 modules	Mid-point - 5 modules
	Beginners

Beginners

One-third point - 3 modules:
```
w2p1-q4-first-attempt < 27.5 days
| w2p1-q4-first-attempt < 4.5 days: Excellent (3251/1194)
| w2p1-q4-first-attempt >= 4.5 days: Good (2339/662)
w2p1-q4-first-attempt >= 27.5 days: Poor (2798/137)
```

Mid-point - 5 modules:
```
w2p1-q4-first-attempt < 27.5 days
| w3p1-q4-first-attempt < 20.5 days
| | w3p1-q4-first-attempt < 7.5 days: Excellent (2654/592)
| | w3p1-q4-first-attempt >= 7.5 days: Good (1390/538)
| w3p1-q4-first-attempt >= 20.5 days: Good (1549/148)
w2p1-q4-first-attempt >= 27.5 days: Poor (2809/138)
```

Intermediate

One-third point - 3 modules:
```
w2p1-q2-first-attempt <= 27 days
| w2p1-q3-first-attempt <= 3 days
| | w2p1-q3-overall-time-inner <= 3.6 days
| | | w2p1-q3-overall-time-outer <= 8.7 min: Excellent (424/104)
| | | w2p1-q3-overall-time-outer > 8.7 min
| | | | w2p1-q1-first-attempt <= 1 days: Excellent (1269/524)
| | | | w2p1-q1-first-attempt > 1 days: Good (285/121)
| | w2p1-q3-overall-time-inner > 3.6 days: Good (199/37)
| w2p1-q3-first-attempt > 3 days: Good (1440/445)
w2p1-q2-first-attempt > 27 days: Poor (1878/31)
```

Mid-point - 5 modules:
```
w2p1-q2-first-attempt < 27.5 days
| w3p1-q2-subs < 1 : Good (1134/121)
| w3p1-q2-subs >= 1
| | w2p2-q2-first-attempt <6.5 days: Excellent (1533/400)
| | w2p2-q2-first-attempt >= 6.5 days: Good (950/424)
w2p1-q2-first-attempt >= 27.5 days: Poor (1880/32)
```

the problem until the student's first attempt. It is the most important feature as it appears in almost all trees and is typically at the root of the tree. It shows the importance of starting early to achieve good grades as this feature is highly negatively correlated with students' final performance. The thresholds sometimes indicate extremes: material released at the start of Week 2 needs to be completed in the remaining 4 weeks, so a student starting Beginners course problem w2p1-q4 27.5 days later is very likely to run out of time.

The number of submissions in problems (*subs*) appears in the Intermediate model with a threshold of 1 submission, indicating whether the student submitted anything. Rather than representing effort or difficulty, this feature suggests that students attempting specific problems are likely to achieve a good outcome, while those who don't will very probably achieve a poor one.

The *overall-time-inner* feature is the time taken to complete a problem, from viewing the problem for the first time to submitting a final answer. It appears only in the Intermediate tree at the one-third stage, where it identifies that students completing the problem quickly (in less than 3.6 days), in conjunction with conditions based on two other time features (*first-attempt* and *overall-time-outer*) will likely achieve an Excellent rather than Good final outcome. By the mid-course point, the feature no longer appears in the decision tree.

The *overall-time-outer* feature is the time from completing the previous problem to completing the current problem. It appears in the Intermediate tree at the one-third stage associated with problem 3 from Week 3, helping to distinguish the Excellent from the Good students.

We were surprised to see other features did not appear in the trees, and their absence may be informative. For instance, the fact that measures of progress through content such as *slides-completed-pc* were absent may suggest that many students do not engage with the material as expected. Exploration of the distributions of these features may reveal important insights for the course developers.

6   Conclusion

We proposed a decision tree-based approach for early prediction of students' final academic performance. The evaluation was conducted on two large-scale programming courses for beginners and intermediate high school students.

We defined appropriate features to characterise student behaviour. Using these features, we build models to predict the final student performance at one-third and mid-course stage, achieving 76% and 82–83% accuracy respectively. Importantly, the generated decision trees are compact and interpretable. They contain a small number of rules and features, and are easy to understand and apply in practice. They provide actionable insights - e.g., highlighting the importance of starting early but also revealing key problems and modules that students should attempt and complete, in order to pass the course successfully.

The results can be used to provide feedback to students if their behaviour in the early stages of the course is likely to be associated with positive or negative final outcomes. This includes identifying at-risk students (failing or dropping out) and also students not achieving their goals, e.g., aiming to achieve 'Excellent' outcome while their current behaviour indicates 'Good'. Students can be provided with personalised feedback and suggested remedial actions.

In future work, we plan to classify the problems into different types, including paired problems, where the solution of one problem is connected to the next problem. Adding more data, for example about the order in which students view and complete the slides and problems, can also help to understand student behaviour better and improve the course design and student engagement.

References

1. Asif, R., Merceron, A., Ali, S.A., Haider, N.G.: Analyzing undergraduate students' performance using educational data mining. Comput. Educ. **113**, 177–194 (2017)
2. Dsilva, V., Schleiss, J., Stober, S.: Trustworthy academic risk prediction with explainable boosting machines. In: Wang, N., Rebolledo-Mendez, G., Matsuda, N., Santos, O.C., Dimitrova, V. (eds.) AIED 2023. LNCS, vol. 13916, pp. 463–475. Springer, Cham (2023). https://doi.org/10.1007/978-3-031-36272-9_38
3. Koprinska, I., Stretton, J., Yacef, K.: Predicting student performance from multiple data sources. In: Conati, C., Heffernan, N., Mitrovic, A., Verdejo, M. (eds.) AIED 2015. LNCS, vol. 9112, pp. 678–681. Springer, Cham (2015). https://doi.org/10.1007/978-3-319-19773-9_90
4. Koprinska, I., Stretton, J., Yacef, K.: Students at risk: detection and remediation. In: Educational Data Mining, pp. 512–515 (2015)
5. McBroom, J., Paassen, B., Jeffries, B., Koprinska, I., Yacef, K.: Progress networks as a tool for analysing student programming difficulties. In: Australasian Computing Education Conference, pp. 158–167 (2021)
6. Pereira, F.D., et al.: Early dropout prediction for programming courses supported by online judges. In: Isotani, S., Millán, E., Ogan, A., Hastings, P., McLaren, B., Luckin, R. (eds.) AIED 2019. LNCS (LNAI), vol. 11626, pp. 67–72. Springer, Cham (2019). https://doi.org/10.1007/978-3-030-23207-8_13
7. Romero, C., Espejo, P.G., Zafra, A., Romero, J.R., Ventura, S.: Web usage mining for predicting final marks of students that use Moodle courses. Comput. Appl. Eng. Educ. **21**(1), 135–146 (2013)
8. Zhang, V., Jefffries, B., Koprinska, I.: Predicting progress in a large-scale online programming course. In: Wang, N., Rebolledo-Mendez, G., Matsuda, N., Santos, O.C., Dimitrova, V. (eds.) AIED 2023. LNCS, vol. 13916, pp. 810–816. Springer, Cham (2023). https://doi.org/10.1007/978-3-031-36272-9_76

MOOCRev: A Large-Scale Data Repository for Course Reviews

Mohammad Alshehri[1]([✉]), Fahd Alfarsi[1], Ahmed Alamri[1], Laila Alrajhi[2], Saman Rizvi[3], Filipe Dwan Pereira[4], Seiji Isotani[5], and Alexandra Cristea[2]

[1] University of Jeddah, Makkah, Saudi Arabia
{malshehri,falfarsi,asalamri4}@uj.edu.sa
[2] Durham University, Durham, UK
{laila.m.alrajhi,alexandra.I.Cristea}@durham.ac.uk
[3] University of Cambridge, Cambridge, UK
ssr2004@cam.ac.uk
[4] Federal University of Roraima, Roraima, Brazil
[5] Harvard University, Cambridge, USA
seiji_isotani@gse.harvard.edu

Abstract. Massive Open Online Courses (MOOCs) have revolutionised the landscape of education, generating unprecedented arrays of diverse data about, from and by learners. However, there are few publicly accessible large-scale text datasets that several stakeholders, including course designers and instructors, can use. This paper presents MOOCRev, a groundbreaking data repository aggregating MOOC reviews from various platforms, leveraging cutting-edge data collection and processing techniques. The dataset encompasses many review characteristics (over 1,250,000 course reviews and ratings), making it, to the best of our knowledge, the *largest publicly accessible MOOC review dataset*. MOOCRev targets the following challenges (1) thorough, *research-based anonymisation of the data to ensure user privacy and data integrity*, (2) *dealing with class (rating) bias*, (3) *addressing language diversity* and *platform-specific rating systems*, (4) *processing emojis and emoticons*, (5) *providing text weights* (6) and *tagging the reviews with the corresponding Parts of Speech* (POS). The repository provides researchers, educators, and platform developers with invaluable opportunities for various Natural Language Processing (NLP) tasks, including learners' sentiments and feedback analysis, aiding future learners in selecting courses aligned with their interests and learning styles based on previous learners' experiences.

Keywords: MOOCs · Course Reviews · Dataset

1 Introduction

MOOCs have redefined the education landscape, offering accessible learning opportunities to a global audience [1, 2]. With the proliferation of MOOCs across various platforms, a substantial amount of user-generated content in the form of reviews and

---

MOOCRev–https://github.com/m-alshehri/MOOCRev.

A. M. Olney et al. (Eds.): AIED 2024 Workshops, CCIS 2151, pp. 124–131, 2024.
https://doi.org/10.1007/978-3-031-64312-5_15

feedback has been amassed [3, 4]. These reviews are a rich data source, reflecting the learners' experiences, preferences, and challenges. However, the potential of this data remains largely untapped, due to the lack of a structured and comprehensive repository, accessible for educational research and analysis. Text analytics, particularly in the education domain, is an essential but complicated task, due to some distinct challenges. This includes (1) the *specific nature* of textual data and the *volume* of information learners generate on online learning platforms [5], (2) annotating learners' expressed opinions, feedback and sentiments in MOOCs, which is known to be a time-consuming and labour-intensive task [6, 7], and (3) the cultural and linguistic variations which can further complicate text annotations and analysis [8, 9]. These challenges underscore the need for a generalisable cross-platform and nonbiased dataset that can be further used to build reliable models in real-world applications. Our ultimate goal is to harness the power of MOOC reviews, by introducing *the largest cross-platform preprocessed MOOC review dataset* to improve online education and make learning more personalised, engaging, and effective for learners.

2   Related Work

Datasets are essential to any data analytics task. Recently, due to the open data movement, supported by major players, such as the US or UK Government, the Open Data Institute[1], the European Data Portal[2] and the World Bank Open Data[3], amongst others, there is an increasing understanding of the importance of data-driven research. Research communities have understood that datasets should not be considered as second-class citizens anymore, and that good datasets, that can lead to excellent research results, are as valuable, if not more, than the research itself, with many top-level conferences offering dataset challenges, and top-level venues sometimes not accepting research without the underlying dataset (or code), or publishing directly datasets. Repositories such as Kaggle[4]; US[5] or UK[6] Government Data; Google Dataset Search[7]; NASA[8]; CERN[9]; British Film Industry[10]; NYCity[11] datasets are just a few examples of the major ones available. However, in the domain of education, and specifically, MOOC related datasets, the landscape is much scarcer. In Table 1 below we list briefly all currently available datasets for MOOCs at the present time, to the best of our knowledge, and compare them briefly with the one we propose, MOOCRev, in terms of dataset size (number of instances), dataset source, number of courses covered.

---

[1] https://theodi.org/.

[2] https://www.europeandataportal.eu.

[3] https://data.worldbank.org/.

[4] https://www.kaggle.com/datasets.

[5] https://data.gov/.

[6] https://www.data.gov.uk/.

[7] https://datasetsearch.research.google.com/.

[8] https://www.earthdata.nasa.gov/.

[9] https://opendata.cern.ch/.

[10] https://www.bfi.org.uk/industry-data-insights.

[11] https://www.nyc.gov/site/tlc/about/tlc-trip-record-data.page.

Table 1. Publicly accessible MOOC datasets versus MOOCRev (CS: Clickstream, Dem: Demographics, TI: Text Inputs, AQE: Assignments/Quizzes/Exams)

Cite	#Instances	Source	#Courses	Type(s)
OULAD [10]	32k	Open University	7	CS; Dem
CCRD	≈458k	Coursera	623	CS; TI
MOOCCube [11]	≈200k	XuetangX	> 700	CS; Dem; AQE
KDDCup [12]	≈120k	XuetangX	39	CS; AQE
MRD	≈40k	iCourse163	39	TI
MOOCRev	**≈ 1.2m**	**Coursera, Udemy, FutureLearn**	**134,200**	**CS; TI**

MRD- https://ieee-dataport.org/documents/mooc-review-data#files.

3   MOOCRev Creation Methodology

3.1   Data Collection

The reviews dataset contains publicly accessible 1,250,830 reviews scraped from 134,200 Coursera, Udemy, and FutureLearn courses, in addition to the learner rating (from 1 to 5). Table 2 shows the distribution of instances per rating, which is highly skewed towards satisfactory ratings of the courses, with reviews with rating 5 marking almost 80% of the collected dataset.

Table 2. Statistics of the original and synthetic instances in the MOOCRev dataset.

Dataset	1	2	3	4	5	Total
Coursera	15,664	13,570	32,073	111,140	595,826	768,273
Udemy	9,172	7,864	22,729	60,042	259,750	359,557
FutureLearn	467	792	3,853	22,088	95,800	123,000
Total	**25,303**	**22,226**	**58,655**	**193,270**	**951,376**	**1,250,830**

3.2   Data Preprocessing

Data preprocessing commenced with automated machine translation of the non-English reviews into English, for language unification and easier processing. Next was data anonymisation since sometimes learners inadvertently reveal information about themselves in their free text (here, the review). Thus, anonymising learners' Personally Identifiable Information (PII) was performed using Presidio[12], a free Microsoft-developed tool for automated text redaction and detailed sensitive data anonymisation [13]. The

---

[12] https://microsoft.github.io/presidio/analyzer/.

last step contained stemming, lemmatisation, removing stop-words, lowering the cases of characters, reforming contractions into the original words, and grammar correction.

Emojis and emoticons tokenisation was performed using UNICODE_EMOJI and EMOTICONS_EMO lexicons which contain 221 emoticons and 3,521 emojis, with their corresponding explanatory words/phrases. Figure 1 depicts an example of converted emojis and emoticons into corresponding explanatory words during the review processing phases.

index	Reviews
0	the course shoud have been better (-_-)
1	babies in mind course is recommended especially for those who arre going to have 😊

index	Reviews
0	the course shoud have been better Shame
1	babies in mind course is recommended especially for those who arre going to have baby

Fig. 1. An example of converting emojis and emoticons into corresponding explanatory words.

Dealing with Bias. The data collected for building the predictive model was obtained from different courses (over 100 thousand courses) from different sources (3 international MOOC platforms), making it not only the largest text dataset collected in the field of MOOC analytics but also the most diverse in terms of course types and disciplines. This would help build a text classifier trained on three of the largest MOOC platforms (Coursera, Udemy, FutureLearn), achieve a higher level of generalisability and reduce the model risk of being overfitting on a single platform.

Having our collected dataset labelled by the learners themselves is another factor for mitigating any potential bias. Human manual labelling for training may introduce, bias because annotators may unintentionally encode their biases. Since learners themselves annotate the data adopted, this typically represents higher accuracy in terms of annotation [14] and promotes the fairness of the dataset.

Weighing Scheme. TF-IDF, the most common weighing scheme for text tokenisation, was used to enrich our collected dataset. Unlike standard vectorisers, TF-IDF assigns a weight to each term in a document based on its frequency in the review and a corpus of documents and generates a matrix of tokenised texts. After that, the vectors that represent the text can be used as input features for prediction.

Part of Speech (POS) Tags. An example of further linguistic features, such as POS tags, is depicted in Fig. 2, which were also measured based on the learners' reviews and included as potential input features (e.g. 'have' is labelled as a 'VBP', i.e. 'verb, singular, present'; and 'interest' as 'noun singular'). This can help (together with other features such as word and character counts) to disclose linguistic patterns associated with learners' reviews and ratings of the courses attended.

For classification, we used ensemble classification models (namely Extremely Randomised Trees (ET), Adaptive Boosting (AdaBoost), Gradient Boosting Machines (GBM), Stochastic Gradient Boosting (XGBoost)). Ensemble models, as confirmed

PRP	VBP	DT	JJ		NN	IN	DT	NN	IN	PRP	VBP	JJ		IN	PRP$	JJ	NN
I	have	a	personal	interest	in	this	course	as	I	'm	pregnant	with	my	first	child

Fig. 2. An illustrative example of POS tagging for a course review.

in previous similar experiments, achieve more generalisable results (being based on several voters (algorithms)) and at the same time address building a reliable model that can reduce variance and avoid overfitting [15–17]. Due to (1) the massive size of MOOCRev dataset and (2) the limited access to computational resources, the best representative 10,000 reviews of each class (rating) of the dataset has been used to build the classification model.

4   Results and Discussion

Our analysis commences with a preliminary visualisation of MOOCRev dataset, to explore students' typing patterns, while reviewing the courses. The word clouds in Fig. 3 present the most frequently occurring trigram phrases from MOOCRev dataset categorised by their associated ratings. The figure is reduced to 3 classes, 1 (left), 3 (middle), and 5 (right), due to paper width restrictions. Each word cloud visualises the trigrams' frequency by varying the size and colour of the text; the larger the text, the more frequently the phrase occurs in the reviews. In the left section, representing a rating of 1, the prominence of phrases such as "worst course ever," "complete waste time," and "waste time money" unsurprisingly suggests strong dissatisfaction among reviewers. These trigrams convey clear negative feedback, indicating that students found the course to be of poor quality, not worth the investment of time and, presumably, money. The middle section, corresponding to a rating of 3, shows a mix of sentiments with trigrams like "course content good," "would recommend course," and "course could better." This suggests that while reviewers found some aspects of the course satisfactory or had a generally neutral opinion, they also saw room for improvement. The right section, associated with the highest rating of 5, displays overwhelmingly positive trigrams such as "really enjoyed course," "highly recommend course," and "best courses." These phrases reflect a strong approval and appreciation for the course, indicating that reviewers found the course to be valuable, enjoyable, and well-structured.

Fig. 3. The most frequent trigram phrases in the reviews with Rating = 1 (left), 3 (middle) and 5 (right).

Table 3 outlines the performance metrics of the ML models used for classifying student reviews, incrementally ordered by the performance of the models as AdaBoost, GBM, XGBoost, and ET. Each model has been evaluated on its ability to correctly classify reviews into one of five ratings. The first five columns from the left represent the *recall* for each class (Rating). F1 (the trade-off between *precision* and *recall*) and *accuracy* (giving the same weighting for all classes by dividing the total number of correctly predicted instances by the total number of instances) are also provided.

Table 3. Model performance results

Model	1	2	3	4	5	F1	Acc
AdaBoost	0.88	0.75	0.85	0.82	0.88	0.83	0.84
GBM	0.88	0.75	**0.89**	0.85	**0.90**	0.85	0.85
XGBoost	0.93	**0.92**	0.81	0.85	0.89	0.88	0.88
ET	**0.94**	0.88	0.84	**0.89**	**0.90**	*0.90*	*0.89*

Regarding the model performance, AdaBoost shows consistent *recall* across most categories but relatively lower in category 2. An F1-score of 0.83 and accuracy of 0.84 suggest good overall performance, although there's room for improvement, especially in categories where precision is lower. GBM demonstrates a slight improvement in categories 3, 4, and 5 compared to AdaBoost. The F1-score and accuracy are marginally higher than AdaBoost, indicating a modest enhancement in the balance between *precision* and *recall*.

Although XGBoost exhibits high *recall* in category 1 and 2 but has a notable drop in category 3. ET demonstrates the highest *recall* across all categories, with notable performance in category 1. With the highest F1-score of all models at 0.90 and an accuracy of 0.89, ET is very competitive with XGBoost, potentially offering a more consistent performance across different classes. The varying performance of the models adopted in this experiment might be due to various factors including, but not limited to, the algorithms' sensitivity to noise and outliers, or the nature of the feature space and how each algorithm constructs the decision boundary. It is also suggested that some categories (Ratings) may be inherently more challenging to predict. This could be due to the nuanced nature of language in different review categories, or a limited representation of features that capture the characteristics of those particular categories.

5 Conclusion

This paper introduces MOOCRev, a large-scale cross-platform MOOC review dataset. MOOCRev's showcases the meticulous attention to data quality, privacy, and representativeness. Our innovative approach to anonymisation protects user privacy without compromising the rich insights contained within over one million reviews. By addressing critical challenges such as rating bias, language diversity, and platform-specific rating systems, MOOCRev sets a new standard for MOOC data repositories. The incorporation of natural language nuances like emojis, emoticons, text weights, and POS tagging provides an unprecedented depth for analysis. The dataset was examined using various ensemble learning methods and yielded high performance accuracies, further underscoring its potential as a versatile tool for stakeholders. For researchers and educators, MOOCRev offers a fertile ground for exploring learner sentiment and feedback, advancing the field of NLP within educational settings. For platform developers, the insights gleaned from this dataset can lead to more sophisticated course recommendation algorithms, enhancing learner engagement and educational outcomes. Ultimately, MOOCRev empowers stakeholders to craft MOOC experiences that are more personalised, engaging, and effective, thereby supporting learners in their quest for knowledge and skills that align with their personal and professional goals.

References

1. Alshehri, M., Alamri, A., Cristea, A.I.: Predicting certification in MOOCs based on students' weekly activities. In: International Conference on Intelligent Tutoring Systems. Springer (2021)
2. Cristea, A.I., et al.: How is Learning Fluctuating? FutureLearn MOOCs Fine-Grained Temporal Analysis and Feedback to Teachers (2018)
3. Alshehri, M., et al.: On the need for fine-grained analysis of Gender versus Commenting Behaviour in MOOCs. In: Proceedings of the 2018 The 3rd International Conference on Information and Education Innovations. ACM (2018)
4. Cristea, A.I., et al.: Earliest predictor of dropout in MOOCs: a longitudinal study of FutureLearn courses (2018)
5. Kastrati, Z., et al.: Sentiment analysis of students' feedback with NLP and deep learning: a systematic mapping study. Appl. Sci. **11**(9), 3986 (2021)
6. Kastrati, Z., Imran, A.S., Kurti, A.: Weakly supervised framework for aspect-based sentiment analysis on students' reviews of MOOCs. IEEE Access **8**, 106799–106810 (2020)
7. Alshehri, M.A., et al.: MOOCSent: a Sentiment Predictor for Massive Open Online Courses. In: Information Systems Development (ISD). Valencia, Spain: Universitat Politècnica de València (2021)
8. Mirza, M., Lukosch, S., Lukosch, H.: Twitter sentiment analysis of cross-cultural perspectives on climate change. In: International Conference on Human-Computer Interaction. Springer (2023)
9. Cristea, A.I., et al.: Can learner characteristics predict their behaviour on MOOCs? In: Proceedings of the 10th International Conference on Education Technology and Computers (2018)
10. Kuzilek, J., Hlosta, M., Zdrahal, Z.: Open university learning analytics dataset. Sci. Data **4**(1), 1–8 (2017)
11. Yu, J., et al.: MOOCCube: a large-scale data repository for NLP applications in MOOCs. In: Proceedings of the 58th Annual Meeting of the Association for Computational Linguistics (2020)
12. Feng, W., Tang, J., Liu, T.X.: Understanding dropouts in MOOCs. In: Proceedings of the AAAI Conference on Artificial Intelligence (2019)
13. Subramani, N., et al.: Detecting personal information in training corpora: an analysis. In: Proceedings of the 3rd Workshop on Trustworthy Natural Language Processing (TrustNLP 2023) (2023)

14. Malko, A., et al.: Demonstrating the reliability of self-annotated emotion data. In: Proceedings of the Seventh Workshop on Computational Linguistics and Clinical Psychology: Improving Access (2021)
15. Asharf, J., et al.: A review of intrusion detection systems using machine and deep learning in internet of things: challenges, solutions and future directions. Electronics **9**(7), 1177 (2020)
16. Alamri, A., et al.: Predicting MOOCs dropout using only two easily obtainable features from the first week's activities. In: International Conference on Intelligent Tutoring Systems. Springer (2019)
17. Alshehri, M., et al.: Towards designing profitable courses: predicting student purchasing behaviour in MOOCs. Int. J. Artif. Intell. Educ. 1–19 (2021)

Handwritten Equation Detection in Disconnected, Low-Cost Mobile Devices

Everton Souza[1], Ermesson L. dos Santos[2], Luiz Rodrigues[2(✉)],
Daniel Rosa[1], Filipe Cordeiro[1], Cicero Pereira[1], Sergio Chevtchenko[3],
Ruan Carvalho[1], Thales Vieira[2], Marcelo Marinho[1], Diego Dermeval[2],
Ig Ibert Bittencourt[2,4], Seiji Isotani[4], and Valmir Macario[1]

[1] Federal Rural University of Pernambuco, Recife, Brazil
valmir.macario@ufrpe.br
[2] Center for Excellence in Social Technologies, Federal University of Alagoas,
Maceio, Brazil
luiz.rodrigues@nees.ufal.br
[3] Western Sydney University, Penrith, Australia
[4] Harvard Graduate School of Education, Cambridge, USA

Abstract. Artificial Intelligence in Education (AIED) implementation in underserved regions faces challenges due to limited digital infrastructure, such as restricted device and internet access. A solution to these challenges lies in AIED Unplugged, a framework designed to address these challenges by tailoring AI solutions to the specific issues prevalent in such regions. AIED Unplugged incorporates principles like Conformity, Disconnect, Proxy, Multi-User, and Unskillfulness, ensuring accessibility by aligning with existing infrastructure, operating offline, simplifying interfaces, and accommodating users' digital skills. Particularly, the framework leverages computer vision to digitalize students' activities and enable AIED-based learning on disconnected, low-cost devices, wherein object detection is crucial to identify which solution areas to digitalize. However, prior research has not assessed the technical feasibility of such applications in the context of AIED unplugged for math education. Therefore, this paper addresses the intersection of "conformity" and "disconnected" principles with an empirical analysis of handwritten equation detection on disconnected, low-cost mobile devices. By optimizing state-of-the-art algorithms for offline inference and considering device constraints, we utilize a dataset of student equations, explore YOLOv8 models, and evaluate its predictive performance. The trained model is converted to Tensorflow Lite for mobile deployment, and a testbed application assesses inference times on diverse low-cost devices, contributing valuable empirical insights to the intersection of AIED Unplugged, Computer Vision, and Education in underserved regions.

Keywords: Computer Vision · Object Detection · YOLO · Unplugged · Mobile

This work was supported by the Brazilian Ministry of Education (MEC), TED 11476.

1   Introduction

Artificial Intelligence in Education (AIED) adoption in underserved regions faces obstacles due to inadequate digital infrastructure, including limited device and internet access, widening the educational divide [3]. AIED Unplugged aims bridge the digital divide by tailoring AI solutions to underserved regions' prevailing challenges [4]. It integrates five key elements: Conformity, Disconnect, Proxy, Multi-User, and Unskillfulness, making AI adaptable to diverse settings. AIED unplugged seeks to ensure accessibility by aligning with existing infrastructure, operating offline, simplifying interfaces, and accommodating users' varying digital skills. This adaptable framework empowers customization for contextual relevance, which is vital for global education enhancement [4].

Most often, approaches towards implementing AIED unplugged suggest students will perform learning tasks on paper sheets, which latter will be digitalized using computer vision techniques so they can be assessed by an intelligent system [12,16]. An example is a case study in which computer vision transcribes handwritten texts so they can be automatically assessed by natural language processing algorithms [4]. Accordingly, a recent review on mathematics intelligent tutoring systems highlighted the importance of computer vision for handling handwritten inputs [13].

Despite their valuable contributions, these approaches overlook a critical aspect of AIED unplugged: the intersection between the *conformity* and *disconnected* principles [4]. While the disconnected principle implies independence from internet availability, conformity emphasizes alignment with existing infrastructure, often limited to low-cost mobile devices. Achieving reliable, offline inferences on such devices requires overcoming resource constraints and optimizing algorithms for reduced computational power without compromising accuracy. However, previous research on computer vision within AIED unplugged has not empirically analyzed it in the context of disconnected, low-cost devices [12,14].

Therefore, this paper presents an empirical analysis of handwritten equation detection in disconnected, low-cost mobile devices. On the one hand, the AIED community has been actively involved in trying to improve math education. Hence, delivering AIED unplugged systems for this domain is of utmost importance. On the other hand, equation detection is a central problem as it is the first step to successfully transcribe handwritten math solutions, so they can be assessed by AIED systems [1,11,16]. Thus, we present an analysis to reveal the technical feasibility of achieving state of the art results, in terms of predictive performance, in equation detection with computer vision models optimized for disconnected, low-cost mobile devices.

2   Related Work

2.1   Object Detection

Object detection is a pivotal computer vision task, addressing the identification of instances of visual objects within digital images. Object detection frameworks are divided into two main categories: (1) two-stage and (2) one-stage object detectors [6]. The two-stage framework divides the object detection process into object localization and classification stages. While offering high detection accuracy, two-stage detectors suffer from relatively slower detection speeds [6]. Conversely, one-stage object detection frameworks concurrently locate and categorize objects using unified deep learning models, offering faster results suitable for real-time applications. Popular one-stage architectures include Single Shot Detection (SSD) [8] and Only Look Once (YOLO) [5], with YOLOv8 [5] currently representing the state-of-the-art for real-time detection.

Recent deep learning techniques led to significant breakthroughs for object detection across various real-world applications, such as autonomous driving [15], medicine [19] and text detection [17]. Object detection for the handwriting localization task is the first stage to achieve handwriting-to-text recognition, used in AIED [18]. Rosa et al. [14] used the object detector YOLOv8 [5] to locate and identify mathematical equations of vertical addition and subtraction. Zhu et al. [22] proposed a three-stage approach for arithmetic exercise correction, incorporating an Feature Pyramid Network (FPN) [7] object detector as the initial stage in an AIED system, but its integration into mobile devices incurs high costs. Despite recent applications of deep learning models in education, there is, to the best of our knowledge, a gap in literature regarding the investigation of handwritten equation detection in disconnected, low-cost mobile devices.

YOLO has also been integrated on mobile devices in practical applications. Yue et al. [21] introduced a preventive embedded system against COVID-19, utilizing a YOLOv4-tiny based system quantified and deployed on Jetson Nano. Morioka et al. [10] proposed a YOLO model-based Android system for ancient text recognition, implemented by communicating with a server equipped to recognize AI models. While these applications showcase the promising use of YOLO for low-cost mobile device development, the technical feasibility of Handwritten Mathematical Equation Detection remains unexplored in the existing literature.

2.2   AIED Unplugged

Recent research has sought to implement AIED unplugged across varied domains. In the writing domain, studies have employed a multi-step process to transcribe handwritten texts, including binarization, segmentation, and text recognition [2,4,12]. Notably, because the application performs this procedure using an external, cloud-based API, it does not comply to the AIED unplugged's principles of conformity and disconnected.

There also are attempts to implement AIED unplugged in the math domain. Authors have proposed a framework for Intelligent Tutoring Systems unplugged, including the need for transcribing math equations from an activities list [16]. However, this study is based on a prototype, not implementing the transcription feature. In [11], the authors use handwriting recognition to digitalize solutions to math tasks. Nevertheless, their study is concerned with equation recognition, not detection, and their application is based on an external API [11].

Differently, other authors used the SSD Mobilenet pre-trained on COCO dataset, a convolutional neural network developed for object detection [1]. They also converted this model, using TensorFlow Lite, and integrated it into a mobile application. Hence, this application complies to the idea of having a disconnected model running on a mobile device. However, despite the overall positive results (accuracy over 95%), a detailed analysis revealed that most of their model's misclassifications were made on handwritten digits. This is a key limitation because the math problems in their dataset are printed, with only students' answers being handwritten [1]. Consequently, most of their application's success does not concern our problem of interest: handwritten equation detection.

Possibly, these findings [1] might be attributed to directly using a pre-trained model. In contrast, recent research has been successful in handwritten equation detection by optimizing state of the art methods, such as YOLO models [9,14, 20]. However, those have not been optimized, deployed and tested in the context of disconnected, low-cost mobile devices. Therefore, there is an empirical gap in terms of the performance of handwritten equation detection models built upon state of the art methods and optimized for the unplugged context. Thus, this paper expands the literature with an empirical analysis that fulfills that gap.

3   Method

The dataset employed in our project originated from manually extracted segments of mathematical questions. Those were provided to us by elementary school teachers, which applied those questions in lessons as part of their pedagogical practices. Each paper sheet contained five questions and responses mainly involved elementary-level mathematical operations, such as addition or subtraction, presented in various formats, including horizontal and vertical equations, as well as isolated numerical answers. Regarding dataset composition, each element consisted of a cropped response zone, capturing both the question text and the student's written response. Annotated bounding boxes provided precise coordinates within these response zones, indicating regions of interest corresponding to what the student wrote on the sheet in response to a particular question. This approach not only allowed us to train models that detect equations, but to teach them about textual and erased content. Hence, an AI system using this model would be able to differentiate true regions of interest (i.e., student equations) from other elements, such as equations written on the question's prompt, among others. In total, the dataset comprises 1063 images, featuring a class distribution of 173 instances of horizontally presented equations, 126 instances of vertically presented equations, and 940 instances of isolated numerical responses. Note that each question allowed for multiple response elements (e.g., a student could an equation, erase it, write another one, and so on), so each image could feature multiple bounding boxes.

As our main goal was to develop an effective model for equation detection, the training process was focused on maximizing the model's ability to correctly identify equations' bounding boxes. Assume an AI system that needs to assess a given student's equation. It might infer the equation's orientation (e.g., vertical or horizontal) from its transcription. However, it can do so without properly detecting its location. Therefore, while knowing an equation's class might be helpful, this research concentrated on the detection task of equation detection as we considered it of higher relevance.

Given that context, the process of model development and validation was based on three steps. **First**, we split our dataset using the hold-out approach to maximize our model's robustness, saving 10% (n = 108) and 20% (n = 211) of our data for validation and testing, respectively. The validation set was used to assess the training process and prevent overfitting. The testing set was only analyzed after finishing the training process as a means to assess the model's generalizability by evaluating it on unseen data. **Second**, to develop our model, we used the Ultralytics' hub, which allows one to explore pre-trained, state-of-the-art models. Here we chose to build our model upon the pre-trained weights of YOLOv8n as it exhibits an interesting balance between speed and accuracy, aligning to the unplugged context. Then, we set the development procedure trained the pre-trained YOLOv8n model using the native tools of the Ultralytics Hub, consisting of 300 epochs, with an early stopping patience of 100 epochs, a batch size of 64, and image dimensions set at 320×320. **Third**, we evaluated our trained model. For this assessment, we considered the precision and recall metrics, utilizing the trained model to extract these insights from the test dataset, which the model had not encountered during its training.

Aligned with our main goal, the study's last step concerned deploying and assessing our trained model's capabilities on disconnected, low-cost mobile devices. For this, the steps were the following. **First**, we converted our trained model so that it was compatible with mobile devices using Tensorflow Lite (TFLite), which is designed to convert models to a format that is compatible with mobile devices through quantization. **Second**, we developed a testbed application to assess the model's performance on mobile-devices, implementing a simple application with the Kotlin programming language as it offers support for TFLite. This testbed application would load our the TFLite version of our trained model, load a set of images from our test set, use the model to predict the images' bounding boxes, and save the time it took to run the predictions. **Third**, we executed our testbed application for 100 trials. Despite YOLOv8 models might be used to make real-time predictions, we chose to run it based on static images to mitigate any operating system processes that could affect inference times. For this, we used two devices of less than 200 USD: a Xiaomi Redmi *Note 10* (2×2.2 GHz Kryo 460 Gold + 6×1.7 GHz Kryo 460 Silver, 128 GB, 4 GB) and a Xiaomi Redmi *Note 8* (Octa-core Max 2.01 GHz, 64 GB, 4 GB).

4   Results and Discussion

This study presents two main findings. First, our results demonstrate that, after training and converting our model for mobile compatibility and offline execution, it yielded a moderate predictive performance, achieving a precision of 62.71% and a recall of 56.74% on unseen data. This finding reveals that our mobile-optimized model achieved an interesting, yet below state-of-the-art performance. Second, we found that devices *Note 10* and *Note 8* took an average of 171 (standard deviation, SD = 31) and 222 (SD = 2) milliseconds to perform inferences for the whole testing set (n = 211). As expected, inferences were faster on the newest of the two devices. Nevertheless, these finding suggest that - even in the worse case analyzed - our mobile-optimized model is likely to able to perform nearly five inferences per second. Hence, our results provide promising evidence on the suitability of deploying and using such a model for real-time inference, as discussed next.

On the one hand, our model's performance is below that of related work. Studies on object detection often report near perfect predictive performance [10,21], including results such as 95% of accuracy [14]. Compared to those, our model's performance was modest (Precision = 0.63; Recall = 0.57). Nevertheless, we limited our model's architecture during training and, then, quantized it using TFLite, which reduces its number of parameters as well as reduces their precision, respectively. Accordingly, one should expected that such a model would yield reduced predictive performance compared to larger, more complex ones.

On the other hand, our model proved to be able to perform real-time inferences in disconnected, mobile devices. Particularly, our results demonstrated it was able to perform predictions within the range of milliseconds, generating multiple bounding boxes within a second while simply relying on the mobile device's resources. In contrast, related work relies on an API, which hosts the model, to perform such inferences [2,11]. This implies that the device must be connected to the internet and, consequently, deal with connection issues.

Given that context, we argue our approach provides a positive trade-off between predictive performance and inference time. While our optimized model is not as accurate as state-of-the-art ones, it is likely to yield way more predictions than more robust, internet-based alternatives. Consequently, the user might benefit from adjusting the camera based on real-time feedback and readily achieve a sweet stop to capture the desired equations. Conversely, an API-based approach would require the user to take the picture, wait for the result (even that for a second), and possibly take another photograph. Alternatively, such an API-based approach could be used to realize a real-time feature, but the waiting time could lead to inconsistent results in case the user slightly moved the camera.

Therefore, this paper contributes empirical evidence on the technical feasibility of performing equation detection in disconnected, low-cost mobile devices. Prior research often relies on large models, demands internet connection, and has not explored equation detection in offline settings [1,2,11]. On the other hand, this paper presented the development, validation, evaluation, and deployment of a model able to perform real-time inferences in disconnected, low-cost mobile devices. Thus, by demonstrating this approach's technical feasibility, this paper addresses a crucial gap in prior research, offering a practical solution with implications for users in underserved regions with limited internet access.

This research holds significant implications for the AIED field. Based on the trade-off between predictive performance and real-time inference, we offer a practical solution for equation detection on disconnected, low-cost mobile devices. This evidence enables crucial advancements on areas such AIED unplugged, which aim to increase access to AIED systems in underserved regions. Thereby, our study has potential societal impact, enabling individuals in regions with limited resources to access and utilize advanced AIED systems. Furthermore, by presenting a feasible and optimized model, this research highlights the feasibility of developing such solutions, which could serve as inspiration for further developments, such as equation assessment and similar applications for other domains.

References

1. Davis, S.R., DeCapito, C., Nelson, E., Sharma, K., Hand, E.M.: Homework helper: providing valuable feedback on math mistakes. In: Bebis, G., et al. (eds.) ISVC 2020. LNCS, vol. 12510, pp. 533–544. Springer, Cham (2020). https://doi.org/10.1007/978-3-030-64559-5_42

2. Freitas, E., et al.: Learning analytics desconectada: Um estudo de caso em análise de produçoes textuais. In: Anais do I Workshop de Aplicações Práticas de Learning Analytics em Instituições de Ensino no Brasil, pp. 40–49. SBC (2022)

3. Gasevic, D., et al.: Learning Analytics for the Global South. Foundation for Information Technology Education and Development, Quezon City, Philippines, published edn. (2018)

4. Isotani, S., Bittencourt, I.I., Challco, G.C., Dermeval, D., Mello, R.F.: AIED unplugged: leapfrogging the digital divide to reach the underserved. In: Wang, N., Rebolledo-Mendez, G., Dimitrova, V., Matsuda, N., Santos, O.C. (eds.) Artificial Intelligence in Education. Posters and Late Breaking Results, Workshops and Tutorials, Industry and Innovation Tracks, Practitioners, Doctoral Consortium and Blue Sky. AIED 2023. CCIS, vol. 1831, pp. 772–779. Springer, Cham (2023). https://doi.org/10.1007/978-3-031-36336-8_118

5. Jocher, G., Chaurasia, A., Qiu, J.: YOLO by Ultralytics, January 2023. https://github.com/ultralytics/ultralytics

6. Kaur, R., Singh, S.: A comprehensive review of object detection with deep learning. Digit. Signal Process. **132**, 103812 (2023)

7. Lin, T.Y., Dollár, P., Girshick, R., He, K., Hariharan, B., Belongie, S.: Feature pyramid networks for object detection. In: Proceedings of the IEEE Conference on Computer Vision and Pattern Recognition, pp. 2117–2125 (2017)

8. Liu, W., et al.: SSD: single shot MultiBox detector. In: Leibe, B., Matas, J., Sebe, N., Welling, M. (eds.) ECCV 2016. LNCS, vol. 9905, pp. 21–37. Springer, Cham (2016). https://doi.org/10.1007/978-3-319-46448-0_2

9. Mahdavi, M., Zanibbi, R., Mouchere, H., Viard-Gaudin, C., Garain, U.: ICDAR 2019 CROHME+ TFD: competition on recognition of handwritten mathematical expressions and typeset formula detection. In: 2019 International Conference on Document Analysis and Recognition (ICDAR), pp. 1533–1538. IEEE (2019)

10. Morioka, T., Aravinda, C., Meng, L.: An AI-based android application for ancient documents text recognition. In: ATAIT, pp. 91–98 (2021)
11. Patel, N., et al.: Equitable access to intelligent tutoring systems through paper-digital integration. In: Crossley, S., Popescu, E. (eds.) Intelligent Tutoring Systems. ITS 2022. LNCS, vol. 13284, pp. 255–263. Springer, Cham (2022). https://doi.org/10.1007/978-3-031-09680-8_24
12. Portela, C., et al.: A case study on AIED unplugged applied to public policy for learning recovery post-pandemic in Brazil. In: Wang, N., Rebolledo-Mendez, G., Dimitrova, V., Matsuda, N., Santos, O.C. (eds.) Artificial Intelligence in Education. Posters and Late Breaking Results, Workshops and Tutorials, Industry and Innovation Tracks, Practitioners, Doctoral Consortium and Blue Sky. AIED 2023. CCIS, vol. 1831, pp. 788–796. Springer, Cham (2023). https://doi.org/10.1007/978-3-031-36336-8_120
13. Rodrigues, L., et al.: Mathematics intelligent tutoring systems with handwritten input: a scoping review. Educ. Inf. Technol. 1–27 (2023)
14. Rosa, D., et al.: Recognizing handwritten mathematical expressions of vertical addition and subtraction. In: 2023 36th SIBGRAPI Conference on Graphics, Patterns and Images (SIBGRAPI), pp. 1–6. IEEE (2023)
15. Vahab, A., Naik, M.S., Raikar, P.G., Prasad, S.: Applications of object detection system. Int. Res. J. Eng. Technol. (IRJET) 6(4), 4186–4192 (2019)
16. Veloso, T.E., et al.: Its unplugged: leapfrogging the digital divide for teaching numeracy skills in underserved populations. In: Towards the Future of AI-augmented Human Tutoring in Math Learning 2023 - Proceedings of the Workshop on International Conference of Artificial Intelligence in Education co-located with the 24th International Conference on Artificial Intelligence in Education (AIED 2023). Springer (2023)
17. Wojna, Z., et al.: Attention-based extraction of structured information from street view imagery. In: 2017 14th IAPR International Conference on Document Analysis and Recognition (ICDAR), vol. 1, pp. 844–850. IEEE (2017)
18. Wu, Y., Hu, Y., Miao, S.: Object detection based handwriting localization. In: Barney Smith, E.H., Pal, U. (eds.) ICDAR 2021. LNCS, vol. 12917, pp. 225–239. Springer, Cham (2021). https://doi.org/10.1007/978-3-030-86159-9_15
19. Xiao, Y., et al.: A review of object detection based on deep learning. Multimed. Tools Appl. **79**, 23729–23791 (2020)
20. Yuan, Y., et al.: Syntax-aware network for handwritten mathematical expression recognition. In: Proceedings of the IEEE/CVF Conference on Computer Vision and Pattern Recognition, pp. 4553–4562 (2022)
21. Yue, X., Li, H., Meng, L.: Ai-based prevention embedded system against COVID-19 in daily life. Procedia Comput. Sci. **202**, 152–157 (2022)
22. Zhu, Q., Luo, Z., Zhu, S., Jing, Q., Xu, Z., Xue, H.: Fate: a three-stage method for arithmetical exercise correction. Neural Comput. Appl. **35**(32), 23491–23506 (2023)

Predicting Academic Performance: A Comprehensive Electrodermal Activity Study

Guilherme Medeiros Machado[(✉)] and Aakash Soni

LyRIDS, ECE Research Center, Paris, France
{gmedeirosmachado,aakash.soni}@ece.fr

Abstract. Electrodermal Activity (EDA) provides insights into the Sympathetic Nervous System (SNS) activation, primarily responsible for "fight-or-flight" response, including stress. Stress, in turn, significantly affects academic performance. Despite EDA's widespread application in various fields, its use in educational research remains underexplored. Existing studies on EDA's relationship with academic outcomes often focus on isolated aspects of the signal, thus narrowing their scope and findings. This paper presents a comprehensive analysis of the entire EDA signal, employing it as input for regression models for predicting students' grades. By augmenting the dataset to compensate for sample size limitations, we achieved low prediction errors (MSE of 0.002 and RMSE of 0.045) and high explanatory power (up to 80% of R^2). Our findings indicate a preference for regression models capable of capturing non-linear relationships with minimal complexity. This research underscores the potential of using EDA as a tool for educators to identify students whose academic performance may be adversely affected by stress, paving the way for targeted interventions.

Keywords: Grades Prediction · Sympathetic Nervous System · Electrodermal Activity · Regression Models · Dataset Augmentation

1 Introduction

Electrodermal activity has been widely used to measure the sympathetic nervous system arousal, owing to the direct link between the skin's conductance and the SNS. The SNS orchestrates the body's involuntary responses, notably initiating the "fight-or-flight" reaction in stressful situations [8]. Stress, a key response governed by the SNS [4] [9], has been identified to adversely affect educational outcomes by diminishing academic achievement and increasing dropout rates [5], impairing student retention and progression [7], and negatively impacting memory retrieval [11].

Given EDA's capability to reflect physiological responses, including stress among various stimuli induced by the SNS, this raises the question: Can we use EDA signals to predict student achievement? Addressing this inquiry necessitates processing the complex EDA signal to decipher its components while acknowledging its sensitivity to a broad spectrum of SNS stimuli.

In our literature review, we identified a limited number of studies employing EDA for predicting student performance. Rafiul et al. [6] achieved up to 80% accuracy in performance prediction using the tonic component of the EDA signal as input for classifiers across three exams among ten students. This methodology, however, has several constraints: it simplifies grade data into binary categories (good or bad), focuses solely on the tonic portion of the EDA signal, and draws on a small sample size. Another study by Gahyun et al. [10] explored EDA's relationship with learning gains, finding inconclusive evidence of EDA arousal correlating with hands-on learning outcomes but suggesting a potential link between EDA arousal during passive learning, engagement, and improved learning gains. This research, though more comprehensive, limited its analysis to the phasic component of the EDA signal. Notably, neither study conducted a thorough examination of features derivable from both EDA signal components nor explored their utility in understanding and predicting student grades.

Analysing both the tonic and phasic components of EDA offers a detailed perspective on an individual's stress response. The tonic component sheds light on general stress levels and long-term stress patterns, identifying characteristics of students under chronic stress. Conversely, the phasic component reflects immediate stress reactions and emotional responses, crucial for detecting short-term stress episodes, such as a student's response to challenging exam questions. Neglecting any part of the EDA signal results in an incomplete analysis of its impact on student achievement. Given the identified research gaps, we propose a novel methodology to utilise EDA signals for predicting student performance, specifically their grades. Utilising the public dataset provided by Rafiul et al. [6], we expanded it to address the issue of insufficient sample size. This study not only focuses on the prediction of student grades but also emphasises careful processing of the EDA signal to extract features from both its tonic and phasic components. Our findings suggest that employing the entire EDA signal enables the generation of very accurate models for grade prediction - with 0.002 of MSE and 0.045 of RMSE in a regression task - and facilitates the differentiation between students achieving good and bad grades, even prior to the actual prediction task.

The remainder of this paper is structured as follows: Sect. 2 introduces the basic principles underlying the EDA signal. Section 3 describes the methodology employed for EDA signal processing and the development of prediction models. Section 4 discusses our findings. Lastly, Sect. 5 summarizes our conclusions and outlines directions for future research.

2   EDA Signal

EDA refers to variation in sweat gland activity, serving as a reflection of an individual's emotional intensity due to its intimate connection with the SNS [1].

EDA signals are characterised by a tonic component (called Skin Conductance Level (SCL)), which changes slowly, and a phasic component (called Skin Conductance Response (SCR)), which varies more rapidly. The phasic component manifests as bi-exponential peaks, reflecting a burst of neural activity to the sweat glands. The frequency, amplitude and duration of SCR peaks serve as markers of sympathetic arousal, with higher rates and amplitudes correlating with increased arousal levels. The tonic component, although influenced by factors like temperature and humidity, also encapsulates overall arousal variations.

3   Methodology

The data utilised in this study were collected by [6], employing the Empatica E4 device to acquire EDA signals. The data were collected from a group of 10 students during three exams (midterm 1 and midterm 2, each lasting 90 min, and a final exam lasting 180 min). The dataset also includes signals for heart rate, blood volume pulse, skin surface temperature, interbeat interval, and accelerometer data. However, for this study, only the EDA was taken into account. One limitation of the EDA signal produced by Empatica is its low collection frequency of 4 Hz and the placement of the sensor on the wrist, whereas the optimal location for collecting EDA is on the palm of the hand. For further details regarding the data acquisition process and experimental design, please refer to [6].

EDA Data Processing and Feature Extraction: In the EDA data analysis, the first step involves the separation of the tonic component and the sequence of neural impulses underlying the phasic component. For that purpose, we used the Biopac Acqknowledge algorithm [2]. It basically involves low-pass and high-pass filtering to isolate the tonic and phasic components, respectively, using a Butterworth filter. The filter cut-off frequencies used were the same as described in [6], ranging between 0.0002 Hz and 0.002 Hz.

For feature extraction, the SCL and SCR components are processed separately. The SCR component is initially used to identify peaks and compute their characteristics, including peak amplitudes and durations. Only the SCR peaks exceeding a predefined threshold (0.03 μS, a commonly used value [2]) are considered significant. Additionally, the number of peaks and area under the curve (AUC) of SCR are computed. These features allow capturing information about instantaneous sympathetic arousal during the exam, such as reactions to specific events when a student recognises a familiar (or unfamiliar) question [3]. The SCL component is used to calculate common tonic features, including mean SCL, representing the average conductance over a specific time period. However, since SCL may differ between participants, a high mean value for a particular participant may not necessarily be high for another participant who typically has a higher SCL. So, we additionally compute the rate of change of SCL. This provides information about how rapidly/slowly sympathetic arousal increases/decreases during the exam. For instance, a well-prepared student may experience a gradual decline in sympathetic arousal upon experiencing subjective ease while looking through the exam questions.

Data Augmentation and Grades Prediction: Upon completing feature extraction, we employed three data windowing strategies to segment and summarize EDA signals into intervals of 30, 60, and 120 s. These strategies facilitate efficient data management and enable the detection of trends and periodicity. To address the limited sample size, we implemented a Conditional Probabilistic Auto-Regressive Neural Network (CPAR) [12], adept at modelling dependencies within multi-sequence data, maintaining consistent context across sequences. This neural network, flexible in adjusting its parameters and loss function to suit the data type, employs a variant of binary cross-entropy for both continuous and discrete data, leveraging Gaussian and binary distributions for probability calculations.

Following training with the PAR algorithm, we synthesised data for 90 new student IDs, including summarised EDA signals within a 120-s framework. This augmented dataset was then analyzed using Supervised Regression algorithms to predict student grades based on EDA features, assessing prediction accuracy through Mean Squared Error (MSE), Root Mean Squared Error (RMSE), R-squared (R^2), and Akaike Information Criterion (AIC).

4   Results and Discussions

For feature analysis, we choose segments from the beginning and end of the exam based on an assumption that sympathetic arousal may be significant at the initial stage of the exam due to the personal sense of discovering the exam questions and at the end of the exam due to the concerns regarding the amount of time left. Moreover, for visualisation purposes, the feature dataset is divided into two parts based on the student grades in respective exams. We arbitrarily choose the first quartile of student grades as the cutoff for data split since it showed significant differences in feature variations. Based on this criteria, we have 7 students for midterm 1 and final exams and 6 students for midterm 2 who are above the cutoff (high grades).

Figure 1 illustrates the global difference in EDA features between students with high grades (HG) and low grades (LG). The bar plots depict the mean feature values for each student group, with error bars representing the 95% confidence interval. A consistent trend is observed in the sum of the number of SCR peaks across all exams. For LG students, it is relatively higher at the beginning of the exam compared to HG students, decreasing significantly towards the end, whereas it shows minimal change for HG students.

The variation (or rate of change) in EDA tonic is higher at the beginning of the exam and decreases towards the end for all students. While the absolute value of EDA tonic for HG students decreases (hence the **-ve variation**) at the beginning and increases towards the end of the exam, there is no specific trend observed for LG students, and the variations are unusual for different exams.

The AUC of EDA phasic exhibits a similar trend to the number of SCR peaks since they are directly related. A higher number of peaks typically leads to a higher AUC of the EDA phasic component. However, the characteristics of these peaks, such as amplitude and duration, may impact the AUC. Therefore, they must be analysed separately.

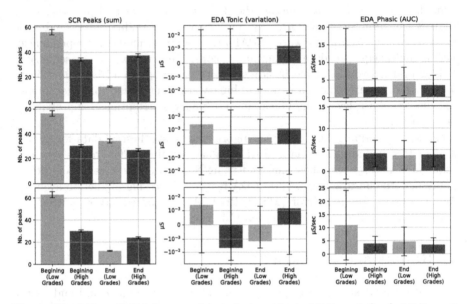

Fig. 1. Variability of EDA features for student grades in Midterm 1 (top), Midterm 2 (middle), and Final (bottom) exams.

Figure 2 compares the trends in amplitude and duration of SCR peaks among HG and LG students. For comparison, SCR peaks are grouped based on their duration (x-axis) with a step of 0.01 sec. The mean amplitude of the corresponding SCR peaks is plotted on the y-axis. For midterms 1 and 2, a consistent trend is observed: for a given duration of SCR peak, the amplitude of SCR peaks for HG students is almost always higher than that of LG students throughout the exams, except when there is no SCR peak at all in HG students, of the given duration. Additionally, SCR peaks with higher absolute amplitude values are observed at the beginning of the exam compared to the end. For the final exam, the trend in relative amplitudes of HG and LG students seems to be reversed; however, the absolute amplitude values are smaller than in the other two exams.

In data augmentation experiments, we employed the CPAR algorithm, which treats student ID as contextual information. This approach not only generates new student IDs but also replicates EDA features within the same timeframe as the original dataset. This setup enabled us to explore three distinct windowing strategies. The effectiveness of our data augmentation was assessed by comparing the synthetic data against actual data, with findings summarised in Table 1. The "KS + TV Complements" column averages the outcomes of two statistical tests: the Kolmogorov-Smirnov (KS) test and the Total Variation Distance (TV), with higher values indicating a closer match between the distributions of the original

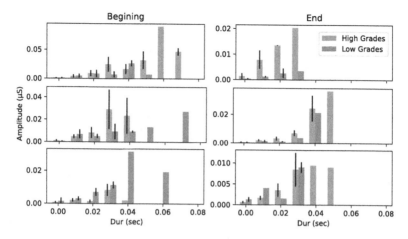

Fig. 2. SCR peak amplitude and duration trends for student grades in Midterm 1 (top), Midterm 2 (middle), and Final (bottom) exams.

and augmented data. The "Column Pairs" column reflects the performance of the algorithm in capturing inter-variable relationships, averaging the differences in Pearson correlations (and using contingency tables for integer data) between original and augmented datasets- again, higher values signify a more accurate representation of the original data. Based on these metrics, with a comprehensive mean presented in the last column, the 120-s window was selected for further experimentation. The improved performance observed with the 120-s windowing strategy may be attributed to its resulting in a smaller dataset size compared to other windowing strategies. This reduced dataset size likely requires less learning effort from neural network, leading to better results.

Table 1. Comparative Analysis on Quality of Data Augmentation

	Wind. Size	KS + TV Complements	Column Pairs	Mean
Midterm 1	**30**	0.77	0.88	0.825
	60	0.76	0.87	0.815
	120	0.82	0.90	0.86
Midterm 2	**30**	0.77	0.88	0.825
	60	0.77	0.86	0.815
	120	0.82	0.87	0.845
Final	**30**	0.65	0.85	0.75
	60	0.70	0.86	0.78
	120	0.72	0.84	0.78

Wind. Size = Window Size (it is given in seconds). KS + TV Complements is the mean of both metrics. Column Pairs is a mean between Pearson Correlations and Contingency Tables results.

Table 2 delineates the performance of various regression models, evaluated based on their prediction accuracy (MSE, RMSE), ability to explain EDA arousal variations (R^2), and model simplicity to preclude overfitting (AIC), aiming for low MSE, RMSE, and AIC values alongside high R^2. After optimising hyperparameters via five-fold cross-validation and grid search, these models were assessed on a test set comprising 20% of the data. The diversity of EDA responses across different exams, attributed to variables such as stress levels, exam difficulty, and student fatigue, underscores the absence of a universally superior model.

However, the Random Forest (RF), Stacking Regressor (SR), and Multi-layer Perceptron (MLP) emerged as the most effective in predicting and accounting for grade variability. Both SR and VR utilised RF, Gradient Boosting (GB), and MLP for their predictions. A notable observation was the generally low R^2 scores but minimal errors for "Midterm 2", indicating these models' predictive capability without adequately explaining grade variance for this specific exam. The findings suggest a preference for nonlinear yet uncomplicated regressors, as indicated by lower AIC scores signifying superior performance. Deep learning methods were also explored but excluded from final analyses due to their inferior performance compared to simpler machine learning approaches. Further details about our results and algorithms implementation are available in this GitHub repository: https://github.com/guimedeiros1/eda_stress.git.

Table 2. Comparative Analysis of Performances Across the Regression Models

		LR	RR	GB	RF	SVR	MLP	VR	SR
Midterm 1	MSE	0.003	0.003	0.002	0.002	0.003	0.002	0.008	0.005
	RMSE	0.055	0.057	0.053	0.045	0.058	0.054	0.093	0.072
	R^2	0.663	0.645	0.690	0.774	0.630	0.682	0.061	0.430
	AIC	−97.5	−96.4	−99.2	−105.5	−95.6	−98.6	−76.9	−86.9
Midterm 2	MSE	0.028	0.028	0.026	0.019	0.029	0.025	0.019	0.018
	RMSE	0.167	0.168	0.161	0.139	0.173	0.158	0.138	0.136
	R^2	−0.086	−0.103	−0.011	0.249	−0.161	0.022	0.259	0.276
	AIC	−53.4	−53.1	−54.9	−60.8	−52.1	−55.6	−61.1	−61.6
Final	MSE	0.004	0.003	0.005	0.004	0.004	0.002	0.006	0.006
	RMSE	0.066	0.054	0.077	0.064	0.069	0.051	0.077	0.081
	R^2	0.674	0.780	0.562	0.695	0.651	0.802	0.559	0.512
	AIC	−90.2	−98.1	−84.3	−91.5	−88.8	−100.2	−84.1	−82.1

LR: Linear Regression/RR: Ridge Regression/GB: Gradient Boosting/RF: Random Forest/SVR: Support Vector Regression/MLP: Multi-layer Perceptron/VR: Voting Regressor/SR: Stacking Regressor/RNN: Recurrent Neural Network/LSTM: Long Short-Term Memory

5 Conclusion

Our findings demonstrate that EDA signal can allow identifying stress-related physiological changes and prediction of exam grades with high accuracy and low error margins. Such insights hold potential for future applications, like creating systems to assist educators in identifying stress-impacted under-performances and enabling tailored interventions. However, these conclusions are derived from a specific demographic and should be approached with caution. Given these promising outcomes with an external dataset. The subsequent stages of this research involve acquiring an extended dataset encompassing students from diverse backgrounds, enrolled in various countries and academic fields. This new dataset holds the potential to be correlated with examination questions, thereby establishing a foundation for investigating potential correlations between exam content and EDA arousal. Such correlations will contribute to a deeper exploration of exam content and its impact on sympathetic nervous system activation. Consequently, the study will be better positioned to discern the intricate manifestations of stress across diverse contexts, fostering a more comprehensive and inclusive analysis of the results, bolstered by enhanced statistical validation.

Acknowledgements. We would like to thank the ECE-Paris and the Omnes Education Group for financing this project and our results.

References

1. Amin, M.R., Faghih, R.T.: Identification of sympathetic nervous system activation from skin conductance: a sparse decomposition approach with physiological priors. IEEE Trans. Biomed. Eng. **68**(5), 1726–1736 (2021)
2. Braithwaite, J.J., Watson, D.G., Jones, R., Rowe, M.: A guide for analysing electrodermal activity (EDA) & skin conductance responses (SCRs) for psychological experiments. Psychophysiology **49**(1) (2013)
3. Gorson, J., Cunningham, K., Worsley, M., O'Rourke, E.: Using electrodermal activity measurements to understand student emotions while programming. In: Proceedings of the 2022 ACM Conference on International Computing Education Research - Volume 1, pp. 105–119. ICER '22, NY, USA (2022)
4. Kyrou, I., Tsigos, C.: Stress hormones: physiological stress and regulation of metabolism. Curr. Opin. Pharmacol. **9**(6), 787–93 (2009)
5. Pascoe, M., Hetrick, S., Parker, A.: The impact of stress on students in secondary school and higher education. Int. J. Adolesc. Youth **25**, 104–112 (2020)
6. Rafiul Amin, M., Wickramasuriya, D.S., Faghih, R.T.: A wearable exam stress dataset for predicting grades using physiological signals. In: 2022 IEEE Healthcare Innovations and Point of Care Technologies (HI-POCT), pp. 30–36 (2022)
7. Robotham, D., Julian, C.: Stress and the higher education student: a critical review of the literature. J. Furth. High. Educ. **30**, 107–117 (2006)
8. Scott-Solomon, E., Boehm, E.D., Kuruvilla, R.: The sympathetic nervous system in development and disease. Nat. Rev. Neurosci. **22**, 685–702 (2021)
9. Smith, S.M., Vale, W.: The role of the hypothalamic-pituitary-adrenal axis in neuroendocrine responses to stress. Dialogues Clin. Neurosci. **8**, 383–395 (2006)

10. Sung, G., Bhinder, H., Feng, T., Schneider, B.: Stressed or engaged? addressing the mixed significance of physiological activity during constructivist learning. Comput. Educ. **199**, 104784 (2023)
11. Vogel, S., Schwabe, L.: Learning and memory under stress: implications for the classroom. NPJ Sci. Learn. **1** (2016)
12. Zhang, K., Patki, N., Veeramachaneni, K.: Sequential models in the synthetic data vault. arXiv preprint arXiv:2207.14406 (2022)

How Can Artificial Intelligence Be Used in Creative Learning?: Cultural-Historical Activity Theory Analysis in Finnish Kindergarten

Jeongki Lim[1,2]([⊠]) [iD], Teemu Leinonen[1] [iD], and Lasse Lipponen[3] [iD]

[1] School of Arts, Design and Architecture, Aalto University, Espoo, Finland
{jeongki.lim,teemu.leinonen}@aalto.fi
[2] Parsons School of Design, The New School, New York, USA
jeongki@newschool.edu
[3] Department of Education, University of Helsinki, Helsinki, Finland
lasse.lipponen@helsinki.fi

Abstract. Recent advances in artificial intelligence (AI) present unprecedented challenges and opportunities in creative education that focus on developing the ability to make something new and useful. In this study, we conducted an exploratory experiment in a Finnish Kindergarten examining the affective relations between AI and children. AI was embodied using a robotic arm and text-to-image model and positioned as a peer to the children in a series of counter-compositional design exercises. The resulting artifacts and observation data were analyzed using the Cultural-Historical Activity Theory, focusing on speaking as an associative action for creative learning. The study's results indicate that AI can be perceived as a member of the community instead of a tool in the learning activity system. The presence of an AI peer tended to foster more speech among the children and teachers, which can lead to creative learning. However, the AI was not as effective as the presence of a human peer due to the contradiction between the actual and perceived objectives of the AI in the learning system. In addition, we identified the importance of the role of the teacher and storytelling in creating the conditions for creative learning. We theorized a human-AI hybrid system for learning where the pedagogy of the teachers and the narrative and empathetic capabilities of the AI play an essential role in the children's creative capacity development.

Keywords: Human-AI Hybrid System for Creative Learning ·
Artificial Intelligence · Cultural-Historical Activity Theory

1 Introduction

Artificial intelligence in education research (AIED) focuses on developing and integrating the techniques from Artificial Intelligence (AI) research in various

A. M. Olney et al. (Eds.): AIED 2024 Workshops, CCIS 2151, pp. 149–156, 2024.
https://doi.org/10.1007/978-3-031-64312-5_18

learning activities [6]. Recent advancements in generative AI can build on existing research domains like intelligent tutoring systems by increasing access to personalized learning [3] and open up new domains by allowing students to create digital artifacts instantly. However, students may become dependent on technology instead of developing their original thinking and skills. We see a promising approach examining human and machine collaboration [11], particularly the relational perception of these technologies may lead to capacity development [8]. In this study, we designed an exploratory experiment in a Finnish Kindergarten. We examined the potential role of AI as a community member in the learning system using the Cultural-Historical Activity Theory (CHAT). To explore the phenomena, we formulated the following three research questions: *R1: Did children perceive AI as a peer in a creative learning activity system?; R2: If the first question is true, did AI, as a community member, make the activity system more conducive to creative learning?; R3: Was AI as effective as a human peer in fostering creative learning?* We hope this research will contribute to the wider AIED research for designing a new human-AI hybrid system for creative learning.

2   Background

While creativity as a multidisciplinary research subject has varying definitions, many share two common characteristics: novelty (new) and value (usefulness) [14]. In education, creativity, as the ability to create new products and services and solve problems, is considered essential in the twenty-first-century knowledge economy [15]. As the latest generative AI models can automate the creation of statistically new outputs, we see the importance of learning activities in the creative capacity development. One of the promising approaches of understanding activities is CHAT [13]. CHAT is considered one of the most effective methods of understanding learning and technology amongst the human-computer interaction research community [7]. Based on the traditions of Russian psychology, the theory places the importance of historical, cultural, and social artifacts and the community in understanding a person or organization by analyzing a situation using following elements in a theoretical system. Human activities are oriented toward objective goals (Object). A Subject is a person or group acting to achieve an Object. The Tools are physical or abstract resources that the Subject uses to achieve their Object. The Subject can work with the Community in a Division of Labor and Rules that exist as explicit or implicit norms. While it shares some characteristics with other theories like distributed cognition, CHAT emphasizes the importance of motive in the subject and considers people and artifacts fundamentally different [12]. CHAT holds that the constant movement between the internal and external is the foundation of a human activity system. This perspective aligns with the notions of creative learning that acknowledge the intra-psychological and inter-psychological activities that can create new and meaningful understandings for oneself and others [1]. All activities, whether internal monologue, verbalization, articulation, making connections, active listening, vocalizing concepts, and communication with peers, contribute to creative capacity development (Fig. 1).

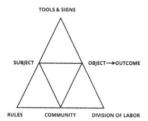

Fig. 1. The cultural-historical diagram is a popular tool for representing the interaction and contradictions amongst the units within single or multiple activity systems [5].

3   Methodological Approach

In this study, we are using select elements of Educational Design Research [10]. Although the experiment was conducted in a kindergarten environment and included a common activity among children, the situation was partly controlled. The aim was to engage in use-inspired research, with elements from basic research: looking for fundamental theoretical understanding and developing theory from an experiment in an authentic environment.

Demography. We conducted the exploratory research experiment over three days at a kindergarten in Vantaa, Finland. There were 14 participants in the study, 12 of whom were children (female = 7), and two were their teachers (female = 2). All the children were six years old. As a kindergarten serving various immigrant communities, we had a diverse group of children participants. All children were familiar with Finnish, while some had other primary languages spoken at home (Arabic, Russian, Ukrainian). Two teachers were Finnish.

Experiment Activities and Data Collection. The experiment consisted of sessions engaging in counter-compositional exercises [2]. The exercise comes from a part of design pedagogy focusing on ontological investigation, where the concepts and forms are explored through variations. First, the child was situated at a table with a marker and whiteboard next to a teacher. The teacher asked the child to draw an object based on their interpretation. (e.g., 'Draw a person.') Afterward, the teacher asked the child to change one part of the drawing. (e.g., 'Now, change one part of your drawing') Once the change was complete, the teacher asked the child to reflect and explain if the drawing was still the same object. (e.g., 'Tell us what you have drawn here. Is this still a person?')

The exercise focused on the compositional aspects of an object, where the children investigated what constitutes the core aspect of a concept, articulated, and exchanged their rationale with their peers and teachers.

This counter-compositional exercise was held in six different groups of children. Each group had six sessions with varying participant compositions to understand the potential of AI as a peer in comparison to humans. The session compositions were: each child alone, two children, each child with AI, and two children with AI. Every session had a teacher present and lead the activity with the children. As we designed the experiment, AI was manifested as the robotics arm [16] executing machine-made images for the children to interact with AI in the context of embodied learning [9]. We wanted the children to engage with AI that is physically present and doing the same activity as their human peers would. We used the lightweight six-axis industrial robot arm with a pen installed at its end-effector, whose desktop size was chosen to not intimidate the children and the range of motion enabled the drawing at scale on par with the children. The robotics arm was operated by the researchers in the room to draw the image developed prior to the sessions. We generated the images using an open-sourced text-to-image generation model, Craiyon, formerly known as DALL · E mini [4]. We chose the model that was significantly smaller than the original DALL · E (0.4 billion parameters vs. 12 billion parameter version of GPT-3), and was trained for only three days to increase the accessibility and potential impact of the study to the educators with limited resources. We prompted the model to draw the same subjects that would be asked of the children (e.g., 'A drawing of a small car.') and a variation of the subject (e.g., 'A drawing of a small car whose tires are made out of long legs.'). We were prescriptive in the prompt for the system to draw visually obvious outputs for the children. In addition, we selected images with simple shapes, considering the optimal conditions for drawing on a physical whiteboard surface with the robotic arm. Between each group, the session prompts were randomly assigned.

There were 36 drawing sessions (six groups of students, each engaged in six sessions). We documented the activities with contemporary note-taking, audio recording, and observational video (a total of 7 h and 37 min), and took photos of the drawings from each session (a total of 132 images).

CHAT Application. We analyzed the research data from a CHAT perspective by identifying the specific actions by the subjects in the system as an unit of analysis. Amongst those, we chose "speaking" by the children and teachers as an associative action for creative learning activities. Verbalizing one's thoughts signals the articulation of ideas and exchange with other community members in the system (inter-psychological learning). When the teacher speaks, the student may actively listen (intra-psychological learning). Using the algorithmic speech detection techniques and reviewing the suggested outputs manually, we quantified the amount of speech in each session. Afterward, we focused on the following four scenarios: how the presence of AI may result in increased speaking in the sessions with 1) a non-talkative child and a teacher, 2) non-talkative children and a teacher, and 3) two talkative children and a teacher. We determined the talkativeness of the students by comparing the amount of time they talked across the sessions. Lastly, we examined 4) how the presence of a human peer is more effective in increasing speaking than AI. The application of the CHAT model in each of the scenarios led to the results that address the research questions.

4   Results and Analysis

R1: Did Children Perceive AI as a Peer in a Creative Learning Activity System?
We determined that the children perceived AI as a peer community member in the activity system based on the following four activity patterns. First, children acted around AI as if they were human peers. They observed its movements, listened to its sounds, commented on its drawings, and asked their teachers about their work. Second, the children took ideas from AI on the assignment. As the teacher gave the same drawing prompts to the children and AI, the children paid attention to how AI engaged with the assignment. In one incident, the student observed the AI's action and borrow the concepts from AI to complete their drawings. Third, the children engaged their emotions with AI. They expressed various emotions in the same room with AI, from boredom to excitement. Lastly, AI's peer-like actions contributed to the inclusive social dynamics of the system. Sitting side by side with AI and watching it draw just like they were asked to do, the children entered into tangential discussions that were not directly related to the assignment. The divergent conversations provided additional context about their ideas and communities. AI's presence as a peer contributes to creating a condition for socializing within the learning system. Based on these activity patterns, we interpreted AI as a community member in the activity system where it functions as a relational artifact. This is a significant as often in the Activity Theory studies, the computer systems are considered as mediating tools. Our

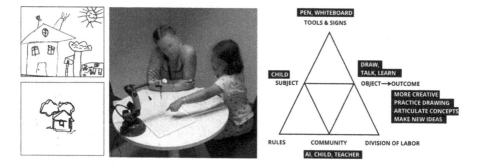

Fig. 2. Drawings and a screenshot from Group 3 Session D (left). The child actively compared her drawing with that of AI. Afterward, she implemented the ideas from the AI into her drawing. This exemplifies one of the activity patterns that led to the application of the CHAT model, where we positioned AI as a community member, unlike the common approach of assigning it as a tool (right).

application of CHAT in the context of AI captured the dynamics where the children interacted with the actions and artifacts of the AI, similar to how they treated their human peers (Fig. 2).

R2: Did AI, as a Community Member, Make the Activity System more Conducive to Creative Learning? When AI was present in the creative learning activity system, we divided 36 sessions into two groups. We compared the sessions with AI to the sessions with the same composition of children and teachers except AI (e.g., participant 1, teacher 1 vs. participant 1, teacher 1, and AI). Out of the 18 sessions, 72% had an increased amount of speaking (13 sessions; mean = 156; standard deviation = 275.74). We found this result consistent with the activity patterns of AI being considered as a peer community member because there is a higher likelihood of the children attempting to communicate with and understand its meaning if there is an additional community member.

R3: Was AI as Effective as a Human Peer in Fostering Creative Learning? We found that AI was less effective than a human peer can be in fostering creative learning. Out of the total 36 sessions, we focused on 12 sessions to compare the session with AI to the sessions with the same child, teacher, and another child (e.g., participant 1, teacher 1, and AI vs. participant 1, teacher participant 2). Out of the 12 sessions, 75% had an increased amount of speaking (9 sessions; mean = 107; standard deviation = 172.67). To better understand AI's ineffectiveness, we analyzed the observation and noticed emerging activity patterns of disconnect and passivity in the system. We found that often, the children were aware of the AI's presence but did not know that they were doing the same task. They are not engaging in substantive ideas and actions that can be exploratory and contribute to collective learning. The disconnect worsened in the groups where none of the children were talkative. The addition of AI did not alter the solitary dynamics of learning in the system. Based on our CHAT analysis, we theorize that the children (Subjects) were experiencing disconnect due to the contradictions between the perceived and actual objective (Contradicting Objects) of the AI (Community Member) in the learning system. The children assumed that AI, as a peer member of the community, might socialize, share stories, and learn with them like their human peers. Nevertheless, AI was executing the commands of translating input images onto a whiteboard. The children could not understand the AI's ideas or stories behind its drawings as they only observed its drawing actions and artifacts. When the child was only to do the exercise, not to socialize with their peers or teacher, then AI functioned like a mechanical spectacle in the background. For creative learning to emerge from the system, there is a need for active inter-psychological learning actions among the community members. AI is an ineffective community member when there is a lack of communication mechanisms. We found that the presence of AI itself does not contribute to creative learning. Communicating ideas and emotionally engaging with peers and teachers is critical for a community member to contribute and achieve its objectives in the learning system. Using this contradiction as a diagnostic tool of dysfunction within the system, we began to theorize how the system can improve its creative learning capacity with AI as a community member.

5 Discussion and Conclusion

Based on the analysis, we see two opportunities to make the creative learning activity system more effective. The first is supporting teachers with the knowledge of AI's capabilities and pedagogical resources. In all the sessions, the teachers were the primary driver of engaging the children to participate in speaking and critically examining the assignment. They directed the students' attention to the AI's drawings and actions even when the AI was not verbally communicative. The teachers were the connectors between the subject and community members. The second is creating the conditions for children to engage in narrative building. We observed that the sessions with exponential growth in speaking had stories with substantive topics, emotions, ideas, and details that the teacher and another community member could build on to develop new ideas. The conditions to enter into storytelling require empathy and supportive relationships. There was a sense of trust amongst the community members, and the children sensed there was permission to go off the tangent and be playful. Incorporating these factors, we theorize a human-AI hybrid activity system for creative learning. The teachers are familiar with the capabilities of AI and direct the children to engage in dialogue and share ideas. The children actively engage in the learning activities, asking questions and building on ideas from their peers. AI can simultaneously communicate its rationale and build on its concepts to the students during the session. It can convey a sense of agency, emotion, and story that engage with the children and teachers to accept them fully as a member of their community in the learning system. In addition, we identified a series of limitations that can be improved in future studies. The study can include more participants to provide a more statistically robust finding. In analyzing speaking, extracting the verbal features like the pause duration, word counts, and semantics while deleting the actual recording will enhance the privacy safeguards of the study especially with the children research participants.

In this study, we examined the role of AI in a Finnish kindergarten from a CHAT perspective. We found that children perceived AI as a peer community member, and their presence led to more speaking, which we considered an associative action for creative learning. Still, the AI peer was less effective than the human peer. This was due to inherent contradictions between the actual and perceived objectives of AI. In response, we theorized that empowering teachers with the pedagogical potentials of AI and developing the computer system with empathetic and narrative-building capabilities. These initiatives can help overcome the challenges of potential dependency on generative tools and contribute to research initiatives that expand a new human-AI hybrid system for creative learning.

Acknowledgements. This work is supported by the LG AI Research, Parsons School of Design and the Office of Research Support at the Provost's Office at The New School.

References

1. Beghetto, R.A.: Creative learning: a fresh look. J. Cogn. Educ. Psychol. **15**(1), 6–23 (2016). num Pages: 18 Place: New York, United States Publisher: Springer Publishing Company
2. Bennett, J., Bühlmann, V., Harman, G., Weizman, I., Witt, A., Sutela, J.: Add metaphysics. Aalto University publication series. Crossover, 1/2013, Aalto University School of Arts, Design and Architecture, Helsinki (2013)
3. Chiu, T.K.F., Xia, Q., Zhou, X., Chai, C.S., Cheng, M.: Systematic literature review on opportunities, challenges, and future research recommendations of artificial intelligence in education. Comput. Educ. Artif. Intell. **4**, 100118 (2023)
4. Dayma, B., et al.: Dall·e mini, July 2021. https://github.com/borisdayma/dalle-mini
5. Engeström, Y.: Activity theory and learning at work. In: Malloch, M., Cairns, L., Evans, K., O'Connor, B.N. (eds.) The SAGE Handbook of Workplace Learning, pp. 86–104. Sage, London (2010)
6. Holmes, W., Tuomi, I.: State of the art and practice in AI in education. Eur. J. Educ. **57**(4), 542–570 (2022)
7. Kaptelinin, V., Nardi, B: Introduction: Activity theory and the changing face of HCI. In: Kaptelinin, V., Nardi, B. (eds.) Activity Theory in HCI. Synthesis Lectures on Human-Centered Informatics, LNCS, pp. 1–13. Springer, Cham (2012). https://doi.org/10.1007/978-3-031-02196-1_1
8. Lim, J., Leinonen, T., Lipponen, L., Lee, H., DeVita, J., Murray, D.: Artificial intelligence as relational artifacts in creative learning. Digit. Creat. **34**(3), 192–210 (2023). publisher: Routledge
9. Macedonia, M.: Embodied learning: why at school the mind needs the body. Front. Psychol. **10** (2019). publisher: Frontiers
10. McKenney, S., Reeves, T.: Conducting Educational Design Research, 2nd edn. Routledge, London (2018)
11. Molenaar, I.: Towards hybrid human-AI learning technologies. Eur. J. Educ. **57**(4), 632–645 (2022)
12. Nardi, B.A.: Studying context: a comparison of activity theory, situated action models, and distributed cognition. In: Context and Consciousness: Activity Theory and Human-Computer Interaction, pp. 69–102. The MIT Press, Cambridge, MA, US (1996)
13. Sannino, A., Ellis, V.: Activity-theoretical and sociocultural approaches to learning and collective creativity: an introduction. In: Learning and Collective Creativity. Routledge (2013). num Pages: 19
14. Sternberg, R.J., Lubart, T.I.: The concept of creativity: prospects and paradigms. In: Handbook of Creativity, pp. 3–15. Cambridge University Press, New York, NY, US (1999)
15. Thornhill-Miller, B., et al.: Creativity, critical thinking, communication, and collaboration: assessment, certification, and promotion of 21st century skills for the future of work and education. J. Intell. **11**(3), 54 (2023). number: 3 Publisher: Multidisciplinary Digital Publishing Institute
16. WLKATA Robotics, W.: Wlkata robotics documents center (2021). https://document.wlkata.com/

Enhancing the Analysis of Interdisciplinary Learning Quality with GPT Models: Fine-Tuning and Knowledge-Empowered Approaches

Tianlong Zhong[✉][ID], Chang Cai[ID], Gaoxia Zhu[ID], and Min Ma[ID]

Nanyang Technological University, 50 Nanyang Ave, Singapore 639798, Singapore
tianlong001@e.ntu.edu.sg

Abstract. Assessing the interdisciplinary learning quality of student learning processes is significant but complex. While some research has experimented with ChatGPT for qualitative analysis of text data through crafting prompts for tasks, the in-depth consideration of task-specific knowledge, like context and rules, is still limited. The study examined whether considering such knowledge can improve ChatGPT's labeling accuracy for interdisciplinary learning quality. The data for this research consists of 252 online posts collected during class discussions. This study utilized prompt engineering, fine-tuning, and knowledge-empowered approaches to evaluate student interdisciplinary learning and compare their accuracy. The results indicated that unmodified GPT-3.5 lacks the capability for analyzing interdisciplinary learning. Fine-tuning significantly improved the models, doubling the accuracy compared to using GPT-3.5 with prompts alone. Knowledge-empowered approaches enhanced both the prompt-based and fine-tuned models, surpassing the researchers' inter-rater reliability in assessing all dimensions of student posts. This study showcased the effectiveness of combining fine-tuning and knowledge-empowered approaches with advanced language models in assessing interdisciplinary learning, indicating the potential of applying this method for qualitative analysis in educational settings.

Keywords: GPT · Prompt engineering · Fine-tuning · Interdisciplinary learning · Qualitative coding

1 Introduction

Interdisciplinary learning is an educational approach that synthesizes perspectives, strategies, and methodologies from multiple disciplines to comprehensively address complex issues beyond the scope of a single field [1]. In interdisciplinary learning, students gain experience tackling complex real-world problems, thereby enhancing their problem-solving skills and employability [2]. One challenge in this research area is evaluating interdisciplinary learning quality based on the

A. M. Olney et al. (Eds.): AIED 2024 Workshops, CCIS 2151, pp. 157–165, 2024.
https://doi.org/10.1007/978-3-031-64312-5_19

learning process data, which involves much qualitative coding of text and artifacts [6]. Qualitative coding refers to the process of making sense of qualitative data such as online discussions, interviews, observations, and essays by tagging them with particular labels related to research aims [4]. Analyzing interdisciplinary learning quality is a qualitative coding task involving coding process data from multiple dimensions such as diversity, cognitive advancement, disciplinary grounding, and integration [15]. However, automated qualitative coding remains a thorny topic. While some studies employ machine learning methods to accomplish this, they still encounter challenges like the need for large datasets with ground truth and issues with low transferability [3].

ChatGPT (Chat Generative Pre-trained Transformer), a chatbot based on GPT foundation models (i.e., GPT-3.5 and GPT-4) developed by OpenAI [10], shows its potential to support qualitative coding. There are two types of coding: inductive coding (labels emerging from context) and deductive coding (labels adapted from predefined frameworks or codebooks) [5]. This study focuses on the latter one, which inherently functions as a classification task, with researchers categorizing text according to predefined criteria in codebooks. Large language models (LLMs) like GPT can classify text into predefined labels [8]. For instance, Xiao and colleagues [14] combined LLMs with an expert-drafted codebook to label children's curiosity-driven questions, which demonstrated fair to substantial agreements with expert-coded results. Qualitative analysis requires a deep understanding of task-specific knowledge (e.g., context and rules)—capabilities unmodified GPT models lack. However, few studies provide a clear definition of knowledge within the context of LLMs. In our research, we use "knowledge-empowered approaches" to describe the integration of external information, such as domain knowledge and codebook rules, into prompts to improve LLMs' performance. Furthermore, while strategies like prompt engineering and fine-tuning are recognized as beneficial for improving LLMs' performance [7,11], their application in the context of qualitative coding with GPT is still rare. This research seeks to experiment with whether incorporating prompt engineering, fine-tuning, and task-specific knowledge into GPT can help automatically analyze students' interdisciplinary learning quality.

We adopted the codebook and employed prompt engineering, fine-tuning, and knowledge-empowered approaches to compare the agreement between GPT models' outputs and ground truth (human labels). The study revealed that generic LLMs like GPT-3.5, trained on broad, non-specialized texts, are not immediately equipped for assessing interdisciplinary learning. For effective evaluation, a combination of fine-tuning and knowledge-empowered approaches is essential. Fine-tuning methods doubled GPT-3.5's performance compared to using prompts alone. Moreover, incorporating knowledge-empowered approaches enabled fine-tuned GPT models to surpass researchers' inter-rater reliability across all dimensions of student posts. The contribution of this research is twofold. First, this study introduces a novel combination of fine-tuning and knowledge-empowered approaches to enhance the capabilities of GPT models in qualitative analysis of interdisciplinary learning—a challenge inadequately tackled by unmodified GPT models. This approach offers a valuable methodology for advancing qualitative analysis. Second, this study provides a practical framework for the automated analysis of interdisciplinary learning quality, which serves as a beneficial resource for both practitioners and researchers in the interdisciplinary learning area.

2   Methods

2.1   The Dataset

This research collected data after getting approval from the institutional review board at the researchers' institutions. The study was conducted in an interdisciplinary digital literacy class at a Southeast Asian university in 2023. The participants were 130 first- and second-year undergraduate students. This study collected learning process data: student posts on the Miro platform, an online cooperation platform (https://miro.com), during weekly learning.

In the end, 252 posts on the Miro platform were included in this study. The posts were written during the activity in the Artificial Intelligence module about preparing a debate on "Artificial Intelligence or the Internet, which has more profound implications for our society?" The posts were about examples, analysis frameworks, and arguments about the debate topic.

The dataset was divided into a training dataset and a testing dataset so that we could do fine-tuning, which is explained in detail in Sect. 2.3. The frequency of each code for each dimension in the training and the testing dataset is displayed in Table 1. The dimensions refer to the elements of interdisciplinary learning quality, which are elaborated in Sect. 2.2.

Table 1. The frequency of each code in each dimension

Dimension	Training data			Test data		
	Level 0	Level 1	Level 2	Level 0	Level 1	Level 2
Diversity	54	89	59	10	31	10
Cognitive advancement	79	63	60	27	15	9
Disciplinary grounding	67	134	1	10	39	2
Integration	163	32	7	41	8	2

2.2   Human-Labeled Interdisciplinary Learning Quality

In this study, we adopted a codebook [15], which consists of four dimensions of interdisciplinary learning: diversity, cognitive advancement, disciplinary grounding, and integration. Based on interdisciplinarity, each dimension is divided into three levels. For instance, level 0 of diversity refers to the text containing no disciplinary perspective, as opposed to level 1 and level 2, in which the text contains one or more disciplinary perspectives respectively. Subsequently, two coders independently labeled students' posts, with each post as an analysis unit. The coders, who have over a year of interdisciplinary learning and research experience, are familiar with qualitative analysis. The inter-rater reliability between human raters on each dimension is evaluated by Cohen's Kappa score (Diversity: 0.68, Cognitive advancement: 0.75, Disciplinary grounding: 0.67, Integration 0.54, overall: 0.75). They then discussed and resolved their disagreements to reach a consensus on every item, which was considered the ground truth for interdisciplinary learning quality.

2.3 Strategies for Enhancing GPT Model Performance

Prompt Engineering

Tailored Prompt. To optimize the use of GPT for the online posts and essay datasets, we systematically crafted the prompts by integrating the latest prompt engineering methods and tailoring them to fit the specific context in the following stages. To begin with, we adopted a codebook drawing upon educational theories [15]. Subsequently, we built a template to convert natural language in the codebook into a format that GPT can interpret. For example, we used a standardized form like the "if... then..." structure to express rules in the codebook. Additionally, system messages and task instructions were incorporated to enhance GPT's comprehension of the task. In the end, each tailored prompt was formed based on the template and consists of the following elements (see Table 2): (1) A system message that refers to a persona description for GPT; (2) A customized task instruction that provides information and requirements about the task; (3) A rule sourced from the codebook containing a set of guidelines that includes concise explanations of various examples that are relevant to different levels of a particular dimension.

Chain-of-Thought Prompting. In this work, we used CoT prompting to direct GPT by providing step-by-step tasks. Drawing from the CoT, the prompt consists of three primary components (see Table 2): First, the task clarification provides information about the tasks, including background, requirements, and expected output, assisting GPT in understanding the core of the task. Second, the main task is split into several smaller ones in the task breakdown section. By combining these elements, we developed a systematic, structured prompting framework that should be able to handle complex or multi-layered questions with acceptable precision and depth in responses.

Fine-Tuning Methods. A GPT model can be fine-tuned by adjusting it to make it more effective for specific purposes [7]. In our study, we fine-tuned the GPT-3.5 baseline model and created four fine-tuned models on four dimensions of the dataset. Then, we applied these fine-tuned models to the validation dataset and calculated Cohen's Kappa scores between GPT labels and ground truth (human labels) on these dimensions to compare the performance.

Table 2. Elements of prompts

Prompt	Element	Example
Tailored Prompt	System messages	"You are an advanced researcher that can precisely follow the user's instructions"
	Customized task instructions	"Please evaluate the cognitive advancement level of students' posts, and then return ONLY numerical values 0, 1 and 2"
	Rules sourced from the Codebook	"Return 1 if the content has explanations, reasons, relationships, or mechanisms mentioned without explanation in detail; or elaborations of terms, phenomena"
Chain-of-Thought Prompting	Task clarification	"Please read the post and check the following levels about the cognitive advancement"
	Task breakdown	"First, please check if the content has basic explanations, reasons, relationships, or mechanisms. If not, please return 0. Second, Return 2 if the content provides extended reasoning with details, mechanisms, and examples. The explanations tend to be longer than average and contain logical words such as "thus", "because", "however", and "but". Return "1" on other conditions"

To deal with imbalanced data (see Table 1), we used ChatGPT to generate additional samples, thereby augmenting the imbalanced dataset. The specific method is to allow ChatGPT to alternatively substitute the original samples with synonyms while maintaining the sentence's structure and meaning. Through oversampling, we ensure that each data category is represented equally in the training dataset.

Knowledge-Empowered Strategies. The knowledge in our study refers to external information that can help GPT better understand task-specific expertise. We incorporated two types of knowledge into the prompt: dictionary-based and rule-based knowledge.

Dictionary-Based Knowledge. In automated labeling tasks, a dictionary refers to a collection of words linked to a specific category [13]. To improve the performance of LLMs like GPT, we associate words with labels in the prompt.

For instance, in coding the diversity dimension, the objective is to identify the range of disciplines represented in student work. However, when students mention words such as "COVID", GPT struggles to categorize them under a specific discipline. Therefore, we integrated dictionary-based knowledge in the following format: "When students discuss topics like WORD (COVID), it indicates coverage of the LABEL (Clinical, Pre-Clinical, and Health)." Similarly, dictionaries were also applied to other dimensions of interdisciplinary learning quality.

Rule-Based Knowledge. While dictionary-based knowledge primarily depends on matching text with dictionary labels, rule-based knowledge draws mechanisms from the rules derived from tasks. In qualitative analysis, codebooks and inherent links between the dimensions can serve as a source for these rules. For instance, if the text does not specifically fall into any discipline (diversity = 0), then the disciplinary grounding should also be 0 (no disciplinary knowledge nor methods). However, we found that the GPT models could not easily find these inherent links in the codebook without explicit prompts. Thus, we added rules to prompts based on knowledge from the codebook in the following format: "IF DIMENSION A (Diversity) is 0, then DIMENSION B (Disciplinary Grounding) is likely to be 0; "IF DIMENSION A (Diversity) is smaller than 2 (no or just one discipline mentioned), then DIMENSION C (Integration) is very likely to be 0 (no connecting nor comparing ideas across disciplines)". This rule-based knowledge, which considers the interactions between different dimensions, has the potential to enhance GPT model performance beyond using text input alone.

3   Results

We conducted experiments with GPT-3.5 to assess the interdisciplinary learning quality of student posts and essays in four conditions: prompts, fine-tuning, knowledge-empowered prompts, knowledge-empowered fine-tuning. In the prompts condition, we directly utilized prompts in the GPT-3.5 model. In the fine-tuning condition, we applied the same prompts but with an additional step of fine-tuning the GPT models. The knowledge-empowered prompts condition integrated knowledge into the prompts without fine-tuning. Lastly, in the knowledge-empowered fine-tuning condition, we employed these knowledge-empowered prompts and also fine-tuned the GPT models. Figure 1 illustrates Cohen's Kappa scores for students' interdisciplinary learning quality regarding posts, reflecting the agreement between GPT-generated labels and the ground truth (human labels) in each condition. The inter-rater reliability among human raters for each dimension serves as the benchmark.

The results indicated that, by using only the prompt, GPT-3.5 shows minimal agreement with human labels, as reflected in Cohen's Kappa scores (all lower than 0.35). After using fine-tuning approaches, Cohen's Kappa scores show a twofold increase (ranging from 0.39 to 0.74), achieving moderate agreement with the ground truth. However, Cohen's Kappa scores on the disciplinary grounding and integration dimensions are still fairly lower than human raters (for disciplinary grounding: 0.61 versus 0.67, and for integration, 0.39 versus 0.54).

Additionally, knowledge-empowered approaches enhance both prompt-based and fine-tuning methods. By incorporating knowledge into prompts, the performance on each dimension increases. Similarly, after combining fine-tuning and knowledge-empowered approaches, GPT models' performance shows a notable increase. The efficacy of integrating fine-tuning with knowledge-enhanced methods has been validated across all dimensions in student posts: diversity (0.80 vs. 0.68), cognitive advancement (0.79 vs. 0.75), disciplinary grounding (0.71 vs. 0.67), and integration (0.70 vs. 0.54). The results suggest that knowledge-empowered fine-tuning approaches are outperforming the proficiency of human judges. We additionally compared the models between imbalanced data and balanced data with oversampling in the knowledge-empowered fine-tuning condition. The result showed that models trained by GPT-augmented balanced data exceed the models trained by imbalanced data on all the four dimensions: diversity (0.80 vs. 0.70), cognitive advancement (0.79 vs. 0.73), disciplinary grounding (0.71 vs. 0.61), and integration (0.70 vs. 0.57).

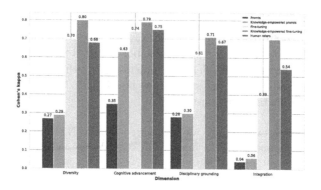

Fig. 1. Cohen's Kappa scores derived from prompt-based, fine-tuned, and knowledge-powered GPT models versus human raters in student posts

4   Discussion and Conclusion

This study developed a method for evaluating interdisciplinary learning with GPT models. Starting with human experts evaluating students' work to establish a benchmark, we then tested the GPT 3.5 model's performance on four conditions: prompts, fine-tuning, knowledge-empowered prompts, and knowledge-empowered fine-tuning. The findings indicate that knowledge-empowered approaches in prompts and fine-tuning can significantly enhance the effectiveness of GPT 3.5 and achieve accuracy surpassing that of human experts.

Our results align with previous research [7] on fine-tuning's benefits for natural language tasks like open-domain question answering and table-to-text generation. Moreover, we found oversampling beneficial for handling imbalanced data, resonating with previous studies [12]. Our findings indicate that knowledge-empowered strategies can improve both prompt-based and fine-tuned GPT models. Echoing previous work on the value of external knowledge for prompt engineering in automatic essay analysis [13], our research shows that even small-scale domain-specific dictionaries or rules can significantly improve LLM effectiveness. This contrasts with studies relying on large knowledge graphs to solve knowledge-driven problems [9]. Furthermore, our findings highlight the importance of incorporating knowledge in the fine-tuning process, which hasn't received much attention.

This study still has some limitations. First, the results of this study are based on a limited number of online notes generated by a particular cohort of undergraduate students. Whether the methods can be generalized to other datasets needs further research. Secondly, this study focuses solely on measuring agreement levels; a more thorough error analysis of disagreement data points is essential for a deeper comprehension of the model's capabilities. Third, our study exclusively evaluated the GPT-3.5 model, leaving out numerous other LLMs like LLaMA and Gemini from future investigation.

Our research offers a practical framework for the automated analysis of interdisciplinary learning quality, paving the way for large-scale studies that were previously unfeasible due to the intensive labor required for qualitative analysis. The precision approach is beneficial for understanding the complex dynamics of interdisciplinary education and sets a new benchmark for the use of artificial intelligence in educational research.

Acknowledgement. This study was supported by the NTU Edex Teaching and Learning Grants (Grant No. NTU EdeX 1/22 ZG). We are indebted to the students and instructor who participated in this study.

References

1. Boix-Mansilla, V.: Learning to Synthesize: The Development of Interdisciplinary Understanding, pp. 288–306. Oxford University Press, Oxford (2010)
2. Brassler, M., Dettmers, J.: How to enhance interdisciplinary competence-interdisciplinary problem-based learning versus interdisciplinary project-based learning. Interdiscip. J. Probl.-Based Learn. 11(22) (2017)
3. Chejara, P., et al.: EFAR-MMLA: an evaluation framework to assess and report generalizability of machine learning models in MMLA. Sensors 21(8), 2863 (2021)
4. Elliott, V.: Thinking about the coding process in qualitative data analysis. Qual. Rep. 23(11) (2018)
5. Fereday, J., Muir-Cochrane, E.: Demonstrating rigor using thematic analysis: a hybrid approach of inductive and deductive coding and theme development. Int. J. Qual. Methods 5(1), 80–92 (2006)
6. Gvili, I.E.F., et al.: Development of scoring rubric for evaluating integrated understanding in an undergraduate biologically-inspired design course. Int. J. Eng. Educ. (2016)
7. Liu, J., et al.: What makes good in-context examples for gpt-3? arXiv preprint arXiv:2101.06804 (2021)

8. Liu, P., et al.: Pre-train, prompt, and predict: a systematic survey of prompting methods in natural language processing. ACM Comput. Surv. **55**(9), 195:1–195:35 (2023)
9. Liu, W., et al.: K-BERT: enabling language representation with knowledge graph. In: Proceedings of the AAAI Conference on Artificial Intelligence, vol. 34, no. 03, pp. 2901–2908 (2020)
10. OpenAI: Openai. https://openai.com/
11. Reynolds, L., McDonell, K.: Prompt programming for large language models: beyond the few-shot paradigm. In: Extended Abstracts of the 2021 CHI Conference on Human Factors in Computing Systems, pp. 1–7 (2021)
12. Shelke, M.S., et al.: A review on imbalanced data handling using undersampling and oversampling technique. Int. J. Recent Trends Eng. Res. **3**(4), 444–449 (2017)
13. Ullmann, T.: Automated Analysis of reflection in writing: validating Machine learning approaches. Int. J. Artif. Intell. Educ. **29**(2), 217–257 (2019)
14. Xiao, Z., et al.: Supporting qualitative analysis with large language models: combining codebook with GPT-3 for deductive coding. In: Companion Proceedings of the 28th International Conference on Intelligent User Interfaces, pp. 75–78 (2023)
15. Zhong, T., et al.: The influences of chatgpt on undergraduate students' perceived and demonstrated interdisciplinary learning. OSF preprint (2023)

Knowledge Distillation of LLMs for Automatic Scoring of Science Assessments

Ehsan Latif[1,2], Luyang Fang[1,3], Ping Ma[3(✉)], and Xiaoming Zhai[1,2(✉)]

[1] AI4STEM Education Center, Athens, GA, USA
[2] Department of Mathematics, Science, and Social Studies Education, University of Georgia, Athens, GA, USA
[3] Department of Statistics, University of Georgia, Athens, GA, USA
{pingma,xiaoming.zhai}@uga.edu

Abstract. This study proposes a method for knowledge distillation (KD) of fine-tuned Large Language Models (LLMs) into smaller, more efficient, and accurate neural networks. We specifically target the challenge of deploying these models on resource-constrained devices. Our methodology involves training the smaller student model (Neural Network) using the prediction probabilities (as soft labels) of the LLM, which serves as a teacher model. This is achieved through a specialized loss function tailored to learn from the LLM's output probabilities, ensuring that the student model closely mimics the teacher's performance. To validate the performance of the KD approach, we utilized a large dataset, 7T, containing 6,684 student-written responses to science questions and three mathematical reasoning datasets with student-written responses graded by human experts. We compared accuracy with state-of-the-art (SOTA) distilled models, TinyBERT, and artificial neural network (ANN) models. Results have shown that the KD approach has 3% and 2% higher scoring accuracy than ANN and TinyBERT, respectively, and comparable accuracy to the teacher model. Furthermore, the student model size is 0.03M, 4,000 times smaller in parameters and x10 faster in inferencing than the teacher model and TinyBERT, respectively. The significance of this research lies in its potential to make advanced AI technologies accessible in typical educational settings, particularly for automatic scoring.

Keywords: large language model (LLM) · BERT · knowledge distillation · automatic scoring · education technology

1 Introduction

Artificial Intelligence (AI) in education has evolved from a theoretical concept to a practical tool, significantly impacting classroom assessment practices and adaptive learning systems [4,6]. AI for personalized learning and assessment provides opportunities for more tailored and effective educational experiences [13]. Integrating Large Language Models (LLMs) from domains on AI like BERT [1] into education has been a significant milestone in enhancing learning experiences,

A. M. Olney et al. (Eds.): AIED 2024 Workshops, CCIS 2151, pp. 166–174, 2024.
https://doi.org/10.1007/978-3-031-64312-5_20

providing personalized learning content and support, and facilitating automatic scoring [9–11,14]. Despite their potential, the deployment of LLMs in educational settings is constrained by their considerable size (714MB for 178 million parameters and 495MB for 124 million parameters) and computational requirements (16 Tensor Processing Units), presenting a challenge for widespread adoption in resource-constrained educational environments such as mobiles/tablets and school-provided laptops with no GPUs or TPUs and limited memory [5].

To bridge this gap, our study explores the feasibility of distilling the knowledge of LLMs into smaller neural networks, referred to as *student models*, with fewer parameters and hidden layers. By training a smaller student model using soft labels provided by a fine-tuned LLM (i.e., *teacher model*), we aim to achieve a similar scoring performance to that of LLMs, but with reduced model size.

The significance of this research lies in its potential to make advanced AI technologies accessible in typical educational settings. The study addresses the technical challenges of deploying AI models in resource-constrained environments and highlights the potential of AI to transform educational assessment practices. By enabling the deployment of efficient automatic scoring systems on less powerful hardware available in school settings, we contribute to the democratization of AI in education. The key contributions of this paper are:

- We demonstrate the successful application of a novel knowledge distillation (KD) strategy that, while inspired by [5], is uniquely adapted and optimized for the context of educational content.
- Our approach achieves a significant reduction in model size and computational requirements without compromising accuracy. The student model, distilled from a fine-tuned BERT teacher model, exhibits a model size that is 4,000 times smaller and demonstrates an inference speed that is ten times faster than that of its teacher counterpart.
- Through comprehensive evaluations using a large dataset of 10k student-written responses to science questions, our work not only validates the effectiveness of our KD method against state-of-the-art models like TinyBERT [7] and generic ANN models [3], but also highlights its superior performance.

2   Proposed Knowledge Distillation

KD is a technique to transfer knowledge from a trained large model (teacher) to a more compact and deployable model (student). We take inspiration from the prominent KD approach, introduced by [5], which involves using the class probabilities generated by the pre-trained large model as *soft labels* for training the smaller model, effectively transferring its predictive and generalization capabilities. Building on this concept, we develop a method for applying KD in the context of automated scoring systems, aiming to improve the process of evaluating educational content using AI.

Specifically, for each data point $\mathbf{x}_i$ in the training sample $\mathcal{D}$, the teacher model predicts the class probability $\boldsymbol{p}_i = (p_{i1}, \ldots, p_{iK})^T$, where p_{ij} represents the predicted probability that the ith data point belongs to class j. The student model is trained using both the training sample $\mathcal{D}$ and the corresponding *soft labels* $\boldsymbol{p} = \{\boldsymbol{p}_i\}_{i=1}^{N}$ produced by the teacher model. We represent the student model by a neural network $f(\cdot, \boldsymbol{\theta})$. The discrepancy between the student and teacher models is measured as

$$\tilde{\mathcal{L}}(f(\cdot, \boldsymbol{\theta}); \mathcal{D}, \boldsymbol{p}) = \frac{1}{N} \sum_{i=1}^{N} \mathrm{CE}(\boldsymbol{p}_i, f(\mathbf{x}_i, \boldsymbol{\theta})),$$

$$= -\frac{1}{N} \sum_{i=1}^{N} \sum_{k=1}^{K} p_{ik} \log \left(f_k(\mathbf{x}_i; \boldsymbol{\theta}) \right), \tag{1}$$

which is the sample mean of the cross-entropy $\mathrm{CE}(\boldsymbol{p}_i, f(\mathbf{x}_i, \boldsymbol{\theta}))$ across i. To leverage the information from both the training data and the teacher model's predictions, KD aims to solve

$$\boldsymbol{\theta}_{\mathrm{KD}}^{*} = \arg \min_{\boldsymbol{\theta} \in \mathbb{R}^{d}} \left\{ \mathcal{L}^{\mathrm{KD}}(f(\cdot, \boldsymbol{\theta}); \mathcal{D}, \boldsymbol{p}, \lambda) \right\}$$

$$= \arg \min_{\boldsymbol{\theta} \in \mathbb{R}^{d}} \left\{ \mathcal{L}(f(\cdot, \boldsymbol{\theta}); \mathcal{D}) + \lambda \tilde{\mathcal{L}}(f(\cdot, \boldsymbol{\theta}); \mathcal{D}, \boldsymbol{p}) \right\}, \tag{2}$$

where the minimized KD loss $\mathcal{L}^{\mathrm{KD}}(f(\cdot, \boldsymbol{\theta}); \mathcal{D}, \boldsymbol{p}, \lambda)$ is the linear combination of two loss terms in KD loss and (1). The first term of the KD loss equation measures the discrepancy between the predictions of the student model and the actual labels. The second term assesses the prediction discrepancy between the student and teacher models. In this context, λ serves as a constant that balances the impact of these two aspects of the loss. Setting $\lambda = 0$ reduces the KD loss to the conventional empirical risk loss.

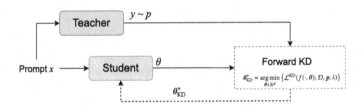

Fig. 1. The architecture of the proposed KD approach uses prediction probabilities as soft labels from the teacher model and forces the student model to achieve these prediction probabilities through the fitting loss function.

KD enables the student model to attain performance comparable to the teacher model while considerably reducing the computational resources required for training. The teacher model's predicted probability outputs $\boldsymbol{p}$ provide valuable insights into its data interpretation. By minimizing the discrepancy between the outputs of the student and teacher models, the student model can effectively adopt the knowledge and insights of the teacher model. Consequently, despite its simpler architecture and reduced computational resources, the student model can match the performance of the more complex teacher model.

In Fig. 1, we present the architecture of the proposed KD method. With a well-performing, fine-tuned, large teacher model and a new dataset, we run the teacher model on the dataset and extract knowledge to guide the training of a more compact student model. In this study, we extract the class probabilities predicted by the teacher model as the knowledge to be transferred to the student model. sing both the knowledge from the teacher model and the data from the dataset, we train the student model based on optimization, as outlined in Eq. (2).

3 Experimental Setup

Our study investigates whether a significantly smaller neural network can effectively mimic the capabilities of a fine-tuned LLM through the proposed KD strategy. Additionally, the study explores how this approach can enhance model performance. We apply our proposed methodology across diverse datasets to train a compact model to achieve this goal. This model is then compared with the SOTA TinyBERT [7] and a trained smaller ANN [3] to evaluate the performance in terms of accuracy and efficiency. TinyBERT stands out due to its specific design to compress the size and computational demands of BERT through a sophisticated distillation process, making it an ideal benchmark for assessing the efficiency and effectiveness of our distillation approach. On the other hand, ANN represents a broader category of neural networks that, while not specialized in natural language processing tasks to the same extent as TinyBERT, provides a contrasting baseline to evaluate the general applicability and performance of our distilled model in automatic scoring. The comparison against these models allows us to validate the superiority of our approach not only against a state-of-the-art distilled model like TinyBERT, which is directly relevant to our domain but also against more generic neural network architectures.

3.1 Data Collection and Preprocessing

The study utilized a meticulously categorized dataset of student-written responses to a science question and three mathematical assessment items, each falling under the multi-class category for automatic scoring. Each student response in datasets is graded by a human expert for automatic scoring, and human scores are used for validation. On average, each student's written textual response contains 15 words. The detailed composition of each assessment item's dataset is presented in Table 1.

Table 1. Sample size and the teacher and student model used for each dataset. The number of parameters for each model is shown in parentheses.

Dataset	Sample size	Classes	Teacher	Student
7T	6,684	10	SciEdBERT (114M)	E-LSTM (0.03M)
Bathtub	1,145	5	BERT_base (110M)	E-LSTM (0.03M)
Falling Weights	1,148	4	BERT_base (110M)	E-LSTM (0.03M)
Gelatin	1,142	5	BERT_base (110M)	E-LSTM (0.03M)

Table 2. Accuracy and F-1 score performance comparison of teacher, TinyBERT [7], ANN [3], and KD model for benchmark datasets. The mean accuracies and standard deviations are displayed.

	Accuracy			
	Teacher	TinyBERT	ANN	KD
7T	0.891±0.016	0.752±0.003	0.716±0.002	0.757±0.001*
Bathtub	0.938±0.014	0.833±0.019	0.831±0.021	0.852±0.012*
Falling Weights	0.904±0.013	0.856±0.015	0.865±0.014	0.888±0.008*
Gelatin	0.871±0.018	0.735±0.010	0.739±0.011	0.780±0.014*
	F-1 Score			
7T	0.842±0.017	0.749±0.009	0.706±0.001	0.751±0.005*
Bathtub	0.914±0.021	0.832±0.069	0.830±0.024	0.851±0.011*
Falling Weights	0.893±0.018	0.855±0.015	0.864±0.014	0.886±0.009*
Gelatin	0.804±0.016	0.731±0.008	0.733±0.014	0.766±0.017*

* KD has shown higher accuracy and F-1 score than TinyBERT and ANN, and is comparable to the Teacher model for each dataset.

3.2 Dataset

The 7T dataset is a large dataset consisting of seven tasks from the SR1 dataset, including short constructed student responses and human-expert graded scores. Overall, the 7T dataset consists of 6,684 labeled student responses from [12], similar to the dataset used for SciEdBERT by Liu et al. [9]. We utilized three multi-class assessment tasks from the Mathematical Thinking in Science (MTS) project [2] responded to by high-school students: Bathtub, Falling Weights, and Gelatin containing 1,145, 1,148, and 1,142 student-written responses respectively. Each dataset contains a different number of classes (scores assigned by human experts) for student-written responses. More specific details about scoring and assessment items can be found in [8]. This comprehensive dataset facilitated a nuanced analysis of the capacity of compact scoring models for student-written responses, ensuring robust and broadly applicable study findings. We processed each dataset by excluding empty responses and ensuring text-formatted student responses and ranged labels.

3.3   Training Scheme

Model Setup. This study uses SciEdBERT [9] with 114M parameters as a specialized Science Education BERT model, and the standard BERT base model [1] contains 110M parameters as the teacher model. These models have been shown to perform brilliantly in processing textual data. For performance comparison, we used TinyBERT [7] with 67M parameters and small ANN [3]. For the KD method, we constructed a compact neural network with an embedding layer with an output dimension of 32 and a bidirectional LSTM layer with 16 units (significantly fewer parameters than transformers), followed by a GlobalMax-Pooling1D layer. Further, it includes two dense layers, with the first having 16 neurons and 'relu' activation, and the final layer is equipped with a softmax activation for multi-class classification. Additionally, dropout layers are integrated for regularization, and the model is optimized with Adam.

Evaluation and Validation. We partition each dataset into training, validation, and testing sets in a $7 : 1 : 2$ ratio. The model optimization employs cross-entropy loss, and to prevent overfitting, an early stopping callback that monitors the validation loss is utilized. We present the prediction accuracy on the test set to assess the model's performance.

The summary of the dataset and the teacher and student (KD) models used for each dataset is detailed in Table 1. We provide the number of parameters for each model in parentheses. The student model is much smaller than the teacher model.

3.4   Results

The comparative analysis of model accuracy across four datasets is presented in Table 2. Results reveal the efficacy of KD in enhancing the performance of a student model as compared to the SOTA TinyBERT [7] and ANN [3] for text classification, in terms of both accuracy and F-1 score. Furthermore, it also provides close accuracy and F-1 score as the complex teacher model. The Falling Weights dataset serves as a typical example, with KD providing performance comparable to the teacher model, suggesting that even models with much smaller sizes can achieve similar performance to the large teacher model. We observed that KD outperforms TinyBERT and ANN in accuracy by 2.5% and 3.2%, respectively, and in F-1 score by 2.2% and 3.0%, respectively. This observation highlights the superiority of KD over SOTA model distillation approaches. Considering both accuracy, F-1 score (shown in Table 2), and model size (shown in Table 1), results highlight the practicality and applicability of the KD approach for automatic scoring on resource constrained-devices.

Despite the success of KD, it is essential to recognize that the student models, although improved, usually do not reach the performance benchmarks set by the teacher models. This is notably apparent in the 7T dataset; the integration of KD leads to better performance compared to the ANN and TinyBERT but still does not match the teacher models' accuracy. Such a discrepancy can be attributed to the inherent limitations of the student models, which possess simpler architectures and are trained on smaller datasets with far fewer training parameters.

The results demonstrate that the KD strategy is a powerful tool in model training, beneficial for applications such as automatic scoring. By effectively condensing the knowledge of a large, pre-trained model into a more compact one, KD not only improves performance but also facilitates the deployment of such models in resource-constrained environments.

3.5 Sensitivity Analysis

We investigated the impact of the hyperparameter λ in Eq. (2) on scoring accuracy to learn more about the resilience of the KD approach. We assessed the KD approach using a step size of 0.02 and a range of λ values from 0.08 to 0.02 for the Bathtub dataset. Although there were little variations in the accuracy of the KD technique with the modification of λ, consistently outperformed the baseline models. These findings suggest that the KD approach is comparatively resistant to the selection of $\lambda = 0.2$, demonstrating consistent performance across a range of hyperparameter values.

4 Discussion

The results of this study highlight the revolutionary possibilities of KD in educational technology, especially in light of the limitations of standard school computing resources. The use of KD in education represents a substantial breakthrough, particularly in automated grading systems. Nevertheless, like any emerging technology, it is important to recognize its limitations as well as its potential for development in the future. In the traditional education system, automatic evaluation is yet a point of discussion [10]. Therefore, our proposed solution is a supplementary tool designed to support and not replace traditional assessment methods established in the education system. Further studies and educational policy adaptations are necessary to fully integrate such technologies into formal school environments.

The most noteworthy application of KD in education is the creation of accurate and efficient automatic scoring systems. A major challenge in many educational contexts is that traditional scoring systems can demand extensive processing resources to function successfully on school-setting devices such as entry-level laptops and tablets. This problem is addressed by KD, which enables the creation of "student models" with significantly lower processing requirements while preserving much of the accuracy and efficiency of larger "teacher models.

Furthermore, KD models are ideally suited for integration into tablet- and smartphone-based learning apps due to their smaller size and reduced processing requirements. The capacity to run complex AI models on these devices, which are increasingly prevalent in educational contexts, creates new opportunities for interactive and adaptable learning experiences.

5 Conclusion

This study effectively illustrates how KD can be used to optimize LLMs for usage in educational technology, especially on low-processor devices. We maintain great accuracy with a much smaller model size (0.03M parameters) and processing requirements by condensing the knowledge of LLMs into smaller neural networks. The distilled models perform better than SOTA TinyBERT and ANN models on various datasets, demonstrating the efficacy of this approach even though their parameter sizes are up to 100 times less than teacher models. This work has important applications since it provides a method to incorporate cutting-edge AI tools into conventional school environments, which frequently have hardware constraints. The learning process and accessibility of personalized education technology can be significantly improved by the capacity to implement effective and precise AI models for uses such as autonomous scoring. Essentially, this work establishes the foundation for future developments in the field and validates the viability of KD in educational contexts, underscoring the significance of ongoing research and innovation in AI for education. In the future, we will work on processing soft-labels and prompt processing to avoid amplification of faults of teacher models by employing more sophisticated techniques.

Acknowledgment. This work was funded by the National Science Foundation(NSF) (Award Nos. 2101104, 2138854) and partially supported by the NSF under grants DMS-1903226, DMS-1925066, DMS-2124493, DMS-2311297, DMS-2319279, DMS-2318809, and by the U.S. National Institutes of Health under grant R01GM152814.

References

1. Devlin, J., Chang, M.W., Lee, K., Toutanova, K.: Bert: pre-training of deep bidirectional transformers for language understanding. arXiv preprint arXiv:1810.04805 (2018)
2. ETS-MTS, T.: Learning progression-based and ngss-aligned formative assessment for using mathematical thinking in science, November 2023. http://ets-cls.org/mts/index.php/assessment/
3. Ghiassi, M., Olschimke, M., Moon, B., Arnaudo, P.: Automated text classification using a dynamic artificial neural network model. Expert Syst. Appl. **39**(12), 10967–10976 (2012)
4. González-Calatayud, V., Prendes-Espinosa, P., Roig-Vila, R.: Artificial intelligence for student assessment: a systematic review. Appl. Sci. **11**(12), 5467 (2021)
5. Hinton, G., Vinyals, O., Dean, J., et al.: Distilling the knowledge in a neural network. arXiv preprint arXiv:1503.02531 (2015)
6. Holmes, W., Tuomi, I.: State of the art and practice in AI in education. Eur. J. Educ. **57**(4), 542–570 (2022)
7. Jiao, X., et al.: Tinybert: Distilling bert for natural language understanding. arXiv preprint arXiv:1909.10351 (2019)
8. Latif, E., Zhai, X.: Fine-tuning chatgpt for automatic scoring. Comput. Educ. Artif. Intell. 100210 (2024)
9. Liu, Z., He, X., Liu, L., Liu, T., Zhai, X.: Context matters: a strategy to pre-train language model for science education. arXiv preprint arXiv:2301.12031 (2023)

10. Selwyn, N.: Should Robots Replace Teachers?: AI and The Future of Education. John Wiley & Sons, Hoboken (2019)
11. Zhai, X.: Chatgpt user experience: Implications for education (2022). SSRN 4312418
12. Zhai, X., He, P., Krajcik, J.: Applying machine learning to automatically assess scientific models. J. Res. Sci. Teach. **59**(10), 1765–1794 (2022)
13. Zhai, X., Yin, Y., Pellegrino, J.W., Haudek, K.C., Shi, L.: Applying machine learning in science assessment: a systematic review. Stud. Sci. Educ. **56**(1), 111–151 (2020)
14. Guo, S., Zheng, Y., Zhai, X.: Artificial intelligence in education research during 2013–2023: a review based on bibliometric analysis. Educ. Inf. Technol. 1–23 (2024)

Exploring the Usefulness of Open and Proprietary LLMs in Argumentative Writing Support

Reto Gubelmann[1]([✉]), Michael Burkhard[2], Rositsa V. Ivanova[1],
Christina Niklaus[1], Bernhard Bermeitinger[1], and Siegfried Handschuh[1]

[1] ICS-HSG, University of St Gallen, St Gallen, Switzerland
{reto.gubelmann,rositsa.ivanova,christina.niklaus,bernhard.bermeitinger,
siegfried.handschuh}@unisg.ch
[2] IBB-HSG, University of St Gallen, St Gallen, Switzerland
michael.burkhard@unisg.ch

Abstract. In this article, we present the results of an exploratory study conducted with our self-developed tool **Artist**. The goal of the tool is to give formative feedback to develop students' argumentation skills. We compare the feedback that two different LLMs, an open-sourced one by META and one of OpenAI's fully proprietary ones, give to students' argumentative writing. We find that, overall, students find the feedback provided by both LLMs helpful (7.51 vs. 7.65 on a scale from 1 to 10), and they rate the quality of the feedback as good to very good. We take this as a very encouraging provisional result that invites larger and more extensive studies on the topic.

Keywords: Writing Support · Argument Quality · ChatGPT · Large Language Model · Computer-Supported Learning · Feedback Provision

1 Introduction and Relevant Previous Research

Argumentative writing support aims to improve students' argumentative skills. By learning how to develop strong arguments, students also practice and improve critical thinking skills, counting among the so-called 21st-century skills [16].

Developing these critical thinking skills requires a close supervision by an experienced teacher: Receiving regular, high-quality feedback can help students to improve their argumentative writing skills. However, this is resource-intensive and, for some settings, including massive open online courses, simply impossible. Therefore, it is desirable to support teachers in the task of providing such feedback to students. This is one of the core tasks of argumentative writing support, a subfield of writing support, which in turn belongs to Natural Language Processing (NLP).

For an overview of recent developments in this field, see [6]. For a recent study of scaling attempts of writing feedback, see [12]. For a recent study on

© The Author(s), under exclusive license to Springer Nature Switzerland AG 2024
A. M. Olney et al. (Eds.): AIED 2024 Workshops, CCIS 2151, pp. 175–182, 2024.
https://doi.org/10.1007/978-3-031-64312-5_21

the effects of computer-generated feedback on overall writing quality, see [14]. Within this field, our approach focuses on a text's argumentative structure, and it builds on argumentative analysis approaches that decompose and/or classify text units and their most important components, usually claims and premises, and then assess the quality of the essay based on this structure. For a recent approach using this paradigm, see [2]. For the most important recent survey of the field, see [18]. For the field's connection to natural language inference (NLI), see [8].

With the advent of highly successful generative large language models (generative LLMs) such as *gpt-3.5-turbo* and *gpt-4*, which power *ChatGPT*, a very promising new path to this goal of aiding students in developing their argumentative writing abilities has entered the field. LLMs are a kind of neural network methods in NLP, as opposed to rule-based or Good Old-Fashioned (GOFAI) approaches. Inspired by the transformer architecture [17], researchers developed a number of influential natural language understanding (NLU) architectures as well as training routines, pioneered by the *BERT* architecture [5]. The transformer architecture has also inspired *GPT-3* [3], which in turn grounds the models powering *ChatGPT*.

Unfortunately, OpenAI, the company that has released *ChatGPT* in November 2022, has decided to contradict current practice in the field and refused to publish any important details on its models and its training, let alone the model weights themselves, while it has communicated that its models are based on the GPT-architecture [1].

Furthermore, with regard to OpenAI's proprietary models, concerns over data protection have alerted watchdogs in many countries, including Poland, the Netherlands, Canada, and Italy.[1] Recently, these privacy worries have been complemented by lawsuits[2] around potential intellectual property violations. Finally, the sheer size of the LLMs served by OpenAI implies that every request sent to *ChatGPT* also exerts a considerable carbon footprint: all other things being equal, a larger model requires more resources to process a request[3].

This creates an uncomfortable situation for universities: it is their mission to equip young people with cutting-edge knowledge and competencies, but they are also required to comply with national laws, including data protection laws, to foster open science, and to reduce their carbon footprint. Fortunately, with the recent surge in smaller open generative LLMs such as the ones released by Meta AI (see below, Sect. 3), a new option has entered the field. On paper, these models are extremely promising. They perform comparably to OpenAI's models at standardized benchmarks, but unlike those, they are relatively small, openly available, and pose no privacy issues, as they can be run on local hardware.

---

[1] See Reports by Reuters on Poland, the Netherlands and Canada, by the Financial Times on Italy.

[2] See Report by AP News. All links last consulted on January, 19th 2024.

[3] Unfortunately, it is impossible to even venture an educated guess on the specific extent of the carbon footprint of one request sent to *ChatGPT*, see [10].

In this paper, we provide a first exploration of the promise of open, locally-run LLMs in the task of argumentative feedback support provision. We provided first-year university students who are enrolled in an academic writing class with the opportunity to obtain feedback on argumentative texts from two different LLMs. We wanted to assert whether (1) students considered this feedback helpful at all, and (2) whether the small, open, and locally deployed LLM is perceived as equally helpful as OpenAI's *ChatGPT*. Answering both of our research questions affirmatively constitutes an important first insight on the performance of open, smaller, and locally run LLMs compared to OpenAI's models in the wild. Rather than evaluating on generic benchmarks, we tap directly into a real-world classroom scenario and let the students compare the two models side-by-side.

2   Our Argumentative Feedback Tool: Artist

For our experiment, we rely on our tool **Artist**[4] (see [4]). The main purpose of **Artist** is to provide students with insight into their argumentative texts. Based on results gathered from a prior user survey, we split the provided analysis into three categories for a better overview. When a user opens **Artist**, they are prompted to type their argumentative text in a designated field or to select one of three example texts used for demo and survey purposes. In the next step, they may choose from three different analytic dimensions:

(1) *Argument Structure Analysis*: uses a random forest classifier to identify argumentative components and visualize their structure in the form of a graph (this part of the tool builds on [19]);
(2) *Discourse Structure Analysis*: uses an RST parser [7] to analyze the rhetorical structure of the text;
(3) *Improvement Suggestions*: lets the user send their texts to an LLM and receive suggestions on how the argumentative quality could be improved. We have incorporated two options:
   (*i*) *llama-2*, a self-hosted instance of *meta-llama/llama-2-70b-chat-hf*,
   (*ii*) *gpt-3.5* that sends the request to *gpt-3.5-turbo* via OpenAI's API.
   The responses from the models are presented to the user in a textual form.

The *Improvement Suggestions* part of the tool is the main focus of this paper. The experiment takes largely place in this part of the tool.

In sum, our approach seamlessly embeds both a commercial LLM accessed with an API and an open LLM running on our infrastructure into a tool that aims at aiding students in their argumentation by providing formative feedback.

3   Set-Up of Exploratory Experiment

Technical Aspects. For the experiment we are relying on the self-developed tool **Artist** that is available via a web interface[5]. The focus of our experiment was

---

[4] Code and screenshots available at https://gitlab.com/ds-unisg/aied2024.
[5] https://artist.datascience-nlp.ai.

a new addition to our tool, namely the ability to receive text-specific feedback regarding the argumentative quality from two different LLMs. In order to reduce the bias of the participants for the study, we simply call them "Model 1" and "Model 2" respectively.

From among OpenAI's proprietary models, we evaluated the version of so-called *gpt-3.5-turbo* (in what follows *gpt-3.5*) available via the API during our experiments in October 2023. In stark contrast to OpenAI, the AI research group of META (formerly Facebook AI) has decided to publicly release its latest series of LLMs [15], allowing for reproducible and rigorous scientific experimentation with these models. We use their model called *meta-llama/llama-2-70b-chat-hf* (in what follows abbreviated by *llama-2*), which we serve using the very efficient serving method vLLM [9]. To the best of our knowledge, we are reporting on the first experiment to deploy this framework in a real-life educational setting. We give an overview on the differences between the two models tested in Table 1.

Table 1. Comparison of the two models used in our explorative experiment.

Model	Size	Serving Method	Open?	Privacy Concerns?
llama-2	70B	locally on 8 GPUs of a NVIDIA DGX-2 via vLLM	yes	none (runs locally)
gpt-3.5	175B	remotely via OpenAI API	no	multiple (see Sect. 1)

We use two different prompts to interact with the models, as they react differently to the same prompts. Following is the prompt for OpenAI's *gpt-3.5*:

"Please give two short suggestions for improving the argumentative quality of the following Essay:" + *input text*

Following is the prompt for *llama-2*:

"[INST] You are an argumentation expert and an experienced teacher that loves to give helpful and encouraging advice to students. You always respond in short, concise, well-formed sentences, and you are also creative. You receive the student's text from me, which has already been analyzed with Discourse Structure Analysis. Referring to very specific elements of the text, you give the student two specific tips on what they could improve about the text in terms of argumentation. Please try to be as specific and supportive as possible giving two formative and instructive feedbacks and nothing more in 2-3 sentences. Here is the text: " + *input text* + "[/INST]"

As can be seen, the prompt used for *llama-2* is much more sophisticated than the one for *gpt-3.5*. We found that more detail is necessary to obtain good results from *llama-2*. However, we applied the same purely formal routine to determine the prompt. We started with identical prompts and then continued to develop the prompts until the following formal requirements had been met by returned feedback: (1) the language of the response is English, (2) the entire response fits into the space allocated to the answer window of 240 tokens, (3) the model gives exactly two suggestions (where we did not investigate the quality of the suggestions, just the count of two).

While *gpt-3.5* fulfilled these three formal criteria with the original version of the prompt, *llama-2* required more information to do so. We hypothesize that this is due to the more extensive reinforcement-based fine-tuning that went into OpenAI's product.

Participants. A total of 63 students participated in our study. All of them were first-year university students who had almost completed the course for academic writing. The course includes several cycles of peer review, thus the students were experienced in both giving and receiving formative feedback. All of the students were enrolled in the same general study program that then allows them to study for a variety of degrees in business administration and social sciences. 66% of the participants were male, 32% female, and 2% preferred not to state their gender. The first language of 89% was German and for 3% English. In regards to the age, we observed a higher variability. The youngest participant(s)'s age was 17 and the oldest(s)'s 37 with an overall (rounded-up) mean of 20.

Table 2. Results of LLM-specific questions (name of LLM that has a higher percentage of strongly or somewhat disagree (question 1) or agree (questions 2 and 3) printed in **boldface**, all values in [%]).

Question	Model	Strongly disagree	Somewhat disagree	Neither nor	Somewhat agree	Strongly agree
Loading answer took too long?	**llama-2**	41.27	31.75	14.29	9.52	3.17
	gpt-3.5	46.03	25.40	17.46	9.52	1.59
Understood the recommendation?	**llama-2**	1.61	3.23	9.68	37.10	48.39
	gpt-3.5	0.00	4.84	9.68	50.00	35.48
Was the feedback useful?	llama-2	1.59	4.76	12.70	46.03	34.92
	gpt-3.5	0.00	3.17	11.11	44.44	41.27

Details on the Process. The experiment was conducted in class on October 20th, 2023, in a course that introduces students to the basics of scientific writing, such as correct referencing, composition, topic selection, and argumentation. The entire class lasted for 90 min and was dedicated to the topic of argumentation. The experiment itself was conducted within this class and took approximately 15 min. The participants were asked to complete a questionnaire about the utility of the entire tool with a clear focus on the recommendations of the two LLMs. Participants in the experiment had one web browser window open with the questionnaire and another one with the tool. To have more control over the variables of the experiment, we randomly chose three texts from a well-known dataset [13] that the students used to obtain the LLMs' feedback.

4   Results and Discussion

Results. In Table 2, we give the results of the LLM-focused questions of our experiment. The table shows that *llama-2* receives slightly better scores with the first two questions (note that the difference regarding the second question is only 0.01%), while *gpt-3.5* significantly outperforms its competitor in question 3. Figure 1 depicts the distribution of participants' responses to the question: *How do you perceive the received feedback quality of Model 1 and Model 2? Rate the two Models on a scale of 1 to 10, where 10 is the highest value and 1 is the lowest value.* The figures show that (1) both models are perceived as helpful, and (2) the specific scores of the two models are very close (average for *llama-2* 7.51, for *gpt-3.5* 7.65). This means that the quality of the feedback provided by the two models was perceived as good to very good, and almost on a par.

Discussion. We emphasize three aspects of the results of our study. First, the perceived quality of the feedback is remarkable. It was not to be expected that general-purpose LLMs with no specific fine-tuning for giving feedback on argumentative texts would perform so well. The task is very difficult, as it requires a combination of strict, logical modes of linguistic abilities with associative, topical modes of abilities. The vast majority of participants — over 80% at the least — understood the recommendation and found it useful. Furthermore, on a scale of 1 to 10, students ranked the quality of the feedback above 7.5 on average. Bearing in mind that even feedback by experienced human teachers would not get a straight 10,[6] it is clear that this is a very good score.

Second, it is surprising that *llama-2* is competitive with *gpt-3.5* throughout the experiment, and that our serving method outperforms OpenAI's API in terms of response time. Given that OpenAI's model is 2.5 times the size of *llama-2*, and given that only the former has been extensively fine-tuned using human feedback and a special flavor of reinforcement learning [11], it would not have been rational to expect that the models are almost on a par at this complex task. From a research-political as well as from an environmental perspective, this is encouraging: open-source models can compete with highly resourced proprietary models even at very challenging tasks. And the fact that the model is 2.5 times smaller will, roughly, reduce its carbon footprint per processed request by the same factor. Where the vast resources invested in OpenAI's models might show is in the ease with which one can extract *formally satisfactory* feedback from

---

[6] For instance, Weaver [20], finds that only 18% of business and design students always find the (human) feedback that they receive during their studies clear and easy to read.

gpt-3.5. While getting *llama-2* to give feedback in the requested form required a carefully engineered, rather long prompt, the prompt that was used to interact with *gpt-3.5* was short and ready in a matter of seconds.

Third, we wish to point out three limitations of our study. As the demographics in Sect. 3 show, the majority of our participants' mother tongue is not English, but rather German. While they all have to be able to take classes in English, native speakers might perceive the usefulness of the feedback provided differently. Note, however, that students across the world have to learn to write academic texts in English as a second language. In these settings, our results are directly applicable. The second limitation is inherent in the design of our study: to make the results as comparable as possible, we pre-defined the texts that the students used to interact with the tool. It is possible that using different texts would lead to different results. However, by choosing three texts at random from a well-respected argumentative writing dataset, we tried to keep this probability as low as possible. Lastly, we wish to emphasize once more that the size and scale of this study means that it can only offer preliminary results that have to be confirmed in larger, more comprehensive settings.

Fig. 1. *llama-2* vs. *gpt-3.5*: perceived quality of argumentative feedback.

5 Conclusion

In this article, we have tested the capacities of an AI-based tool that is intended to support and improve students' argumentative writing skills. We have focused on using two different LLMs to provide students with case-specific formative feedback to improve their argumentative texts. Our provisional findings are overall very encouraging. Students perceive the quality, comprehensibility, and helpfulness of the feedback by the two LLMs as good or very good. We are particularly encouraged by the fact that students rate the much smaller open-source LLM *llama-2* almost as highly as OpenAI's *gpt-3.5* (7.51 vs. 7.65 on a scale from 1 to 10). As a consequence, we plan to experiment with more flexible settings and larger test groups to confirm our findings. Furthermore, as the field of LLMs is evolving at an impressive pace, we would like to explore the promise of even smaller and more efficient models to further reduce the carbon footprint of our method.

Acknowledgements. We would like to thank the Swiss National Science Foundation (SNSF) for the grant to support our project "Next Generation of Digital Support for Fostering Students' Academic Writing Skills: A Learning Support System based on Machine Learning (ML)", a collaboration project between the University of St. Gallen in Switzerland and the Mahidol University in Thailand.

Disclosure of Interests. The authors declare that they have no competing interests that could influence the content of the research presented here.

References

1. Achiam, J., et al.: GPT-4 technical report. arXiv: 2303.08774 (2023)
2. Alhindi, T., Ghosh, D.: 'Sharks are not the threat humans are': argument component segmentation in school student essays. In: BEA@EACL, pp. 210–222 (2021)
3. Brown, T., et al.: Language models are few-shot learners. In: Advances in Neural Information Processing Systems, vol. 33, pp. 1877–1901 (2020)
4. Burkhard, M., et al.: Computer supported argumentation learning: design of a learning scenario in academic writing by means of a conjecture map. In: CSEDU (1), pp. 103–114 (2023)
5. Devlin, J., et al.: BERT: pre-training of deep bidirectional transformers for language understanding. In: NAACL 2019, pp. 4171–4186 (2019)
6. Strobl, C., et al.: Digital support for academic writing: a review of technologies and pedagogies. Comput. Educ. **131**, 33–48 (2019)
7. Feng, V.W., Hirst, G.: A linear-time bottom-up discourse parser with constraints and post-editing. In: Proceedings of the 52nd Annual Meeting of the Association for Computational Linguistics (Volume 1: Long Papers), pp. 511–521 (2014)
8. Gubelmann, R., et al.: Capturing the varieties of natural language inference: a systematic survey of existing datasets and two novel benchmarks. J. Log. Lang. Inf. **33**, 21–48 (2023). https://doi.org/10.1007/s10849-023-09410-4
9. Kwon, W., et al.: Efficient memory management for large language model serving with pagedattention. In: SOSP 2023: Proceedings of the 29th Symposium on Operating Systems Principles, pp. 611–626 (2023)
10. Patterson, D., et al.: Carbon emissions and large neural network training. arXiv: 2104.10350 (2021)
11. Proximal policy optimization. OpenAI (2017). https://openai.com/blog/openai-baselines-ppo/
12. Rapp, C., Kauf, P.: Scaling academic writing instruction: evaluation of a scaffolding tool (thesis writer). Int. J. Artif. Intell. Educ. **28**, 590–615 (2018)
13. Stab, C., Gurevych, I.: Annotating argument components and relations in persuasive essays. In: COLING 2014, pp. 1501–1510 (2014)
14. Stevenson, M., Phakiti, A.: The effects of computer-generated feedback on the quality of writing. Assessing Writ. **19**, 51–65 (2014)
15. Touvron, H., et al.: Llama 2: open foundation and fine-tuned chat models. arXiv: 2307.09288 (2023)
16. Van Laar, E., et al.: The relation between 21st-century skills and digital skills: a systematic literature review. Comput. Hum. Behav. **72**, 577–588 (2017)
17. Vaswani, A., et al.: Attention is all you need. In: Advances in Neural Information Processing Systems, vol. 30 (2017)
18. Wachsmuth, H., et al.: Computational argumentation quality assessment in natural language. In: EACL 2017, pp. 176–187 (2017)
19. Wambsganss, T., et al.: AL: an adaptive learning support system for argumentation skills. In: Proceedings of the 2020 CHI Conference on Human Factors in Computing Systems, pp. 1–14 (2020)
20. Weaver, M.R.: Do students value feedback? Student perceptions of tutors' written responses. Assess. Eval. High. Educ. **31**(3), 379–394 (2006)

Good Fit Bad Policy: Why Fit Statistics Are a Biased Measure of Knowledge Tracer Quality

Napol Rachatasumrit[(✉)][ID], Daniel Weitekamp[ID], and Kenneth R. Koedinger[ID]

Carnegie Mellon University, Pittsburgh, PA 15213, USA
{napol,weitekamp,koedinger}@cmu.edu

Abstract. Knowledge tracers are typically evaluated on the basis of the goodness-of-fit of their underlying student performance models. However, for the purposes of supporting mastery learning the true measure of a good knowledge tracer is not its goodness-of-fit, but the degree to which it optimally selects next problem items. In this context, a knowledge tracer should minimize under-practice to ensure students master learning materials and minimize over-practice to reduce wasted time. Prior work has suggested that fit-statistic-based measures of knowledge tracer quality may misrank the relative quality of knowledge tracers' item selection. In this work, we evaluate this claim by measuring over- and under-practice directly in synthetic data drawn from ground-truth learning curves. We conduct an experiment with 3 well-known student performance models: Performance Factor Analysis (PFA), BestLR, and Deep Knowledge Tracing (DKT), and find that in 43% of the synthetic datasets, the models with higher measures of overall predictive performance (e.g. AUC and MSE) were worse than a comparison model with a lower predictive performance at minimizing over-practice and under-practice. These results support the hypothesis that overall fit statistics are not a reliable measure of a knowledge tracer's ability to optimally select next items for students, and bring into question the validity of traditional methods of knowledge tracer comparison.

Keywords: Knowledge Tracing · Mastery Learning · Intelligent Tutoring System

1 Introduction

Student performance models estimate the probability that students will correctly answer the next question items given their prior correct and incorrect responses and serve both online and offline roles in education. In an online setting student performance models can be used as knowledge tracers to adaptively select next problems based on students' current abilities. In an offline setting, they can be

---

N. Rachatasumrit and D. Weitekamp—These authors contributed equally to this work.

A. M. Olney et al. (Eds.): AIED 2024 Workshops, CCIS 2151, pp. 183–191, 2024.
https://doi.org/10.1007/978-3-031-64312-5_22

used to reveal patterns in student's learning which can be used to make data-driven improvements to instructional materials [3].

Knowledge tracing is the online use of a student performance model to actively estimate students' mastery of individual knowledge components (KCs)—the pre-specified facts, skills and principles which students must understand in order to have mastered a particular domain [6]. Mastery of a KC is typically characterized as the point when a student's predicted chance of correctly answering future question items associated with the KC exceed some preset mastery threshold, typically chosen in the range 85–95% [2]. A knowledge tracer leverages its student performance model to estimate which KCs are mastered and which are not so that it can select the next practice items for students which correspond to unmastered KCs. Thus, the challenge of knowledge tracing is to actively adapt to students as they practice to optimize their use of time—giving them enough practice problems for each KC to ensure full domain mastery, but not more than this to avoid wasting time better spent practicing new material. Thus, the ideal knowledge tracer jointly minimizes over-practice, the number of prescribed practice problems given after the student has reached mastery, and under-practice, the number of practice problems which a student would still need to solve in order to achieve mastery.

Unfortunately, over- and under-practice are not directly measurable quantities. Instead, the relative quality of knowledge tracers is typically compared on the basis of the overall fit of their underlying student performance models to student data. Overall fit statistics take the form $\pi(\hat{y}, y)$ and measure the degree to which the continuous student model predictions $\hat{y}$ are a good approximation of the discrete sequence of binary correctness values $y = y_0, ..., y_n$ (correct=1, incorrect=0) collected from student transaction logs. Prior work has used a variety of fit statistics for knowledge tracer comparisons including Mean-Square Error (MSE), prediction accuracy, log-likelihood, AIC [1], BIC [7], and Area under the receiver operating characteristic curve (AUC).

In this work, we demonstrate that overall fit statistics can in fact be a biased basis for knowledge tracer comparison since there are circumstances where a model's total predictive performance can be improved without any corresponding change in the behavior of a knowledge tracer utilizing that model. A model can fit better without producing any corresponding reduction in the number of over- and under-practice problems experienced by students.

A similar concern, yet one unrelated to the claims of this work, is the debate over interpretable versus non-interpretable student performance models. The last decade of knowledge tracing research has been inclusive of a broader machine learning community which have eschewed traditional models based on Item-Response Theory (IRT) [5], hidden markov models, and logistic regression for uninterpretable yet often performant, deep-learning models. Since black-box models possess more parameters than can be practically interpreted, they are less amenable to generating defensible and actionable insights about student data. Thus proponents of deep-knowledge tracers have typically placed a greater emphasis on the practical use of their models for online knowledge tracing over their use as tools of offline analysis.

In this work, our main objective is not to make an argument for interpretable or uninterpretable blackbox student models, but to bring into question an assumption held in common by both sides of that debate. We show through simulation that it is possible for a student performance model to fit better than a baseline model but be worse at knowledge tracing. And we contend that this raises serious doubts about whether the collective research project of trying to produce better-fitting student models is necessarily leading to knowledge tracers which are better at mastery-based item selection.

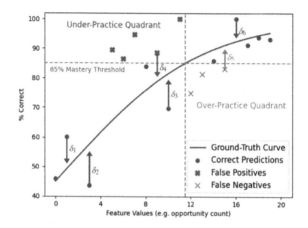

Fig. 1. Illustration of over-practice and under-practice attempts

2   Over-Practice and Under-Practice

Although counts of over- and under-practice are not directly measurable from student data, they can be defined relative to a notion of a student's ground-truth learning curve—their true probability of answering next question items correctly at each practice opportunity. Framed in non-stochastic terms, a student's ground truth curve for a given KC represents the degree to which that KC has been mastered at each learning opportunity. It captures the progression of complex cognitive factors beyond the scope of what statistical performance models typically capture. A point along the curve captures the degree to which a student has partially constructed knowledge—a notion that statistical models typically estimated solely from binary observations of correct and incorrect performance.

By reference to a ground-truth learning curve and a choice of mastery threshold, a model's instances of under-practice are those where the performance model predicts performance to be above the mastery threshold when the ground truth is below it, and the model's instances of over practice are those where it predicts performance to be below mastery when the ground-truth is above the mastery threshold (Fig. 1).

Student performance modeling can be framed as estimating students' ground-truth learning curves from the noisy sampling of performance data collected from tutoring system transactions. The logic of comparing knowledge tracers by their overall goodness-of-fit to data is motivated by the idea that an optimal recreation of the ground-truth learning curve should produce an optimal prediction of student mastery. However, this perspective conflates the logic of offline statistical modeling, in which goodness-of-fit can be used to justify hypotheses about students' learning trajectories and their relationship to learning materials, with the narrower aims of online item selection. In this context, a knowledge tracer's purpose is simply to make one critical decision: after a student completes each problem it decides whether to continue prescribing new practice problems with particular KC requirements or not. Thus, certain variations in the predictions of a student performance model simply have no bearing on the real-world quality of their knowledge tracer.

Figure 1 demonstrates how this can be the case by offering an illustration of a hypothetical set of performance model predictions relative to a ground-truth learning curve. The intersection of the ground-truth curve with the mastery threshold divides the figure into 4 quadrants. Predictions in the top-left and bottom-right quadrants are instances where the model would cause under- or over-practice. The dots and x's in Fig. 1 represent the predictions $\hat{y}_A$ of a baseline model A. Consider that there is also a comparison model B with predictions $\hat{y}_B = \hat{y}_A + \delta$ perturbed by some δ which brings B closer to the ground truth than A. With these perturbations B's expected overall fit to a sample of the ground-truth curve should be better than model A's. However, only a subset of the shown perturbations would produce improvements in mastery prediction, only those perturbations which move predictions out of the over- and under-practice quadrants (e.g. like δ_4 and δ_5).

A core hypothesis of this work is that the prediction differences between different types of student performance models mostly do not correspond to differences in expected over- and under-practice like perturbations δ_4 and δ_5. Instead, we hypothesize that the majority of model improvements are like δ_1, δ_2, δ_3, and δ_6: inconsequential to levels of over- and under-practice, and generally outside the neighborhood of the ground-truth mastery threshold. One reason to expect this result is that the more data that models have about students the more similar their predictions are likely to be. We expect models to have the greatest difference in their predictions under uncertain circumstances, particularly in early practice attempts when evidence about the student's knowledge is sparse.

To test this hypothesis we utilize synthetic student data to establish ground-truth learning curves. Then we fit various student performance models on the synthetic data and utilize the ground-truth curves to measure over- and under-practice. We evaluate whether the student performance models which produce the least over- and under-practice are also the best fitting models with respect to overall performance statistics like AUC and MSE. Finally, we graph MSE as a function of ground-truth probability to evaluate whether differences in model fit tend to be greatest within or outside the neighborhood of the mastery threshold.

3   Related Works

Prior work has argued that comparing knowledge tracers on fit statistics alone fails to estimate their relative efficacy on a substantive scale [14]. For instance, to claim that model A achieves a 5% improvement in MSE over model B fails to capture the time savings or post-test performance improvements that would be achieved by utilizing that model for adaptive item selection. Prior attempts to estimate this relationship by simulation [14] and analytically [12], have supported the conclusion that relatively small overall model improvements can yield large reductions in over- and under-practice. However, Weitekamp et al. [12] point out that in theory, it is indeed possible for a better-fitting student performance model to actually perform worse at item selection than a baseline model. The key idea is that the only predictions which matter for item selection are those in the ground-truth neighborhood of the mastery threshold—the region where a knowledge tracer makes its critical decision: to stop prescribing problems for a particular KC or not. Overall fit statistics may produce a biased sense of knowledge tracer quality because they capture the goodness of fit of a performance model on early student transactions which are unambiguously part of the unmastered region. By contrast prediction differences between models in the neighborhood of the mastery threshold are likely to be small since there is typically more supporting evidence from the student transactions preceding it.

4   Methods

We utilize 3 models for synthetic data generation and evaluation: BestLR [4], DKT [9], and PFA [8]. For each dataset, we use each model to create a simulated dataset and evaluate each generated dataset with all 3 models to create a 3×3 experiment. In all cases, we use implementations from Gervet et. al. [4].

Our synthetic data generation works by (1) fitting a generation model to the real data, (2) predicting an error rate for each transaction with a fitted model, and using the predicted value as a ground truth for an error rate in synthetic data, (3) sampling a synthetic outcome for each transaction in the synthetic data based on the corresponding error rate. In this work, we use the same 7 real-world datasets from Gervet et. al. [4], so we generated 21 synthetic datasets for our experiment using 3 generation models For each synthetic dataset, we use random cross-validation splitting by students. The data of 90% of the students are used for training and the data of the other 10% are reserved for the test set. We resample and retrain 5 times for each condition, examining the relative counts of over- and under-practice on the test set between models, and compare this to their relative AUC scores on the test set. We report the average and standard deviation for each metric across replicates.

Table 1. Average numbers of over- and under-practice for each dataset and model

Dataset	Generate	BestLR	DKT	PFA
algebra05	BestLR	**4.577 ± 0.235**	7.261 ± 0.199	10.843 ± 0.433
	DKT	13.164 ± 5.184	**8.300 ± 0.327**	32.522 ± 1.957
	PFA	9.067 ± 0.672	13.116 ± 0.835	**5.028 ± 0.563**
assistments09	BestLR	**3.355 ± 0.078**	4.488 ± 0.181	5.393 ± 0.151
	DKT	7.280 ± 0.151	**4.000 ± 0.107**	9.597 ± 0.265
	PFA	4.258 ± 0.136	5.706 ± 0.246	**3.309 ± 0.184**
assistments15	BestLR	**2.398 ± 0.045**	4.388 ± 0.323	2.961 ± 0.056
	DKT	8.096 ± 0.107	**3.963 ± 0.063**	8.233 ± 0.167
	PFA	**2.377 ± 0.118**	4.997 ± 0.186	2.425 ± 0.043
assistments17	BestLR	**2.638 ± 0.045**	3.567 ± 0.060	5.297 ± 0.085
	DKT	6.334 ± 0.226	**2.808 ± 0.027**	3.614 ± 0.098
	PFA	4.663 ± 0.280	4.738 ± 0.395	**3.495 ± 0.581**
bridge_algebra	BestLR	**3.936 ± 0.094**	5.494 ± 0.217	6.405 ± 0.132
	DKT	14.033 ± 0.368	**6.751 ± 0.165**	22.319 ± 0.712
	PFA	4.762 ± 0.300	6.539 ± 0.200	**3.759 ± 0.218**
spanish	BestLR	**2.447 ± 0.022**	4.213 ± 0.160	3.173 ± 0.083
	DKT	10.798 ± 0.222	**4.600 ± 0.194**	12.701 ± 0.345
	PFA	2.397 ± 0.041	4.324 ± 0.145	**2.109 ± 0.036**
statics	BestLR	**3.962 ± 0.205**	4.263 ± 0.185	10.559 ± 0.442
	DKT	10.379 ± 0.415	**5.095 ± 0.235**	19.067 ± 0.843
	PFA	8.333 ± 0.687	7.565 ± 0.589	**3.743 ± 0.457**

5   Results and Discussion

Table 1 shows the average instances of over- and under-practice and Table 2 shows the average AUC for each dataset and evaluation model pair. Conventional evaluations assume that between two models the one with the higher predictive performance (e.g. higher AUC) will be the better model—the one expected to make fewer over- and under-practice errors. However, our results demonstrate that this assumption is not always true. We find that in 43% of the synthetic datasets, there are pairs of models where the higher AUC model commits more over- and under-practice errors than the lower AUC model. These results support the hypothesis that overall fit statistics are not a reliable measure of a knowledge tracer's ability to optimally select next items for students, and challenge the credibility of conventional approaches to comparing knowledge tracers.

Table 2. Average and SD of AUC for each dataset and evaluation model

Dataset	Generate	BestLR	DKT	PFA
algebra05	BestLR	**0.794 ± 0.002**	0.728 ± 0.004	0.716 ± 0.004
	DKT	**0.808 ± 0.003**	0.764 ± 0.007	0.737 ± 0.002
	PFA	0.689 ± 0.004	0.645 ± 0.004	**0.705 ± 0.002**
assistments09	BestLR	**0.712 ± 0.003**	0.636 ± 0.006	0.653 ± 0.003
	DKT	**0.736 ± 0.005**	0.696 ± 0.004	0.670 ± 0.004
	PFA	0.629 ± 0.003	0.565 ± 0.007	**0.653 ± 0.003**
assistments15	BestLR	**0.721 ± 0.005**	0.702 ± 0.006	0.713 ± 0.005
	DKT	0.658 ± 0.001	**0.674 ± 0.002**	0.656 ± 0.001
	PFA	**0.659 ± 0.002**	0.630 ± 0.001	0.659 ± 0.003
assistments17	BestLR	**0.734 ± 0.004**	0.717 ± 0.005	0.654 ± 0.004
	DKT	0.702 ± 0.002	**0.728 ± 0.001**	0.617 ± 0.001
	PFA	0.636 ± 0.002	0.619 ± 0.002	**0.639 ± 0.002**
bridge_algebra	BestLR	**0.834 ± 0.031**	0.780 ± 0.033	0.780 ± 0.034
	DKT	**0.774 ± 0.003**	0.747 ± 0.008	0.705 ± 0.004
	PFA	0.699 ± 0.005	0.645 ± 0.002	**0.715 ± 0.003**
spanish	BestLR	**0.820 ± 0.003**	0.764 ± 0.001	0.811 ± 0.004
	DKT	0.808 ± 0.006	**0.813 ± 0.006**	0.788 ± 0.003
	PFA	0.813 ± 0.006	0.763 ± 0.006	**0.814 ± 0.006**
statics	BestLR	**0.799 ± 0.007**	0.785 ± 0.010	0.661 ± 0.010
	DKT	**0.804 ± 0.005**	0.801 ± 0.004	0.665 ± 0.005
	PFA	0.661 ± 0.005	0.647 ± 0.004	**0.670 ± 0.004**

6   Conclusion and Future Works

In this work, we have utilized synthetic data generated by popular knowledge tracers to test whether models with the highest overall fit statistics necessarily produce the best predictions of student mastery. Our method allows us to answer questions of the nature: what is the quality of knowledge tracer X's item selection assuming student learning behaves like model Y? Varying models X, Y, and datasets we find that in 43% of the synthetic datasets, models with higher measures of overall predictive performance (i.e. AUC) were worse than a comparison model with a lower predictive performance at minimizing over-practice and under-practice. We conclude that traditional measures of overall performance (e.g. AUC) are in fact not reliable proxies for rates of over- and under-practice. These results raise serious doubts about whether the field of knowledge tracing follows a sound logic of justification when it comes to model comparison.

As in prior works that have utilized synthetic data for analyses of student performance models [11], our method relies upon a theoretical commitment to an underlying model for generating ground-truth curves. Thus our method is not a stand-in replacement for traditional metrics of model fit which evaluate models directly on datasets. Yet, methods which draw comparisons between statistical models and synthetic ground truths have the potential to enable deeper evaluations than the simple notion of that which fits best is best. In this work, we have used statistical performance models as ground-truth generators, but more theory-driven generators such as computational models of learning [10,12,13] could be used in their place, to serve as more precise, predictable, and explainable generators of ground-truth learning curves and synthetic data.

Utilizing more controlled theory driven models for data generation could enable more concrete analyses of the sensitivities of different student performance models to individual student differences and domain types, and models' behavior under uncertainty. For instance, while no model in our analyses stood out as decidedly better than the others, in some cases certain models performed better in terms of over- and under-practice on certain datasets. Future work may also include further analysis of the nature of the unproductive predictions that each model commits. For example, investigating the conditions when the models commit those errors and how extreme those errors are could show interesting insights that lead to a better evaluation metric for knowledge tracers.

References

1. Akaike, H.: Akaike's information criterion. Int. Encycl. Stat. Sci. 25–25 (2011)
2. Arlin, M.: Time, equality, and mastery learning. Rev. Educ. Res. **54**(1), 65–86 (1984)
3. Cen, H., Koedinger, K., Junker, B.: Learning factors analysis – A general method for cognitive model evaluation and improvement. In: Ikeda, M., Ashley, K.D., Chan, T.-W. (eds.) ITS 2006. LNCS, vol. 4053, pp. 164–175. Springer, Heidelberg (2006). https://doi.org/10.1007/11774303_17
4. Gervet, T., Koedinger, K., Schneider, J., Mitchell, T., et al.: When is deep learning the best approach to knowledge tracing? J. Educ. Data Min. **12**(3), 31–54 (2020)
5. Harvey, R.J., Hammer, A.L.: Item response theory. Couns. Psychol. **27**(3), 353–383 (1999)
6. Koedinger, K.R., Corbett, A.T., Perfetti, C.: The knowledge-learning-instruction framework: bridging the science-practice chasm to enhance robust student learning. Cogn. Sci. **36**(5), 757–798 (2012)
7. Neath, A.A., Cavanaugh, J.E.: The Bayesian information criterion: background, derivation, and applications. Wiley Interdiscip. Rev. Comput. Stat. **4**(2), 199–203 (2012)
8. Pavlik, P.I., Jr., Cen, H., Koedinger, K.R.: Performance factors analysis-a new alternative to knowledge tracing. In: Artificial Intelligence in Education: Building Learning Systems that Care: From Knowledge Representation to Affective Modelling, Proceedings of the 14th International Conference on Artificial Intelligence in Education, AIED 2009, July 6-10, 2009, Brighton, UK (2009)
9. Piech, C., et al.: Deep knowledge tracing. In: Advances in Neural Information Processing Systems, vol. 28 (2015)

10. Rachatasumrit, N., Carvalho, P.F., Li, S., Koedinger, K.R.: Content matters: a computational investigation into the effectiveness of retrieval practice and worked examples. In: Wang, N., Rebolledo-Mendez, G., Matsuda, N., Santos, O.C., Dimitrova, V. (eds.) International Conference on Artificial Intelligence in Education, pp. 54–65. Springer, Cham (2023). https://doi.org/10.1007/978-3-031-36272-9_5
11. Rachatasumrit, N., Koedinger, K.R.: Toward improving student model estimates through assistance scores in principle and in practice. In: Proceedings of the 14th International Conference on Educational Data Mining (EDM 2021) (2021)
12. Weitekamp, D., Koedinger, K.: Computational models of learning: deepening care and carefulness in AI in education. In: Wang, N., Rebolledo-Mendez, G., Dimitrova, V., Matsuda, N., Santos, O.C. (eds.) International Conference on Artificial Intelligence in Education, pp. 13–25. Springer, Cham (2023). https://doi.org/10.1007/978-3-031-36336-8_2
13. Weitekamp III, D., Harpstead, E., MacLellan, C.J., Rachatasumrit, N., Koedinger, K.R.: Toward near zero-parameter prediction using a computational model of student learning. In: Proceedings of The 12th International Conference on Educational Data Mining (EDM 2019) (2019)
14. Yudelson, M., Koedinger, K.: Estimating the benefits of student model improvements on a substantive scale. In: Educational Data Mining (2013)

Writing Analytics and AI for Special Education: Preliminary Results on Students with Autism Spectrum Disorder

Akira Borba Colen França[1]([⊠]) ⓘ, Eliseo Reategui[1] ⓘ, Joseph Mintz[2] ⓘ,
Ricardo Radaelli Meira[1] ⓘ, and Regina Motz[3] ⓘ

[1] PGIE, Universidade Federal do Rio Grande do Sul, Porto Alegre, RS, Brazil
akirabcf@outlook.com
[2] Institute of Education, University College London, London, England
j.mintz@ucl.ac.uk
[3] Facultad de Ingeniería, Universidad de la República, Montevideo, Uruguay
rmotz@fing.edu.uy

Abstract. This article discusses the utilization of writing analytics in Special Education, with a particular focus on students with Autism Spectrum Disorder (ASD). Research increasingly supports the use of data mining and Artificial Intelligence (AI) to analyze and support students' writing processes, showcasing the potential of these systems to enhance student engagement and the accuracy of automated feedback. However, concerns persist regarding potential biases and ethical implications. The literature highlights limitations in applying Writing Analytics and AI to atypical students since most research and tools are designed with typical students in mind, reflecting societal biases. Autistic students often encounter challenges in writing performance due to factors such as rigid style, limited vocabulary, and difficulties expressing thoughts. This paper presents a study involving the analysis of 2643 essays from secondary education students, including a subset with ASD, using text-to-network tools and NLP analysis to compare texts and examine computational linguistics metrics and text mining patterns. Preliminary findings suggest the necessity for tailored evaluation and interventions for ASD students. While AI offers opportunities for personalized interventions, further research is essential to effectively adapt current tools for atypical students.

Keywords: Writing Analytics · Autism Spectrum Disorder · Special Education · Text Mining · Natural Language Processing

1 Introduction

The interest in different forms of Artificial Intelligence (AI) within academic literature has grown exponentially, particularly since 2015 [24]. It is important to notice that AI is not a singular entity, but an umbrella term encompassing a set of modeling capabilities that are expanding on a daily basis. In this context, this paper explores Writing Analytics (WA) in the educational context, with preliminary results of its potential and limitations in Special Education with students with Autism Spectrum Disorder (ASD).

© The Author(s), under exclusive license to Springer Nature Switzerland AG 2024
A. M. Olney et al. (Eds.): AIED 2024 Workshops, CCIS 2151, pp. 192–199, 2024.
https://doi.org/10.1007/978-3-031-64312-5_23

Previous research states that WA applications can have high accuracy in evaluation tasks [25] and improve students' engagement in the writing process [26]. However, the predominant focus of research and the development of tools and applications, tends to revolve around the needs of typical students. Students with learning disabilities, when compared to their typically achieving peers, exhibit reduced planning time, generate less coherent ideas, and dedicate less effort to revising for meaning and content [9]. Concerning autistic students, it's widely acknowledged that they often demonstrate writing performance below their expected intellectual capabilities [10]. This discrepancy is attributed to factors such as a more rigid writing style, limited vocabulary, challenges in expressing thoughts through writing, difficulties with theory of mind, and weaker coherence. This paper presents a study with 2643 secondary education essays, aiming to bridge this gap.

2 Writing Analytics

The term writing analytics emerged to describe a field of study that focuses on analyzing and assessing texts to understand writing processes and products in educational contexts [13]. Writing Analytics' primary objective is supporting students to improve their writing skills. Patout [20] highlights the importance of these systems in providing context-aware feedback and scaffolding improvements in writing skills. However, Roscoe [23] emphasizes the need for these systems to offer appropriate and formative feedback, suggesting the use of pedagogically-guided algorithms. Lee [14] further supports the long-term effectiveness of these systems, demonstrating their positive impact on writing development over time. Another approach that is of interest in the field is that of verifying how precise can technological advances be in automated essay scoring (AES), when compared with human raters. While some earlier works have found only modest correlations between automated and human scores in MOOC context, suggesting some validity, other researchers advocate better results, describing AES approaches that may provide more specific feedback to writers and may be relevant to other natural language computations, such as the scoring of short answers in comprehension or knowledge assessments [17]. More recently, with the advent of generative AI, it has been also demonstrated how the integration of generative artificial intelligence with computational linguistics measures can enhance text scoring accuracy [18].

In this context, Natural Language Processing (NLP) plays an important role, providing techniques for textual classification based on linguistic characteristics [17]. Among the various applications of these technologies, the assessment of the readability level and complexity of texts has a certain prominence [27]. However, fewer studies focus on the use of textual metrics to evaluate textual quality, especially in languages other than English. Even fewer studies focus on the use of data-based instruction (DBI) for children with disabilities, although some research has provided preliminary support for the effectiveness of a combined research-based intervention and DBI approach in addressing the intensive writing needs of students, particularly those with disabilities [11].

3   Text Analytics in Portuguese

As this study focuses on the analysis of texts written in Portuguese, we found it relevant to explore previous research in which Computational Linguistics and Natural Language Processing have been employed for the analysis of Portuguese texts. One of these studies was carried out by Evers and Finatto [6]. The authors used machine learning and natural language processing to distinguish levels of proficiency in Brazilian Portuguese. The work showed how a set of metrics, obtained using Coh-Metrix-Port, demonstrated non-random behavior for the established categories. A decision tree capable of classifying works into these categories was implemented, however, demonstrating a low classification capacity. Martins [15] carried out a study on the correlation between indicators of textual complexity and academic performance. The results indicated that, in relation to lexical complexity, there was a positive correlation between the frequency of nouns and progression, and a negative correlation between the frequency of verbs and progression. With regard to syntactic complexity, the correlation varied according to the type of subordinate clause in the two sets of data analyzed. Oliveira et al. [19] used regression models based on conventional machine learning methods and pre-trained deep learning language models to automatically estimate the cohesion of essays in Portuguese and English. They analyzed 4,570 essays in Portuguese and 7,101 in English, finding that a deep learning model performed best, with a moderate Pearson correlation with human-rated cohesion scores. Cavalcanti et al. [4] sought to identify the most influential textual characteristics in feedback texts written by teachers, seeking to predict their quality. A classifier was trained and evaluated, in Portuguese, at various levels of feedback. The most important variable identified was the number of paragraphs, directly linked to the amount of corrective information intended for students. More recently, Meira et al. [16] conducted a study focusing on university students' texts. The findings revealed a substantial correlation between most metrics and teachers' assigned grades, providing insights into textual quality indicators.

4   Writing with ASD and AI Bias

The research presented in this paper distinguishes itself from previous studies by directing its focus specifically towards exploring how writing analytics can be effectively employed to support writing instruction for students with Autism Spectrum Disorder (ASD). Previous research has demonstrated that students with ASD present specific difficulties in written expression, including differences in text length, legibility, size of handwriting, writing speed, spelling, and general text structure [8]. Although the research on ASD, education and technology is growing, the majority of research is interested in early education and young children [7]. Such research is focused mainly on support for handwriting, grammar and spelling [5]. Some studies delved in more intricate aspects of writing, though. For instance, Hilvert [10] presented a linguistic analysis of the expository writing abilities of school-age children with ASD in comparison to neurotypical children. The study revealed that, in contrast to their neurotypical peers, children with ASD produced shorter expository texts containing more grammatical errors, requiring additional assistance during the writing assessment. In that sense, it is possible to say that the vast majority of Writing Analytics and AI tools availables are strongly biased. WA and AI tools are highly derivative from NLP metrics and text analysis that have been carried with neurotypical students as a target audience for the tools. This is a case of both data and societal bias in AI [12]. In that sense, analysis and tools built without ASD students in mind may not be suited to the individual needs of those students.

5 Method

For this study, 2643 essays produced by 1416 secondary school students were collected and analyzed using basic NLP and Network Analysis. The essays were obtained through the support of RevisãoOnline's team, a peer review system to support the process of producing and reviewing argumentative and dissertative texts, particularly tailored to the ENEM examination (National Secondary School Exam). For this preliminary study, 12 texts written by 10 students with ASD were identified and compared with the remaining 2631 texts written by their typical peers. The students with ASD were formerly diagnosed as such.

Basic NLP was done to the whole corpus of essays, comparing students with ASD numbers and the general population. Besides, the texts were converted to a network of terms using Sobek [21]. This process was carried out so network analysis could be done to the corpus, allowing the use of graph theory and the study of the relationships between terms and the inference of text characteristics.

Due to the small sample of students with ASD, this paper will focus on general measurements of network analysis, such as mean degree, distance and transitivity of those networks. All data was processed using R, version 4.2.2.

6 Basic NLP Results

A few NLP metrics were selected to understand patterns in students' writing behavior.

Number of Sentences: The general average was 10.17 sentences in each text, while the median was 10 sentences. Standard deviation was found to be 5.02. Three essays from students with ASD had the number of sentences above average.

Word Count: The group mean was 256.9 words, with a median of 261 words, and standard deviation of 119.81. Once again students with ASD had values somewhat distant from those.

Average Sentence Length: The mean of average sentence length across the group was 27.05 words, with a median of 24.625 words. Standard deviation was 12.95. On this metric, 3 essays written by students with ASD had above average values.

Sentiments Mean: Sentiment analysis was used to analyze the essays. Due to the argumentative nature of the texts, results were expected to be close to zero. In fact, the general sentiment mean was 0.0002, with median 0 and standard deviation 0.005.

Basic NLP Results showed that the autistic students in this study showed variations in their writing patterns when compared to averages from their neurotypical peers. However, these patterns are not easily identifiable and seem to be highly individual. Number of sentences, word count and sentence length appear to be the metrics with a more noticeable pattern, since in all 3 of those, 9 out of 12 essays had values below average. This could suggest a tendency of writing short and fewer sentences, and generally using words with a more negative connotation. However, this data is not sufficient for drawing strong conclusions since no definitive pattern was found.

7 Network Analysis Results

The following analysis was conducted through the conversion of essays to a web of terms using text mining tool Sobek [21].

Average Degree: This network measure indicates how many connections to different nodes there are in a specific node. A higher average degree suggests a more densely connected network of terms, indicating stronger associations between terms in the text. A lower average degree indicates a sparser network with fewer connections between terms. The average degree was found to be 3.04, with median value of 3.125 and standard deviation of 0.836. Five essays of students with ASD had average degree above general mean, the highest being 3.78, while 7 essays had values below mean, the lowest being 0.79.

Average Distance: This metric refers to the shortest path between two nodes. It provides insights into the overall connectedness and information flow within the text. A lower average distance indicates that terms in the text are closer to each other, which suggests an interconnected chain of ideas in a text. The general average distance across networks was 2.12, with median 2.08 and standard deviation 0.6. Three essays of students with ASD had average distance above the general mean, the highest being 3.61, and the other 9 essays had values below mean, the lowest being 1.

Transitivity: Network transitivity refers to the probability that two nodes that are connected to a common node are also connected to each other, i.e. that triangles are formed in the network. The measure can indicate the extent to which terms in the text are clustered together, with a higher transitivity indicating a more tightly clustered network and therefore a text with clustered and dense ideas, understanding the overall structure of the text network. General average transitivity was found to be 0.23, with median of 0.18 and standard deviation of 0.16. Five essays written by students with ASD had values above average, with the highest network transitivity being 1, while 7 essays written by students with ASD had below-average values, the lowest being 0.

General Network Analysis Results were similar to those of basic NLP analysis. It was evident that for the measurements selected (degree, distance and transitivity), students with ASD had a range of atypical values, but with no evident pattern could be traced. Network Average Distance was the metric with higher congruence, with 9 out of 12 essays with below-average values. More data is necessary to better understand and possibly generalize these results.

8 Conclusions and Future Directions

Natural Language Processing (NLP) and Network Analysis (NA) have shown potential in detecting atypical patterns in the writing styles associated with Autism Spectrum Disorder (ASD). While these tools successfully identify deviations from typical linguistic and structural norms, they fall short of providing a comprehensive understanding of the unique writing patterns specific to individuals with ASD. The complexity of ASD expression and individual's characteristics and inconsistencies poses a challenge for NLP and NA to pinpoint a singular direction in interpreting the ASD writing, and at the moment is not possible to verify if such singular direction exists. Many AI systems rely on measures derived from NLP and NA. In that sense, the identified atypicalities raise concerns regarding the suitability of existing systems for ASD students. The reliance on generic linguistic and structural measures may contribute to the generation of frustrating experiences for ASD students and unhelpful tools for educators, as these systems may not be able to accommodate their distinctive communication styles and limitations. To address this challenge, it is imperative to tailor AI systems explicitly to the nuanced needs of ASD students, ensuring that these technologies enhance rather than hinder their learning experiences. This would require further research and AI systems that do not rely on generalization of metrics as a measurement of writing success.

The current limitations in adapting AI systems for ASD students highlight a significant bias in the existing AI landscape. The predominant focus on samples from neurotypical writers in the development of AI systems has created a bias that may render these technologies less suitable for the distinctive needs of individuals with ASD and other atypical students, that are still to be fully understood. The lack of representation from ASD writers in training datasets may contribute to the difficulty of these systems in accurately capturing the nuances of atypical writing patterns. Recognizing and addressing this bias becomes paramount in the pursuit of developing inclusive and effective AI tools for individuals with ASD. Finally, it is essential to acknowledge the limitations of the current findings, primarily stemming from the utilization of a small sample size of students with ASD and their essays. Generalizing the outcomes to a broader population of ASD students demands further research with larger samples. More extensive research is crucial to understand how AI systems can be effectively adapted to the highly individualized needs and limitations of atypical students.

Acknowledgments. This research received funding from CAPES/PROEX, Brazil, and the Agencia Nacional de Investigación y Innovación, Uruguay, Proj. FSED_2_2021_1_169701.

References

1. Bigolin, M., Santanna, M., Albilia, C., Reategui, E.B., Barcellos, P.S.C.C.: RevisãoOnline: Sistema de revisão por pares para apoio ao processo de produção de textos dissertativo-argumentativos. Anais do XXVII Ciclo de Palestras sobre Novas Tecnologias na Educação, pp. 163–173 (2019)
2. Caldeira, S.M.G., Petit Lobão, T.C., Andrade, R.F.S., Neme, A., Miranda, J.G.V.: The network of concepts in written texts. Europ. Phys. J. B **49** (2006)

3. Carley, K.M.: Network text analysis: the network position of concepts. In: Roberts, C.W. (ed.) Text Analysis for the Social Sciences: Methods for Drawing Statistical Inferences From Texts and Transcripts. Routledge (1997)
4. Cavalcanti, A.P., et al.: How good is my feedback? In: LAK, 2020, Frankfurt, Germany. Tenth International Conference on Learning Analytics & Knowledge. New York: ACM (2020). https://doi.org/10.1145/3375462.3375477
5. Coffin, A.B., Myles, B.S., Rogers, J., Szakacs, W.: Supporting the writing skills of individuals with autism spectrum disorder through assistive technologies. In: Cardon, T. (eds) Technology and the Treatment of Children with Autism Spectrum Disorder. Autism and Child Psychopathology Series. Springer, Cham (2016). https://doi.org/10.1007/978-3-319-20872-5_6
6. Evers, A., Finatto, M.J.B.: Corpus linguistics, textual lexicon-statistics and natural language processing: perspective for vocabulary studies in textual productions. GTLex Magazine, [sl], vol. 1, no. 2, p. 271 (2016). https://doi.org/10.14393/lex2-v1n2a2016-3
7. Ferreira, W., et al.: Panorama das Publicações Nacionais sobre Autismo, Educação e Tecnologia. In: Brazilian Symposium on Computers in Education (Simpósio Brasileiro de Informática na Educação - SBIE), p. 913 (2018). https://doi.org/10.5753/cbie.sbie.2018.913
8. Finnegan, E., Accardo, A.L.: Written expression in individuals with autism spectrum disorder: a meta-analysis. J. Autism Develop. Disord. **48**, 868–882 (2018). https://doi.org/10.1007/s10803-017-3385-9
9. Gillespie, A., Graham, S.: A meta-analysis of writing interventions for students with learning disabilities. Except. Child. **80**(4), 454–473 (2014). https://doi.org/10.1177/0014402914527238
10. Hilvert, E., Davidson, D., Scott, C.M.: An in-depth analysis of expository writing in children with and without autism spectrum disorder. J. Autism Dev. Disord. **49**(8), 3412–3425 (2019). https://doi.org/10.1007/s10803-019-04057-2
11. Jung, P.-G., McMaster, K. L., delMas, R.C.: Effects of early writing intervention delivered within a data-based instruction framework. Except. Child. **83**(3), 281–297 (2016)
12. Khosravi, H., et al.: Explainable artificial intelligence in education. Comput. Educ.: Artif. Intell. **3** (2022) https://doi.org/10.1016/j.caeai.2022.100074
13. Lang, S., Aull, L., Marcellino, W.: A taxonomy for writing analytics. J. Writ. Anal. [sl] **3**(1), 13–37 (2019). https://doi.org/10.37514/JWA-J.2019.3.1.03
14. Lee, Y.-J.: The long-term effect of automated writing evaluation feedback on writing development. English Teach. **75**(1), 67–92 (2020). https://doi.org/10.15858/engtea.75.1.202003.67
15. Martins, M.G.D.C.: Textual complexity and school progression in two registers: a correlation study based on a quasi-longitudinal corpus. Doctoral thesis, University of Lisbon (2016). https://repositorio.ul.pt/handle/10451/23963. Accessed on 29 Oct 2023
16. Meira, R.R., Weiand, A., Reategui, E., Bigolin, M., Motz, R.: A Analítica da Escrita para Identificação de Indicadores de Qualidade Textual. Revista Novas Tecnologias na Educação, Porto Alegre, vol. 21, no. 2, pp. 342–351 (2023). https://seer.ufrgs.br/index.php/renote/article/view/137756
17. McNamara, D.S., et al.: Automated Evaluation of Text and Discourse with Coh-Metrix. Cambridge University Press, Cambridge (2014)
18. Mizumoto, A., Eguchi, M.: Exploring the potential of using an AI language model for automated essay scoring. Res. Methods Appl. Linguist. **2**(2), 100040 (2023). https://doi.org/10.1016/j.rmal.2023.100050
19. Oliveira, H., et al.: Towards explainable prediction of essay cohesion in Portuguese and English. In: LAK, 2023, Arlington, TX. 13th International Learning Analytics and Knowledge Conference. New York: Association for Computer Machinery (2023). https://doi.org/10.1145/3576050.3576152

20. Patou, P.-A., Cordy, M.: Towards context-aware automated writing evaluation systems. In: Proceedings of the 1st ACM SIGSOFT International Workshop on Education through Advanced Software Engineering and Artificial Intelligence, August 2019, pp. 17–20 (2019). https://doi.org/10.1145/3340435.3342722

21. Reategui, E., Bigolin, M., Carniato, M., dos Santos, R.A.: Evaluating the performance of SOBEK text mining keyword extraction algorithm. In: Holzinger, A., Kieseberg, P., Tjoa, A.M., Weippl, E. (eds) Machine Learning and Knowledge Extraction. CD-MAKE 2022. Lecture Notes in Computer Science, vol. 13480, p. 15. Springer, Cham. (2022). https://doi.org/10.1007/978-3-031-14463-9_15

22. Reilly, E.D., Stafford, R.E., Williams, K.M., Corliss, S.B.: Evaluating the validity and applicability of automated essay scoring in two massive open online courses. Int. Rev. Res. Open Distrib. Learn. **15**(5) (2014). https://doi.org/10.19173/irrodl.v15i5.1857

23. Roscoe, R.D., Varner, L.K., Crossley, S.A., McNamara, D.S.: Developing pedagogically-guided algorithms for intelligent writing feedback. Int. J. Learn. Technol. **8**(4), 362–381 (2014)

24. U.S. Department of Education, Office of Educational Technology: Artificial Intelligence and Future of Teaching and Learning: Insights and Recommendations. Washington, DC (2023). https://tech.ed.gov

25. Ullmann, T.D.: Automated analysis of reflection in writing: validating machine learning approaches. Int. J. Artif. Intell. Educ. **29**, 217–257 (2019). https://doi.org/10.1007/s40593-019-00174-2

26. Wilson, J., Czik, A.: Automated essay evaluation software in English Language Arts classrooms: Effects on teacher feedback, student motivation, and writing quality. Comput. Educ. **100** (2016). https://doi.org/10.1016/j.compedu.2016.05.004

27. Xia, M., Kochmar, E., Briscoe, T.: Text readability assessment for second language learners. In: BEA@ACL, 2016, San Diego, CA. 11th Workshop on Innovative Use of NLP for Building Educational Applications. Association for Computational Linguistics (2016)

Towards Automated Slide Augmentation to Discover Credible and Relevant Links

Dilan Dinushka[✉], Christopher M. Poskitt[✉], Kwan Chin Koh[✉],
Heng Ngee Mok[✉], and Hady W. Lauw[✉]

Singapore Management University, Singapore, Singapore
{dinushkasa,cposkitt,kckoh,hnmok,hadywlauw}@smu.edu.sg

Abstract. Learning from concise educational materials, such as lecture notes and presentation slides, often prompts students to seek additional resources. Newcomers to a subject may struggle to find the best keywords or lack confidence in the credibility of the supplementary materials they discover. To address these problems, we introduce SLIDE++, an automated tool that identifies keywords from lecture slides, and uses them to search for relevant links, videos, and Q&As. This interactive website integrates the original slides with recommended resources, and further allows instructors to 'pin' the most important ones. To evaluate the effectiveness of the tool, we trialled the system in four undergraduate computing courses, and invited students to share their experiences via a survey and focus groups at the end of the term. Students shared that they found the generated links to be credible, relevant, and sufficient, and that they became more confident in their understanding of the courses. We reflect on these insights, our experience of using SLIDE++, and explore how Large Language Models might mitigate some augmentation challenges.

1 Introduction

Students go through considerable changes in their learning patterns once they commence their tertiary education. Rather than following a managed learning path, students must transition to an independent learning approach that relies on their own research. Self-learning becomes even more important when the provided course materials are concise or limited, especially when they are the sole reference materials for certain lectures. Thus, it would be beneficial to supplement such materials with credible and relevant links to additional resources.

As an initial attempt to support students in their self-learning journeys, we developed SLIDE++[1], a system for viewing and managing augmented slides (Fig. 1). Our platform enables students to access slide sets uploaded by their university instructors, automatically enriched with complementary resources curated from the web. SLIDE++ uses statistical methods to identify the most significant keywords inside slides, then uses them to query the web for relevant

---

[1] https://slideplusplus.preferred.ai/.

A. M. Olney et al. (Eds.): AIED 2024 Workshops, CCIS 2151, pp. 200–208, 2024.
https://doi.org/10.1007/978-3-031-64312-5_24

articles, videos, and discussions. Instructors can further refine the results by adding/removing resources and 'pinning' the most important recommendations.

We trialled SLIDE++ in four undergraduate computing courses, then conducted surveys and focus groups at the end of term. Students shared that they found the generated links to be credible, relevant, and sufficient, and increased their confidence in their understanding of the courses. We also explored the use of Large Language Models (LLMs), finding that GPT4 can potentially complement SLIDE++'s statistics-based approach on image-heavy slides.

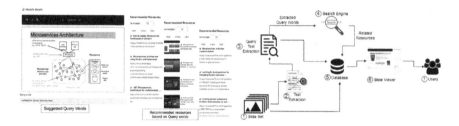

Fig. 1. SLIDE++ website preview (left) and architecture (right): 1. upload slide set; 2. extract & preprocess; 3. query text extraction; 4. search links; 5. store in database; 6. SLIDE++ website; 7. end users (students)

2   Related Work

Although we are not aware of another tool similar to SLIDE++, there are some works that analyse and augment lecture slides for other purposes. Wolfe [9], for example, creates annotated slides that can be used by 'conversational agents' that students can interact with. Similarly, Shimada et al. [7] analyse lecture slides to generate summarised versions of the materials to read as previews. Like SLIDE++, this tool also automatically analyses a slide set, but instead of providing additional references, it provides a summarised version of its content.

A related approach to recommending links in slides is recommending content personalised to the student [8]. For example, Oliveira et al. [4] generate such recommendations in learning management systems: they take into account the learning profiles of students in order to suggest activities and materials that better suit their learning styles. Barria-Pineda et al. [2] integrate personalised recommendations into a Java programming practice system, with the system able to explain its recommendations based on the student's prior interactions. These approaches are quite different to SLIDE++ in that existing catalogues of content, activities, or exercises are utilised as the recommendations.

Recently, practitioners are beginning to explore the potential pedagogical benefits of AI tools such as ChatGPT [1,3,5], e.g. for generating new content, explaining examples, or analysing error messages. Our preliminary tool is effective at generating keywords using simpler statistical methods, but its approach could also be used in conjunction with advanced AI models, e.g. leveraging Chat-GPT to generate better query terms to improve the the relevance of curated resources.

3   SLIDE++ Tool

The SLIDE++ website's primary functionality centres around recommending three main categories of resources: web pages, videos, and discussion (Q&A) threads on forums such as Stack Overflow as shown in Fig. 1 (left).

The overall architecture and content augmentation process is shown in Fig. 1 (right). To initiate the content augmentation process, instructors upload their teaching slides to SLIDE++ for student viewing. A statistics-based keyword extraction mechanism (detailed shortly) is used to generate potential query words for each slide, which in turn are used to extract and display resources from the web. Instructors can subsequently refine these keywords based on their expertise, and also 'pin' any important links retrieved by the process to highlight those resources to students.

We implemented a logging mechanism to capture student behaviour such as their login patterns, engagement with course content (including slide set viewing and navigation) and their interactions with the provided supplementary materials such as web pages, videos, and Q&A links.

Query Word Extraction. In SLIDE++, we employ a blend of statistical methods to automate the extraction of query words. The aim is not to generate flawless keywords for each slide but rather to serve as an assisting mechanism for instructors to properly define the query words for a given slide.

We initially transform the text using the TF-IDF technique [6], which gauges the relative importance of a word within the given corpus. (In our context, a single document corresponds to a slide in a slide deck, and a term is a word present in the slide.) The Eqs. 1, 2 and 3 provide the related definitions for the representations utilised in our query text extraction mechanism.

$$tf(t, i) = f_{(t,i)} \tag{1}$$

$$idf(t, D_m) = log\left(\frac{|D_m| + 1}{df(t, D_m) + 1}\right) + 1 \tag{2}$$

$$x_t = tfidf(t, i, D_m) = tf(t, i) * idf(t, D_m) \tag{3}$$

The term $f_{(t,i)}$ denotes the frequency of word t in the document i. In our scenario, it can be considered as a count of a word's appearance in a slide in the given slide set. Term $df(t, D_m)$ specifies the number of documents in the corpus D_m in which a word t appears uniquely. The x_t term provides the final representation of word t, with respect to the considered corpus. Beyond word occurrences, to leverage the structural details in PowerPoint slide sets, we use bullet levels inside of a slide and consider it as a weighting factor for word representations. Below are the relevant equations to calculate the weighted values for the considered tokens in a given slide set based on order and level information.

$$penalty_{(t,l)} = tf(t,i) * e^{-log(\beta_l)*l} \tag{4}$$

$$penalty_{(t,o)} = \begin{cases} tf(t,i) * e^{-log(\beta_o)*o} & \text{if } W \geq \omega \\ 1 & \text{otherwise} \end{cases} \tag{5}$$

$$weight_{t,i} = penalty_{(t,l)} * w_l + penalty_{(t,o)} * (1 - w_l) \tag{6}$$

Here, t, l, i, o, and W are symbols denoting the considered token, token level, document (slide), token order value in the slide, and the total number of words in the document respectively. Furthermore, w_l denotes a parameter that adjusts the values coming from level penalty and order penalty, and ω is a threshold parameter we set that limits the effect of penalties on slides with low text content.

In addition to considering the importance of word ordering within a slide, we utilise the importance of slide titles as an additional factor along with a Gaussian window-based method to incorporate information from neighbouring slides.

$$token_{(t,i)} = weight_{(t,i)} * \alpha + title_{(t,i)} * (1 - \alpha) \tag{7}$$

$$x_{(t,i)} = token_{(t,i)} * P(i \mid \mathcal{N}(0, \sigma^2)) \tag{8}$$

Here, $x_{(t,i)}$ represents the value of term t in document i. The Gaussian distribution $\mathcal{N}(0, \sigma^2)$ represents a predefined variance σ^2 and zero mean. We calculate $x_{(t,i)}$ for a given slide text content around its neighbouring slides (a hyperparameter) to calculate the final representation values for all vocabulary tokens.

For a given slide we select the highest ranking tokens based on the above weights and display those in the SLIDE++ tool as the selected query words. Note that these query words are *suggestions*, in that they can be overridden by the instructor, but they help to reduce the load on instructors.

4 Study and Discussion

4.1 Study Design

To evaluate the effectiveness of SLIDE++, we deployed the tool in a field study involving four undergraduate computing courses (running in parallel) at our institution. Instructors of the four courses agreed to participate by utilising the website to upload their slides and verifying the generated keywords and supplementary resources. In the first lesson of the term, instructors demonstrated the SLIDE++ website, and invited students to use it as a learning resource throughout the course. Students were not forced to use the tool (slides were also distributed through the usual learning management system). At the end of the term, students were invited to participate in an online survey and a focus group session. This participation was entirely optional for the students, but came with a small monetary token of appreciation.

We trialled SLIDE++ on two programming courses (*Data Structures & Algorithms, Web Application Development*) as well as two software engineering courses (*Software Project Management, IT Solution Lifecycle Management*). Based on information recorded in the system, over 400 unique student accounts were registered on the website throughout the term, and over 40 slide sets were uploaded by instructors across the four courses.

For the survey, we were able to gather 49 student responses, with the following distribution across the four courses respectively: 33.33%, 11.11%, 22.22% and 33.33%. Among the respondents, 57% of them reported their gender as female, 39% as male, and the remaining 4% preferred not to say. (Note that our student cohort is approximately 40% female and 60% male.) Furthermore, 73% of the students shared that they were taking our Information Systems degree, with the others all taking Computer Science. Of these 49 students, 13 students accepted our invitations to join one of three focus group sessions.

Based on the data we collected from SLIDE++ over the considered term, the 'activity count' on the website hit around 6,500 actions (clicking links, opening slide sets, traversing slides) per week in total. Among the students who clicked at least one recommended link, the overall average number of clicks rises to 7.64, with a maximum of 41. This data indicates a reasonable amount of tool usage and provides motivation to explore the use-cases further.

4.2   Relevancy, Credibility, and Sufficiency of SLIDE++

In our survey, we focused on questions to explore the participants' thoughts on the generated query words and recommended web resources (web links, videos, and Q&A). We used a 7-point Likert scale to measure each participant's evaluation of the relevance, credibility, and sufficiency of the recommended resources.

Figure 2 summarises the responses received to these questions. Overall, students agreed that SLIDE++ was able to provide relevant, credible and sufficient additional resources. Web links received the highest percentage of positive responses, whereas Q&A links and query words received slightly lower responses. We believe this is because good Q&A links (i.e. specific technical discussions on Stack Overflow) may not exist for the content of every slide, especially when compared to the links curated from around the broader web.

During the focus group sessions we discovered that even though students appreciated the links generated by the SLIDE++, they still preferred to have an explicit indication of credibility regarding the suggested contents. The reasoning behind this request seems to be originating from their experiences with the 'hallucination' effects of novel generative models. In focus group sessions, students mentioned issues with tools like ChatGPT and it seems that such technologies have made them suspicious of anything automatically generated:

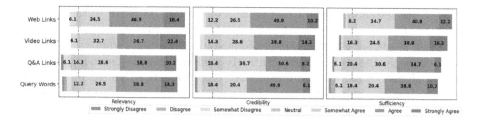

Fig. 2. Extent students agree that resources are relevant, credible, and sufficient

"...I actually don't use ChatGPT; instead I use Bing AI. It's just based on ChatGPT, but because it provides links and sources to what it's talking about, [I can verify] if it's hallucinating or not."

4.3   Perceived Benefits and Impact of SLIDE++

In our survey, we asked students for their feedback on how useful they found the SLIDE++ tool for four objectives. Figure 3 (left) summarises how useful they perceived the tool for those objectives. The overall responses lean positive, with perceived usefulness for finding additional explanations being the highest.

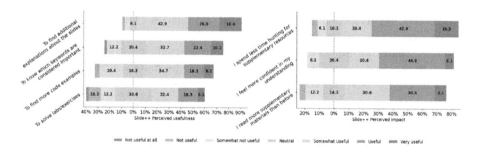

Fig. 3. Perceived usefulness and impact of SLIDE++

During the focus groups, some students shared that the tool was particularly helpful for more 'subjective' topics (e.g. the agile software process 'scrum' as taught in Software Project Management) and appreciated having links to additional articles with different perspectives to those of their instructors.

While the curated resources themselves were found useful, the focus group participants conveyed that they did not find it helpful to see the query words that were generated. We suspect that this is because SLIDE++ currently displays them as a list of keywords, rather than as a description of the slide.

Finally, we investigated the impact of SLIDE++ on students' confidence in their understanding, as well as how the tool has changed their approach to finding and using supplementary material. Figure 3 (right) summarises the responses.

Overall, the responses suggest that students feel more confident in their under-standing of course materials and seemingly spend less time finding resources.

A key lesson we learnt in running this study is the importance of ensuring new pedagogical tools integrate seamlessly with the learning methods and sys-tems students normally use. For example, some students suggested to integrate SLIDE++ into the university learning management systems, or even directly in slides via a plug-in (as they prefer to annotate slides directly as part of their learning process). We hope to explore such possibilities in the future.

5 Generative Model-Based Query Sentence Generation

SLIDE++ currently employs an efficient statistics-based approach, but we also wanted to assess whether LLMs may help to improve the query words generated—especially for image-heavy slides that lack keywords to extract.

We implemented a proof of concept query word generator using the OpenAI GPT4-VISION API, and used it to extract key descriptions of slide contents. Instead of just using the text inside of a slide, we incorporated the full slide image as an input. Furthermore, while the statistics-based approach generates a list of keywords, we wanted to explore whether GPT4 could present those keywords in a more intelligible *sentence* describing the content of the slide (while also being useful to extract resources from the web).

Table 1 shows the course-wise BERTScore [10] comparisons between the statistics-based approach and GPT4-based summaries. The score differences between summary values are not significant, indicating that the statistics-based approach performs well, despite being a more cost-efficient method. However, the query sentences generated by GPT4 appear to be more readable than those in the statistics-based approach as can be seen from the examples in Table 2.

Table 1. BERT F1-scores of statistics-based and LLM-based summaries

Course Name	Statistics-based	LLM-based
Web Application Development	0.77	0.81
Data Structures and Algorithms	0.83	0.80
IT Solution Lifecycle Management	0.80	0.80
Software Project Management	0.83	0.81

We will be developing SLIDE++ further to explore combinations of these pos-sibilities. For example, the statistics-based approach could be used for text-heavy slides, with the LLM-based approach for image-heavy ones. LLMs may also be helpful for generating summaries of the whole slide set, similar to the approach of Shimada et al. [7].

Table 2. Examples of query text from statistics-based and LLM-based approaches

Statistics-based	LLM-based
type linked declaring array arrays elements initialize java	Initializing Java arrays with types and sizes
analysis algorithm exists proof	Quantifiers in logical statements described
dom getElementsByTagName javascript array index	Accessing elements in HTML by tag name
granularity service team size microservices	Microservice size varies, consider business needs and two-pizza rule

6 Conclusion

In order to provide students with an enhanced learning experience, we developed a website, SLIDE++, that offers functionality for automatically generating query words and links to supplementary materials. We trialled it in four undergraduate computing courses, inviting students to participate in a survey and focus group discussion at the end of the term. We found that students deemed the generated links to be relevant, credible, and sufficient, and that SLIDE++ helped to improve their confidence in their understanding of the course content. Students conveyed the usefulness of the tool in a variety of scenarios, e.g. generating links to explain diagrams in slide sets, finding interesting articles for class reflection exercises, and discovering different (but credible) perspectives to those of their instructors.

In the future, we plan to further explore LLM-based approaches for more advanced query extraction (especially on image-heavy slides). In addition, we would like to explore how the slide augmentation features of SLIDE++ might be integrated into common educational tools such as learning management systems, in order to reduce the friction currently imposed by requiring the use of a standalone website.

Acknowledgements. This research/project is supported by the Ministry of Education, Singapore under its Tertiary Education Research Fund (MOE Reference Number: MOE2021-TRF-013). Any opinions, findings and conclusions or recommendations expressed in this material are those of the author(s) and do not reflect the views of the Ministry of Education, Singapore. Hady W. Lauw gratefully acknowledges the support by the Lee Kong China Fellowship awarded by Singapore Management University.

References

1. Adeshola, I., Adepoju, A.P.: The opportunities and challenges of ChatGPT in education. Interact. Learn. Environ. **1**, 1–14 (2023)
2. Barria-Pineda, J., Akhuseyinoglu, K., Želem-Ćelap, S., Brusilovsky, P., Milicevic, A.K., Ivanovic, M.: Explainable recommendations in a personalized programming practice system. In: Roll, I., McNamara, D., Sosnovsky, S., Luckin, R., Dimitrova,

V. (eds.) AIED 2021. LNCS (LNAI), vol. 12748, pp. 64–76. Springer, Cham (2021). https://doi.org/10.1007/978-3-030-78292-4_6

3. Montenegro-Rueda, M., Fernández-Cerero, J., Fernández-Batanero, J.M., López-Meneses, E.: Impact of the implementation of ChatGPT in education: a systematic review. Computers **12**(8), 153 (2023)

4. Oliveira, A., Teixeira, M.M., Neto, C.D.S.S.: Recommendation of educational content to improve student performance: an approach based on learning styles. In: CSEDU (2), pp. 359–365 (2020)

5. Ouh, E.L., Gan, B.K.S., Shim, K.J., Wlodkowski, S.: ChatGPT, can you generate solutions for my coding exercises? An evaluation on its effectiveness in an undergraduate Java programming course. In: ITiCSE (1), pp. 54–60. ACM (2023)

6. Salton, G., McGill, M.J.: Introduction to Modern Information Retrieval. McGraw-Hill, Inc. (1986)

7. Shimada, A., Okubo, F., Yin, C., Ogata, H.: Automatic summarization of lecture slides for enhanced student preview-technical report and user study. IEEE Trans. Learn. Technol. **11**(2), 165–178 (2018)

8. Verbert, K., et al.: Context-aware recommender systems for learning: a survey and future challenges. IEEE Trans. Learn. Technol. **5**(4), 318–335 (2012)

9. Wölfel, M.: Towards the automatic generation of pedagogical conversational agents from lecture slides. In: Fu, W., Xu, Y., Wang, S.-H., Zhang, Y. (eds.) ICMTEL 2021. LNICST, vol. 388, pp. 216–229. Springer, Cham (2021). https://doi.org/10.1007/978-3-030-82565-2_18

10. Zhang, T., Kishore, V., Wu, F., Weinberger, K.Q., Artzi, Y.: BERTScore: evaluating text generation with BERT. In: ICLR. OpenReview.net (2020)

Learner Agency in Personalised Content Recommendation: Investigating Its Impact in Kenyan Pre-primary Education

Chen Sun[1]([⊠]), Louis Major[1], Nariman Moustafa[2], Rebecca Daltry[3], and Aidan Friedberg[4]

[1] University of Manchester, Manchester, UK
chen.sun@manchester.ac.uk
[2] Open Development & Education, Barnet, UK
[3] Jigsaw, Richmond, London, UK
[4] EIDU, Berlin, Germany

Abstract. There is a lack of understanding regarding how pre-primary learners exercise their agency in their learning processes when interacting with AI-powered digital personalised learning (DPL) tools. This study aims to address the gap by investigating the interaction between pre-primary learners' agency and a DPL tool in a Kenyan classroom setting. A total of 76,479 pre-primary learners participated in a two-month experiment, where each learner was randomly assigned to two partitions. Learners in the control partition followed the learning content designated by an algorithm within an adaptive DPL tool. In the experimental partition, learners received two additional learning units to choose from as well as the default content unit. Learning outcomes were assessed through six summative test units measuring literacy and numeracy skills. The results revealed that learners who were provided with a choice scored significantly higher in four out of the six test units. This study highlights the potential that pre-primary learners can exercise some degree of agency and direct their own learning within a structured set of choices provided by a DPL tool. Future research is needed for a comprehensive understanding of pre-primary learner agency.

Keywords: Learner agency · digital personalised learning · human-AI collaboration · pre-primary education · content recommendation · low- and middle-income country

1 Introduction

The application of AI in education can benefit learners in a variety of areas, such as increasing engagement, providing personalised learning experiences, offering timely feedback, and ultimately improving learning outcomes [18]. Digital personalised learning (DPL), which is often powered by AI, has demonstrated positive effects on learning outcomes, by tailoring content to students' characteristics and learning needs [15, 19]. However, there is a potential concern that the automation inherent in DPL might reduce human autonomy in the learning process [8].

A. M. Olney et al. (Eds.): AIED 2024 Workshops, CCIS 2151, pp. 209–216, 2024.
https://doi.org/10.1007/978-3-031-64312-5_25

To ensure a beneficial integration of DPL in educational environments, it is important to maintain a balance between DPL automation and learner agency [14]. Learner agency refers to the degree of freedom and control that learners have to make choices, exert influence, and take responsibilities for their learning when interacting with AI-powered learning systems [13, 14]. It has been identified as a key feature to promote learner engagement, motivation, and effective learning [1, 6]. Identifying the optimal moments and methods for integrating learner agency into DPL is challenging but essential for its effective deployment and to unleash its full potential [17].

Providing learners with choices in learning paths, content, and pace, is considered as an important feature of DPL, as it may enhance learner agency and motivate them through their learning processes [1, 16]. A key strategy for activating learner agency involves enabling learners to navigate freely through learning content, such as instructional videos and game levels, rather than confining them to a fixed sequence of materials [9, 13]. However, it is worth noting that unrestricted freedom in determining learning paths does not necessarily lead to improved learning outcomes [13]. Therefore, careful design and rigorous investigation of learner agency should be conducted to gauge the degree of agency that effectively enhances learning within the Artificial Intelligence in Education (AIED) field.

The majority of agency research focuses on post-primary learners, from lower secondary level up to university. However, understanding early-grade learners' agency in the AIED field plays a key role in incorporating their perspectives into the design and implementation of educational technologies, including who makes the choice, what options are available to them, and the context in which choices occur [3]. [4] emphasised the importance of examining how young learners engage with AI-powered content recommender systems, considering factors such as their age and developmental readiness. This suggests a notable gap in our understanding of how agency is expressed when pre-primary learners are interacting with AI-powered learning environments.

Although not specifically within the AIED field, studies have shown that early-grade learners have the ability to manifest agency in their learning. In traditional classroom settings, these early-grade learners recognise the need to actively engage in their learning, and understand how various pedagogical activities can facilitate or hinder agency among diverse learners [11, 12]. Listening to pre-primary learners is crucial because their perspectives can help construct learning environments that provide meaningful opportunities for them to make decisions about their education [11]. Building on this foundation, this current study investigated how pre-primary learners interact with and make choices about learning content recommended by a DPL system.

1.1 Contributions

This study contributes to the literature on integrating agency within DPL systems by providing learners opportunities to select learning content from a collection of algorithm-generated choices. The study has a particular focus on how pre-primary learners exercise this described agency in a DPL tool within a low- and middle-income country (LMIC). The research informs the AIED community about the plausible, scalable implementation of DPL in typical school systems, taking into consideration learner agency in determining their own learning paths. The main research question addressed in the paper is: What's the impact of providing learners with choices in DPL on pre-primary numeracy and literacy outcomes in Kenya?

2 Method

2.1 EIDU DPL Tool

EIDU is a classroom-based intervention in government run schools, providing digitised structured pedagogy resources for teachers and DPL literacy and numeracy gamified learning exercises for students, which are curriculum aligned. The software is accessed through low-cost android devices with each learner having an account linked to their class. There are one to two devices per class, with teachers setting up a dedicated table in the classroom for learners to engage individually with the devices for 5-min sessions. This can be during regular classroom instruction or during break times. After a learner finished a session the next learner's profile appears to facilitate peer to peer handover without any teacher involvement. Over the study period, learners engaged with the device for an average of 30 min per week. EIDU is aligned with the Kenyan curriculum in domains (literacy and numeracy), strands (e.g., Classification), and substrands (e.g., Matching & Grouping) and all content has been approved by the Kenyan Institute for Curriculum Development (KICD). Teachers select which area of the curriculum they want their learners to focus on prior to providing them with the device. EIDU's personalisation system then selects a content unit (a predefined game made up of 3 to 4 exercises) for a given learner based on that learner's performance history.

2.2 Personalisation

The adaptive feature of the DPL tool is achieved by providing a sequence of learning content that tailors to learners' performance. A learner's historical binary outcome on previous content units (successful or non-successful completion) is provided as a sequence to a Long Short-Term Memory (LSTM) neural network, similar to the deep knowledge tracing model described by [10]. Given the low-connectivity environment, the model needs to be available locally on devices through TensorFlow Lite (https://www.tensorflow.org/lite).

For every available content unit in the teacher selected area, the model takes the learner history at time t, then predicts performance at $t + 1$, and then simulates the learner's knowledge state at $t + 2$ given this predicted performance. Knowledge state is defined as the average predicted success rate on the vector of available content units in a teacher-selected area. The content unit expected to provide the largest increase in knowledge state at $t + 2$, taking into account probability of completion at $t + 1$, is then selected as the next optimal unit for a learner.

As part of their product development process EIDU runs A/B tests on new features to explore impacts on learning outcomes. Since the method described above returns a ranked list of content units for a given area, it is an arbitrary decision to provide the learner with only one unit. To explore whether providing learners with a greater degree of agency in selecting content would be beneficial, this study implemented an A/B test in the EIDU platform. In one partition, learners were shown a thumbnail preview of the top three units the personalisation algorithm expected to have the greatest impact, while in the other partition the default of only displaying a single unit, also with a thumbnail preview, was maintained.

2.3 Data Collection and Sampling

Prior to releasing any A/B test EIDU went through a development process where teachers were first consulted in discussion groups, prototypes are provided to individual teachers and learners. If feedback was positive, there was a release to a small group of dedicated Beta testing schools, who provided extensive feedback. Once the feature was deemed safe to release, schools where teachers and local government authorities have provided EIDU with digital gatekeeper consent at sign-up were included in the A/B test. This test lasted from August 28 to October 27, 2023. A total of 76,479 pre-primary learners participated. Learners were assigned into the two partitions using simple randomisation. All personal identifying data is always encrypted locally on devices with anonymised usage data being uploaded to EIDU servers when an internet connection is available. A data-sharing protocol was established to facilitate sharing of anonymised data for this study.

2.4 Summative Assessment

As learners engaged with the learning units, they were occasionally provided with dedicated assessment units instead of learning units. There were six assessment units used to evaluate this A/B test: initial sound identification, letter sounds, number discrimination, number identification, word sounds, word sounds Swahili. These assessment units are digitised versions of established assessment batteries such as EGMA, EGRA or MELQO [5]. Learners engage with no more than one assessment unit per day. This routine of continuous assessment creates longitudinal learning data which can be used to evaluate the impact of feature changes through A/B testing. Summative assessment scores were calculated based on the correctness – as the percentage of correctly answered items over the total number of items within a test unit. The possible range of the score was from 0 to 1.

2.5 Data Cleaning

Duplicated data points were removed before any in-depth check of the dataset. Data was excluded where an assessment was tagged as having ended through consecutive timeouts. This happens when a learner has either disengaged with the content and stopped touching the screen, or a teacher has removed the device mid-assessment without ending it. This represents approximately 0.1% of total assessments, so has negligible impact on analysis. The assessment outcomes that were larger than 1 were removed as these are generated due to data errors. The cleaning process resulted in a total of 288,339 assessment outcomes the default partition and 280,435 outcomes in the strand Choice partition. The latest assessment for each learner in each test unit type was selected resulting in 194,751 assessment outcomes in the default partition and 203,169 outcomes in the Choice partition. The loss of information is justified in this initial analysis to avoid unbalanced repeated measures being present as part of this initial analysis.

2.6 Analysis

Descriptive statistics, including the mean and standard deviation, were calculated on the correctness of summative assessment for each test unit. To understand how learners interact with the available choices, we computed the percentage of time a learner selected what would be the default option, which is the unit the personalisation model considered optimal. The primary objective was to ascertain whether the intervention—characterised by learner autonomy in content selection—significantly impacted educational outcomes in a government-run school setting.

In the context of our study, where assessments are undertaken sporadically across time, we opted for an ANOVA model after filtering for only the latest result per learner and per test unit. This choice was to ensure a robust evaluation of group differences taking into account the longitudinal nature of our data, where learners engaged with assessment units across different time points during the A/B test. Levene's test for equality of variances was performed to assess the assumption of homoscedasticity across groups, yielding a statistic of $F = 7.85$, $p = .005$. ANOVA is robust to the homogeneity assumption when group sizes are large and similar [2]. Secondary analysis using a regression model examined the impact of choice on learning outcomes by using the percentage of default option selections as a predictor for each test unit. This analysis was performed for the choice partition only.

3 Results

The descriptive statistics in Table 1 show that on average learners have low scores across all assessments, apart from Initial Sound Identification and Number Discrimination. This could be representative of general low levels of learning outcomes in this context but could also be due to higher levels of disengagement with digital assessments [5]. Learners in the Choice partition score marginally higher than learners in the Default partition in all test units apart from in Number Discrimination. Across both partitions there were at least 30,000 learners who completed each test unit type.

Our ANOVA analysis (Table 2) reveals that learners in the choice partition scored significantly higher in all test units apart from Number Discrimination and Word Sounds Swahili. This suggests that providing learners with increased agency in this pre-primary context did lead to higher learning outcomes in specific numeracy and literacy domains.

Our secondary analysis explored how the degree of agency impacted assessment outcomes in the Choice partition only. On average, learners selected the default choice 35% (SD 15%) of the time when offered with other options in literacy-related content. In contrast, for numeracy-related content, the default choice was selected 32% (SD 15%) of the time. This frequency of default choice per learner was added as a variable to the analysis. This suggests choosing the default option significantly and positively predicted the score of all test units (Table 2). This implies that although increasing agency in the choice partition led to higher overall outcomes it may be beneficial to restrict this choice further in future A/B tests for example by only providing 2 rather than 3 choices.

Table 1. Descriptive statistics of assessment score and sample size per partition per test unit

Test unit name	Assessment score Mean (SD)		Learner count	
	Default	Choice	Default	Choice
Initial Sound Identification	0.474 (0.313)	0.480 (0.315)	31513	30084
Letter Sounds	0.246 (0.171)	0.253 (0.173)	33843	32139
Number Discrimination	0.524 (0.249)	0.522 (0.249)	35856	34685
Number Identification	0.231 (0.262)	0.241 (0.266)	34788	33648
Word Sounds	0.273 (0.198)	0.278 (0.201)	33569	32123
Word Sounds Swahili	0.201 (0.169)	0.202 (0.172)	33600	32072

Table 2. Regression coefficients on group comparisons and agency effects per test item

	Group comparison	Choice partition only
Test unit name	Parition_id (coefficient)	Agency (content related) (coefficient)
Initial Sound Identification	.006**	.127***
Letter Sounds Short	.006***	.046***
Number Discrimination	−.002	.112***
Number Identification	.010***	.117***
Word Sounds Short	.005**	.061***
Word Sounds Swahili	.002	.085***

Note: ** $p < .05$, *** $p < .001$

4   Discussion and Future Work

This study investigated how providing pre-primary learners with an increased choice in sequencing their learning experience versus a restricted AI-powered sequence impacted learning outcomes. The results showed that providing learners with increased agency significantly improved learners' scores in numeracy and literacy domains compared with the learners who followed the system-generated learning content. This is in line with previous research from [7] showing positive effects of DPL on literacy and numeracy development. Importantly though this work enhances these findings by exploring how learners even at a young age can effectively act as humans-in-the-loop when it comes to AI systems. The fact that high use of the default option was associated with higher learning outcomes despite the primary results finding agency increased learning outcomes, suggests that a more restricted number of choices may be beneficial.

Although the study shows promising results in increasing agency of learners in digital spaces in the Kenyan pre-primary context, a few limitations exist that offer directions for future research. In particular, why certain content units were chosen by learners over others or how the expected gain in knowledge state across different choices modulated the effectiveness of learner agency. Exploring what other strategies foster learner agency could also provide insights for how DPL systems can be optimised to benefit learners [3].

Acknowledgments. This study was funded by the Bill and Melinda Gates Foundation (grant number 035197).

Disclosure of Interests.   Author Aidan Friedberg is an employee of EIDU, no other authors have competing interests.

References

1. Alamri, H., Lowell, V., Watson, W., Watson, S.L.: Using personalized learning as an instructional approach to motivate learners in online higher education: learner self-determination and intrinsic motivation. J. Res. Technol. Educ. **52**(3), 322–352 (2020). https://doi.org/10.1080/15391523.2020.1728449
2. Blanca, M., Alarcón, R., Arnau, J., et al.: Effect of variance ratio on ANOVA robustness: might 1.5 be the limit? Behav. Res. Meth. **50**, 937–962 (2018). https://doi.org/10.3758/s13428-017-0918-2
3. Brod, G., Kucirkova, N., Shepherd, J., et al.: Agency in educational technology: interdisciplinary perspectives and implications for learning design. Educ. Psychol. Rev. **35**, 25 (2023). https://doi.org/10.1007/s10648-023-09749-x
4. Brusilovsky, P.: AI in education, learner control, and human-AI collaboration. Int. J. Artif. Intell. Educ. **34**, 122–135 (2024). https://doi.org/10.1007/s40593-023-00356-z
5. Friedberg, A.: Can A/B testing at scale accelerate learning outcomes in low- and middle-income environments? In: Wang, N., Rebolledo-Mendez, G., Dimitrova, V., Matsuda, N., Santos, O.C. (eds.) Artificial Intelligence in Education. Posters and Late Breaking Results, Workshops and Tutorials, Industry and Innovation Tracks, Practitioners, Doctoral Consortium and Blue Sky, pp. 780–787. Springer Nature Switzerland, Cham (2023). https://doi.org/10.1007/978-3-031-36336-8_119
6. Høgheim, S., Reber, R.: Supporting interest of middle school students in mathematics through context personalization and example choice. Contemp. Educ. Psychol. **42**, 17–25 (2015). https://doi.org/10.1016/j.cedpsych.2015.03.006

7. Major, L., Francis, G.A., Tsapali, M.: The effectiveness of technology-supported personalised learning in low- and middle-income countries: a meta-analysis. Br. J. Edu. Technol. **52**(5), 1935–1964 (2021). https://doi.org/10.1111/bjet.13116
8. Molenaar, I.: Towards hybrid human-AI learning technologies. Eur. J. Educ. **57**(4), 632–645 (2022). https://doi.org/10.1111/ejed.12527
9. Nguyen, H., Harpstead, E., Wang, Y., McLaren, B.M.: Student agency and game-based learning: a study comparing low and high agency. In: Penstein Rosé, C., Martínez-Maldonado, R., Hoppe, H.U., Luckin, R., Mavrikis, M., Porayska-Pomsta, K., McLaren, B., du Boulay, B. (eds.) Artificial Intelligence in Education, pp. 338–351. Springer International Publishing, Cham (2018). https://doi.org/10.1007/978-3-319-93843-1_25
10. Piech, C., et al.: Deep knowledge tracing. In: Cortes, C., Lawrence, N., Lee, D., Sugiyama, M., Garnett, R. (eds.) Advances in Neural Information Processing Systems. vol. 28. Curran Associates, Inc. (2015)
11. Ruscoe, A., Barblett, L., Barratt-Pugh, C.: Sharing power with children: Repositioning children as agentic learners. Australas. J. Early Childhood **43**(3), 63–71 (2018). https://doi.org/10.23965/AJEC.43.3.07
12. Sirkko, R., Kyrönlampi, T., Puroila, A.M.: Children's agency: opportunities and constraints. Int. J. Early Childhood **51**, 283–300 (2019). https://doi.org/10.1007/s13158-019-00252-5
13. Taub, M., Sawyer, R., Smith, A., Rowe, J., Azevedo, R., Lester, J.: The agency effect: the impact of student agency on learning, emotions, and problem-solving behaviors in a game-based learning environment. Comput. Educ. **147**, 103781 (2020). https://doi.org/10.1016/j.compedu.2019.103781
14. Tsai, Y.S., Perrotta, C., Gašević, D.: Empowering learners with personalised learning approaches? Agency, equity and transparency in the context of learning analytics. Assess. Eval. High. Educ. **45**(4), 554–567 (2020). https://doi.org/10.1080/02602938.2019.1676396
15. Van Schoors, R., Elen, J., Raes, A., Depaepe, F.: An overview of 25 years of research on digital personalised learning in primary and secondary education: a systematic review of conceptual and methodological trends. Br. J. Edu. Technol. **52**(5), 1798–1822 (2021). https://doi.org/10.1111/bjet.13148
16. Van Schoors, R., Elen, J., Raes, A., et al.: The charm or chasm of digital personalized learning in education: teachers' reported use. Perceptions Expect. Tech. Trends **67**, 315–330 (2023). https://doi.org/10.1007/s11528-022-00802-0
17. Walkington, C., Bernacki, M.L.: Appraising research on personalized learning: definitions, theoretical alignment, advancements, and future directions. J. Res. Technol. Educ. **52**(3), 235–252 (2020). https://doi.org/10.1080/15391523.2020.1747757
18. Zhang, K., Aslan, A.B.: AI technologies for education: recent research future directions. Comput. Educ.: Artif. Intell. **2**, 100025 (2021). https://doi.org/10.1016/j.caeai.2021.100025
19. Zheng, L., Long, M., Zhong, L., et al.: The effectiveness of technology-facilitated personalized learning on learning achievements and learning perceptions: a meta-analysis. Educ. Inf. Technol. **27**, 11807–11830 (2022). https://doi.org/10.1007/s10639-022-11092-7

Interactive AI-Generated Virtual Instructors Enhance Learning Motivation and Engagement in Financial Education

Thanawit Prasongpongchai[1](✉), Pat Pataranutaporn[2](✉),
Chonnipa Kanapornchai[1](✉), Auttasak Lapapirojn[3](✉),
Pichayoot Ouppaphan[3](✉), Kavin Winson[3](✉), Monchai Lertsutthiwong[3](✉),
and Pattie Maes[2](✉)

[1] Beacon Interface, Nonthaburi, Thailand
{thanawit.p,chonnipa.k}@kbtg.tech
[2] MIT Media Lab, Massachusetts Institute of Technology, Cambridge, MA, USA
{patpat,pattie}@media.mit.edu
[3] Kasikorn Labs, Nonthaburi, Thailand
{auttasak.l,pichayoot.o,kavin.w,monchai.le}@kbtg.tech

Abstract. AI-generated virtual instructors powered by large language models (LLMs) have the potential to augment online learning by making it more interactive and personalized. In this work, we focus specifically on applying this technology to financial literacy education, a critical life skill that is unfortunately undertaught in most school systems. To investigate this potential, we developed three conditions for delivering financial education: A) a Passive Virtual Instructor, B) an Interactive Virtual Instructor powered by LLMs to enable two-way conversation with the learner, and C) a Personally Selected Interactive Virtual Instructor, allowing learners to choose their preferred AI instructor. In a study with 90 participants aged 18–25, we found that incorporating interactivity significantly improved learning motivation, perceived learning facilitation, engagement, and humanness compared to the passive agent. The ability to personally select an instructor provided little additional benefit for most measures. These results highlight the potential of interactive virtual instructors for enhancing motivational and experiential factors, even if comprehension itself is unaffected.

Keywords: AI-Generated Virtual Instructor · Human-AI Interaction · Personalized AI · AI for Education · Pedagogical Agent · Financial Literacy Education

1 Introduction

With the ongoing increase in remote education, "virtual instructors" or "pedagogical agents" have emerged as an alternative delivery mode for video-based

---

T. Prasongpongchai and P. Pataranutaporn—These authors are contributed equally to this work.

A. M. Olney et al. (Eds.): AIED 2024 Workshops, CCIS 2151, pp. 217–225, 2024.
https://doi.org/10.1007/978-3-031-64312-5_26

instruction [19]. Prior research has explored the potential of using virtual instructors based on different personas to improve the learning experience [11,15–17,21]. Neuroimaging also reveals enhanced socio-emotional activity regions during learning with a virtual agent, suggesting social processes aid the experience [12].

These insights motivate developing virtual instructors specifically for financial literacy education. Financial literacy refers to the skills needed to make sound financial decisions, including concepts like budgeting, saving, investing, and managing credit [24]. Despite its critical significance, the level of financial literacy differs significantly across countries globally. This disparity leaves a considerable number of individuals worldwide without the necessary skills to make knowledgeable financial choices [9].

In response to the pressing need for improved financial literacy, we were motivated to explore how we might use generative artificial intelligence (AI) to create virtual instructors to support financial literacy education. Specifically, we are focused on evaluating the potential for interactivity and personalization to boost engagement. We developed a web-based financial literacy education platform with three conditions: A) Passive Virtual Instructor: non-interactive pre-generated video lessons, B) Interactive Virtual Instructor: learners can ask questions during lessons and get personalized answers in return, C) Personally Selected Interactive Virtual Instructor: allowing learners to select an interactive virtual instructor of choice.

This paper presents two novel contributions: 1) An AI-powered financial literacy education system with added two-way interactivity and personalization features where the virtual instructor occasionally poses questions to the learner, and learners can also ask questions back to the virtual agent. 2) A comparative study across 90 participants investigating learning motivation, experience, and comprehension between the three conditions with the aim of understanding how interactivity and character personalization change outcomes.

2   Related Work

Our work on understanding the role of AI-generated interactive virtual instructors is situated within three different contexts of research as follows:

Learning Motivation and Personalization: The concept of learner-centric, personalized educational experiences has gained increasing traction in recent years [1]. The central strategy, *context personalization*, aligning content with the student's interests and background, has been shown to impact learners' attention and engagement. [18,25]. However, past efforts required extensive manual effort and failed to scale broadly [7,10,14]. With its capacity for generating text for virtually any topic, generative AI offers a scalable method for dynamically producing tailored content. We intend to explore these issues to pave the path forward for AI-enabled personalized learning.

AI-Generated Characters: Advances in generative AI have facilitated the development of virtual characters using techniques like generative adversarial networks (GANs) [8,27], neural radiance fields [6,13,26], and self-supervised motion transfer networks [22]. These digital avatars can be based on contemporary, historical, or imaginary personalities and used in education, wellness, and entertainment [2,15,16]. Additionally, simulated conversations have become possible thanks to natural language processing (NLP) and large language models (LLMs), built on widely available transformer architectures [23].

Virtual Pedagogical Agents: Digitally synthesized "virtual instructors" or "pedagogical agents" have been developed as an additional method to help facilitate video-based learning [19]. Studies have suggested socio-emotional processes are in play during an interaction with a virtual instructor [11,12] which may explain the impact on the learning experience. Research has also explored various characteristics of these agents, e.g., age and appearance [21], admired people [17], and historical figures [15], and their effects on the learning experience.

These insights led us to investigate enhancing existing virtual instructors with interactivity, powered by LLMs, and personalization through instructor selection, which we hypothesize could improve the learning experience and outcomes for financial literacy education.

3   Methodology

To evaluate the potential of interactivity and personalization in AI-generated instruction for financial literacy, we developed an AI-based web platform with three conditions: *A) Passive Virtual Instructor* that presents pre-generated video lessons. This condition has no chat interface for interactive conversations with the virtual instructor and acts as our control condition representing traditional online learning, *B) Interactive Virtual Instructor* featuring two-way question-and-answer capabilities with personalized responses powered by an LLM, *C) Personally Selected Interactive Virtual Instructor* which is Condition B with the added ability to select a preferred character as their instructor.

3.1   Functional Design

Upon accessing the web system, learners are prompted to select a generated character as their "AI virtual instructor." Then, the lesson is delivered via a slideshow, accompanied by a pre-generated talking head video of the selected instructor, utilizing an AI-generated characters pipeline [16]. Throughout each lesson, this AI instructor interjects at pre-scripted points to pose relevant questions that learners can answer through a chat interface. The AI instructor is pre-prompted to ask follow-up questions to encourage the learner's self-reflection, as shown in Fig. 2. Moreover, learners can always invoke the "Raise Hand" button to ask questions through a pop-up chat window. Once each conversation ends, the AI instructor will ask the user to resume the video lesson. These dynamic responses are powered by a GPT-4-based model through a back-end API. This two-way interactivity was designed to foster an active, bidirectional engagement throughout the lessons. The functional design of the proposed web-based system is illustrated in Fig. 1. This description features the Personally Selected Interactive Virtual Instructor condition. Reduced versions for the other two conditions are crafted by removing the chat-based interactivity and instructor selection.

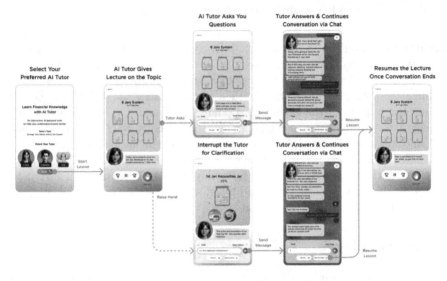

Fig. 1. The overall system of AI-generated virtual instructors for financial literacy.

Fig. 2. Examples of lesson scripts for each AI instructor.

3.2   Lesson Content and Personalized Virtual Instructors

For this initial investigation, the Six Jars Method for basic money management, developed by T. Harv Eker [5], was adapted into our lesson format. The method suggests budgeting one's income into six "jars," namely, Necessities, Financial Freedom, Long-Term Savings, Education, Play, and Give, using certain recommended proportions. The topic was chosen because it gives a simple yet essential component for personal finance skills suitable to our target student demographic.

The scripts were crafted to be conversational to fit with the interactivity of the instructor. We also included an initial greeting to acclimate learners to interactions like chat and hand-raising. In this research, the lessons are provided in the Thai language to best match our target demographics.

To test the effect of personalization by instructor selection, we designed 3 distinct virtual characters: *Luna* (a gentle, cheerful female instructor), *Captain* (a confident, serious male instructor), and *Ura* (a playful, friendly elephant cartoon instructor), as shown in Fig. 2. Each persona has a slightly different scripted speech to simulate their personalities while delivering a similar content structure. For this experiment, *Captain* and *Ura* are only available for selection in the Personally Selected Interactive Virtual Instructor condition while *Luna* is available in all conditions.

3.3   Experiment

We recruited 90 Thai students aged 18-25 and randomly assigned them into three experimental groups for each condition: A) Passive Virtual Instructor, B) Interactive Virtual Instructor, and C) Personally Selected Interactive Virtual Instructor. Afterward, participants completed a 10-question assessment quiz along with several self-report surveys, which measure learning motivation and positive emotion (adapted from [17]), and perceptions of the system regarding learning facilitation, agent credibility, engagement, and human-likeness (adapted from established metrics for pedagogical agents [3,20]). Each survey consists of multiple 5-point Likert scale questions. Each session lasted approximately 15 min.

One-way analysis of variance (ANOVA) was conducted to determine differences between conditions, with virtual instructor type as the grouping variable. Dependent measures were quiz performance, learning motivation, positive emotion, and the pedagogical agent ratings along each subscale. Benjamini-Hochberg correction was then performed to control false discovery rate [4]. Statistically significant main effects were followed up with Tukey HSD post-hoc tests to assess differences between each condition pair.

4   Result

With the students in the three conditions, we aim to analyze two key factors: the impact of a passive versus active learning environment and how a personally selected AI character influences learning.

The results from the study are shown in Table 1 and visualized in Fig. 3. It can be seen that while the added interactivity and ability to personally select a virtual instructor did not yield a significant difference in the test scores from the assessment quiz, the added interactivity improved certain self-reported metrics while the personal selection of the virtual instructors provided little added benefit. According to post-hoc Tukey HSD tests, we found a significant increase in several self-reported metrics in the Interactive Virtual Instructor condition compared to the Passive condition. These include learning motivation (p = .003), learning facilitation (p = .003), engagement (p = .003), and agent humanness (p = .005). The tests revealed no significant improvements in positive emotion and agent credibility between the two groups (p > .05). Additionally, a significant increase in agent humanness was also found when comparing the Personally Selected Interactive condition to the Passive variant (p = .02).

Table 1. Measured mean and standard deviation (in parentheses) for each learner experience metric for the three conditions, with results from the F-Test for statistical significance *(in italics)*, including Benjamini-Hochberg adjusted p-values. Test scores range from 0 to 10. Others are Likert ratings ranging from 1 to 5.

Condition	Test Scores	Learning Movitation	Positive Emotion	Learning Facilitation	Agent Credibility	Engage-ment	Human-ness
A) Passive	8.48	3.75	3.80	3.68	4.07	3.63	2.88
	(1.23)	(0.65)	(0.80)	(0.76)	(0.76)	(0.79)	(1.03)
B) Interactive	8.13	4.41	4.19	4.39	4.32	4.29	3.69
	(1.36)	(0.83)	(0.78)	(0.86)	(0.96)	(0.67)	(1.09)
C) Personalized,	8.21	4.15	3.97	3.98	4.07	3.85	3.55
Interactive	(1.45)	(0.82)	(0.88)	(0.86)	(0.88)	(0.89)	(0.88)
$F(2, 114)$	*0.60*	*5.99*	*1.82*	*5.83*	*0.87*	*5.71*	*6.02*
p-value	*0.551*	*0.004*	*0.168*	*0.004*	*0.422*	*0.005*	*0.003*
p-value, BH-adjusted	*0.551*	*0.012*	*0.235*	*0.010*	*0.492*	*0.008*	*0.024*

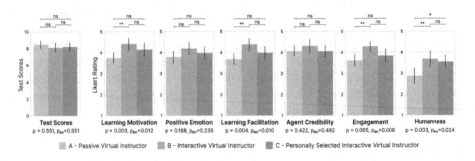

Fig. 3. Statistical analysis of the AI-generated virtual instructor.

5   Discussion and Future Work

This study demonstrates that two-way interactivity—through dynamic, personalized responses—significantly boosts learning motivation, perceived facilitation, engagement, and the AI instructor's "humanness" compared to passive formats. This may be because interactive systems better mimic the dynamics of a class-room, offering a more engaging learning experience. However, instructor selection showed minimal extra benefits. One explanation could be that the options pro-vided were not sufficiently meaningful for learners' experiences. For instance, they did not include liked or admired figures, as was effective in [17]. This calls for further investigation into what types of personalization are most effective for educational AI.

The study also noted no differences in test scores across groups, which may be attributed to participants' initial high performance and/or too-easy lesson and quiz materials. This necessitates difficulty level calibration of the content to potentially reveal more distinct differences in learning outcomes.

Several limitations of the results must also be acknowledged. These included limited participant sample size and demographics, limited lesson content domain, and short interactions with the AI instructor. Future research should explore these factors for greater generalization, including in educational fields other than financial literacy. Longitudinal studies, as well as studies on potential adverse effects and strategies to mitigate them, would also be beneficial.

One crucial ethical consideration is that these AI systems should not be viewed as replacements for human teachers but rather as supplementary tools that assist in areas that may not be extensively covered in traditional education systems, such as financial literacy. This ensures that while students gain access to specialized knowledge areas like financial literacy, they continue to benefit from the unique and invaluable human elements of teaching, such as empathy, ethical guidance, and real-world experience sharing. Thus, the objective is to harmonize AI's capabilities with human qualities, fostering an enriched learning environment that prepares students for both the technical and interpersonal aspects of the real world.

6   Conclusion

Our development and analysis of three AI-driven financial education platforms revealed significant advantages of interactive virtual instruction, enhanced with LLMs for dynamic question-and-answer capabilities, in terms of boosting learn-ers' motivation, engagement, perceived learning facilitation, and the agent's humanness, compared to a non-interactive approach. While personalization via selection of an instructor showed minimal extra benefits, the results highlight the potential of adding interactivity to virtual instructors to enrich remote learning experiences, making them more engaging and emotionally fulfilling without nec-essarily altering comprehension levels. As online learning continues to evolve, further exploration into optimizing interactive virtual instructors may lead to even greater improvements in learner engagement and motivation.

References

1. Alamri, H., Lowell, V., Watson, W., Watson, S.L.: Using personalized learning as an instructional approach to motivate learners in online higher education: learner self-determination and intrinsic motivation. J. Res. Technol. Educ. **52**(3), 322–352 (2020)
2. Ammanabrolu, P., Riedl, M.O.: Situated language learning via interactive narratives. arXiv:2103.09977 (2021)
3. Baylor, A., Ryu, J.: The API (agent persona instrument) for assessing pedagogical agent persona. In: EdMedia+ Innovate Learning, pp. 448–451. Association for the Advancement of Computing in Education (AACE) (2003)
4. Benjamini, Y., Hochberg, Y.: Controlling the false discovery rate: a practical and powerful approach to multiple testing. J. Roy. Stat. Soc. Ser. B (Methodol.) **57**(1), 289–300 (1995)
5. Eker, T.H.: Secrets of the Millionaire Mind. Harper Business (2005)
6. Guo, Y., Chen, K., Liang, S., Liu, Y.J., Bao, H., Zhang, J.: AD-NeRF: audio driven neural radiance fields for talking head synthesis. In: Proceedings of the IEEE/CVF International Conference on Computer Vision, pp. 5784–5794 (2021)
7. Høgheim, S., Reber, R.: Eliciting mathematics interest: new directions for context personalization and example choice. J. Exp. Educ. **85**(4), 597–613 (2017)
8. Hong, F.T., Zhang, L., Shen, L., Xu, D.: Depth-aware generative adversarial network for talking head video generation. arXiv preprint arXiv:2203.06605 (2022)
9. Klapper, L., Lusardi, A., Van Oudheusden, P.: Financial literacy around the world. World Bank. Washington DC: World Bank **2**, 218–237 (2015)
10. Ku, H.Y., Harter, C.A., Liu, P.L., Thompson, L., Cheng, Y.C.: The effects of individually personalized computer-based instructional program on solving mathematics problems. Comput. Hum. Behav. **23**(3), 1195–1210 (2007)
11. Lawson, A.P., Mayer, R.E., Adamo-Villani, N., Benes, B., Lei, X., Cheng, J.: Recognizing the emotional state of human and virtual instructors. Comput. Hum. Behav. **114**, 106554 (2021)
12. Li, W., Wang, F., Mayer, R.E., Liu, T.: Animated pedagogical agents enhance learning outcomes and brain activity during learning. J. Comput. Assist. Learn. **38**, 621–637 (2021)
13. Liu, X., Xu, Y., Wu, Q., Zhou, H., Wu, W., Zhou, B.: Semantic-aware implicit neural audio-driven video portrait generation. arXiv preprint arXiv:2201.07786 (2022)
14. López, C.L., Sullivan, H.J.: Effect of personalization of instructional context on the achievement and attitudes of hispanic students. Educ. Tech. Res. Dev. **40**(4), 5–14 (1992)
15. Pataranutaporn, P., et al.: Living memories: AI-generated characters as digital mementos. In: Proceedings of the 28th International Conference on Intelligent User Interfaces (IUI), pp. 889–901 (2023). https://doi.org/10.1145/3581641.3584065
16. Pataranutaporn, P., et al.: AI-generated characters for supporting personalized learning and well-being. Nat. Mach. Intell. **3**(12), 1013–1022 (2021)
17. Pataranutaporn, P., Leong, J., Danry, V., Lawson, A.P., Maes, P., Sra, M.: AI-generated virtual instructors based on liked or admired people can improve motivation and foster positive emotions for learning. In: IEEE Frontiers in Education Conference (FIE), pp. 1–9. IEEE (2022). https://doi.org/10.1109/FIE56618.2022.9962478

18. Reber, R., Canning, E.A., Harackiewicz, J.M.: Personalized education to increase interest. Curr. Dir. Psychol. Sci. **27**(6), 449–454 (2018)
19. Schroeder, N.L., Adesope, O.O., Gilbert, R.B.: How effective are pedagogical agents for learning? A meta-analytic review. J. Educ. Comput. Res. **49**(1), 1–39 (2013)
20. Schroeder, N.L., Romine, W.L., Craig, S.D.: Measuring pedagogical agent persona and the influence of agent persona on learning. Comput. Educ. **109**, 176–186 (2017)
21. Shiban, Y., et al.: The appearance effect: influences of virtual agent features on performance and motivation. Comput. Hum. Behav. **49**, 5–11 (2015)
22. Siarohin, A., Lathuilière, S., Tulyakov, S., Ricci, E., Sebe, N.: First order motion model for image animation. arXiv preprint arXiv:2003.00196 (2020)
23. Thoppilan, R., et al.: Lamda: language models for dialog applications. arXiv preprint arXiv:2201.08239 (2022)
24. Tomášková, H., Mohelská, H., Němcová, Z.: Issues of financial literacy education. Procedia. Soc. Behav. Sci. **28**, 365–369 (2011)
25. Walkington, C., Bernacki, M.L.: Personalization of instruction: design dimensions and implications for cognition. J. Exp. Educ. **86**(1), 50–68 (2018)
26. Yao, S., Zhong, R., Yan, Y., Zhai, G., Yang, X.: DFA-NeRF: personalized talking head generation via disentangled face attributes neural rendering. arXiv preprint arXiv:2201.00791 (2022)
27. Yin, F., et al.: StyleHEAT: one-shot high-resolution editable talking face generation via pretrained StyleGAN. arXiv preprint arXiv:2203.04036 (2022)

A Turkish Educational Crossword Puzzle Generator

Kamyar Zeinalipour[1]([⊠])(iD), Yusuf Gökberk Keptiğ[1]([⊠])(iD),
Marco Maggini[1]([⊠])(iD), Leonardo Rigutini[2]([⊠])(iD), and Marco Gori[1]([⊠])(iD)

[1] University of Siena, Siena, Italy
{kamyar.zeinalipour2,marco.maggini,marco.gori}@unisi.it,
y.keptig@student.unisi.it
[2] expert.ai, Modena, Italy
lrigutini@expert.ai

Abstract. This paper introduces the first Turkish crossword puzzle generator designed to leverage the capabilities of large language models (LLMs) for educational purposes. In this work, we introduced two specially created datasets: one with over 180,000 unique answer-clue pairs for generating relevant clues from the given answer, and another with over 35,000 samples containing text, answer, category, and clue data, aimed at producing clues for specific texts and keywords within certain categories. Beyond entertainment, this generator emerges as an interactive educational tool that enhances memory, vocabulary, and problem-solving skills. It's a notable step in AI-enhanced education, merging game-like engagement with learning for Turkish and setting new standards for interactive, intelligent learning tools in Turkish.

Keywords: Large Language Models · Educational Puzzles · Interactive Learning

1 Introduction

Crossword puzzles, designed with educational goals, blend puzzle-solving enjoyment with knowledge acquisition in fields like history, linguistics, and sciences, enhancing learners' vocabulary, spelling competence, and cognitive abilities such as memory retention and critical reasoning [1–6]. They are particularly valuable in language learning, aiding in vocabulary building and the assimilation of technical jargon [7,8]. Recent advances in natural language processing (NLP), specifically LLMs, facilitate the creation of educational crosswords, offering high-quality clues and solutions based on user-suggested texts or keywords. This paper describes a novel application that leverages LLMs for generating Turkish educational crossword puzzles, supported by two new datasets: one with hand-crafted

The funding for this paper was provided by the TAILOR project and the HumanE-AI-Net projects, both supported by the EU Horizon 2020 research and innovation program under GA No 952215 and No 952026, respectively.

crossword puzzles and another with categorized Turkish texts, aimed at fostering the development and widespread use of educational crossword puzzles in Turkish learning environments. Through this work, the datasets and all of the developed models[1] will be made available to the scientific community in open-source and free-to-download mode. The structure encompasses a literature review 2, dataset description 3, methodology 4, experiment analysis 5, and concluding remarks 6.

2   Related Works

The development of automatic crossword puzzle generation has evolved over time, incorporating various strategies from exploiting lexical databases and internet text analysis to adopting recent LLM-based techniques like fine-tuning and zero/few-shot learning. Initial attempts by Rigutini et al. leveraged advanced NLP for puzzle creation from online texts [9,10], while other efforts focused on specific languages or themes, such as SEEKH for Indian languages [11] and Esteche's work for Spanish-speaking audiences [12]. Recently, [13–15] shifted from manual crossword design to using pre-trained LLMs, generating puzzles in English, Arabic, and Italian, highlighting the power of computational linguistics in creating culturally diverse puzzles. In [16] they suggest a method for creating educational crossword clues dataset in English. However, the challenge of creating Turkish educational crossword puzzles was unaddressed. This study fills the gap by introducing a novel approach that employs advanced language models for generating engaging Turkish crossword puzzles, marking a significant advancement in educational tool development, especially for Turkish language learning.

3   Dataset

We constructed two datasets: the first consists of clue-answer pairs, and the second, a more elaborate compilation, merges text, answers, categories, and clues into one comprehensive dataset. The next section delves into the methods employed for their acquisition and compilation.

Turkish Answer-clue Pairs Dataset(TAC): [2] In this study, a comprehensive dataset of $252,576$ Turkish answer-clue pairs from online sources[3], including newspapers like Habertürk, was compiled. These pairs, crafted by crossword experts, encapsulate $187,495$ unique entries, $54,984$ unique answers, and $178,431$ unique clues marking it a significant Turkish language resource. It's useful for applications like generating crossword clues for given answers. In Fig. 1 we report a graphical representation of the distribution of answer's lengths.

---

[1] https://huggingface.co/Kamyar-zeinalipour/llama7B_turkish_crossword_clue_gen
    https://huggingface.co/Kamyar-zeinalipour/llama13B_turkish_crossword_clue_gen
    https://github.com/KamyarZeinalipour/CW_Clue_Gen_tr.
[2] https://huggingface.co/datasets/Kamyar-zeinalipour/TAC.
[3] https://www.bulmaca-sozlugum.com
    https://www.bulmacacozumleri.com.

Text for Turkish Answer-clue Pairs (T4TAC) Dataset: [4] The second dataset in our study is crucial for developing models to autonomously produce crossword puzzle clues from the text and answer inputs in specific categories. This has significant potential in education, providing interactive teaching materials. The dataset creation process is outlined in Fig. 2. The workflow in Fig. 2 is divided into stages important for the dataset development, starting with data collection where relevant text is gathered. This leads to data preprocessing to structure the text and keywords for training, followed by categorization to aid crossword clue generation. The Text for Turkish Answer-Clue pairs *T4TAC* dataset is crucial for models that generate Turkish crosswords from text, benefiting educators. We outline the dataset construction phases.

Data Acquisition and Filtering: The extraction of Turkish Wikipedia articles targets the initial sections, emphasizing significant keywords and collecting metadata such as view counts, relevance scores, and URLs[5]. The focus remains on keyword-rich introductory paragraphs. Quality assurance involves filtering pages by popularity, relevance, content length, and keyword specificity, excluding data unsuitable for crossword clues such as keywords outside 3 to 20 characters or containing special characters and numbers.

Prompt Development and Educational Clue Creation: The targeted prompt is crafted to generate informative and engaging Turkish crossword clues, illustrated in Fig. 3. Utilizing the SELF-INSTRUCT framework [17] and *GPT4-Turbo*, our method integrates contextual information with content, keywords, and categories for generating tailored Turkish educational clues. This approach results in clues that are educationally valuable and contextually relevant.

Turkish Answer-Clue Dataset Compilation: From an initial set of 180, 000 Wikipedia pages, 9, 855 pages across 29 themes were selected after filtering, producing over 35, 000 clues using *GPT4-Turbo*. The dataset's content length ranges from 50 to 982 words, with clues generally spanning 5 to 15 words. Figure 4 (i) shows the distribution of word counts for both texts and clues and the lengths of answers. The dataset encompasses a broad spectrum of topics including dominant themes like "Entertainment," "History," and "Science," as illustrated in Fig. 4 (ii).

4   The Crossword Generation System

We developed a system for generating Turkish crossword puzzles from specified answers and texts, useful for educators. There are two main use cases: one where educators already have a list of keywords for creating puzzles without additional text; and another where they wish to generate crosswords from input texts by extracting keywords and clues automatically. We fine-tuned LLMs like *GPT3.5-Turbo* [18], *Llama-2-7b-chat-hf* and *Llama-2-13b-chat-hf* [19] using a specific dataset for clue generation, allowing educators to choose the best options. A

---

[4] https://huggingface.co/datasets/Kamyar-zeinalipour/T4TAC.
[5] https://en.wikipedia.org/wiki/Wikipedia:Lists_of_popular_pages_by_WikiProject.

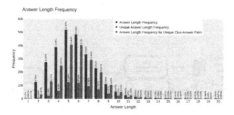

Fig. 1. The dataset entries are showcased visually through the distribution of answer lengths. Blue bars represent all answer-clue pairs, green bars show the frequency of unique answers, and red bars display the frequency of unique answer-clue pairs (Color figure online)

Fig. 2. Diagram of the steps followed in the construction of the T4TAC dataset.

Fig. 3. The prompt utilized in the study.

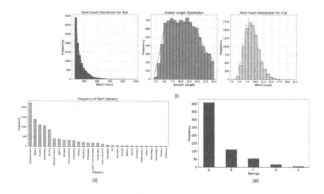

Fig. 4. (i) Word and Character Length Distributions for Contexts, Outputs, and Keywords.(ii) Category Distributions of T4TAC. (iii) Human Evaluation for T4TAC.

specialized algorithm, detailed later, is used to create the crossword layout. The methodology is outlined in Fig. 5.

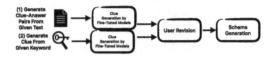

Fig. 5. The scheme of the crossword generation system

Fine-tuning LLMs to Generate Clues: Investigating how to generate crossword clues from answers and text, we improved language models, closely following the dataset described in Sect. 3. For two distinct task: creating crossword clues from provided answers using the *TAC* dataset, and generating crossword clues from given text corresponding to specific categories using the *T4TAC* dataset. We evaluated models like *GPT3.5-Turbo* [18], *Llama-2-7b-chat-hf* and *Llama-2-13b-chat-hf* [19], using a mix of open-source and proprietary versions, including *GPT3.5-Turbo*, after fine-tuning, to assess clue quality.

Schema Generator: The crossword puzzle creation algorithm operates with input criteria like an answer list, puzzle size, and stopping rules to efficiently assemble puzzles. It starts by placing an initial word and adds others strategically, adjusting through removals or resets for an ideal layout. The puzzle's score is judged by the $\text{Score} = (\text{FW} + 0.5 \cdot \text{LL}) \times \text{FR} \times \text{LR}$ where, FW denotes the total words inserted, LL is the count of crossing letters, FR represents the proportion of filled space, and LR the density of intersecting letters. End criteria include reaching necessary puzzle complexity, obtaining a given space ratio, limiting grid adjustments, or maximizing allocated time, all ensuring a targeted and time-bound puzzle output. It emphasizes adaptability in honoring preferred answers to align with specific educational aims or content themes.

5   Experiments

In this section, we describe the experimental evaluation of two methods for generating Turkish crossword clues from keywords and their text categories. We detail the datasets used for training, the experimental setup for each method, and the methodologies implemented. Finally, we showcase a Turkish crossword created using our methodology as an example of its application.

Generating the Crossword Clues From Given Answer: In our study on automated crossword clue generation, described in Sect. 3, we employed a *TAC* dataset subset consisting of 60,000 pairs provided by experts. We fine-tuned *GPT3.5-Turbo* on a chosen subset with a batch size of 16 and a learning rate of 0.01 for three epochs, targeting Turkish crossword clues from keywords. We tested the model's performance using 2,135 academic keywords to assess its capability to produce pertinent and creative clues. The evaluation of generated crossword clues was conducted by two native Turkish language speakers who served as human judges. This process revealed that 51.8% of the clues achieved the set standards for acceptability, highlighting the model's capability to create contextually relevant and innovative clues. This result emphasizes the model's potential for further improvements and its applicability in educational tools to enhance learning in diverse subjects.

Generating Clues From Given Text Answer Category: To analyze specific queries from the text, we used a dataset depicted in Sect. 3, refining it down to 8670 unique, clue-paired texts. We designated 8000 of these for training and 670 for testing. We fine-tuned *GPT3.5-Turbo*, *Llama-2-7b-chat-hf* and *Llama-2-13b-chat-hf* models; *GPT3.5-Turbo* with a batch size of 16, a learning rate of 0.001 over three epochs, *Llama-2-7b-chat-hf* and *Llama-2-13b-chat-hf* using Parameter Efficient Fine-Tuning (PEFT) with r=16, alpha=32, and a learning rate of 0.0001 across three epochs for both models. Model performances for generating clues from the test set were assessed via ROGUE-1, ROGUE-2, and ROGUE-L F1 score metrics, compared to the generated clues from *GPT4-Turbo*, with detailed results in Table 1.

Table 1. Performance of LLMs with and without fine-tuning.

model type	model name	# params	ROUGE-1	ROUGE-2	ROUGE-L
Base LLMs	LLAMA2-CHAT	7B	–	–	–
	LLAMA2-CHAT	13B	–	–	–
	GPT3.5 TURBO	–	24.57	8.48	16.61
Finetuned LLMs	LLAMA2-CHAT	7B	22.50	5.00	14.80
	LLAMA2-CHAT	13B	25.66	6.45	17.24
	GPT3.5 TURBO	–	**39.96**	**18.08**	**23.48**

In our study, we evaluated *GPT3.5-Turbo*, *Llama-2-7b-chat-hf*, and *Llama-2-13b-chat-hf* models on their ability to generate Turkish crossword clues using the *T4TAC* dataset. The base models of *Llama-2-7b-chat-hf* and *Llama-2-13b-chat-hf* initially showed limited capability in processing Turkish. To analyze the impact of fine-tuning on model performance, we conducted a detailed human evaluation using 200 texts from both the base and fine-tuned versions, examining three clues per text, totaling 600 examples. Native Turkish speakers carried out the evaluations to ensure accurate assessment in terms of linguistic and cultural relevance. The results, depicted in Fig. 6, indicate a clear enhancement

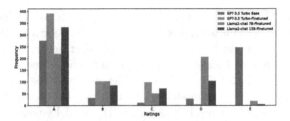

Fig. 6. Comparison of model ratings.

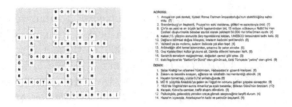

Fig. 7. Crossword crafted using the proposed approach.

in performance post-fine-tuning with the *T4TAC* dataset. Both *Llama* models initially struggled with Turkish crossword clues but improved markedly after fine-tuning, as did the *GPT3.5-Turbo* model.

We examined two crossword clue creation methods: one with set answers and another categorizing text-generated clues. Both support custom clues for educational goals, allowing educators to choose optimal clues for crossword construction (see Sect. 4). A sample Turkish crossword crafted using this system is depicted in Fig. 7.

6   Conclusion

In this study, we introduced the Turkish Educational Generator, a novel tool powered by LLMs, to create dynamic crosswords in Turkish for educational purposes. This system allows educators to easily generate subject-specific crosswords, boosting student engagement and learning retention. Additionally, we enriched Turkish language datasets by providing two comprehensive datasets: one from expert-crafted answer-clue pairs and another from generated clues with corresponding text, answers, and categories, both vital for educational system development and research. Future plans include expanding our tool to more languages and enhancing clue-generation techniques with advanced LLMs, pushing the boundaries of educational technology.

References

1. Orawiwatnakul, W.: Crossword puzzles as a learning tool for vocabulary development. Electron. J. Res. Educ. Psychol. **11**, 413–428 (2013)
2. Bella, Y., Rahayu, E.: The improving of the student's vocabulary achievement through crossword game in the new normal era. Edunesia Jurnal Ilmiah Pendidikan **4**, 830–842 (2023)
3. Mueller, S., Veinott, E.: Testing the effectiveness of crossword games on immediate and delayed memory for scientific vocabulary and concepts. In: CogSci (2018)
4. Zirawaga, V., Olusanya, A., Maduku, T.: Gaming in education: using games as a support tool to teach history. J. Educ. Pract. **8**, 55–64 (2017)
5. Zamani, P., Haghighi, S., Ravanbakhsh, M.: The use of crossword puzzles as an educational tool. J. Adv. Med. Educ. Professionalism **9**, 102 (2021)
6. Dol, S.: GPBL: an effective way to improve critical thinking and problem solving skills in engineering education. J. Eng. Educ. Trans. **30**, 103–13 (2017)
7. Sandiuc, C., Balagiu, A.: The use of crossword puzzles as a strategy to teach maritime English vocabulary. Sci. Bull. Mircea cel Batran Naval Acad. **23**, 236A – 242 (2020)
8. Yuriev, E., Capuano, B., Short, J.: Crossword puzzles for chemistry education: learning goals beyond vocabulary. Chem. Educ. Res. Pract. **17**, 532–554 (2016)
9. Rigutini, L., Diligenti, M., Maggini, M., Gori, M.: A fully automatic crossword generator. In: 2008 Seventh International Conference on Machine Learning and Applications, pp. 362–367. IEEE (2008)
10. Rigutini, L., Diligenti, M., Maggini, M., Gori, M.: Automatic generation of crossword puzzles. Int. J. Artif. Intell. Tools **21**, 1250014 (2012)
11. Arora, B., Kumar, N.: Automatic keyword extraction and crossword generation tool for Indian languages: SEEKH. In: 2019 IEEE Tenth International Conference on Technology for Education (T4E), pp. 272–273. IEEE (2019)
12. Esteche, J., Romero, R., Chiruzzo, L., Rosá, A.: Automatic definition extraction and crossword generation from Spanish news text. CLEI Electron. J. **20**, 6 (2017)
13. Zeinalipour, K., Saad, M., Maggini, M., Gori, M.: ArabIcros: AI-powered Arabic crossword puzzle generation for educational applications. In: Proceedings of ArabicNLP 2023, pp. 288–301 (2023)
14. Zeinalipour, K., Zanollo, A., Angelini, G., Rigutini, L., Maggini, M., Gori, M.: Others Italian crossword generator: enhancing education through interactive word puzzles. arXiv preprint arXiv:2311.15723 (2023)
15. Zeinalipour, K., Iaquinta, T., Angelini, G., Rigutini, L., Maggini, M., Gori, M.: Building bridges of knowledge: innovating education with automated crossword generation. In: 2023 International Conference on Machine Learning and Applications (ICMLA), pp. 1228–1236. IEEE (2023)
16. Zugarini, A., Zeinalipour, K., Kadali, S., Maggini, M., Gori, M., Rigutini, L.: Clue-instruct: text-based clue generation for educational crossword puzzles. arXiv preprint arXiv:2404.06186 (2024)
17. Wang, Y., et al.: Self-instruct: aligning language model with self generated instructions. arXiv preprint arXiv:2212.10560 (2022)
18. Brown, T., et al.: Language models are few-shot learners. In: Advances in Neural Information Processing Systems, vol. 33, pp. 1877–1901 (2020)
19. Touvron, H., et al.: Llama 2: open foundation and fine-tuned chat models. arXiv preprint arXiv:2307.09288 (2023)

An Optimization Approach
for Elementary School Handwritten
Mathematical Expression Recognition

Sergio F. Chevtchenko[3,5]([✉]) [iD], Ruan Carvalho[1]([✉]) [iD], Luiz Rodrigues[2]([✉]) [iD],
Everton Souza[1]([✉]) [iD], Daniel Rosa[1]([✉]) [iD], Filipe Cordeiro[1]([✉]) [iD],
Cicero Pereira[1]([✉]) [iD], Thales Vieira[2] [iD], Marcelo Marinho[1] [iD],
Diego Dermeval[2] [iD], Ig Ibert Bittencourt[2,4] [iD], Seiji Isotani[4] [iD],
and Valmir Macario[1] [iD]

[1] Federal Rural University of Pernambuco, Recife, Brazil
{ruan.carvalho,everton.silvasouza,filipe.rolim,valmir.macario}@ufrpe.br,
dan.rosa10@gmail.com, cicerojrlima@gmail.com
[2] Center for Excellence in Social Technologies, Federal University of Alagoas,
Maceió, Brazil
luiz.r70@gmail.com
[3] Western Sydney University, Penrith, NSW, Australia
[4] Harvard Graduate School of Education, Cambridge, MA, USA
[5] ISI-TICs - SENAI Innovation Institute for Information and Communication
Technologies, Recife, Pernambuco, Brazil
sergio.chevtchenko@sistemafiepe.org.br

Abstract. This study introduces a novel approach to Handwritten
Mathematical Expression Recognition (HMER), focusing on elementary
school mathematical expressions. Recognizing the challenges posed by
limited training data and the unique characteristics of elementary students' handwriting, we present a multiobjective optimization method
tailored for small training datasets. We employ state-of-the-art HMER
methods, including transformer-based and attention mechanism models, and optimize them using a custom dataset comprised of elementary school arithmetic equations. This dataset contains 1237 images and
includes both horizontal and vertical equations and isolated numbers,
featuring common errors in children's handwriting. Additional similar
datasets are also leveraged for training augmentation. Our experimental
results demonstrate the efficacy of the optimization approach, significantly improving the performance of the evaluated models in terms of
expression recognition rate and inference speed. This study contributes
to the field of HMER by providing an effective optimization approach for
SOTA models and by introducing a specialized dataset for elementary
school mathematics. The dataset is available upon request.

Keywords: HMER · Deep Learning · Optimization

1 Introduction

Computer-Assisted Learning (CAL) tools can contribute positively to teaching and learning mathematics, helping with automatic examining scores and

A. M. Olney et al. (Eds.): AIED 2024 Workshops, CCIS 2151, pp. 234–241, 2024.
https://doi.org/10.1007/978-3-031-64312-5_28

feedback [10,14]. A common task of these tools is to use an image of a handwritten problem, so that the system can recognize handwritten characters to verify the solution and give students feedback [11]. One of the main stages of such systems is Handwritten Mathematical Expression Recognition (HMER), which is an image-to-text task to generate the corresponding mathematical symbols from an input image [6]. Although HMER can be compared to traditional handwriting text recognition tasks, it is considered more challenging [19,21]. The incorrect classification of a single number or mathematical symbol changes the result of the operation and evaluation of the expression.

Recent advances in the field use deep neural networks to handle HMER as an end-to-end approach, using an image-to-sequence procedure in an encoder-decoder architecture [2]. Using deep learning (DL) models, end-to-end solutions transform an input representation (image) directly to mathematical notation. Although these techniques demonstrate impressive results, they typically require large amounts of data to train the model. For instance, popular datasets such as CROHME 2019 [9] and HME100k [17] consist of thousands of images of equations in a horizontal format.

In this work, we built a small image dataset of basic arithmetic equations (i.e., addition and subtraction) from elementary school students. To the best of our knowledge, there is no available public dataset with specific focus on elementary mathematics. In contrast, among the many consequences of the Covid 19 pandemic, children face substantial learning deficits in subjects such as numeracy. Thereby, providing datasets for developing HMER solutions targeting basic addition and subtraction is a significant step towards AI systems designed to mitigate these educational setbacks. We highlight that, although the images in our dataset are related to mathematical expressions, such as algebraic equations [9,17], it is a different domain because the equations are written by elementary school children and include both horizontal and vertical equations, as well as isolated numbers. Thus, in order to obtain good results on the presented dataset, the existing SOTA models have to be trained from scratch.

The main contributions of this paper are three-fold. 1) A new dataset of elementary school mathematical equations is presented; 2) The limitations of SOTA methods on our dataset are evaluated; 3) A multiobjective optimization approach is shown to significantly improve upon the default SOTA models.

2   Related Work

In recent years, encoder-decoder architectures have shown increased performance in various image-to-text tasks, using DL models [2,3,16]. Using a different neural architecture, BTTR [20] was proposed to first use the transformer decoder for solving HMER tasks and perform bidirectional language modelling with a single decoder. Zhao and Gao propose CoMER [19] to introduce the coverage mechanism into the transformer decoder and proposes an Attention Refinement Module. Wang et al. introduced a multi-modal attentional network (MAN) [15], which uses a decoder that employs two unidirectional Gated Recurrent Units (GRUs) with a multi-modal attention mechanism. Although SOTA research in HMER have been extensively evaluated in large public datasets, such as CROHME 2019 [9] and HME100k [17], their performance has not been evaluated in small-sized datasets. For instance, Neri et al. [11] propose a small custom dataset of linear equations, but use a classical approach (i.e., not end-to-end) and do not compare with SOTA methods. Zhu et al. [22] propose a model composed of detection, recognition, and correction components, dubbed as FATE. This model is evaluated on a dataset of 5000 sheets of paper and 77,097 arithmetical tasks, known as AEC-5k. The equations are often composed of printed initial parts, followed by a handwritten solution. The experimental results suggest that the FATE recognition model outperforms other SOTA baselines, although without hyperparameter tuning for the specific dataset. An evaluation of SOTA methods for a small mathematical dataset is done by Chevtchenko et al. [4]. However the optimization method only considers a single objective and the dataset is limited to vertical equations. Liu et al. [8] recently proposed a HMER network that includes a semantic aware module (SAM), designed to improve recognition on expression that contain similar symbols. While the SAM network is tailored towards more complex mathematical expression, it can also be optimized and evaluated for elementary school expressions using our proposed approach.

In summary, while existing works predominantly focus on large-scale datasets featuring complex mathematical expressions, the current study underscores the potential improvements SOTA models can achieve through hyperparameter optimization when applied to small-scale datasets. Therefore, we evaluate four recently proposed models, contrasting the optimized configurations with those presented in their original publications. Furthermore, this research fills a gap in existing literature by providing a public dataset in elementary school mathematics that encompass both vertical and horizontal equation formats. This dataset will help developing models that are sensitive to the nuances of children's handwriting and the simplified mathematical constructs typical of elementary-level education.

3   Methodology

The present work explores the optimization of hyperparameters of four SOTA networks as an approach to improve their performance on the proposed small-scale dataset: Counting Aware Network (CAN) [7], Syntax-Aware Network (SAN) [18], Bidirectionally Trained Transformer (BTTR) [20] and CoMER [19].

3.1   Datasets

As our dataset is smaller than what is usually required to train large models, two other datasets are used as an addition during the training stage: a small dataset of basic vertical summation and subtraction [12] and a simplified version of the HME100K dataset [17]. This is akin to the data augmentation technique in machine learning, where the goal is to expand and diversify the training set.

Vertical Summation and Subtraction. This dataset, first presented by Rosa et al. [12], contains basic sum and subtraction arithmetic equations written in a vertical format. Figure 1 shows image samples from this dataset, which contains a total of 232 images, created by three human annotators. In addition to the different writing styles, image writing also varies using a pencil or pen.

 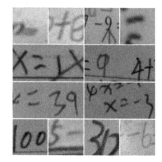

Fig. 1. Examples of images from Rosa et al. [12] and the simplified HME100K dataset.

Reduced HME100K Dataset. The original HME100K dataset contains a large amount of symbols and operators. Meanwhile, the scope of the present work is limited to evaluation of SOTA HMER models on simple mathematical expressions. Thus, the initial dataset is reduced to 7337 images, containing only digits from 0 to 9, the letter 'x', summation and subtraction, as illustrated in Fig. 1.

Custom Dataset. We built a custom dataset that combines characteristics from both of the previously introduced HME datasets. It contains 1237 images, separated in isolated numbers (949), vertical (119) and horizontal (169) equations. Notably, the dataset comprises equations handwritten by elementary school students and, thus, the digits are drawn with errors that are common to children that are learning to write, enhancing the dataset's validity for experimentation. Figure 2 presents some equations from this dataset.

3.2   Metrics

Intelligent tutoring systems can be employed on smartphones and other similar devices with possible constrains in terms of internet connection or computational resources. Thus, it is important to consider the trade-off between the inference time of HMER and a recognition rate. Similar to prior research [4,7,18], the following metrics are used to evaluated the models in this work: **ExpRate:** the rate of correct mathematical expressions recognized by the model. Defined as percentage of predictions of mathematical expressions that match the LaTeX

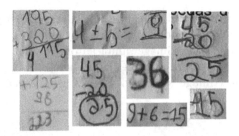

Fig. 2. Examples of images from our dataset.

label; **E1:** the proportion of correctly recognized expression, allowing for one error or less. **Inference time** – the inference time of the model in seconds. The images are presented to the model one at a time and the final metric is averaged across all images in the test subset. It is worth noting that ExpRate is the most challenging metric, as it requires the model to get the entire expression right, without missing or adding new symbols. By definition, the E1 metric will provide a higher score for the same model, when compared to ExpRate. Finally, because our dataset contains an imbalanced number of types of equations, the final ExpRate and E1 metrics are reported as an average of the individual metrics for each class: isolated numbers, vertical and horizontal equations.

3.3   Model Training and Hyperparameter Optimization

For the proposed experimental methodology, our custom dataset is partitioned in three subsets: **training**, composed of 741 images; **validation**, with 245 images used for early stopping of the training process and as the optimization target; and **test**, containing 251 images, which are separated for an unbiased comparison between the optimized models. We perform hyperparameter optimization based on that division. Nevertheless, note that the training set comprises our custom dataset's training partition, augmented by the other ones described in Sect. 3.1. This is a popular approach in machine learning, and aims to improve the model's predictive performance [13].

DL models often have a large number of hyperparameters and their performance can be significantly affected by hyperparameter selection. A manual approach becomes less preferable as search space increases. Thus, a widely used optimization framework, *Optuna* [1], is employed to find a good set of hyperparameters for the four benchmark models previously described. Note that the search space[1] includes the default configurations, nevertheless these are separately compared to the optimized ones in Sect. 4. In order to provide a fair comparison in terms of inference time, all models are evaluated on the same Linux system with Python 3.9, Intel i5 processor and an RTX3060 GPU.

---

[1] https://osf.io/bf86z/?view_only=249760c785b246709a6cc20a3987e201.

4 Experimental Results

This section presents the optimization approach and experimental results on our custom dataset, described in Sect. 3.1. Due to nonlinear relation between Expression and E1 rates, we adopt a multiobjective approach for optimization on the custom dataset. The NSGAIII [5] algorithm with population size of 10 is used to optimize both of these objectives concurrently. Each of the four models is optimized independently for 100 trials, following the protocol described in Sect. 3.3. In addition to the training set from the proposed dataset (741 images), the networks are trained with the 232 images from the vertical summation and subtraction images, described in Sect. 3.1, as well 200, 400 or 600 images, randomly selected from the reduced HME100K dataset, described in Sect. 3.1. The choice between the number of HME100K images to be added for training is considered as an additional optimization hyperparameter.

After optimization, five configurations that are closest the the theoretical optimum point are selected for each model. The resulting 20 configurations[1] are evaluated on the test subset. As previously, ExpRate, E1 and inference time are evaluated for each configuration on the test subset. The trade-off between ExpRate and E1 is presented in Fig. 3(left). Note that a balanced expression rate is used, which is an average between the three classes of the proposed dataset: isolated, horizontal and vertical. On these two metrics, a model based on BTTR obtains the best results, followed closely by CAN and SAN. Note that most of the optimized models achieved better performance in terms of both metrics that the corresponding default configurations.

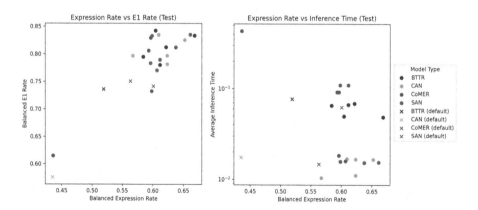

Fig. 3. Evaluation of the trade-off between Expression Rate and E1 Rate (left) and between Expression Rate and Inference Time (right) on the custom dataset.

Complementary to the results above, Fig. 3(right) also presents the trade-off between ExpRate and inference time. Both CAN and SAN models achieve a significantly better performance in terms of inference time. However, when considering the expression recognition rate, there is no clear distinction between these two models. It is worth noting that two of the optimized CAN models also prove a faster inference time than its default configuration: 10 vs. 18 milliseconds, respectively.

5   Conclusion and Future Work

This study presented an approach to optimizing large models for the recognition of handwritten mathematical expressions in elementary school contexts. Our approach leveraged a multiobjective optimization framework to significantly improve the performance of SOTA models on a custom dataset. The presented dataset reflects the unique characteristics of children's handwriting and is available upon request. The optimized configurations of the evaluated models (BTTR, CAN, SAN, CoMER) outperformed their default settings in terms of expression recognition rate, E1 rate, and inference time. This improvement was consistent across different models (see Fig. 3).

In future work we plan to explore the integration of these optimized models into real-world educational tools. Furthermore, we plan to explore a deeper optimization strategy on the most promising models, simultaneously adjusting parameters and structure of the network.

References

1. Akiba, T., Sano, S., Yanase, T., Ohta, T., Koyama, M.: Optuna: a next-generation hyperparameter optimization framework. In: Proceedings of the 25rd ACM SIGKDD International Conference on Knowledge Discovery and Data Mining (2019)
2. Bian, X., Qin, B., Xin, X., Li, J., Su, X., Wang, Y.: Handwritten mathematical expression recognition via attention aggregation based bi-directional mutual learning. In: Proceedings of the AAAI Conference on Artificial Intelligence, vol. 36, pp. 113–121 (2022)
3. Cheng, Z., Bai, F., Xu, Y., Zheng, G., Pu, S., Zhou, S.: Focusing attention: towards accurate text recognition in natural images. In: Proceedings of the IEEE International Conference on Computer Vision, pp. 5076–5084 (2017)
4. Chevtchenko, S., et al.: Algoritmos de reconhecimento de dígitos para integração de equações manuscritas em sistemas tutores inteligentes. In: Anais do XXXIV Simpósio Brasileiro de Informática na Educação, pp. 1442–1453. SBC (2023)
5. Deb, K., Jain, H.: An evolutionary many-objective optimization algorithm using reference-point-based nondominated sorting approach, part I: solving problems with box constraints. IEEE Trans. Evol. Comput. $18(4)$, 577–601 (2013)
6. Deng, Y., Kanervisto, A., Ling, J., Rush, A.M.: Image-to-markup generation with coarse-to-fine attention. In: International Conference on Machine Learning, pp. 980–989. PMLR (2017)
7. Li, B., et al.: When counting meets HMER: counting-aware network for handwritten mathematical expression recognition. In: Avidan, S., Brostow, G., Cissé, M., Farinella, G.M., Hassner, T. (eds.) Computer Vision–ECCV 2022: 17th European Conference, Tel Aviv, Israel, October 23–27, 2022, Proceedings, Part XXVIII, pp. 197–214. Springer, Cham (2022). https://doi.org/10.1007/978-3-031-19815-1_12

8. Liu, Z., Yuan, Y., Ji, Z., Bai, J., Bai, X.: Semantic graph representation learning for handwritten mathematical expression recognition. In: International Conference on Document Analysis and Recognition, pp. 152–166. Springer, Cham (2023). https://doi.org/10.1007/978-3-031-41676-7_9
9. Mahdavi, M., Zanibbi, R., Mouchere, H., Viard-Gaudin, C., Garain, U.: ICDAR 2019 CROHME+ TFD: competition on recognition of handwritten mathematical expressions and typeset formula detection. In: 2019 International Conference on Document Analysis and Recognition (ICDAR), pp. 1533–1538. IEEE (2019)
10. Meeter, M.: Primary school mathematics during the COVID-19 pandemic: no evidence of learning gaps in adaptive practicing results. Trends Neurosci. Educ. **25**, 100163 (2021)
11. Neri, M.C.G., Villegas, O.O.V., Sánchez, V.G.C., Domínguez, H.D.J.O., Nandayapa, M., Azuela, J.H.S.: A methodology for character recognition and revision of the linear equations solving procedure. Inf. Process. Manage. **60**(1), 103088 (2023)
12. Rosa, D., et al.: Recognizing handwritten mathematical expressions of vertical addition and subtraction. arXiv preprint arXiv:2308.05820 (2023)
13. Shorten, C., Khoshgoftaar, T.M.: A survey on image data augmentation for deep learning. J. Big Data **6**(1), 1–48 (2019)
14. Verbruggen, S., Depaepe, F., Torbeyns, J.: Effectiveness of educational technology in early mathematics education: a systematic literature review. Int. J. Child Comput. Interact. **27**, 100220 (2021)
15. Wang, J., Du, J., Zhang, J., Wang, Z.R.: Multi-modal attention network for handwritten mathematical expression recognition. In: 2019 International Conference on Document Analysis and Recognition (ICDAR), pp. 1181–1186. IEEE (2019)
16. Xu, K., et al.: Show, attend and tell: neural image caption generation with visual attention. In: International Conference on Machine Learning, pp. 2048–2057. PMLR (2015)
17. Yuan, Y., et al.: Syntax-aware network for handwritten mathematical expression recognition. In: Proceedings of the IEEE/CVF Conference on Computer Vision and Pattern Recognition, pp. 4553–4562 (2022)
18. Yuan, Y., et al.: Syntax-aware network for handwritten mathematical expression recognition. In: Proceedings of the IEEE/CVF Conference on Computer Vision and Pattern Recognition, pp. 4553–4562 (2022)
19. Zhao, W., Gao, L.: CoMER: modeling coverage for transformer-based handwritten mathematical expression recognition. In: Avidan, S., Brostow, G., Cissé, M., Farinella, G.M., Hassner, T. (eds.) Computer Vision–ECCV 2022: 17th European Conference, Tel Aviv, Israel, October 23–27, 2022, Proceedings, Part XXVIII, pp. 392–408. Springer, Cham (2022). https://doi.org/10.1007/978-3-031-19815-1_23
20. Zhao, W., Gao, L., Yan, Z., Peng, S., Du, L., Zhang, Z.: Handwritten mathematical expression recognition with bidirectionally trained transformer. In: Lladós, J., Lopresti, D., Uchida, S. (eds) Document Analysis and Recognition–ICDAR 2021: 16th International Conference, Lausanne, Switzerland, September 5–10, 2021, Proceedings, Part II 16, pp. 570–584. Springer, Cham (2021). https://doi.org/10.1007/978-3-030-86331-9_37
21. Zhelezniakov, D., Zaytsev, V., Radyvonenko, O.: Online handwritten mathematical expression recognition and applications: a survey. IEEE Access **9**, 38352–38373 (2021)
22. Zhu, Q., Luo, Z., Zhu, S., Jing, Q., Xu, Z., Xue, H.: Fate: a three-stage method for arithmetical exercise correction. Neural Comput. Appl. **35**(32), 23491–23506 (2023)

Can Large Language Models Generate Middle School Mathematics Explanations Better Than Human Teachers?

Allison Wang[1]([✉])[iD], Ethan Prihar[3][iD], Aaron Haim[2][iD], and Neil Heffernan[2][iD]

[1] University of California, Berkeley, Berkeley, CA 94720, USA
`allison-wang@berkeley.edu`
[2] Worcester Polytechnic Institute, Worcester, MA 01609, USA
`{ahaim,nth}@wpi.edu`
[3] Swiss Federal Institute of Technology in Lausanne, Rte Cantonale, 1015 Lausanne, Switzerland
`ethan.prihar@epfl.ch`

Abstract. The rapid development of large language models has led to increased interest in evaluating the efficacy of models for education technology. In this paper, we are particularly interested in GPT created by OpenAI. Specifically, we are interested in comparing GPT-3.5 and GPT-4 in their ability to generate explanations for math problems. Computer-generated explanations could be readily utilized in online tutoring systems as a supplement to existing teacher-created explanations. Conclusions from prior work regarding GPT-3.5 generated explanations found a high mathematical error rate of 90% and 57% respectively for two methods tested and a poor quality rating when assessed by teachers. In this work, we present improved methods using GPT-4 that significantly reduced errors in generated explanations to 6%. Additionally, a preregistered study with evaluators rated the GPT-4 explanations as higher in quality compared to human explanations. Our main interest is looking at the quality of generated explanations, so other AIED systems can begin using GPT-4 to create feedback to better foster student learning.

Keywords: Large Language Models · Online Tutoring · GPT-3 · GPT-4 · Distance Learning

1 Introduction

In this work, we explore the effectiveness of using LLMs to generate explanations for mathematics problems within the ASSISTments online learning platform [4]. In prior work, we generated explanations using two different methods to prompt GPT-3.5, which resulted in high rates of mathematical error [5]. An analysis of evaluation from teachers also found that GPT-3.5 explanations performed statistically worse than human explanations in perceived quality.

With the continual development of LLMs, it is likely that newer and more powerful GPT-4 can generate mathematical explanations better, even though

A. M. Olney et al. (Eds.): AIED 2024 Workshops, CCIS 2151, pp. 242–250, 2024.
https://doi.org/10.1007/978-3-031-64312-5_29

they have not been specifically fine-tuned [1]. In this work, prompting methods for generating explanations with GPT-4 to the same problems from prior work with GPT-3.5 were explored as a means of comparing whether the new model was more reliable in generating correct explanations. GPT-4 explanations were then generated with the most reliable prompting method to compare to teacher explanations in a single-blinded study where explanations were evaluated by undergraduate math students.

To assess the mathematical explanations, we define the term *perceived-effectiveness* as the metric of measure when judged and rated by humans. The term *error-rate* refers to explanations that had math errors. Mathematical error was only judged by one person, compared to the multiple raters for the more subjective *perceived-effectiveness*.

To reiterate, this paper addresses the following research questions:

1. GPT-3.5 Study: (Prihar et al.)[1]
   (a) What is the *perceived-effectiveness* of explanations generated by GPT-3.5 compared to human-written explanations?
2. GPT-4 Study: (Wang et al.)[2]
   (a) What is the *perceived-effectiveness* of GPT-4 explanations compared to GPT-3.5 explanations?
   (b) What is the *perceived-effectiveness* of GPT-4 generated explanations compared to human-written explanations?

2   Background

2.1   ASSISTments

ASSISTments is an online learning platform focusing on helping students with K-12 mathematics [4]. As students complete problems, they can request an explanation, which are currently written by teachers. This work aims to investigate whether the quality of GPT-4 explanations generated is comparable to current human-written explanations and better than GPT-3.5 explanations. As the GPT-4 engine is significantly more costly to run, the analysis is essential to determine whether the investment in GPT-4 generated explanations is ultimately worth it to scale ASSISTments.

For this paper, we sampled problems from the two most popular Open Educational Resource math curriculum used in the US, called EngageNY (Eureka Math) and Illustrative Math (Open Up Resources).

---

[1] Results have been previously reported. In Prihar et al. [5], a greater emphasis was placed on evaluating how explanations were generated.

[2] Results were presented in an earlier form at a workshop and have not been published prior.

2.2   GPT-3.5 Vs. GPT-4

Many other works have already shown that GPT-4 has improved mathematical capabilities compared to GPT-3.5. When comparing model performance on the GSM-8K benchmark data set, a dataset of 8.5k grade school math problems [3], GPT-4 performed significantly better than GPT-3.5.

The increase in GPT-4 accuracy provides solid reasoning that the model will show improvement in generating math explanations for middle school math problems. It is important to note that the focus of this paper is not a demonstration of the inherent increases in capability that come with larger LLMs. This work differentiates itself from previous work showing increases in capability due to an additional emphasis on assessing the perceived quality of explanations for student learning outcomes alongside a measure of correctness.

Compared to a single prompt with GPT-3.5, prompting GPT-4 requires an additional overall system prompt, and a user prompt that interacts with the system for specific requests.

3   Prior Work

3.1   Methodology

The methodology used with GPT-3.5 was focused on comparing two different methods of generating explanations. GPT-3.5 was either asked to summarize the advice from a tutoring chat log or prompted to generate new explanations when provided teacher-written examples to similar problems in a few-shot learning approach.

For the summarization approach, tutoring chat logs from ASSISTments were split between a development and evaluation data set. The chain-of-thought prompting used to instruct GPT-3.5 is reported in the prior work of Prihar et al. [5]. For the few-shot learning approach, problems in the ASSISTments database from the two curriculum described above were used. The same problems used for evaluating the summarization approach were used for few-shot explanation generation for comparison purposes. More information can also be found in the prior work [5].

3.2   Results

Both methods of prompting GPT-3.5 to generate explanations had a high rate of mathematical error present, with an *error-rate* of 57% for the summarization approach and 90% for few-shot learning. The valid explanations that remained were combined with existing ASSISTments explanations for the same problems, and a blinded survey was given to 5 middle school math teachers to evaluate GPT-3.5 compared to ASSISTments explanations. A mixed-effects model to aggregate ratings concluded that explanations from both methods were rated as significantly worse than current ASSISTments explanations [5].

4 Methodology

4.1 Optimization

To investigate the differences between the GPT-3.5 and GPT-4 models, a preliminary comparative study was first conducted to determine whether an alternative method of zero-shot prompting without examples with GPT-4 would be less error-prone, as it has been shown to been successful for other uses [6].

The dataset of problems that were given to GPT-4 included all problems from the summarization and few-shot learning approaches in the previous GPT-3.5 work. Duplicate problems or problems with images were removed. After this, 33 problems remained to be given to GPT-4 for explanation generation. This was fewer than the 42 problems from the GPT-3.5 study because problem duplicates were removed.

Explanations were generated using the system-level prompt, which was found by attempting to generate explanations to three mathematics problems that all tested different skills. The wording was altered until the explanations generated were of adequate quality for all three problems. The Temperature was 0.31 and Maximum Length was 256 tokens. Temperature was chosen arbitrarily within a range that consistently generated correct explanations.

"The user will provide a middle school math problem that a student is currently struggling on. The student requests for an explanation of how to find an answer to the problem. Provide a step-by-step explanation as a middle school math teacher that is easy enough for a student to understand, and that they will learn from. Problem explanations must be under 170 words and very concise, and easy to follow. Respond in a direct and factual tone in third person. Value efficiency in finding the answer using the least number of steps rather than a single-step mathematical operation. Find a creative solution."

Problems were given to the model as specific user requests in the original HTML format from the website to accurately account for math symbols.

Evaluation Method and Results. After explanations were generated, they were manually evaluated based on if they were valid (structured as an explanation, and no math errors present). There were 2 out of the 33 that were invalid, equal to an approximately 6% *error-rate.*

The other 31 valid explanations generated by GPT-4 were appended to the survey from the prior work that included the 42 GPT-3.5 explanations and 31 newly generated valid explanations from GPT-4. The three math undergraduate students that were evaluators were instructed independently to rate the 73 explanations on a scale of 1–5 (1 = Very Bad, 5 = Very Good) based on if they believed they would help students.

All ratings were aggregated, and a mixed-effects model was fit with random effects for the problem and rater, and fixed effects for the source of the explanation. This model was identical to the model used in prior work [5]. The fixed effects for the source of the explanations were used to measure the difference in quality of explanations, as these effects can be interpreted as the average rating of an explanation generated by the corresponding source after factoring out confounding from different raters strictness and the difficulty of explaining specific problems.

Figure 1 shows the graph of the mean ratings and 95% confidence interval for explanations created by humans, both methods from GPT-3.5, and GPT-4. The explanations from the ASSISTments source were drawn from ASSISTments problem databases, and are written by teachers.

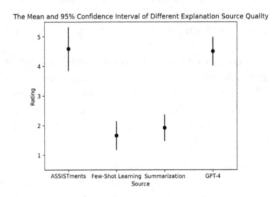

Fig. 1. The mean and 95% confidence interval of explanation ratings for 4 sources of explanation generation.

The GPT-4 explanations were rated much higher than both methods of GPT-3.5 generation, demonstrating the new method for explanation generation used with GPT-4 had both a lower mathematical *error-rate* and the explanations were more highly perceived. The main study following will utilize the new zero-shot GPT-4 explanation generation method.

4.2   Improved GPT-4 Study

The analysis plan has been preregistered on OSF and can be found at https:// osf.io/x3qrh?view_only=03b19d094a9440a0ae5df4177907a1d1.

Using the same zero-shot method from the GPT-4 optimization section above, the study randomly sampled 100 problems and explanations from the ASSISTments database that were text-based. Only the first problems in a multi-problem were sampled from so that each problem required no prior context to solve. Problems were provided to GPT-4 in HTML format to account for special symbols, the temperature was set to 0.5, and the responses had no maximum token length.

A system prompt was given to GPT-4 that applied to each individual problem request. The system prompt was modified from the optimization method based on picking 5 problems with different skills from the entire ASSISTments dataset and altering the prompt wording until the explanations generated were satisfactory.

> You are a middle-school math teacher. A student is completing an online math assignment. Provide the student with a very concise explanation that teaches them, step-by-step, how to solve for the answer to the following problem. The explanation should be easy for a middle-school student to understand and learn from. If there are efficient shortcuts or rules of thumb that can be used to solve the problem, include them in the explanation. Return only the explanation formatted as HTML starting with <p>.

Then, the following user prompt was given to GPT-4:

> Problem HTML:
> [Problem HTML]
> Acceptable Answer(s):
> [First Acceptable Answer]
> [Second Acceptable Answer]
> ...
> [Last Acceptable Answer]

Evaluation Method. After explanations for the 100 sampled problems were generated by GPT-4, they were combined with the existing teacher-written explanations in the ASSISTments database for the same problems. The 200 problem and explanation pairs were combined into a survey. Explanations were evaluated using the same 1–5 scale as the preliminary study, and a new group of 10 undergraduate mathematics students were given the same verbal instructions for how to rate explanation's *perceived-quality*. The source of the explanation (GPT-4 or human) was blinded and the order of the explanations was randomized for each rater to reduce ordering effects.

To compare the *perceived-quality* of GPT-4 generated explanations to human-written explanations, the same model from the optimization study that was used to judge the ratings between GPT-3.5 and GPT-4 was used, except random effects for the interactions between raters and problems were also included to help remove any additional confounding from the interactions. The fixed effects of GPT-4 generated explanations and human-written explanations were again measures of the average quality of each source's explanations.

5 Results

Of the 100 explanations generated, 4 of them contained errors in the explanation structure or utilized a problem approach different to the one specified in the problem, which is a 4% *error-rate*. This error rate can be interpreted as more generalized compared to the 6% *error-rate* in the GPT-4 optimization section, as the 6% *error-rate* only applied to a subset of ASSISTments problems.

While there are still instances of explanations that were classified as invalid, all 4 errors still led to the correct answer, so we are confident the explanations could be implemented into the ASSISTments system without harming student learning. The distribution of ratings is shown in Fig. 2. The GPT-4 explanations were rated higher than the human created explanations, with an average rating of 4.3 compared to human-written explanations with a rating of 3.7.

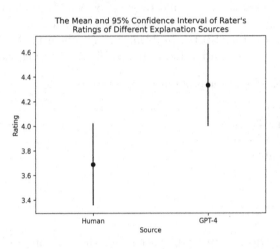

Fig. 2. Distribution graph comparing ratings for GPT-4 and human created explanations, showing the mean and 95% confidence intervals.

The lack of overlap in the 95% confidence intervals indicates that it is highly likely the GPT-4 explanations were preferred with higher ratings. Such results are quite surprising, as we assumed the current human-created explanations would be the most preferred. After the survey was completed and the purpose was revealed, evaluators noted that they preferred GPT-4 explanations because it had clearer step-by-step approach using relevant concepts compared to ASSISTments explanations.

It is also important to note that with Fig. 2 above compared to Fig. 1 in the methodology, the difference between the averages and confidence intervals for human and GPT-4 ratings are not the same. This is likely due to the different evaluators for the two experiments. In the study comparing GPT-4 and humans, the evaluators also shared how they were more inclined to rate the GPT-4 explanations higher because they were longer and more thorough. With the increased volume of explanations, it is possible many did not assess every explanation and extrapolated quality only on amount, and not necessarily correctness, of detail.

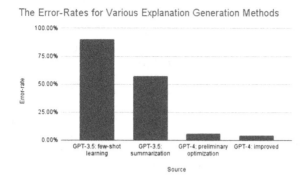

Fig. 3. Graph comparing rates of mathematical errors in different generation methods.

6   Conclusion

This work concludes that GPT-4 explanations were preferred over both GPT-3.5 and humans in *perceived-effectiveness*, and also had a lower *error-rate* of explanation correctness. Even though the *error-rates* are not zero as shown in Fig. 3, the large magnitude in the drop in *error-rate* is large enough to begin using explanations in ASSISTments, and also to begin investing in GPT-4 generated explanations at large.

6.1   Limitations and Future Work

A major limitation to this study is the changing nature of LLMs. Prior work has demonstrated both positive and negative changes in LLM performance with mathematics tasks for GPT-3.5 and GPT-4 when evaluations were conducted only months apart [2]. While it would be best to compare at the same point in time, the continuously changing nature of both GPT-3.5 and GPT-4 makes such a task incredibly difficult, so this limitation is inevitable.

While experienced raters preferred GPT-4 generated explanations, there is no guarantee that the explanations cause more learning, as the ratings were only a measure of perceived quality. We recognize the limitations posed by such weakness. In January 2024, we launched a randomized controlled trial comparing 276 problems where we had both teacher-created and GPT-created explanations inputted into ASSISTments and given to students randomly. The study investigating whether the messages are effective for student learning has been preregistered and can be found at https://osf.io/b7p6v/?view_only=03b19d094a9440a0ae5df4177907a1d1.

References

1. Bubeck, S., et al.: Sparks of artificial general intelligence: early experiments with GPT-4. arXiv preprint arXiv:2303.12712 (2023)
2. Chen, L., Zaharia, M., Zou, J.: How is chatgpt's behavior changing over time? (2023)
3. Cobbe, K., et al.: Training verifiers to solve math word problems. arXiv preprint arXiv:2110.14168 (2021)
4. Heffernan, N.T., Heffernan, C.L.: The assistments ecosystem: building a platform that brings scientists and teachers together for minimally invasive research on human learning and teaching. Int. J. Artif. Intell. Educ. **24**, 470–497 (2014)
5. Prihar, E., et al.: Comparing different approaches to generating mathematics explanations using large language models, pp. 290–295, June 2023. https://doi.org/10.1007/978-3-031-36336-8_45
6. Wu, W., Yao, H., Zhang, M., Song, Y., Ouyang, W., Wang, J.: Gpt4vis: What can gpt-4 do for zero-shot visual recognition? (2023)

System Dynamics Through Qualitative Modeling in DynaLearn: Results from Research Conducted in a Brazilian High School Context

Paulo Vitor Teodoro[1]([✉]) [iD], Paulo Salles[2] [iD], and Ricardo Gauche[2] [iD]

[1] Federal University of Uberlândia, Ituiutaba, MG, Brazil
paulovitorteodoro@ufu.br
[2] University of Brasília, Federal District, Brazil

Abstract. In this paper, we present the results of research aimed at demonstrating that high school students are capable of understanding complex phenomena through the exploration of qualitative simulation models. To achieve this, we adopted the approach of Learning by Modeling using techniques developed in an area of Artificial Intelligence (AI) called Qualitative Reasoning (QR). As a tool for constructing and simulating qualitative models, we used the DynaLearn platform, which diagrammatically displays possible causal relationships between the components of a system and representations of mathematical functions, allowing for the calculation of qualitative values of system properties. Thirty-nine students from the first grade of high school participated in this study, from a federal public school, in Brazil. Data collection took place through questionnaires and semi-structured interviews. The data showed that qualitative modeling is an important tool for students to think about dynamic systems, as opposed to a one-sided view. Indeed, students understood the structure and functioning of the system, distinguishing the main elements that make up the system, as well as recognizing concepts of System Dynamics (SD), such as feedback mechanisms. The data allow us to consider that the construction and manipulation of qualitative simulation models by students contribute to the understanding of complex systems in the classroom.

Keywords: Systems Dynamics · Qualitative Modeling · DynaLearn

1 Introduction

The classroom space is a learning environment that extends beyond the mere transmission of information by the teacher. This includes, for instance, the implementation of innovative pedagogical methods, as we will present in this study [1]. It emphasizes that the efficacy of the learning environment is not solely based on its physical structure or available resources, but also on how it is utilized by the participants, namely, teachers and learners. Therefore, it is crucial to have the support of appropriate educational materials for the complex activities of teaching and learning. Teaching, coupled with effective pedagogical materials, can develop in students the ability to continuously reflect on understanding a changing world [1].

© The Author(s), under exclusive license to Springer Nature Switzerland AG 2024
A. M. Olney et al. (Eds.): AIED 2024 Workshops, CCIS 2151, pp. 251–258, 2024.
https://doi.org/10.1007/978-3-031-64312-5_30

In the literature, numerous texts illustrate experiences of using various teaching and learning materials in Science Education [1, 2]. However, they generally aim to review specific subjects, promote socialization, and motivate learners [1]. All these points are crucial for the educational process [1]; however, they are still limited when considering pedagogical strategies that truly promote student autonomy and protagonism, critical formation, the development of a systemic worldview, knowledge acquisition, and opportunities for decision-making.

Therefore, one possible strategy would be to adopt tools that allow the teacher, together with the learners, to develop their own educational materials [2]. These materials can be created through the construction of simulation models, which enable learners to be inserted as protagonists in the learning process through knowledge representation [2]. Indeed, modeling can be used to understand complex Systems Dynamics (SD), a fundamental subject for scientific comprehension. That's because modeling allows the student to learn through 'doing' [3], in an approach known as Learning by Modeling (LbM), that they create models, tests hypotheses, predicts outcomes, and adopts a more critical stance towards the analyzed problem situations [4].

These characteristics allow models to be not only a way to represent knowledge but also a resource to be used as suitable educational material for understanding complex systems [4]. However, there are few examples of the use of simulation models in basic education. One limiting factor is that simulations almost always rely on complex mathematical functions, which are usually inaccessible to high school students.

An interesting solution to overcome these limitations is qualitative simulation models, developed by researchers [4] in an area of Artificial Intelligence (AI) known as Qualitative Reasoning (QR), which focuses on describing continuous properties of the world based on a discrete system of symbols to support automated reasoning [5]. Through modeling techniques and with the AI tools, it is possible to use qualitative knowledge to answer questions about highly complex systems, even when quantitative information or data is absent or incomplete.

With qualitative models, it is possible to identify central and peripheral information, contextualize, interpret, interrelate different areas of knowledge, problematize, make inferences, explain mechanisms based on the functioning of system parts, plan, analyze solutions, and even predict possible outcomes [6]. Through qualitative models, it is possible to acquire conceptual knowledge qualitatively and, from this, predict relevant solutions and conclusions.

Research conducted with students from various countries in Europe, North America, and Asia has shown that QR has potential to represent Qualitative System Dynamics and also assist in students' scientific development. As a result, the DynaLearn software was developed through the efforts of European researchers in collaboration with other countries [6], including Brazil. DynaLearn is a modeling platform that uses QR techniques and diagrammatically shows possible causal relationships between the topics covered and the prediction of simulation results, without the use of numbers. Therefore, this work aims to investigate how the construction of simulation models based on QR in DynaLearn can facilitate high school students' understanding of complex systems.

2 Methodology

We developed a pedagogical intervention in two classes (class A and class B) of the first year of high school at a federal public institution located in Brazil. Student participation in the project occurred through voluntary enrollment during the school's extracurricular activities. The students engaged in a series of modeling activities during the extracurricular sessions, totaling 54 sessions, with each session lasting 50 min. In the research, 25 participants from class A and 23 from class B participated. Of these, 22 from class A (88%) and 17 from class B (74%) completed the modeling activities.

During the modeling intervention, we utilized the Qualitative Process Theory (QPT) proposed by Forbus [7]. In this theory, the world is modeled by objects, and any change in the system is attributed to the action of a process. Accordingly, a process characterizes changes in objects over time. The world is modeled by objects whose properties are described by quantities [7]. Thus, the combination of objects, their properties, the relationships between them, as well as the processes that occur, defines a system.

Processes can affect objects in various ways, and many of these effects can be modeled by changes in some properties of the objects. As these properties are represented by quantities, changes that objects undergo due to the action of processes are modeled with direct influences (represented by I + or I-), corresponding to changes in quantity values [7]. In this sense, changes cause alterations in the values of various quantities, and these changes propagate to other parts of the system through indirect influences (also called qualitative proportions - P+ or P−). After all changes are computed, a new state will have been characterized. The simulation, which begins with the description of an initial scenario, now presents a new state of the system. Similarly, the simulation allows the conditions for some mechanism of change (process) to become active and, therefore, initiates possible changes. This generates a new state, and so on, until no further change is possible, and the simulation ends.

Thus, the concepts through QPT guide modeling activities, which are suitable for representing complex systems: processes, rates, state variables, direct influences, qualitative proportions, and feedback mechanisms [7]. Through the DynaLearn platform, learners developed models at increasing levels of complexity, broadening their understanding of system behavior. Our intention was to ensure that students had active participation in representing SD through model construction, aligned with the LbM approach. Consequently, from the first session, learners began exploring the DynaLearn platform by constructing simpler models, such as conceptual maps (LS1 in DynaLearn). From there, progressively more complex models were built (LS2, LS3), until we reached more specific concepts of SD, such as the notion of processes, feedback mechanisms, in LS4. In total, 55 models were constructed by the students and explored, with the vast majority, 32, being developed in LS4 (the main level for addressing complex systems). For all the models used in the school, we conducted a concept test using the DynaLearn software. The models were evaluated by experts (teachers and researchers in the Natural Sciences fields from a federal public institution in Brazil) with both conceptual and operational assessments.

It's worth emphasizing that we acknowledge the intrinsic presence of biases and limitations in data collection in qualitative research. For this reason, we utilized two data collection instruments: questionnaires and semi-structured interviews. The cross-referencing of the collected data allowed us to minimize bias risks. The study participants were minors (aged between 15 and 16 years). Therefore, the parents (or guardians) of the students signed the Informed Consent Form, which included a guarantee to maintain the anonymity of the participants.

3 Results and Discussion

In total, the students built 55 models using DynaLearn. An example can be seen in Fig. 1, which illustrates how pollutant emissions in the city influence the number of people on the shore.

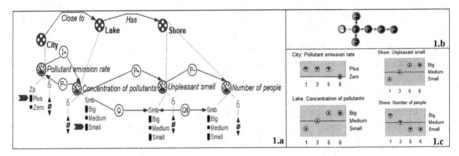

Fig. 1. Model scenario "Pollutant emission in the city" (a), State graph (b) and VHD (c).

This model illustrates that there is a 'City' close to a 'Lake', and the 'Lake' has a 'Shore'. The entire system is comprised of Entities, namely 'City', 'Lake', and 'Shore', each with its configuration (such as "Close to" and "has"). Each Entity also has its properties, known as Quantities, which include *Pollutant emission rate, Concentration of pollutants, Unpleasant smell*, and the *Number of people*. The value of a quantity is represented by the Quantitative Space (QS) < Magnitude, Derivative >. The magnitude represents the 'size of the thing' (e.g., 'Small', 'Big'). The derivative represents the direction of change of the value of this variable, with values from the QS = { 'Positive', 'Zero', or 'Negative'}, indicating that the variable is increasing, stable, or decreasing, respectively. Changes in values represent the behavior of a quantity. Therefore, if a variable has a magnitude value of 'Small' and is increasing, < Small, Positive > will be its value. Thus, changes in variable values indirectly indicate the passage of time.

In this scenario (Fig. 1a), the process is represented by the *Pollutant emission rate*, which influences the *Concentration of pollutants* with I +. This process initiates a chain of causality, causing changes in the entire system. The *Concentration of pollutants* increases up to a certain limit, controlled by the balancing feedback mechanism with a P- (*Pollutant emission rate ← Concentration of pollutants*), indicating that these processes are controlled and do not continue indefinitely. As the *Concentration of pollutants* increases, *Unpleasant smell* also increases, in the same direction, due to their connection by a qualitative proportionality (P +), with the same qualitative values of magnitude and derivative, as they have the correspondence Q between them. Consequently, if *Unpleasant smell* increases, the *Number of people* decreases, as they are connected by P-. It's important to note that *Unpleasant smell* and *Number of people* vary inversely due to the inverse correspondence (Q↓), resulting in inverse qualitative value changes.

During the simulation of this model, the State Graph is generated, showing the change of states as depicted in Fig. 1b. These represent all the possible changes from the initial scenario. The values assumed by the quantities represented in the model during the simulation are shown in Value History Diagrams (VHD) as illustrated in Fig. 1c. The VHDs enable us to comprehend the changes that occurred in the transitions between states: [1] → [3] → [5] → [6].

On the chosen trajectory, the *Pollutant emission rate* has a 'Plus' magnitude, but a decreasing derivative, remaining in states (1, 3, 5) until reaching 'Zero' and a stable magnitude in state 6. This rate stabilizes due to being controlled by the balancing feedback mechanism. The *Concentration of pollutants* in the 'Lake', influenced by the I + of the rate, begins the trajectory with a 'Small' and increasing value, passing through state 3 with a 'Medium' value, and reaching 'Big' at state 5, where it stabilizes in the final state of this trajectory (6). The *Unpleasant smell* changes during the trajectory, mirroring the values of its influencer, *Concentration of pollutants*, as they are connected by the Q correspondence. If *Unpleasant smell* increases, the *Number of people* decreases, as they are connected by P- and exhibit inverse qualitative value changes (Q↓).

3.1 Results from Research Conducted in a Brazilian High School Context

Based on the set of activities conducted at the school and the questionnaire answered by the participants in this study, we found that almost all students understand the system structure concerning objects, their properties (Quantities), Configurations, and Quantity Space. Thus, the data showed that the participants mentioned finding the difference between Entities and Quantities 'easy' (60% in class B and 46.67% in class A) or 'very easy' (53.33% in class A and 30% in class B). The only exception was one participant from Class B who did not provide an opinion agreeing that it is 'easy' or 'very easy'. Regarding configuration, around 73.33% of students (in both classes A and B) mentioned finding it 'easy' or 'very easy' to understand. The quantitative space was classified, in the majority of both classes, as easy (53.33% in class A; 80% in class B).

Besides that, regarding the structure of the system, we also considered correspondences, influences, and feedback mechanisms. Concerning correspondences (Q), most participants in class A (53.33%) and 50% in class B indicated the concept as 'easy' or 'very easy.' To represent influence, we have direct influence (I + or I-) and indirect influence (P + and P-). The distinction between direct influences (Is) and qualitative proportionalities (Ps) was considered 'difficult' by most participants (60%, in both classes).

As for the open-ended questions, we observed that students had more autonomy to write about their understanding of the concepts addressed in the intervention. Understanding the concepts related to direct influences is crucial for comprehending the dynamic aspects of the system. The first open-ended question in the questionnaire ('What do you understand by Direct Influence?') allowed us to gather information about the students' understanding of this subject. From the collected data, we categorized three groups of responses to refer to direct influences: a) represent processes (40% in class A, 30% in class B); b) influence the state variable (60% in class A, and 80% in class B); c) add or subtract value from the stock (30% in class A).

Direct influences represent the effects of processes through rates and state variables. Rates depict flows and are always related to a period of time [7]. State variables, on the other hand, can be understood as the representation of a Stock, which can increase or decrease depending on the type of influence from the Rate. Thus, we can perceive that the students' responses touched on the correct meaning of direct influences. However, the responses were shorter, and most of them only encompassed one of the categories.

Indeed, in class B, only one participant mentioned two of the mentioned categories: 'It is I + and I- used to represent processes. When a rate directly influences a state variable' (one student, class B). In this case, the student mentioned influences as a representation of processes and also that this representation occurs through rates that influence state variables. The other students in class B always pointed out only one of the mentioned categories. Most responses, like in class A, related direct influences as the representation of rates influencing state variables (Stocks, in the students' language), for example: 'That only influences stocks' or 'It is what influences the stock directly'.

In class A, we also had conceptions that encompassed more than one category mentioned in the graph. For example: 'There are two types: I + which means adding for a certain time and I- which means subtracting for a certain time. They represent the processes of the system.' (one student, class A). In this situation, the response was counted for the categories 'Represents processes' and 'Adds or removes values.' Additionally, only in class A was it mentioned in relation to direct influences being used to add or remove values from the state variable. For example: 'That direct influence is proposed to add or remove values in its stock.' And indeed, influences can be positive or negative (I + or I-), where I + means that values are added to the state variable (stock), and I- implies deducting values from the state variable. Thus, the state variable is responsible for describing the system's state.

From the students' responses regarding the differences between Rate and Stock, it was possible to elaborate four distinct groups: a) Different units (86.67% in class A, and 60% in class B); b) Influence on the State Variable (26.67% in class A, and 40% in class B); c) Quantitative spaces (30%, in class B); d) Others (10%, in class B). Considering that some students mentioned more than one of these groups, the numerical value of the percentage exceeded 100%.

As can be noted, most students in both classes mention the difference between rate and state variable in relation to Units. In this category, students present more direct responses, for example, 'The units are different' (one student, class A), or even: 'The difference is in the units' (one student, class B). But other participants relate the difference in units because the rate is related to time, and the stock is not. For illustration, we can show: 'The rate adds something over time, and the stock receives what was added, always for a period of time.', or also: 'Rate is a value per unit of time. The stock or state variable has the same unit of measure but without time.' (one student, class A).

In class B, there were also responses related to Quantitative Spaces. The participant points out that 'Rate uses the quantitative space of magnitude 'Zero' and 'Plus'. Stock uses the quantitative space of large, medium, and small magnitudes.' Two other students also mention: 'Rate is used in the quantitative space of magnitude zero and plus. Stock is used in other quantitative spaces.', or in a more concise way, 'Quantitative Space'. Finally, we created a last group of answers, which we titled 'Other.' In this case, we had one participant from each class who mentioned answers that did not fit into one of the three groups, for example: 'Rate is the process that is being executed with a certain frequency and that will insert or remove values from the stock. The stock is the result of the processes of the rates.' (student, class A). The student provides a confusing answer, and perhaps lacked rethinking to express these concepts better. For example, the participant says that 'Rate is the process [...]', while, in fact, the rate represents a process. The student uses the word frequency, possibly relating to time. And still, she emphasizes that 'The stock is the result of the processes of the rates', and in fact, the stock is the result of flows per unit of time. And in class B, a student mentions 'I don't know the difference', which clearly shows that she does not recognize the difference between rate and state variable. Regarding the three groups of answers mentioned (Different Units, Influences on the state variable, Quantitative Spaces), we consider the answers consistent and correct, indicating that through the questionnaires, it was possible to perceive that most students recognize the differences between rates and state variables.

3.2 Important Concept of SD: Conceptions About the Feedback Mechanism.

The concept of feedback is also crucial for understanding SD and developing system thinking, as we showed in the Review [8]. According to students' conceptions of this concept, we elaborated three response groups: Feedback as Control Mechanism (CM), around 66.67% in classes A and B; Feedback as Return of Process Effects (RPE), with 26.67% in class A, and 70% in class B; and types of feedback, Balancing and Reinforcing (BR), with 60% in class A, and 40% in class B.

Most learners mentioned feedback as a system control mechanism. A significant portion of class B indicated feedback as the return of process effects, and most of class A mentioned both types of feedback. Although it is a complex concept, it was possible to observe that most students were consistent in their formulated answers. Some of them mentioned only one of the listed categories with more concise responses, such as 'It is the system control mechanism' (one student, class A), which belongs to the CM category; or 'It is the return of the process' (one student, class B), as RPE.

However, most participants explicitly stated more than one of these categories, like 'It is when there is balancing or reinforcing control in the system' (one student, class A). In this case, the student points out system control and also mentions both types of feedback. Therefore, the response was placed in the CM and BR categories. Another participant also mentions 'Control mechanisms in which the action of objects generates reactions that ultimately act on the same object that caused the initial reaction' (one student, class B). In this case, belonging to the CM category is evident, as the student begins by explicitly stating 'Control mechanisms,' and in the end, she points out that the action of objects 'ultimately acts on the same object that caused the initial reaction.' In this part of the response, the student refers to the return of the effect of processes, also belonging to the RPE category.

4   Conclusion

This study aimed to demonstrate that high school students are capable of understanding the structure and functioning of complex systems through the construction of qualitative simulation models built in DynaLearn. During the activities at school, learners constructed models, valuing the approach chosen in this study: LbM. This approach emphasizes the students' protagonism in the learning process, as they build models, make adaptations, and analyze simulation results.

A set of actions, based on concepts of System Dynamics (SD), was important to enable students to use modeling language to represent phenomena at increasing levels of complexity and to have solidity in knowledge representation through systems, recognizing, even, that in the world, 'things' are interconnected. The results obtained from the questionnaire provided evidence that modeling, based on SD, promotes understanding of the structure and functioning of systems, from the simplest to the most complex. Students demonstrated recognition of modeling language and concepts established by SD for representing real-world phenomena in qualitative models. Additionally, students also expressed changes in their worldview.

This text points to enriching pathways for a broader education, where students are able to visualize and interpret operative relationships, phenomena of educational and scientific interest. It is worth noting that DynaLearn is a powerful tool that enables collaborative work among students, although this aspect was not investigated in this paper. Therefore, it could be a suggestion for future research.

References

1. Souza, P.V.T.: Modelos de simulação qualitativos como estratégia para o ensino de Ciências. Doctoral thesis, University of Brasília, Brasília/DF (2019)
2. Bredeweg, B., Liem, J., Beek, W., Mioduser, D.: DynaLearn – an intelligent learning environment for learning conceptual knowledge. AI Mag. **34**(4), 46–65 (2013)
3. Borkulo, S.P. The assessment of learning outcomes of computer modeling in secondary science education. Doctoral Thesis, University of Twenty, The Netherlands (2009)
4. Bredeweg, B., Forbus, K.: Qualitative modelling in education. AI Mag. **24**, 35–46 (2003)
5. Bredeweg, B., & Salles, P. Mediating conceptual knowledge using qualitative reasoning. In S. E. Jorgensen, T. C. Chon, & F. Recknagel (Eds.), Handbook of Ecological Modelling and Informatics (pp. 351–398). Southampton, UK: WIT Press (2009)
6. Mioduser, D., et al.: Final report on DynaLearn evaluation studies. DynaLearn, EC FP7 STREP project 231526, Deliverable D7.4 (2012)
7. Forbus, K.D.: Qualitative process theory. Artif. Intell. **24**(1), 85–168 (1984)
8. Richmond, B.: Systems thinking: critical thinking skills for the 1990s and beyond. Syst. Dyn. Rev. **9**(2), 113–133 (1993)

Unhappiness and Demotivation Among Students in Gamified Tutoring Systems: Toward Understanding the Gender Gap

Kamila Benevides[1](✉), Jário Santos[1], Geiser Chalco[2], Marcelo Reis[1], Álvaro Sobrinho[3], Leonardo Marques[1], Diego Dermeval[1], Rafael Mello[4,5], Alan Silva[1], Seiji Isotani[6,7], and Ig Ibert Bittencourt[1,7]

[1] Federal University of Alagoas, Maceió, Brazil
{kbas,alanpedro,ig.ibert}@ic.ufal.br, leonardo.marques@cedu.ufal.br,
diego.matos@famed.ufal.br
[2] Federal Rural University of the Semiarid, Mossoró, Brazil
geiser@alumni.usp.br
[3] Federal University of the Agreste of Pernambuco, Garanhuns, Brazil
alvaro.alvares@ufape.edu.br
[4] Federal Rural University of Pernambuco, Recife, Brazil
rafael.mello@ufrpe.br
[5] CESAR School, Recife, Brazil
[6] University of São Paulo, São Paulo, Brazil
sisotani@icmc.usp.br
[7] Harvard Graduate School of Education, Cambridge, USA

Abstract. In an experimental study, we observed that male-stereotyped elements, such as avatars and trophies, positively impacted men's self-efficacy using gamified tutoring systems based on artificial intelligence. Regardless of gender stereotypes, men achieved higher flow state scores than women when using these systems. Thus, we conducted a qualitative study involving Brazilian high school students to understand these effects. Our findings revealed that male students experience increased self-efficacy when game elements align with their gender, making them feel more challenged and personally connected to the system. Considering the emotional aspects of dejection, agitation, cheerfulness, and quiescence, women displayed slightly higher levels of cheerfulness in neutral and female-stereotyped environments. Moreover, student motivation varied when using a gamified platform, and there were differences in the prevention motivation within the male-stereotyped and neutral environments. These findings are relevant because they can be used to develop recommendations and guidelines for creating gamified intelligent tutoring systems that promote gender equity.

Keywords: Gamification · Self-efficacy · Flow State · Tutoring Systems

1   Introduction

Gamification has gained widespread popularity in educational settings, with numerous studies demonstrating its effectiveness as a relevant tool for engaging learners in education and fields such as sales and sports [6]. Thus, gamification involves integrating common gaming elements into non-gaming environments, where features typically found in games are adapted for different contexts.

Research has shown that gamification effectively enhances students' motivation, engagement, and overall performance [10]. However, as the adoption of gamification in education, particularly in virtual learning environments, continues to grow, studies reveal a potential issue related to the indiscriminate use of gamified educational technologies: gender stereotype threats [1].

The flow state is relevant in this scenario [5], which occurs when an individual is fully immersed in a task, striking a balance between challenge and boredom. Nine dimensions may contribute to an individual achieving a state of flow: having clear goals, maintaining focus and concentration, reducing self-consciousness, experiencing a distorted perception of time, receiving immediate feedback, finding a balance between boredom and anxiety, exercising self-control, experiencing a sense of reward, and having an autotelic experience.

Bizman et al. [4], Higgins et al. [7], Roney et al. [11], and Strauman and Higgins [12] demonstrated that inducing a promotion focus can impact emotions linked to dejection. Conversely, inducing a prevention focus can influence emotions associated with agitation. This suggests that individuals can adjust their motivational orientation in response to various stimuli, including negative stereotypes. Furthermore, individuals may be motivated to counteract the detrimental effects of negative stereotypes by shifting their focus towards promotion-oriented motivations. A study by Keller and Dauenheimer [8] supports this idea, revealing that stereotype threat can lead to lower expected performance and elicit feelings of dejection in affected individuals when a promotion focus is activated.

In this research, we experimented to investigate how gender stereotypes impact a gamified tutoring system based on artificial intelligence designed for teaching logic. Our primary focus was on understanding the role of dejection as a psychological factor. We also examined aspects related to the flow experience and the learning performance of university students.

We conducted a controlled experiment that involved using questionnaires to assess levels of flow and dejection. Additionally, we administered a logic test through a gamified platform, which had three different versions: one with female avatars and purple colors, another with male avatars and blue colors, and a third with mixed avatars and grey colors. We aimed to evaluate participants' flow, motivation, dejection, and performance levels, which were determined by the number of points they earned while solving logic puzzles.

2   Methods

This study relies on the following research questions to identify significant differences between the experimental and control groups: (RQ1) Does student motivation vary when using a stereotyped gamified educational technology? (RQ2) Does student learning vary using a stereotyped gamified educational technology? (RQ3) Do students experience differing flow levels when using a stereotyped gamified educational technology? (RQ4) Are there variations in the students' level of dejection when using a stereotyped gamified educational technology?

Thus, we defined the ten hypotheses: (H1) Flow experience does not vary between participants in different conditions; (H2) Flow experience does not vary between participants in different environments and of different genders; (H3) Dejection, cheerfulness, agitation, or quiescence do not vary between participants in different conditions; (H4) Dejection, cheerfulness, agitation, or quiescence do not vary between participants in different environments and of different genders; (H5) Prevention motivation does not vary between participants in different conditions; (H6) Prevention motivation does not vary between participants in different environments and of different genders; (H7) Promotion motivation does not vary between participants in different conditions; (H8) Promotion motivation does not vary between participants in different environments and of different genders; (H9) Promotion motivation does not vary between participants in different environments and of different genders; and (H10) Performance does not vary between participants in different environments and of different genders.

To assess the impact of dejection, flow levels, and learning performance in a gamified intelligent tutoring system for logic education, we examined various facets, including dejection, cheerfulness, agitation, and quiescence. We conducted a controlled experiment involving higher education students using a customized educational platform with three distinct settings. The ST-F setting was intentionally designed with a bias towards a female gender perspective. Female ranking systems and avatars were prominently featured, and the system and badges were predominantly colored purple. The ST-M setting was intentionally developed with a male gender bias. Male ranking systems and avatars were integrated, and the system and badges were predominantly colored blue. The ST-N (Default) setting was designed with neutral attributes, striving for a gender-balanced approach. The system and badges were colored in grey. This configuration served as the control group for the experiment, providing a baseline for comparison.

The experiment followed a 2×3 factorial design, considering two main factors: the version of the gamified educational technology (ST-F, ST-M, ST-N) and the student's gender (male or female). The experiment involved forming three distinct groups: the Boost Group, the Threat Group, and the Control Group.

The experiment encompassed seven stages: (1) reviewing the informed consent form; (2) employing the Dispositional Flow Scale (DFS) [3]; (3) administering the Regulatory Focus Questionnaire (RFQ) As a second pre-test measure [9]; (4) presenting one of the three versions of the educational technology to the student; (5) administering the post-test questionnaire [8]; (6) administering the Flow State Scale (FSS) questionnaire after completing the motivation questionnaire [2]; and (7) administering a socioeconomic questionnaire.

The experimental tool used in this study was initially designed by Albuquerque et al. [1], but questionnaires were substituted for specific research purposes. The tool was developed using JavaScript, Angular, and Bootstrap.

The recruitment for the experiment involved announcing it to university students through email and instant messaging channels, primarily targeting students within study groups. A significant portion of the participants belonged to programs related to science, technology, engineering, and mathematics (STEM).

The study comprised 107 participants, categorized by gender. The sample size was selected based on convenience, and the participants were approximately 20 years old. Most of them were single and either undergraduates or graduates in their academic status. The predominant racial group among participants was White, followed by Brown, Black, and some who chose not to specify their race.

This study relied on the ANCOVA and ANOVA tests, considering the experiment's 2×3 covariate factorial design. A winsorization method was also applied to reduce the influence of outliers, with values beyond the 5% to 95% probability range being adjusted to maintain data symmetry. Additionally, steps were taken to eliminate noise, ensuring an analysis with minimal bias from external values.

Furthermore, data normality and symmetry assumptions were assessed using the Shapiro-Wilk test. Following applying the statistical tests (ANCOVA and ANOVA), pairwise comparisons were conducted using Estimated Marginal Means (EMMs) to identify statistically significant differences between groups defined by independent variables. The p-values obtained from these comparisons were adjusted using the Bonferroni method to account for multiple comparisons.

3   Results

3.1   Effects on Participants' Flow

We applied the DFS questionnaire twice to assess the participants' state of flow: once before participants engaged with the gamified environment and again after they had completed the logic puzzle.

Hypothesis H1. The analysis focused on FSS, DFS, and condition (inThreat, inBoost, control) to test hypothesis H1. After confirming the DFS's linearity of covariance, an ANCOVA test was conducted to compare the groups defined by the condition variable.

Regarding the FSS (dependent variable), significant effects were observed concerning the DFS factor, yielding an $F(1.103)$ value of 8.985 and a p-value of 0.003, with an effect size (ges) of 0.08. However, no statistically significant differences emerged between the groups when conducting pairwise comparisons. The p-values were as follows: $p = 0.143$ for the comparison between the control and inBoost groups, p=0.409 for the control and inThreat groups, and $p = 0.588$ for the inBoost and inThreat groups.

Hypothesis H2. We also considered the variables FSS, DFS, type of test (stMale, stFemale, default), and gender (male, female) to test hypothesis H2. An ANCOVA test with these independent variables was also employed to compare the different types of tests and genders.

Significant effects of the DFS factor were identified, yielding an F(1.100) value of 8.473 and a p-value of 0.004, with an effect size (ges) of 0.078. Subsequent pairwise comparisons, based on the dependent variable FSS, revealed that the means for the female gender (M-adjusted = 3.263 and SD = 0.929) differed significantly from those of the male gender (M-adjusted = 3.762 and SD = 0.655), with a p-adjusted value of 0.027.

3.2   Effects on Participants' Dejection

To assess the impact of dejection on participants, we administered an emotions questionnaire that gauges various aspects related to dejection, cheerfulness, agitation, and quiescence. This questionnaire was given to participants after they had engaged with the gamified system.

Hypothesis H3. The analysis included dejection, cheerfulness, agitation, quiescence, and the condition. An ANOVA test was conducted to compare the different conditions as the independent variable. The results indicated no significant effects for any of the tested variables, with p-values of 0.925 (dejection), 0.157 (cheerfulness), 0.599 (agitation), and 0.777 (quiescence).

Hypothesis H4. The analysis conducted to test hypothesis H4 considered the variables dejection, cheerfulness, agitation, quiescence, type of test, and gender. An ANOVA was performed to investigate significant differences in the dependent variables dejection, cheerfulness, agitation, and quiescence concerning the independent variables' type of test and gender. The results indicated the following significant variations:

- Dejection: There was a significant difference according to gender, with an F(1.86) value of 6.402, p = 0.013, and an effect size (ges) of 0.069.
- Cheerfulness: Significant variations were observed based on gender, with an F(1.100) value of 10.362, p = 0.002, and effect size (ges) of 0.094.
- Quiescence: Significant differences were identified according to gender, with an F(1.98) value of 6.34, p = 0.013, and an effect size (ges) of 0.061.

Pairwise comparisons indicated significant differences in cheerfulness, with p-adjusted values of 0.002 in the Default environment and 0.028 in the stMale environment. For the dependent variable quiescence, the mean for the Male gender exhibited a significant difference (M-adjusted = 5.214 and SD = 0.963) with a p-adjusted value of 0.028.

3.3   Effects on Participants' Promotion and Prevention Motivation

To evaluate the promotion motivation of participants when solving the logic puzzle, a Regulatory Focus Questionnaire was administered before participants engaged in the logic puzzle within the gamified environment. This questionnaire includes items assessing promotion and prevention motivation, providing insight into the participants' motivational orientation.

Hypothesis H5. The analysis relied on the prevention (pre- and post-test) and condition variables to test hypothesis H5. An ANCOVA test was employed to compare the variable prevention.pre across different conditions.

However, for the dependent variable prevention.pos, we did not identify any significant differences. The p-values were 0.331 for the variable prevention.pos concerning prevention.pre and 0.929 for prevention.pos about the condition.

Hypothesis H6. The analysis considered the variables prevention, type of test, and gender to investigate hypothesis H6. An ANCOVA test was conducted to compare the variable prevention.pre among different types of tests and genders.

Hypothesis H7. To explore hypothesis H7, we considered the variables prevention, type of test, and gender. An ANCOVA test was employed to compare the variables prevention.pre among different types of tests and genders.

Hypothesis H8. To examine hypothesis H8, we considered variables such as promotion, type of test, and gender. We first ensured that the promotion.pre variable's covariance linearity was controlled. Subsequently, ANCOVA tests were performed to determine if there were statistically significant differences in the dependent variable, promotion.pos, based on the independent variables: type of tests and gender.

Regarding the promotion.pos variable, we found significant effects related to the promotion.pre factor, indicated by an $F(1.96) = 13.157$, $p < 0.001$, and an effect size (ges) of 0.121. Additionally, gender also showed significant effects, with an $F(1.96) = 27.554$, $p < 0.001$, and an effect size (ges) of 0.223.

Further analysis through pairwise comparisons focused on the promotion.pos variable demonstrated that there were significant differences between genders. The mean score for females (M-adjusted = 3.7 and SD = 1.222) differed from the mean score for males (M-adjusted = 2.509 and SD = 0.843), with a p-adjusted value of < 0.001. Additionally, the mean score for females (M-adjusted = 3.393 and SD = 0.869) was also significantly different from the mean score for males (M-adjusted = 2.33 and SD = 0.646), with a p-adjusted value of < 0.001.

3.4   Participants' Performance

To assess how well participants performed in solving logic puzzles, we conducted statistical tests using the total points earned by each participant.

Hypothesis H9. We used the variables of points and conditions (inThreat, inBoost, control) to examine hypothesis H9. An ANOVA test focused on the independent variable, condition (inThreat, inBoost, control). However, it is important to note that no statistically significant effects were observed for the points variable, as indicated by a p-value of 0.148.

Hypothesis H10. Finally, we focused on variables related to the number of points, type of test (stMale, stFemale, default), and gender (male, female) to investigate hypothesis H10. ANOVA tests with independent variables were used to assess whether participants' scores, represented by the number of points earned, varied significantly based on the types of tests and gender.

Results showed a significant difference in participants' scores (measured in points) related to the gender factor, as indicated by an $F(1.94) = 10.378$, $p = 0.002$, and an effect size (ges) of 0.099. Further examination through pairwise comparisons revealed that the number of points earned by female participants (M-adjusted = 8.115 and SD = 2.221) significantly differed from male participants (M-adjusted = 6.538 and SD = 1.614), with a p-adj value of 0.014.

4 Discussion and Conclusion

From the results and analyses provided, it is clear that student motivation varies when using a gamified platform. There were specific differences in prevention motivation within the male-stereotyped and neutral environments. Upon closer examination through group comparisons, these differences were primarily associated with the male group in neutral and male-stereotyped environments.

In learning with a stereotyped gamified educational technology, women's performance showed differences across various environments, including female-stereotyped and male-stereotyped settings. In both environments, women achieved higher average scores than men, with a particularly pronounced difference observed in the default environment. This difference in the default environment was statistically significant compared to the other environments.

Regarding students' flow experience on the gamified platform, the findings indicated a difference between the DSF and FSS conditions within the default environment. In the default environment, women displayed lower levels of flow in comparison to men, and this pattern persisted across both environments.

For men, their flow levels remained relatively consistent regardless of the environment. In contrast, women exhibited fluctuations in their flow experience across all environments, with the most significant fluctuations occurring in the neutral one. This suggests that men may be less affected by the type of environment compared to women regarding their flow experience.

Regarding the emotional aspects related to dejection, such as dejection itself, agitation, cheerfulness, and quiescence, women displayed slightly higher levels of cheerfulness in the neutral and female-stereotyped environments. However, these differences were not statistically significant.

The neutral environment was associated with improved learning performance compared to women in different environments. Despite decreased dejection, the female-stereotyped environment decreased learning performance compared to the neutral environment, particularly for female participants. This suggests that women exhibit lower learning performance in competitive settings, as the female-stereotyped environment implies. The levels of dejection may influence performance, with better learning experiences observed in neutral environments, indicating that a lack of competitive pressure can enhance learning outcomes.

References

1. Albuquerque, J., Bittencourt, I.I., Coelho, J.A., Silva, A.P.: Does gender stereotype threat in gamified educational environments cause anxiety? an experimental study. Comput. Educ. **115**, 161–170 (2017)
2. Bittencourt, I.I., et al.: Validation and psychometric properties of the Brazilian-Portuguese flow state scale 2 (FSS-BR). PLOS ON (under review)
3. Bittencourt, I.I., et al.: Validation and psychometric properties of the Brazilian-Portuguese dispositional flow scale 2 (DFS-BR). PLoS ONE **16**(7), e0253044 (2021)
4. Bizman, A., Yinon, Y., Krotman, S.: Group-based emotional distress: an extension of self-discrepancy theory. Pers. Soc. Psychol. Bull. **27**(10), 1291–1300 (2001)
5. Csikszentmihalyi, M.: Flow. The Psychology of Optimal Experience. New York (HarperPerennial) 1990. (1990)
6. Ferreira, G., Pereira, S.: Atividade gamificada em saúde: entendo as viroses e seus métodos de transmissão e prevenção como atividade lúdica no ensino de ciências e biologia. In: Proceedings Congresso internacional ABED de educação a Distância (2017)
7. Higgins, E.T., Shah, J., Friedman, R.: Emotional responses to goal attainment: strength of regulatory focus as moderator. J. Pers. Soc. Psychol. **72**(3), 515 (1997)
8. Keller, J., Dauenheimer, D.: Stereotype threat in the classroom: dejection mediates the disrupting threat effect on women's math performance. Pers. Soc. Psychol. Bull. **29**(3), 371–381 (2003)
9. Kroth, G.L., et al.: O impacto do foco regulatório na procrastinação e nas escolhas intertemporais. Ph.D. thesis, Universidade Federal de Santa Maria (2019)
10. Landers, R.N., Bauer, K.N., Callan, R.C.: Gamification of task performance with leaderboards: a goal setting experiment. Comput. Hum. Behav. **71**, 508–515 (2017)
11. Roney, C.J., Higgins, E.T., Shah, J.: Goals and framing: how outcome focus influences motivation and emotion. Pers. Soc. Psychol. Bull. **21**(11), 1151–1160 (1995)
12. Strauman, T.J., Higgins, E.T.: Automatic activation of self-discrepancies and emotional syndromes: when cognitive structures influence affect. J. Pers. Soc. Psychol. **53**(6), 1004 (1987)

A Web Application for a Cost-Effective Fine-Tuning of Open-Source LLMs in Education

Victor Diez-Rozas[(✉)], Iria Estevez-Ayres[iD], Carlos Alario-Hoyos[iD], Patricia Callejo[iD], and Carlos Delgado Kloos[iD]

Dept Telematic Engineering, Universidad Carlos III de Madrid, Leganés, Madrid, Spain
100451534@alumnos.uc3m.es, {ayres,calario,pcallejo,cdk}@it.uc3m.es

Abstract. Many Generative Artificial Intelligence (GenAI) applications built on Large Language Models (LLMs) emerged in 2023 causing a great impact in the educational landscape. However, most of these GenAI applications require a subscription to access the most advanced models and functionalities. Therefore, the search for cost-effective solutions becomes an important concern, especially for instructors and educational institutions with limited resources. This article introduces a Web Application aimed at facilitating instructors in fine-tuning open-source LLMs and subsequently posing questions to them. Instructors only need to upload a dataset into the Web Application to fine-tune the open-source LLM, specifically Llama 2. This web application was developed using open-source tools (Hugging Face and Langchain), and techniques to reduce hardware resource consumption (LoRA and QLoRA). Preliminary results from the experiments conducted show that LLMs provide more accurate responses when fine-tuned for a specific task through the Web Application. These are the first steps in providing cost-effective GenAI solutions for instructors and educational institutions using open-source tools.

Keywords: Generative Artificial Intelligence · Large Language Models · Web Application · Open Source Models

1 Introduction

November 2022 marked a turning point for Generative Artificial Intelligence (GenAI) with the public release of ChatGPT as a part of a free research preview [4]. The tool turned out to be a success, so OpenAI decided to limit the free service and offer upgrades under subscription [8]. ChatGPT disrupted many sectors and in the educational field led to a reflection on how to properly incorporate GenAI and Large Language Models (LLMs) in the teaching and learning processes.

Numerous GenAI-based applications emerged after the success of ChatGPT, and some of them focused on the educational field [11], for example, to help instructors plan their lessons, generate questions [9], or support online courses [19]. There are also GenAI-based chatbots that help students as personal tutors in specific courses (e.g., the CS50 Duck in the CS50 course at Harvard University) [18]. However, many of these applications require a subscription to access the most advanced models and functionalities. Resources are limited in the field of education, and not all institutions can afford to devote resources to pay for GenAI applications. As a result, there is an economic barrier that undermines equity in access to technology.

This paper aims to address this issue by introducing a free, customisable GenAI Web Application that facilitates instructors to fine-tune open-source LLMs and ask them questions afterwards. The Web Application allows instructors to fine-tune Llama 2 [5], and was implemented using Hugging Face [3], Langchain [6], and techniques to reduce hardware resource consumption (LoRA [16] and QLoRA [13]).

The paper is organised as follows: Sect. 2 discusses the technical aspects on how to conduct fine-tuning of LLMs. Section 3 explains the model implementation. Section 4 shows the results obtained by training the base model on two different datasets. Section 5 discusses the preliminary results. The last section draws the conclusions and presents future lines of work.

2   Technical Aspects

LLMs are the backbone of GenAI applications. LLMs are AI algorithms trained on a large collection of datasets to understand and generate human-like language [15]. The release of ChatGPT using the LLM GPT-3.5 led to an acceleration in the number of LLMs developed, including open-source LLMs, such as Llama 2 [5] or Mistral, among others.

Most GenAI applications currently offer a limited free service, requiring a subscription fee to access their most advanced models and functionalities. Nevertheless, there are also free tools that use the existing open-source LLMs; Hugging Face and Langchain are the most popular ones. Hugging Face [3] is a platform that provides open-source LLMs and datasets; it also offers Python libraries for training models. Langchain [6] is an open-source framework for the development of LLMs tools; it primarily facilitates the implementation of the user-model interaction.

The interesting feature of LLMs is that they can be tailored to a specific task by adjusting their parameters using a technique known as fine-tuning [12,21], although this is usually a computationally expensive process. Parameter-Efficient Fine-Tuning (PEFT) [17] is one of the methods that requires the least hardware resources. PEFT has the capability to train a subset of parameters efficiently using just a single GPU, making it a cost-effective method.

PEFT includes several approaches. One of them is LoRA (Low-Rank Adaptation) [16]. The fine-tuned model is the combination of the parameters of the base model and some new parameters, and the idea behind LoRA is to represent these new parameters as the product of two small matrices of lower rank, called update matrices. The parameters of the base model are frozen and unchanged. Those that undergo changes are the parameters of the update matrices. Thus, the number of parameters to be trained is significantly reduced and, consequently, fewer hardware resources are required.

Another approach is QLoRA (Quantized LLMs with Low-Rank Adaptation) [13], which derives from LoRA. The idea behind QLoRA is to convert the base model parameters to 4-bits and keep the training parameters at 16-bits. QLoRA offers the benefits of LoRA while simultaneously decreasing memory usage, all the while preserving performance levels.

The dataset is another element to consider when fine-tuning an LLM. It must follow the instruction dataset format [2,24], consisting of questions and answers (see Table 1). The instruction prompt is formed by combining a context, the question, and its answer. The prompt is customisable as long as it has a question and an answer. There are also two issues to take into account: the dataset impacts the resources used during training (the larger the dataset, the greater the resource consumption), and the instruction prompt used during fine-tuning must also be employed during inference to ensure the accuracy of results (inference is the ability of a model to generate answers from a given input).

Table 1. Three samples taken from a Python code dataset. Each sample is composed of a question -the task to be learned- and its answer -the result of the task-.

Question	Answer
Write a Python code to print a message which content is "Hello, World!"	print("Hello, World!")
Write a Python code to find the maximum value in a list	max_value = max(lst)
Write a Python code to sum all the elements of a list	total = sum(lst)

3 Development of the Web Application

Building applications which use GenAI with open-source tools requires devices equipped with high VRAM. Since not all educational institutions have the required hardware, the Web Application is implemented on virtual machines offered by Google Colaboratory's free plan [14]. This option includes Python as programming language, 12.7 GB of RAM, 15 GB of VRAM, and 78.2 GB of disk storage. While Google Colaboratory's free plan offers considerable VRAM, it may not be sufficient for training all open-source LLMs. The base model chosen is a modified version of Llama 2 [23] that works with such limitations, with potential for extending support to other open-source LLMs if they operate within the hardware constraints.

The Web Application is implemented using Flask, Celery, Redis, and Ngrok. Flask [22] is an open-source Python framework that is used to develop web applications. Celery [20] is an open-source Python framework that creates task queues and manages long-running tasks (e.g., the fine-tuning of an LLM). Redis [20] is an open-source data store. Ngrok [7] is a free tool that allows an application running locally to be accessible on the Internet. The Web Application is hosted on a Google Colaboratory virtual machine. An HTTP address is assigned to test the Web Application.

On the backend, the server consists of three processes: 1) the main process receives requests from the user and keeps the connection alive between client and server; 2) the Celery Worker fine-tunes the model and generates responses; 3) Redis acts as an intermediary in communication between the main process and the Celery Worker. On the frontend, the client connects to the server and has two options: fine-tuning Llama 2 or running inference on the model.

3.1 Fine-Tuning Llama 2

The fine-tuning starts when the instructor sends to the server an instruction dataset (see Sect. 2) and the name of the LLM. The main process in the Web Application receives the information and forwards it to the Celery Worker process via Redis. The Celery Worker then performs the fine-tuning as follows. First, 300 samples are randomly chosen if the size of the dataset is greater than this number. As discussed in Sect. 2, the size of the dataset impacts resource consumption during fine-tuning. Therefore, the number of samples needs to be adjusted to prevent resource overloading. Next, Llama 2 and its tokenizers (with LoRA and QLoRA configurations) are downloaded from the Hugging Face Hub. Then, the small number of parameters of the update matrices (see Sect. 2) are trained. Once the parameters are trained, Llama 2 and its tokenizers (without LoRA and QLoRA configurations) are downloaded from the Hugging Face Hub. The trained parameters are then merged with the original ones to obtain the fine-tuned LLM. Finally, the fine-tuned LLM and its tokenizers are uploaded to the Hugging Face Hub. The Celery Worker sends status updates during fine-tuning to the client for two purposes: to keep the connection alive and to keep the instructor informed about the progress of the training. This explanation is illustrated in Fig. 1.

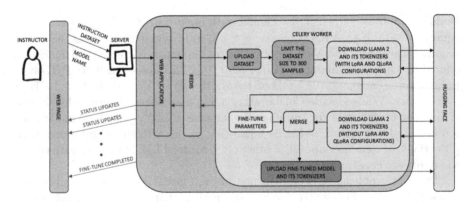

Fig. 1. Fine-tuning steps identified with different colors.

3.2 Running Inference on the Trained Model

Inference begins when the user (instructor or student) sends the name of the fine-tuned LLM to the server. The main process in the Web Application receives and forwards this name to the Celery Worker process via Redis. The Celery Worker then downloads the fine-tuned LLM and its tokenizers from the Hugging Face Hub. Once the download finishes, the Celery Worker notifies the user through Redis and the main process in the Web Application. The user sends questions to the server. The main process in the Web Application receives and forwards them to the Celery Worker, responsible for querying the fine-tuned LLM. The fine-tuned LLM generates responses and the Celery Worker sends them back to the user. This explanation is shown in Fig. 2.

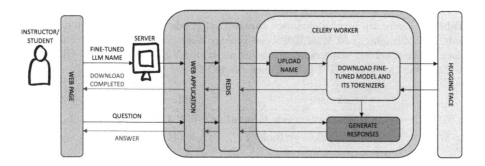

Fig. 2. Inference steps identified with different colors.

4 Results

Two experiments were conducted to evaluate the performance of the Web Application developed. The first experiment analyses the hardware resources consumed during fine-tuning; its results are shown in Table 2. The second experiment analyses the performance of the base model (BM) and the fine-tuned models (FTM) of Llama 2 using two benchmarks; benchmarks results are shown in Table 3. Both experiments were conducted using two different datasets for the fine-tuning process (i.e., using two fine-tuned LLMs). The first dataset included information related to computer programming [1] and the second one included concepts related to biology [10], thus covering two major educational areas: engineering and health sciences. These datasets were chosen due to their availability in Hugging Face, facilitating preliminary tests. As mentioned in Sect. 2, the datasets were modified before fine-tuning to fit the format of the instruction dataset.

5 Discussion

The fine-tuning of an LLM depends on the chosen base LLM, the fine-tuning methods, and the dataset used. The only difference between the two examples

Table 2. Fine-tuning was split into five steps to present the data: S0 -beginning-, S1 -downloading Llama 2 and its tokenizers (with LoRA and QLoRA configurations)-, S2 -fine-tuning parameters of update matrices-, S3 -downloading Llama 2 and its tokenizers (without LoRA and QLoRA configurations), and merging parameters-, and S4 -uploading fine-tune model and its tokenizers to Hugging Face Hub-. This table presents training time in minutes, and hardware resources values -i.e., RAM, VRAM, and disk storage- in GB for each tested dataset.

Fine-tuning steps	Dataset 1: Computer Programming				Dataset 2: Biology			
	RAM (GB)	VRAM (GB)	Disk Storage (GB)	Time (min)	RAM (GB)	VRAM (GB)	Disk Storage (GB)	Time (min)
S0	2.1/12.7	0/15	26.9/78.2	0	2.2/12.7	0/15	26.9/78.2	0
S1	2.2	4	39.5	3	2.2	4	39.5	3
S2	3	6.9	39.5	20	3.1	11.2	39.6	122
S3	3.2	13.4	39.6	2	3.2	13.4	39.6	2
S4	3.3	13.4	52.2	5	3.2	13.4	52.2	5
	Overall:			30	**Overall:**			132

Table 3. There are two benchmarks: HumanEval (Code Generation) related to computer programming, and MMLU College Biology related to biology. The table shows the number of correct answers in each benchmark for Llama 2 -Base Model (BM)- and its fine-tuned versions -FTM (Fine-Tuned Model)-.

Benchmark	BM	FTM
HumanEval (Code Generation)	6/164	**29/164**
MMLU College Biology	0/144	**53/144**

shown in Table 2 is that the number of tokens per sample in the Biology dataset is larger than the one in the Computer Programming dataset. Therefore, it can be concluded that not only the size of the dataset has an impact, but also the length of each sample. Another observation that can be drawn from Table 2 is regarding the benefits of the LoRA and QLoRA approaches. Downloading Llama 2 and its tokenizers without LoRA and QLoRA configurations (step 3) consumed 13.4 GB of VRAM. Fine-tuning the parameters of the update matrices (step 2) increased the VRAM consumption 2.9GB in the first case and 7.2 GB in the second case. Therefore, fine-tuning Llama 2 without PEFT techniques would have been impractical, as the VRAM consumption would have exceeded 15 GB.

The first conclusion that can be drawn from Table 3 is the poor performance of Llama 2 (BM) on both benchmarks. Its inability to provide correct answers for most questions is the primary reason. Instead, it often gives blank responses, repeats, or generates new questions. Llama 2 is an LLM that has limitations and has already been surpassed by other competitors. However, as discussed in Sect. 3, it was chosen because its compatibility with hardware limitations. These constraints also caused that Llama 2 could only be trained on a limited dataset of 300 samples. Nevertheless, the fine-tuned models (FTM) show a significant improvement in the accuracy of the responses and overcome the issues of blank responses, repeated or new questions.

6 Conclusion

The aim of this paper is to present a cost-effective solution for instructors and educational institutions in relation to the fine-tuning and communication with LLMs, in particular with open-source LLMs. This solution is a Web Application that has been implemented using open-source tools (Hugging Face and Langchain), and techniques to reduce hardware resource consumption (LoRA and QLoRA), and that currently supports the fine-tuning of the open-source model Llama 2 with the possibility to support other open-source LLMs if they operate within the hardware constraints. Preliminary results show that it is possible to fine-tune an LLM and pose questions to it through the Web Application. However, there are important limitations, such as the reduced set of LLMs that can be used and the small size of the training dataset that can be used as input in the Web Application. Further research is still needed to test the Web Application with a larger collection of datasets and to address the existing limitations. Nevertheless, we believe that this Web Application can be a useful and innovative solution that responds to the current needs of instructors and educational institutions.

Acknowledgments. The authors acknowledge funding from FEDER/Ministerio de Ciencia, Innovación y Universidades - Agencia Estatal de Investigación through project H2O Learn (PID2020-112584RB-C31). This research has also received partial support from the European Commission through Erasmus+ projects MICROCASA (101081924 ERASMUS-EDU-2022-CBHE-STRAND-2), MICRO-GEAR (101127144 ERASMUS-EDU-2023-CBHE-STRAND-3), POEM-SET (2021-FR01-KA220-HED-000032171) and EcoCredGT (101129122 ERASMUS-EDU-2023-CB-VET). This publication reflects the views only of the authors and funders cannot be made responsible for any use which may be made of the information contained therein.

References

1. Dataset computer programming. https://huggingface.co/datasets/mbpp, Accessed 01 Feb 2024
2. Extended guide: instruction-tune llama 2. https://www.philschmid.de/instruction-tune-llama-2, Accessed 01 Feb 2024
3. Hugging face hub documentation. https://huggingface.co/docs/hub/index, Accessed 01 Feb 2024
4. Introducing ChatGPT. https://openai.com/blog/chatgpt, Accessed 01 Feb 2024
5. Introducing Llama 2. https://llama.meta.com/llama2, Accessed 01 Feb 2024
6. LangChain documentation. https://python.langchain.com/docs, Accessed 01 Feb 2024
7. Ngrok: an introduction and practical use cases. https://medium.com/aia-sg-techblog/ngrok-an-introduction-and-practical-use-cases-2432b9efb87, Accessed 01 Feb 2024
8. OpenAI releases first $20 subscription version Of ChatGPT. https://www.forbes.com/sites/alexkonrad/2023/02/01/openai-releases-first-subscription-chatgpt, Accessed 01 Feb 2024

9. Teaching with AI. https://openai.com/blog/teaching-with-ai, Accessed 01 Feb 2024

10. camel AI: dataset biology. https://huggingface.co/datasets/camel-ai/biology, Accessed 01 Feb 2024

11. Bahroun, Z., Anane, C., Ahmed, V., Zacca, A.: Transforming education: a comprehensive review of generative artificial intelligence in educational settings through bibliometric and content analysis. Sustainability **15**(17), 1–40 (2023). https://doi.org/10.3390/su151712983

12. Church, K.W., Chen, Z., Ma, Y.: Emerging trends: a gentle introduction to fine-tuning. Nat. Lang. Eng. **27**(6), 763–778 (2021). https://doi.org/10.1017/S1351324921000322

13. Dettmers, T., Pagnoni, A., Holtzman, A., Zettlemoyer, L.: QLoRA: efficient fine-tuning of quantized LLMs, pp. 1–26 (2023). https://arxiv.org/abs/2305.14314

14. Google: introducing google colaboratory. https://colab.research.google.com, Accessed 01 Feb 2024

15. Hadi, M.U., Al Tashi, Q., Qureshi, R., et al.: A survey on large language models: applications, challenges, limitations, and practical usage. TechRxiv, pp. 1–29 (2023). https://doi.org/10.36227/techrxiv.23589741.v1

16. Hu, E.J., et al.: LoRA: low-rank adaptation of large language models, pp. 1–26 (2021). https://arxiv.org/abs/2106.09685

17. Lialin, V., Deshpande, V., Rumshisky, A.: Scaling down to scale up: a guide to parameter-efficient fine-tuning, pp. 1–21 (2023). https://arxiv.org/abs/2303.15647

18. Liu, R., Zenke, C., Liu, C., Holmes, A., Thornton, P., Malan, D.J.: Teaching CS50 with AI: leveraging generative artificial intelligence in computer science education. In: Proceedings of the 55th ACM Technical Symposium on Computer Science Education V. 2, pp. 1–27. SIGCSE 2024, Association for Computing Machinery, New York, NY, USA (2024). https://doi.org/10.1145/3626253.3635427

19. Paiva, R., Bittencourt, I.I.: Helping teachers help their students: a human-AI hybrid approach. In: Bittencourt, I.I., Cukurova, M., Muldner, K., Luckin, R., Millán, E. (eds.) AIED 2020. LNCS (LNAI), vol. 12163, pp. 448–459. Springer, Cham (2020). https://doi.org/10.1007/978-3-030-52237-7_36

20. Pankiv, A.: Concurrent benchmark system for web-frameworks on Python. Ph.D. thesis, Ukrainian Catholic University (2019). https://er.ucu.edu.ua/bitstream/handle/1/4562/Andriy-Pankiv.pdf

21. Radiya-Dixit, E., Wang, X.: How fine can fine-tuning be? Learning efficient language models. In: Chiappa, S., Calandra, R. (eds.) Proceedings of the Twenty Third International Conference on Artificial Intelligence and Statistics. Proceedings of Machine Learning Research, vol. 108, pp. 2435–2443. PMLR (2020). https://doi.org/10.48550/arXiv.2004.14129

22. Relan, K.: Beginning with Flask, pp. 1–26. Apress, Berkeley, CA (2019). https://doi.org/10.1007/978-1-4842-5022-8_1

23. TinyPixel: Llama-2-7B-bf16-sharded. https://huggingface.co/TinyPixel/Llama-2-7B-bf16-sharded, Accessed 01 Feb 2024

24. Zhang, S., et al.: Instruction tuning for large language models: a survey, pp. 1–35 (2024). https://arxiv.org/abs/2308.10792

Research on Personalized Hybrid Recommendation System for English Word Learning

Jianwei Li[ID], Mingrui Xu[(✉)][ID], Yuyao Zhou, and Ru Zhang

Beijing University of Posts and Telecommunications, Beijing 100088, China
xumingrui2023@bupt.edu.cn

Abstract. Our study aims to explore a personalized hybrid recommendation system for English word learning. This paper begins with an analysis of the research status and existing problems in the field of personalized English word recommendation, followed by the design of a personalized hybrid recommendation system for English words. The system considers factors such as the user's word mastery level, the forgetting curve, and examination frequency of words. It utilizes three intelligent technologies: word vectors, cosine similarity, and long short-term memory networks, providing an innovative solution for the problem of English word recommendation. The system's efficacy is validated through an offline simulation experiment, comparing its performance with four baseline algorithms. The results reveal that the system outperforms the baseline algorithms in multiple indicators, demonstrating its superior recommendation capabilities. These findings have significant application value for English word learning and contribute valuable insights to personalized English word recommendation.

Keywords: English word learning · Hybrid recommendation system · Word vectors · Cosine similarity · Long short-term memory networks

1 Introduction

The rapid advancement of globalization and informatization has made English increasingly crucial. Mastering a robust English vocabulary is essential for improving English proficiency. In order to meet the growing demand for learning English vocabulary, numerous applications such as Duolingo and Shanbay have emerged in the market. Despite offering various memorization methods like image association and word splitting, these applications often overlook the impact of learners' individual characteristics on word learning. Consequently, users may encounter words that are too difficult or easy, repeat already mastered words, or overlook necessary words, thus reducing their learning efficiency. Therefore, improving the accuracy and personalization of recommendation systems is an urgent problem in the field of English word learning.

© The Author(s), under exclusive license to Springer Nature Switzerland AG 2024
A. M. Olney et al. (Eds.): AIED 2024 Workshops, CCIS 2151, pp. 275–282, 2024.
https://doi.org/10.1007/978-3-031-64312-5_33

2 Status of Research and Existing Problems

2.1 Status of Research

Personalized recommendation systems have been widely applied and have achieved significant success recently. Yet, research in English word learning remains limited. Our study reviews and analyzes recent literature, summarizing several types of personalized recommendation systems for English word learning.

Initially, collaborative filtering-based systems operate based on similarities among users or words. For instance, Wu [1] collected personalized data using tests and questionnaires, and then employed user-based collaborative filtering algorithms to recommend English words. Next, content-based recommendation systems calculate item similarity from attributes, aligning with users' preferences to recommend closely matched items. For example, Zou et al. [2] and Xie et al. [3] utilize content-based systems to recommend words tailored to users' characteristics, preferences, as well as word learning tasks' attributes and effect. Additionally, deep learning-based systems utilize advanced machine learning technologies such as deep neural networks and attention mechanisms. These technologies are advantageous for recommendation systems due to their ability to process large datasets, recognize complex patterns, and make accurate predictions. Researchers such as Xie et al. [4, 5] have implemented these technologies, demonstrating their effectiveness in enhancing recommendation quality. Finally, hybrid recommendation systems combine multiple algorithms or data sources to improve accuracy and diversity of recommendations. Examples include Long [6] and Ru et al. [7], who have effectively utilized these systems. However, research on hybrid recommendation for English words is still scarce.

2.2 Existing Problems

Upon reviewing the current state of recommendation systems, it is clear that personalized recommendation technology positively impacts English vocabulary learning. However, several critical issues remain unresolved.

Firstly, the overlook of high-frequency words crucial for English learning leads to recommendations misaligned with learners' needs. Secondly, the difficulty in recommending easily forgotten words arises from the insufficient integration of psychological principles such as the forgetting curve, resulting in misaligned recommendations with learners' memory states. Thirdly, the challenge of introducing new words, which are essential for expanding and enhancing English learning, arises as current systems overlook learners' personalized characteristics, resulting in poorly matched recommendations to their word mastery levels.

3 Design of a Personalized Hybrid Recommendation System for English Words

To address the aforementioned three problems, our study develops a personalized hybrid recommendation system for English word learning. This system takes into account the learner's word mastery level, the forgetting curve, and examination frequency of words.

It utilizes three intelligent technologies: word vectors, cosine similarity, and long short-term memory networks.

Having outlined the foundational technologies and principles, the following section provides an in-depth exploration of the system's operational process (refer to Fig. 1).

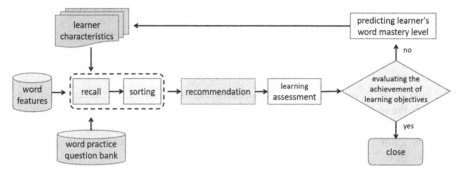

Fig. 1. The operational process diagram of personalized hybrid recommendation system for English word learning

The recommendation process mainly includes recall, sorting, recommendation, learning assessment, evaluating the achievement of learning objectives, and updating learner characteristics. Among these, recall and sorting are the core modules.

3.1 Recall Module Design

The recall module combines word features and learner characteristics to filter a small candidate set from a large database. This set serves as input for further fine-tuning in the sorting component. Word features include level (such as College English Test Level 4 vocabulary, TOEFL vocabulary, etc.), difficulty, and examination frequency. Learner characteristics encompass learning objectives, word mastery level, learning style, and memory ability. The primary learner characteristic used in our study is word mastery level, which is crucial yet challenging to accurately assess. Inspired by deep knowledge tracing research [8], our study automatically constructs a word practice question bank based on the Word2Vec model and cosine similarity algorithm, transforming the recommendation objects from words to questions. Based on deep neural network, the interaction data between learners and questions can be used to accurately capture the learners' mastery levels. This section will introduce the construction of the word practice question bank.

The word exercises consist primarily of objective questions, organized in two distinct ways. The first method presents the stem in English with options in Chinese, while the second reverses this format. Taking the first method as an example, the initial phase is training the Word2Vec model. This training is conducted using a word bank that encompasses a broad range of English words and their corresponding Chinese paraphrases, producing well-trained Chinese and English Word2Vec models.

Subsequently, the trained Chinese Word2Vec model generates several Chinese interpretations for the English stem word. For example, in a single-choice question with four options, the stem is word A, and the options are four Chinese translations including its correct answer and three distractors. These distractors are generated by inputting the correct answer into the Chinese Word2Vec model. The association between the correct answer and the distractors is evaluated using cosine similarity. For instance, if the vector representation of the correct answer B is $[x_1, x_2, ..., x_n]$, and that of the distractor C is $[y_1, y_2, ..., y_n]$, representing their semantic features, the association between B and C is calculated using formula (1).

$$cos(\theta) = \frac{\sum_{i=1}^{n} x_i y_i}{\sum_{i=1}^{n} (x_i)^2 \times \sum_{i=1}^{n} (y_i)^2} \tag{1}$$

The final phase focuses on establishing a threshold M for the number of distractors and selecting the most relevant Chinese interpretations as distractors. The selection is based on their association with the correct answer. In the above example, with M set at 3, the top three Chinese interpretations most closely related to the correct answer are chosen as distractors.

3.2 Sorting Module Design

The sorting module uses the Sorting Algorithm Based on Frequency, Mastery, and Memory (SAFMM). This algorithm assigns a comprehensive score to each candidate word based on examination frequency, mastery level, and memory volume, subsequently sorting them. This process dynamically recommends words to learners in real-time. The algorithm design is as follows:

Word Examination Frequency Statistics and Scoring Weight Mapping. Our study collects past examination papers of a certain English level and creates a word bank by dividing the content into individual words. This simplifies the calculation of word frequency in exams. Next, word examination frequency is mapped to scoring weights using formula (2). For instance, high-frequency words have a scoring weight of 3.

$$P_v = \begin{cases} 3, & \text{high} - \text{frequency} \geq N \\ 2, & M < \text{medium} - \text{frequency} < N \\ 1, & \text{low} - \text{frequency} \leq M \end{cases} \tag{2}$$

Here, P_v represents the frequency scoring weight of word v. The classification of recommended words as high, medium, or low frequency is determined based on their occurrence frequency and the predefined frequency thresholds.

Prediction of Learner's Mastery Level and Scoring Weight Mapping. Our study employs the Long Short-Term Memory (LSTM) model to predict a learner's word mastery level [9]. This model correlates "hidden" states, input vector sequences, and output vector sequences, as shown in Fig. 2. It maps an input vector sequence $x_1...x_T$ to an output vector sequence $y_1...y_T$, determined through the calculation of "hidden" states $h_1...h_T$. The "hidden" states represent a continuous encoding of information related to the user's historical word practice results, facilitating future predictions.

Initially, the LSTM model is trained to develop predictive capabilities using other users' historical word practice results. Subsequently, the trained LSTM model processes

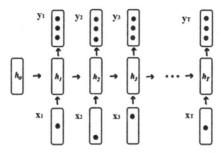

Fig. 2. Principle diagram of LSTM model

the current user's historical word practice feedback to predict the mastery levels of all words in the candidate set. The mastery level is numerically evaluated on a range from 0 to 1, then utilized to determine scoring weights as defined in formula (3).

$$D_{uv} = \begin{cases} 3, & d_{uv} \leq M \\ 2, & M < d_{uv} < N \\ 1, & N \leq d_{uv} \end{cases} \quad (3)$$

Here, d_{uv} is user u's mastery level of word v, which is determined based on the LSTM prediction model. D_{uv} represents user u's scoring weight of word v. For instance, if the user's mastery level of word A is M, the corresponding scoring weight is 3.

Word Memory Volume Statistics and Scoring Weight Mapping. According to the Ebbinghaus forgetting curve, if it takes x (hours) after a user first memorizes a word, the memory retention rate y can be estimated using the formula shown in formula (4).

$$y = 1 - 0.56x^{0.06} \quad (4)$$

The Ebbinghaus forgetting curve can determine a user's memory volume for words at different time intervals, so as to develop a review plan. For example, immediately after memorizing, the memory volume is 100%. After an interval of 20 min, it drops to 58.2%. After 1 h, it drops to 44.2%. Between 8 to 9 h, it drops to 35.8%. After 1 day, it drops to 33.7%. After 2 days, it drops to 27.8%. After 6 days, it drops to 25.4%. This curve indicates that the process of forgetting follows a distinct pattern: rapid at first and then gradually slowing down. Using formula (5), the memory interval for each word is mapped to a corresponding scoring weight.

$$A_{uv} = \begin{cases} 3, & 5 \text{ minutes} < t_{uv} < 20 \text{ minutes} \\ 2.5, & 20 \text{ minutes} < t_{uv} < 1 \text{ hour} \\ 2, & 8 \text{ hours} < t_{uv} < 12 \text{ hours} \\ 1.5, & 1 \text{ day} < t_{uv} < 2 \text{ days} \\ 1, & 2 \text{ days} < t_{uv} < 3 \text{ days} \\ 0.5, & t_{uv} > 4 \text{ days} \\ 0, & \text{in other cases} \end{cases} \quad (5)$$

Here, t_{uv} represents the memory interval of user u for word v. A_{uv} is the scoring weight of user u for word v. According to the memory interval, the user's scoring weight for each word can be determined. For instance, if the memory interval for word A is 1.5 days, the corresponding scoring weight is 1.5.

Comprehensive Scoring Sorting. Based on each word's examination frequency scoring weight P_v, the learner's mastery level scoring weight D_{uv}, and the memory interval time scoring weight A_{uv}, the implicit feedback feature vector $x_v = [P_v, D_{uv}, A_{uv}]$ is easy to calculate. For the implicit feedback I(u) of user u for all words, the cumulative feature contribution is $\sum_{v \in I(u)} x_v$.

The comprehensive score is normalized and calculated as specified in formula (6):

$$R_u = |I(u)|^{-0.5} \times \sum_{v \in I(u)} x_v \qquad (6)$$

Here, R_u represents the comprehensive score. I(u) is the implicit feedback of user u for all words, and x_v is the implicit feedback feature vector determined based on the scoring weights of examination frequency, mastery level, and memory interval time for word v.

Finally, words are sorted according to the comprehensive score, and the highest-scoring words are selected as target words to be recommended to the user.

4 Experimental Design and Result Analysis

The performance of our personalized hybrid recommendation system primarily relies on the sorting algorithm. Therefore, this experiment aims to evaluate the sorting algorithm's effectiveness. The experiment uses an offline simulation method for its efficiency in handling multiple algorithms, making it suitable for academic research. The evaluation metrics include Accuracy, Precision, Recall, F1-Score, and AUC, which are standard in recommendation algorithm assessments. This experiment utilizes real online learning data from the "Public English" course on the English adaptive learning system independently developed by our team. The dataset comprises 323,512 learning records from 2,654 learners, covering 992 word learning tasks. From this dataset, 300 freshmen learners, each of whom has memorized over 500 words, are selected for the word recommendation experiment. The remaining data is divided in an 8:1:1 ratio for the LSTM model's training, validation, and testing, which is designed to predict learners' word mastery levels.

This experiment uses collaborative filtering algorithm, latent factor model, Alphabet, and Random as baseline algorithms for comparison. It selects 300 learners who have memorized over 500 words for word recommendation. The recommended word list is set at 30% of the words studied by learners. The collaborative filtering algorithm employs user similarity to enhance word learning recommendations. User similarity is calculated using the cosine similarity algorithm, which helps identifies a group of "neighbors" with similar learning behaviors. Recommendations are based on these neighbors' word practice results, aligning with the user's needs and preferences. The latent factor model uses matrix decomposition to transform the user-word matrix into user-latent factor matrix and latent factor-word matrix. This separation reveals latent relationships between users and words. Analyzing these connections allows the model to predict a user's performance on unlearned words, recommending suitable words. Next, the Alphabet algorithm recommends words alphabetically. Lastly, the Random algorithm uses simple random sampling to select words.

After recommendation, each word has four possible outcomes. Firstly, the algorithm recommends the word, and the learner does not know it. Secondly, the algorithm recommends the word, but the learner recognizes it. Thirdly, the algorithm does not recommend the word, but the learner does not know it. Fourthly, the algorithm does not recommend the word, and the learner recognizes it.

Table 1 displays the experimental results. SAFMM outperforms the four other algorithms in Accuracy, Precision, Recall, F1-Score, and AUC, with an AUC of 0.7836, indicating optimal performance. The latent factor model shows good effectiveness, the collaborative filtering algorithm exhibits moderate performance, and the Alphabet algorithm has lower recommendation effectiveness. The Random algorithm, with an AUC approximately equal to 0.5, has no recommendation value.

Table 1. Comparison of recommendation performance metrics for different models

Models	Accuracy	Precision	Recall	F1-Score	AUC
Latent factor model	0.7867	0.4824	0.6784	0.5619	0.7473
Collaborative filtering algorithm	0.7464	0.4113	0.5764	0.4782	0.6837
Alphabet	0.6890	0.3093	0.4227	0.3557	0.5892
Random	0.6381	0.2208	0.3107	0.2572	0.5077
SAFMM	**0.8170**	**0.5252**	**0.7355**	**0.6104**	**0.7836**

5 Conclusion

Addressing personalized recommendation challenges for high-frequency, easily forgotten, and new words in English learning, our study introduces a specialized sorting algorithm. This algorithm considers the user's mastery level, forgetting curve, and word examination frequency to ensure accurate recommendations. On this basis, a personalized hybrid recommendation system for English words is designed. This system integrates word vectors, cosine similarity, and long short-term memory networks, offering an innovative English word recommendation solution. Our study conducts an offline simulation experiment to compare the system with four baseline algorithms, verifying its performance. The results show the system surpasses the baseline algorithms in Accuracy, Precision, Recall, F1-Score, and AUC metrics, indicating superior recommendation effects.

Our findings have significant application value for English word learning and contribute valuable insights to personalized recommendation. However, the dataset's size constrains the LSTM model's predictive accuracy, impacting the system's performance.

Therefore, future research could thus be enhanced in these areas: Firstly, expanding the dataset to improve the LSTM model's prediction accuracy and the system's performance. Secondly, incorporating additional learner characteristics (such as media preferences, learning styles, etc.) and word features (such as word images, word audio, etc.) to enrich user and word profiles. Thirdly, exploring more recommendation algorithms and hybrid strategies to increase the accuracy and diversity. Fourthly, considering users' feedback of the recommendation results to achieve dynamic updates and optimization of the recommendation system.

References

1. Wu, S.Y.: Design and development of intelligent recommendation system for English vocabulary learning. Shanghai International Studies University, Shanghai (2021). https://doi.org/10.27316/d.cnki.gswyu.2020.000590 (in Chinese)
2. Zou, D., Xie, H.: Personalized word-learning based on technique feature analysis and learning analytics. J. Educ. Technol. Soc. 21(2), 233–244 (2018)
3. Xie, H., Zou, D., Wang, M., Wang, F.L.: A personalized task recommendation system for vocabulary learning based on readability and diversity. In: Cheung, S., Lee, L.K., Simonova, I., Kozel, T., Kwok, L.F. (eds.) Blended Learning: Educational Innovation for Personalized Learning. ICBL 2019. LNCS, vol. 11546, pp. 82–92. Springer, Cham (2019). https://doi.org/10.1007/978-3-030-21562-0_7
4. Zou, D., Xie, H., Wang, F.L., Wong, T.L., Kwan, R., Chan, W.H.: An explicit learner profiling model for personalized word learning recommendation. In: Huang, T.C., Lau, R., Huang, Y.M., Spaniol, M., Yuen, C.H. (eds.) Emerging Technologies for Education. SETE 2017. LNCS, vol. 10676, pp. 495–499. Springer, Cham (2017). https://doi.org/10.1007/978-3-319-71084-6_58
5. Xie, H., Zou, D., Zhang, R., et al.: Personalized word learning for university students: a profile-based method for e-learning systems. J. Comput. High. Educ. 31, 273–289 (2019). https://doi.org/10.1007/s12528-019-09215-0
6. Long, X.W.: Design and implement of crowd-sensing recommender system for English words. South China University of Technology, Guangzhou (2020). https://doi.org/10.27151/d.cnki.ghnlu.2019.003952 (in Chinese)
7. Ru, S.Y., Deng, J.D., Zhang, Z.W., Zhong, J.Y., Chen, G.X., Tan, Z.J.: Application research of English word learning system based on deep learning. J. Comput. Program. Skills Mainten. 133–134+147 (2020). https://doi.org/10.16184/j.cnki.comprg.2020.02.045 (in Chinese)
8. Pu, S., Yudelson, M., Ou, L., Huang, Y.: Deep knowledge tracing with transformers. In: Bittencourt, I., Cukurova, M., Muldner, K., Luckin, R., Millán, E. (eds.) Artificial Intelligence in Education. AIED 2020. LNCS, vol. 12164, pp. 252–256. Springer, Cham (2020). https://doi.org/10.1007/978-3-030-52240-7_46

The Influence of Aesthetic Personalization on Gamified Learning: A Behavioral Analysis of Students' Interactions

Luiz Rodrigues[1]([✉]) [iD], Cleon X. Pereira Jr.[2] [iD], Emanuel Marques Queiroga[3] [iD], Heder Filho S. Santos[2] [iD], and Newarney T. Costa[2] [iD]

[1] Computing Institute, Federal University of Alagoas, Maceió, Brazil
luiz.rodrigues@nees.ufal.br
[2] Instituto Federal de Educação, Ciência e Tecnologia Goiano (IF Goiano), Iporá, Brazil
[3] Instituto Federal de Educação, Ciência e Tecnologia Sul-rio-grandense (IFSul), Pelotas, Brazil

Abstract. Personalized gamification seeks to address the limitations of the one-size-fits-all approach, mostly by tailoring the selection of game elements to individual preferences. However, there is limited understanding of how aesthetic personalization influences actual student behavior. This paper presents a behavioral analysis of 40 high school students engaged with a Virtual Learning Environment (VLE) over a four-week period. Each participant experienced both the one-size-fits-all and aesthetic personalization conditions for two weeks while submitting homework. Utilizing interaction data, we employed recurrent neural networks and grid search to develop a user model that demonstrated moderate agreement with students' actual behavior. This model was then utilized to examine student behavior over time. We found that, compared to the one-size-fits-all approach, aesthetic personalization appears to be linked with a higher probability of sustained engagement with the VLE during the initial days of interaction, despite this difference becomes inconsistent thereafter. This discovery suggests that while aesthetic personalization might enhance student learning by optimizing engagement with the VLE, it might suffer from the novelty effect.

Keywords: Gamification · Education · Tailoring · Aesthetic

1 Introduction

Gamification is known as the use of game elements in a non-game environment. In an educational context, various game elements can be employed to enhance interaction during the learning process. Nevertheless, it is crucial to recognize the potential adverse effects of gamification, wherein personalization

This Study Was Partially Supported by if Goiano.

can be employed as an alternative to mitigate them [5]. Personalization of gamification can be viewed from two distinct perspectives: i) providing specific game elements, considering that different individuals are motivated by distinct game elements; and ii) tailoring game elements' aesthetics based on individual preferences that may not necessarily be related to the educational context [2]. Most research has focused on analyzing the first case, but recent results provide motivation to investigate aspects of aesthetic-tailoring in the design of game elements [8, 9].

Furthermore, research on personalized gamification, while primarily focusing on tailoring game elements selection, often relies solely on self-reports captured through questionnaires or semi-structured interviews [5], leaving analyses of the user behavior resulting from these interventions still to be investigated [8]. A literature review indicated that research involving personalized gamification is still focused on exploring which characteristics can be taken into account in the customization process and how the school environment should be prepared for such an approach [4]. The research also points out that more empirical studies are needed to examine its motivational effects.

One strategy for analyzing behavioral patterns is investigating log data collected over time, in which the context of Artificial Intelligence in Education (AIED) enables employing techniques to predict distinct student behaviors [7]. As it involves data and behavior over a certain period of time, the use of time series combined with machine learning techniques is an intriguing avenue to pursue new discoveries from personalized gamification, wherein is possible to find studies that use RNN for prediction [6].

Driven by the aim to explore the impact of personalizing game-based elements within an educational setting on student behavior, we conducted an analysis of student interactions with a Virtual Learning Environment (VLE). To conduct the experiment, we collected interaction data from 40 high school students over a four-week period. Each participant experienced both the one-size-fits-all and aesthetic personalization conditions for two weeks while submitting homework. Using Recurrent Neural Network (RNN) and time series dataset, we aimed to predict the possibility of a student interacting with the VLE according to the personalizing game-based elements. Thus, this work expands the literature by investigating student behavior, through interactions in the VLE, in the context of aesthetic-based personalized gamification, revealing insights on the potential and drawbacks of this approach.

2   Method

The goal of this study was to understand how aesthetic-based personalized gamification (personalization hereafter) affects student behavior compared to the standard, one-size-fits-all approach (standard hereafter). For this, aiming for a deeper understanding of student behavior, we relied on user modeling based on time series using the dataset created in [8]. At first, all participants completed a questionnaire regarding their preferences concerning games, TV series, movies, anime, and music to inform the game elements' personalization, as detailed below. Then, the class was randomly divided into 2 groups (A and B). During the first cycle of 2 weeks, a personalized badge was provided to each student in group A upon daily activity submission, while each student in group B received a standard badge. In the second cycle, the dynamics of badges delivery were reversed between the groups. Next, we retrieved interaction logs considering the introduction of the gamification element, whether personalized or standard, and data were prepared using Python. Lastly, predictions were made using RNNs.

The student interaction data used was collected from the Moodle VLE. The types of interactions used in this study were: Comment viewed; Submission status viewed; Submission created; A file has been sent; A submission has been sent; and An online text has been sent. The interaction records provided by Moodle logs include, among other information, the date and time they occurred. These details were used to contextualize them with the insertion of gamification elements (standard vs. personalized).

Regarding personalized gamification through the alteration of the aesthetic aspects of game-based elements, we chose to incorporate affective aspects, similar to [8]. Based on a combination of information collected with a questionnaire, badges were constructed daily and distributed to students according to the configuration (standard or personalized). The standard badge defined for this experiment was emojis, and during the same week, all students who did not receive personalized badges were given emojis. Figure 1 shows examples of badges.

Fig. 1. Badges examples used as acknowledgment game element.

To the preprocessing, *first*, we calculated the number of interaction each student had in each day of the data collection period. *Second*, we standardized the number of interactions to range from 0 to 1 to facilitate model convergence. *Third*, we create the interaction windows, which we used as the RNN's input. The window generation procedure featured four steps. First, we defined the window's time frame, which ranges from the first day of data collection to five weeks after it. Note that we considered five weeks, despite the intervention lasting four weeks only, to accommodate a subsequent week in which students were still allowed to submit the activities. Second, we selected the window sizes to consider, namely seven and 14 days. We chose 7 to create our model based on at least one week of interaction data, whereas 14 was analyzed to investigate how a longer period would contribute to modeling students behavior from our data. Third, for each student, we extracted the number of interactions within a given time period (i.e., the window). During this process, for each time step, we saved whether the student had interacted with the standard (0) or personalized (1) gamification design. This is a key step that aimed to enable the user model to learn how the gamification design affects the changes of interacting with the VLE.

Fourth, we prepared the targets. For this, modeled student interaction as a binary problem (i.e., will the student interact with the VLE in the next day?) to be able to predict the probability they will come back to interact with the system. Accordingly, targets were adjusted to 1 if the number of interactions in the *next day* was bigger than 0 and remained as 0 otherwise. After this procedure, we had our preprocessed dataset (or X), which was composed of two features (i.e., number of interactions and gamification design) with either 7 or 14 time steps each, as well as the target values (y).

The behavior modeling was based on four main steps, according literature recommendations [3]. *First,* we split the data into training (80%) and testing (20%) sets. *Second,* we defined the hyperparameters search space for finding our user model. Concerning the *layer type,* we considered Gated Recurrent Unit (GRU), Long Short-Term Memory (LSTM), and a 1-dimensional Convolutional Neural Network (CNN). GRU and LSTM are advantageous when the temporal dependencies are complex and require a focus on long-term memory, while CNN is preferable for discerning particular features in sequential data. Hence, we considered those alternative would provide a broad yet technically suitable search space for our model's layers.

Additionally, we considered 16, 32, 64, and 128 as the number of neurons/units per layer and 1, 5, and 10 for the number of hidden layers (depth). Aiming to control for model complexity, we also analyzed dropout rates (5%, 10%, and 20%). Lastly, particularly to the CNN layer, we experimented with kernel sizes ranging from 3 up to 13 (to avoid considering the whole sequence), depending on the window size. Overall, these choices aimed to provide a broad search space, while avoiding models that would be too complex for our sample size.

Third, we performed the grid search. All models were optimized using Adam, towards minimizing their binary crossentropy, while saving 20% of the training sample for validation. Then, we used validation loss as the metric responsible to finish training each model. Particularly, we defined a large number of epochs (i.e., 1000) and used early stopping to finish training after 10 epochs with no improvements on validation loss.

Fourth, we selected the best model and assessed its performance on the test set. Similar to related work [3], we used Kappa, Area Under the Curve (AUC), and Accuracy as metrics. Kappa accounts for chance agreement, AUC reveals discriminative power across various thresholds, and Accuracy offers a straightforward measure of overall correctness. Thereby, this combination provides a comprehensive evaluation from different perspectives and allows comparisons to other research. After identifying the best model, we similarly retrained it (but without validation data) and evaluated it on the testing set.

After conducting behavior modeling, the subsequent step involved behavior analysis. This procedure builds upon the best user model, as found through the aforementioned procedure. Fundamentally, the goal is to exploit the user model to understand users behavior with standard/personalized gamification, as well as how it changes over time. For this, we followed three main steps.

First, we simulated behavioral patterns in terms of interacting with both the standard and the personalized gamification. Note that we do not simulate interaction/behavior. Instead, we simulated two patterns focused on the gamification designs. At first, we simulated the student had interacted with standard gamification during the whole window (e.g., 14 days) and, then, started interacting with the personalized design. The second pattern was the opposed, in which we simulated the student had interacted with the personalized design and progressively started interacting with the standard one.

Second, we used the best model to predict user behavior based on those interaction patterns. Note that the model's prediction is a probability. Hence, for each combination, we stored mean and standard deviation (SD) values of all probabilities predicted for X. With this, we were able to understand how students behavior would change depending on how long they had interacted with either standard or personalized gamification design.

Third, we used the statistics generated through the predictions to assess the behavioral patterns. Informed by our user model, we sought to understand how students interact with VLEs featuring either standard or personalized gamification over time based on the probability of interacting in the next day. Accordingly, we present an exploratory analysis based on visualizations and relaxed confidence intervals, as recommended in the literature [13].

3   Results

After the grid search, the most favorable results were obtained with a window size of 14 interactions, where a configuration comprising 16 units, 10 hidden layers, a dropout rate of 0.1, and a kernel size of 13 yielded the lowest validation loss of 0.591. Intriguingly, across both window sizes, the top-10 performing configurations consistently utilized CNNs. However, it's worth noting that while the optimal setting for a window size of 14 interactions achieved a validation loss of 0.591, the lowest loss attained with a window size of seven was 0.654.

Based on these findings, we considered the top-1 setting with window size of 14 as our best model. Considering our main goal was to model user behavior, we chose this model as it seemed to yield a better adjustment to our data, despite its higher complexity compared to models based on a window size of 7. Therefore, we retrained this model with the complete training set and, finally, evaluated its performance on the testing set.

The best model's performance indicated comparable results on both training and testing sets. It achieved AUC values of 0.768 and 0.769, and accuracies of 72,7% and 75% on training and testing sets, respectively. Additionally, the model demonstrated kappa values of 0.439 and 0.484 for training and testing sets, suggesting moderate agreement with ground truth values. These outcomes indicate substantial performance in aligning with actual observations, prompting us to utilize the user model for further analysis of student behavior.

In light of those findings, Fig. 2 demonstrates the results from simulating interaction patterns to understand students behavior. Recall that our user model predicts the probability a given student will interact with the VLE on the next day given i) the number of interactions in the last 14 days and ii) the gamification design they were exposed to during that period. Based on that, for each gamification design (i.e., standard and personalization), Fig. 2 presents the mean probability (i.e., the average of the probabilities for all rows in our preprocessed dataset), along with the standard deviation (shown as error bars).

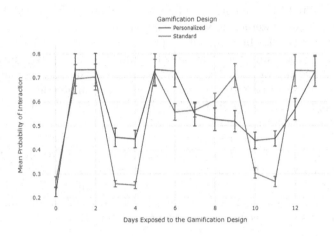

Fig. 2. Analysis of how student behavior changes over time depending on the gamification design they interact with based on our user modeling approach.

The analysis of Fig. 2 highlights mixed outcomes regarding the efficacy of personalized gamification. Initially, within the first week of interaction (up to six days), personalized design exhibited higher or comparable interaction probabilities compared to the standard design, with significantly greater average probabilities (Median = 0.73; Mean = 0.58; SD = 0.21 vs. Median = 0.56, Mean = 0.49; SD = 0.23) at a 90% confidence level. However, results became less consistent during the second week, showing mixed probabilities between the two designs, though the standard design's average probability slightly surpassed that of the personalized one (Median = 0.61, Mean = 0.56, SD = 0.20 vs. Median = 0.53, Mean = 0.54, SD = 0.10). Despite this, the difference was statistically non-significant, possibly due to intermittent periods where personalized design still showed superiority, particularly after 10 and 11 days.

4   Discussion

This paper explored the impact of aesthetic-based personalized gamification on student behavior compared to the standard, one-size-fits-all approach. Unlike prior research on personalized gamification, this study presents three main distinctions. First, we focused on aesthetic-based personalization. In contrast, the literature has been extensively focused on personalization of gamification by tailoring the selection of game elements [2]. Second, despite recent research started exploring this kind of personalization, it concerned empirical studies within laboratory settings [11]. Differently, our study concerns an empirical analysis from data collected during real lessons. Third, our investigation builds upon actual user behavior. Differently, research on aesthetic-based personalization outside laboratory settings is limited to students' self-reports [8]. Therefore, this paper expands the literature by revealing how aesthetic-based personalization affects student behavior based on empirical evidence from actual user behavior collects within the context of real lessons.

The study initially utilized RNN to develop a user model capturing students' behaviors over time, achieving moderate agreement with ground truth data. This model addressed a gap in personalized gamification research by focusing on actual behavior [5]. Subsequently, leveraging this user model, the study investigated student behavior in aesthetic personalized and one-size-fits-all gamification approaches. While aesthetic personalization initially enhanced engagement with the VLE, its impact diminished after a week, suggesting susceptibility to the novelty effect, which is consistent with the literature on one-size-fits-all gamification [1]. Conversely, the one-size-fits-all approach didn't demonstrate a clear advantage, showing fluctuating results over time and closely mirroring the behavior of personalized gamification. This insight might suggest that our aesthetic-based personalization suffered from the familiarization effect, which has been recently discussed in the one-size-fits-all gamification literature [10,12].

For instance, Van Roy et al. [12] found indications that gamification had an initial positive effect on student motivation, which naturally decreased after a few weeks (i.e., the novelty effect). They additionally found that, following the initial decrease, student motivation ended up increasing again, arguing it could be due to a process of familiarization with gamification (i.e., the familiarization effect). Nevertheless, this study is based on a single-sample analysis and relied on self-reports of motivation [12]. Then, Rodrigues et al. [10] conducted a similar analysis based on a 14-week quasi-experimental study. Their findings revealed that during the first four weeks, gamification had a moderate effect on user behavior, which decreased for the next two weeks and presented a small decrease thereafter. Thus, providing further empirical evidence on the novelty and familiarization effects in the context of one-size-fits-all gamification.

Nevertheless, studies on personalized gamification are often limited to short interaction periods or lack an analysis of how its impact changes over time [8,9, 11]. In contrast, this paper presents an analysis of aesthetic-based personalization built upon data collect throughout four weeks of interaction. Moreover, this data concerned interacting with gamification as part of a real learning course. Therefore, our finding expands the literature with empirical evidence suggesting the presence of the novelty effect, and possibly the familiarization effect, in the context of gamification personalized by changing game elements' aesthetics.

References

1. Bai, S., Hew, K.F., Huang, B.: Is gamification "bullshit"? evidence from a meta-analysis and synthesis of qualitative data in educational contexts. Educational Research Review p. 100322 (2020)
2. Hallifax, S., Serna, A., Marty, J.C., Lavoué, É.: Adaptive gamification in education: A literature review of current trends and developments. In: Scheffel, M., Broisin, J., Pammer-Schindler, V., Ioannou, A., Schneider, J. (eds.) Transforming Learning with Meaningful Technologies. pp. 294–307. Springer International Publishing, Cham (2019). https://doi.org/10.1007/978-3-030-29736-7_22
3. Hewamalage, H., Bergmeir, C., Bandara, K.: Recurrent neural networks for time series forecasting: Current status and future directions. Int. J. Forecast. **37**(1), 388–427 (2021)
4. Hong, Y., Saab, N., Admiraal, W.: Approaches and game elements used to tailor digital gamification for learning: A systematic literature review. Computers & Education p. 105000 (2024)
5. Klock, A.C.T., Gasparini, I., Pimenta, M.S., Hamari, J.: Tailored gamification: A review of literature. International Journal of Human-Computer Studies p. 102495 (2020)
6. Murata, R., Okubo, F., Minematsu, T., Taniguchi, Y., Shimada, A.: Recurrent neural network-fitnets: Improving early prediction of student performanceby time-series knowledge distillation. Journal of Educational Computing Research **61**(3), 639–670 (2023)
7. Ouyang, F., Zheng, L., Jiao, P.: Artificial intelligence in online higher education: A systematic review of empirical research from 2011 to 2020. Educ. Inf. Technol. **27**(6), 7893–7925 (2022)
8. Pereira Júnior, C., Santos, H., Rodrigues, L., Costa, N.: Investigating the effectiveness of personalized gamification in enhancing student intrinsic motivation: an experimental study in real context. In: Anais do XXXIV Simpósio Brasileiro de Informática na Educação. pp. 838–850. SBC, Porto Alegre, RS, Brasil (2023)
9. Rodrigues, L., Palomino, P.T., Toda, A.M., Klock, A.C., Pessoa, M., Pereira, F.D., Oliveira, E.H., Oliveira, D.F., Cristea, A.I., Gasparini, I., et al.: How personalization affects motivation in gamified review assessments. International Journal of Artificial Intelligence in Education pp. 1–38 (2023)
10. Rodrigues, L., Pereira, F.D., Toda, A.M., Palomino, P.T., Pessoa, M., Carvalho, L.S.G., Fernandes, D., Oliveira, E.H., Cristea, A.I., Isotani, S.: Gamification suffers from the novelty effect but benefits from the familiarization effect: Findings from a longitudinal study. Int. J. Educ. Technol. High. Educ. **19**(1), 1–25 (2022)
11. Rodrigues, L., Toda, A.M., Palomino, P.T., Klock, A.C., Avila-Santos, A.: Motivation no jutsu: Exploring the power of badge aesthetics in gamified learning. In: Anais do XXXIV Simpósio Brasileiro de Informática na Educação. pp. 631–643. SBC (2023)
12. Van Roy, R., Zaman, B.: Need-supporting gamification in education: An assessment of motivational effects over time. Computers & Education **127**, 283–297 (2018)
13. Vornhagen, J.B., Tyack, A., Mekler, E.D.: Statistical significance testing at chi play: Challenges and opportunities for more transparency. In: Proceedings of the Annual Symposium on Computer-Human Interaction in Play. pp. 4–18 (2020)

Can AI Assistance Aid in the Grading of Handwritten Answer Sheets?

Pritam Sil[(✉)], Parag Chaudhuri, and Bhaskaran Raman

Department of Computer Science and Engineering, IIT Bombay, Mumbai, India
{pritamsil,paragc,br}@cse.iitb.ac.in

Abstract. With recent advancements in artificial intelligence (AI), there has been growing interest in using state of the art (SOTA) AI solutions to provide assistance in grading handwritten answer sheets. While a few commercial products exist, the question of whether AI-assistance can actually reduce grading effort and time has not yet been carefully considered in published literature. This work introduces an AI-assisted grading pipeline. The pipeline first uses text detection to automatically detect question regions present in a question paper PDF. Next, it uses SOTA text detection methods to highlight important keywords present in the handwritten answer regions of scanned answer sheets to assist in the grading process. We then evaluate a prototype implementation of the AI-assisted grading pipeline deployed on an existing e-learning management platform. The evaluation involves a total of 5 different real-life examinations across 4 different courses at a reputed institute; it consists of a total of 42 questions, 17 graders, and 468 submissions. We log and analyze the grading time for each handwritten answer while using AI assistance and without it. Our evaluations have shown that, on average, the graders take 31% less time while grading a single response and 33% less grading time while grading a single answer sheet using AI assistance.

Keywords: AI Assisted Grading · Handwritten Answers · Keyword Highlighting

1 Introduction

Grading of handwritten answer sheets is a time consuming process. Traditionally, during the grading of handwritten answer sheets the grader has to go through each answer and grade it without any external assistance. On the other hand, recent advances in AI have shown a growing interest in using SOTA methods to provide assistance in this area. Presently, most commercial solutions today provide an easy-to-use user interface (UI) that the graders can use to upload their set of answer sheets, add rubrics and grade them. While some strictly restrict the question paper format and answer format in order to provide assistance, others relax these restrictions by only providing an easy-to-use UI (more has been discussed in Sect. 2). Moreover, none of these solutions provide AI-based assistance in grading short and long-answer type questions by highlighting certain keywords that will help them in grading such answers.

A. M. Olney et al. (Eds.): AIED 2024 Workshops, CCIS 2151, pp. 291–298, 2024.
https://doi.org/10.1007/978-3-031-64312-5_35

Hence, this work has designed and deployed a prototype implementation of an AI-assisted grading pipeline (details in Sect. 3) on an existing e-learning platform and verified if keyword highlighting is effective in providing assistance to the grading process. The main contributions of this work are as follows

- Highlighting of important keywords in the answer regions of scanned answer sheets to enable faster grading. Question regions are auto-detected on the question paper.
- An extensive field evaluation on over 5 different real-life examinations across 4 different courses at a reputed institute, consisting of a total of 42 questions, 17 graders, and 468 submissions.
- Our evaluations show that on average, there was a **31%** reduction in grading time for a single response and a **33%** reduction in grading time for a single answer sheet using AI assistance (as discussed in Sect. 4).

2 Related Work

Table 1. Comparison of existing methods with our method

	GS	SAFE	DG	SG	Our method
Permitted Answer Types	Any	Any	Textual	A small phrase or word	Any
Question Paper Format	Any	Any	Any	Restricted	Any
Auto-grouping of Similar Answers	Y	N	N	N	N
Auto Detection of Question Regions	N	N	N	N	Y
Auto Extraction of Roll Number	Y	N	N	Y	Y
Keyword highlighting	N	N	N	N	Y
Field evaluation of methodology	Opinion survey	Opinion based	Not available publicly	Not available publicly	**Grading time based**

The logistics involved in grading handwritten answer sheets can be quite challenging for courses with large enrollments. Gradescope [7] (GS) by Singh et al. allows the grader to manually demarcate the answer and roll number regions on a blank question paper, upload a set of scanned answer sheets and subsequently use this to highlight the whole answer region on the answer sheet. Gradescope supports AI based grouping of similar answers. However, the system does not crop out the answer regions; instead, it just places a bounding box to highlight the answer region for a question as demarcated by the instructor. Additionally, it does not provide any assistance while grading paragraph-type answers.

Smart Authenticated Fast Exams (SAFE) by Chebrolu et al. [5] provides a secure solution to conduct examinations in a proctored environment while following a Bring Your Own Device (BYOD) model. SAFE provides a one stop solution for managing day-to-day class activities. The SAFE system allows upload of answer sheet images and provides a manual grading interface for the graders to correct such uploads. However, it does not assist in grading such images, like our pipeline.

Smartail.ai came up with their flagship product called DeepGrade [1] (DG). DeepGrade can generate a question paper automatically from an instructor created question set. The students upload their answers to the system in the form of text or handwritten answers. For handwritten answers, the interface will automatically perform optical character recognition(OCR) and automatically grade these answers. DeepGrade is effective in grading short answers where the students have to write a short phrase or a short paragraph. The interface provides some feedback on the grammar and spelling errors present in the answer. However, the system does not automatically detect the question regions, or highlight important keywords present in handwritten answers.

Swift Grade [4] (SG) is a solution for assisting in the grading of answer sheets. Their interface allows the instructor to add questions along with their marks and answers and also register students for a course. The interface generates an answer sheet with the student's names on top, which the instructor can download and distribute to the students during the examination. The interface records which bounding box represents the answer to which question. The answer sheets are scanned and uploaded to the system and are automatically graded. Their interface can automatically grade short equations, phrases and numbers. However, Swift Grade requires the grader to use the generated answer sheet format and does not provide much assistance in grading short and long answer type questions.

Unlike some of the others, ours is a human-in-the-loop solution to answer script grading. We believe while there is merit in providing assistance to the task of grading handwritten answers, doing it completely automatically may not be able to assess the variety of answers possible. We observe in our analysis that error or noise in the output of AI models can adversely the effect performance of such systems, hence it is always better to have a human in the loop, so as to not pass on this adverse effect to the students being assessed.

Table 1 compares various aspects of GradeScope, DeepGrade, Swift Grade, SAFE, and our AI-assisted grading pipeline. Regarding field evaluation, only Gradescope and SAFE have made their data publicly available. While Gradescope performed an opinion-based survey and an analysis of grading time at assignment level, SAFE only performed an opinion-based survey. DeepGrade and Swift Grade do not have any publicly available field evaluation data. Our system records the grading time at per question level, and we analyse it as a part of our field evaluation.

3   AI Assisted Grading Pipeline

The AI-assisted grading pipeline, as shown in Fig. 1, assumes that each question will have an answer region provided just below it, and the students will write their answers within this region. Also, the blank question paper must be available as a single PDF file at the start of the process. The first part of this pipeline automatically detects the question regions ([3] [2]) present in the PDF file of a blank question paper uploaded by the instructor. Once the instructor is

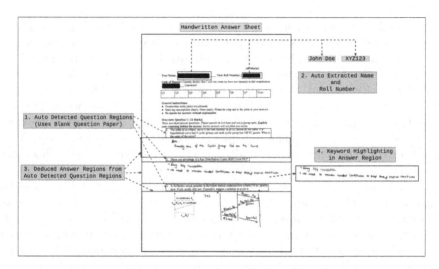

Fig. 1. Overview of AI assistance provided by our pipeline

satisfied with the detected question bounding boxes and the corresponding text, they can be saved to the interface.

Once the examination is over, the instructor will have a set of scanned handwritten answer sheets. Note each such answer sheet is scanned to a separate PDF file. It is difficult to manually create a mapping between each such scanned PDF file to the corresponding student's name and roll number. Our pipeline partially automates this problem using AI. The instructor uploads this set of scanned answer sheets to our interface, and also marks the corresponding bounding box for name and roll number on one PDF. Once done, the system will automatically extract (perform word recognition) the text present in these bounding boxes. For this word recognition step, we use the docTR library as described earlier for question text extraction. Once each answer sheet has been mapped to a student's name and roll number, the mappings are shown to the instructor via a separate user interface. Incorrect mappings, if any, can be manually corrected by the instructor. Once done, the mappings are saved.

Now, the system automatically deduces the handwritten answer regions from each answer sheet. The deduced answer region for a particular question is the region between the bottom part of the bounding box of that question and the top part of the bounding box of the next question. The system automatically crops out the handwritten answer region for a question for a particular student's answer sheet and displays it on the grading interface. This allows the grader to focus on the relevant region where the answer is present. The grader has the option to look at the entire page image, if needed.

It is often seen that graders look for specific keywords while grading a particular answer. However, as seen in see Sect. 2, this has not been investigated by any earlier or currently available system. Our system asks the instructor to input relevant keywords for each answer once in the process and subsequently highlights [6] them on the handwritten answer regions for a particular question.

4 Field Evaluation of Grading Time Reduction

This section describes the field evaluation of the AI-assisted grading pipeline described above.

4.1 Evaluation Method

A prototype implementation of the pipeline is used to grade real-life examinations at a reputed institute[1]. The handwritten answer sheets of the students are scanned and uploaded to the platform, and the answer regions were highlighted using the given keywords. The graders were then asked to grade their respective questions using the manual grading interface, and the time needed to grade each response was automatically recorded by the system. No timer was shown to the grader, but the time was logged by the system in a manner transparent to the grader. This was done to ensure a fair evaluation and not put any pressure on the grader.

Set of All Answer Sheets

Table 2. Details of each examination

Course	# Sub	# Qs	# Graders
Course A	124	21	3
Course B	49	10	3
Course C	50	1	1
Course D	198	5	9
Course E	47	5	1

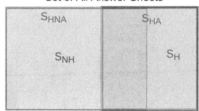

Fig. 2. Grouping of Answer Sheets for field evaluation

The pipeline was evaluated on five examinations for five separate courses. Table 2 displays the number of submissions and the number of questions in which keyword highlighting has been used for each examination. It can be seen that the evaluation dataset is varied and has a significant number of answer sheets. Grading all of them manually would have taken a non-trivial amount of time. Each of these examinations contains a significant number of questions, except the Course C examination, which had only 1 question that could use the keyword highlight feature. Also, all these questions are of different types, including those with numerical answers, short text answers and long paragraph answers. The set of scanned answer sheets was split into two equal parts (see Fig. 2). In the first half, no attempt was made to highlight keywords present in it (S_{HNA}). In the second half, an attempt was made to highlight keywords in it (S_{HA}). Since

---

[1] Name withheld for anonymity.

no handwritten text detector and recognizer is 100% accurate, certain images in S_{HA} did not have any keyword highlighted within them. These are contained in the set S_{NH}(the red rectangle). S_H (the green rectangle) denotes the set of answer sheets where at least one keyword was highlighted. Thus, due to the inaccuracy of the keyword detection model, S_{HA} contains elements from both S_H and S_{NH}.

The main goal of the evaluation was to determine whether graders take less time to grade the student's answer or not when using keyword-highlighting feature[2]. The time taken to grade each response is recorded and analyzed. It should be noted that *no personally identifiable student information was used* for the study.

4.2   Analysis at per Response Level (Course B)

During our evaluation process, the time taken to grade each response was recorded. Both the S_{HNA} and $S_{HighlightAttempted}$ sets contained few grading time instances, which are quite high compared to the mean time recorded. We conjecture that these maybe cases where the grader took a break between grading one question and the next. These are considered to be outliers and are eliminated by removing the top 5% data points in a set.   Fig. 3 shows the average

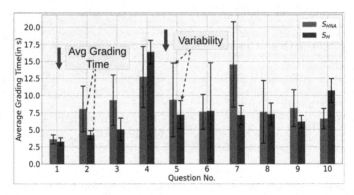

Fig. 3. Average grading time (S_{HNA} vs S_H) for Course B

grading time, for both S_{HNA} and S_H for each question present in Course B. It can be seen that the graders have taken less time to grade all the questions using AI-based keyword highlighting, except questions 4, 6 and 10. Question 4 asks the students to write pseudocode as a solution. It is thus seen that high-lighting keywords in pseudocode is not effective in reducing the grading time. Also, for questions 4 and 6 involved the grader going through the entire expla-nation provided by the student. This suggests that for questions where the logic

---

[2] The keyword highlight feature is used interchangeably with AI-based assistance in this section.

present within the answer is more important than looking at certain keywords, the keyword highlighting feature cannot assist the grader.

We also observe in Fig. 3 that the variability in grading time is significantly reduced while using AI assistance. It is often seen that student's answers are needlessly verbose. We conjecture that because of the keyword highlight feature, the grader can focus on the specific regions within the answer where the relevant parts of the explanation are present. Thus, for each answer the grader has to go through a specific set of sentences rather than the whole answer. Hence, this reduces the variability in grading time while using the keyword feature as each time the grader has to go through approximately same number of lines.

While we have presented results here for one of the five examinations in detail, we note that similar results were found for the other examinations. To summarize this section, we compute the average of individual reductions in grading time at the response level for each question from this data (as shown in Fig. 3 and similar data from other examinations). We compute average reduction in grading time by computing the average of reductions of grading times of each question w.r.t S_{HNA}. The average reduction in grading time per response as computed for Q is 31%. (More Details : https://sites.google.com/view/pritam-sil/aied2024)

4.3   Analysis at per Answer Sheet Level

The time to grade an answer sheet is considered to be the total time taken to grade all the questions in that answer sheet. The average grading per answer sheet for each examination is analysed here.   Fig. 4 shows the average grading

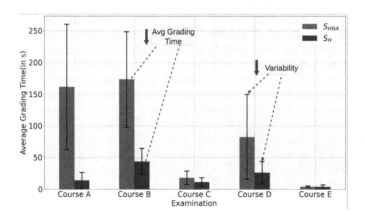

Fig. 4. Average grading time - S_{HNA} vs S_H(Answer Sheet)

time per answer sheet across all examinations, for both S_{HNA} and S_H. It clearly shows the fact that graders took less time while grading the whole answer sheet while using keyword highlighting than when they did not have that assistance.

Overall variability in average grading time can again be observed to be low in Fig. 4 when graders used keyword highlighting. Computing the average of individual reduction in average grading time at the answer sheet level for each examination from this data (as shown in Fig. 4) indicates an overall reduction of 33% in grading time per answer sheet. Similar analysis at question type level revealed a reduction of 23% for long answer type questions, 34% for numerical type and 34% for short answer type questions.

5   Conclusion

We introduce an AI-assisted pipeline for grading of handwritten answer scripts. The system aids the grader by automatically detecting the question regions and text in the blank question paper. Subsequently, it extracts the name and roll number of students from scanned answer sheets and maps the scanned answer sheets to corresponding students. At grading time, it automatically crops out the handwritten answer regions and displays it on our manual grading interface and highlights important keywords in the answer. Extensive field evaluation of this pipeline reveals a 31% reduction in average grading time for each hand-written response and a 33% reduction in average grading time for entire answer sheets, when using the keyword highlight feature. Thus, AI-based assistance can definitely aid in the grading of handwritten answer sheets.

The current work only highlights keywords if there is an exact match. Future work involves but is not limited to using techniques from Natural Language Processing (NLP) to detect different forms or even synonyms of the keywords and eventually automatically grade such handwritten answers.

Acknowledgements. We would like to thank the TIH Foundation for IoT and IoE at IIT Bombay for funding this project (RD/0121-TIH0000-001).

References

1. DeepGrade. https://smartail.ai/deepgrade/, Accessed 25 Jan 2024
2. docTR. https://mindee.github.io/doctr/, Accessed 25 Jan 2024
3. pdfminer. https://pdfminersix.readthedocs.io/, Accessed 25 Jan 2024
4. Swift grade. https://goswiftgrade.com/, Accessed 25 Jan 2024
5. Chebrolu, K., et al.: SAFE: smart authenticated fast exams for student evaluation in classrooms. In: Proceedings of the ACM SIGCSE Technical Symposium on Computer Science Education, pp. 117-122. ACM, New York, NY, USA (2017)
6. Kuang, Z., et al.: MMOCR: a comprehensive toolbox for text detection, recognition and understanding. CoRR **abs/2108.06543** (2021)
7. Singh, A., Karayev, S., Gutowski, K., Abbeel, P.: Gradescope: a fast, flexible, and fair system for scalable assessment of handwritten work. In: Proceedings of the Fourth (2017) ACM Conference on Learning @ Scale, pp. 81–88, April 2017

From Guidelines to Governance: A Study of AI Policies in Education

Aashish Ghimire$^{(\boxtimes)}$ and John Edwards

Utah State University, Logan, UT 84322, USA
{a.ghimire,john.edwards}@usu.edu

Abstract. Emerging technologies like generative AI tools, including ChatGPT, are increasingly utilized in educational settings, offering innovative approaches to learning while simultaneously posing new challenges. This study employs a survey methodology to examine the policy landscape concerning these technologies, drawing insights from 102 high school principals and higher education provosts. Our results reveal a prominent policy gap: the majority of institutions lack specialized guidelines for the ethical deployment of AI tools such as ChatGPT. Moreover, we observed that high schools are less inclined to work on policies than higher educational institutions. Where such policies do exist, they often overlook crucial issues, including student privacy and algorithmic transparency. Administrators overwhelmingly recognize the necessity of these policies, primarily to safeguard student safety and mitigate plagiarism risks. Our findings underscore the urgent need for flexible and iterative policy frameworks in educational contexts.

Keywords: AI in Education · Administrator's attitude · AI Policy

1 Introduction

With the rapid advancement of technology, generative artificial intelligence (Gen AI) tools, particularly Large Language Models (LLMs) like ChatGPT, are increasingly being adopted in various sectors, including education. These technologies offer promising avenues for pedagogical innovation, personalized learning, and administrative efficiency. However, their integration into educational settings is not without challenges, particularly concerning ethical considerations. Issues related to student privacy, data security, algorithmic transparency, and accountability are growing areas of concern.

While the application of these tools offers numerous advantages, the absence of comprehensive policy frameworks governing their ethical use in education can lead to unintended negative consequences. Inadequate policies may expose students to risks such as data misuse, algorithmic bias, and academic dishonesty. Educational institutions, thus, find themselves at a crossroads, balancing the potential benefits of emerging technologies against ethical and legal ramifications.

The research focuses on addressing the following questions:

© The Author(s), under exclusive license to Springer Nature Switzerland AG 2024
A. M. Olney et al. (Eds.): AIED 2024 Workshops, CCIS 2151, pp. 299–307, 2024.
https://doi.org/10.1007/978-3-031-64312-5_36

RQ1 What is the current landscape of policies related to Generative AI in educational settings and what do these policies cover?

RQ2 What are the perceived needs for future policy formulation in relation to Generative AI, and what recommendations can be made for an effective ethical framework?

To answer these questions, this study surveyed over 100 educational administrators in the United States.

2   Related Work

The integration of artificial intelligence (AI) in education is evolving rapidly, necessitating a multidimensional understanding of its applications, ethical considerations, governance frameworks, and pedagogical impacts.

The ethical implications of AI in education are complex, involving considerations of fairness, transparency, and privacy. Holmes et al. [15], Akgun and Greenhow [3], and Adams et al. [1] discuss the ethical challenges in deploying AI in educational settings. Halaweh et al. [14] and Sullivan et al. [22] propose frameworks for responsible implementation, emphasizing the need for policies that ensure student safety and academic integrity. Chiu [6] and Kooli [16] highlight the lack of policy considerations, calling for a balanced approach to leveraging AI's benefits while mitigating its risks. Ghimire et al. studied the AI in education from students' [10] and teachers' perspective. [11,12]

The governance of AI in education involves balancing technological benefits with ethical risks. Garshi et al. [9] and Filgueiras [8] explore and propose frameworks for accountability and human rights in smart classrooms. Li and Gu [17] present a risk framework for Human-Centered AI, emphasizing accountability and bias. Memarian and Doleck [18], and Gillani et al. [13] discuss the challenges of fairness and transparency, necessity of security and privacy, ethical concerns, advocating for human-centered and politically aware governance models. Uunona and Goosen focus on leveraging ethical values in AI-powered online learning applications, particularly in the Namibian educational context [23].

The development of AI-specific policy guidelines is critical for ethical integration into educational systems. Miao et al. [19] and Chan [5] have contributed to guiding policymakers, though existing technology policies provided by CoSN and broader guidelines from organizations like IEEE and the European Commission [2,21] lack the granularity needed for AI and fall short in addressing AI's unique challenges. A multidisciplinary approach is vital for understanding AI's impact on education. Dwivedi et al. [7] and Baidoo-Anu et.al [4] combine insights from various fields, addressing the capabilities and challenges of AI. This underscores the need for more detailed and AI-focused educational policies.

3   Methodology

To gain insights into the current policy landscape regulating the use of AI tools such as ChatGPT in educational settings, as well as to understand the attitudes of educational administrators toward these policies, this study employed a survey. This survey, administered across a diverse array of educational institutions, consists of a mix of multiple-choice questions, Likert-scale questions, and free-form text entries. The survey was specifically designed to discover the current landscape of policies related to Generative AI in educational settings and the perceived needs for future policy formulation in relation to Generative AI. Influenced by prior research such as Nguyen et al. [20] and Adams et al. [1], the survey covered commonly identified policy areas and offered respondents the opportunity to express additional concerns and policy suggestions through free-form text. Survey questions are available at https://github.com/GitAashishG/G2GSurvey

Data Collection. We utilized a publicly available directory (from state education boards' websites) to identify and reach out to high school principals. For higher education institutions, we manually curated the mailing list by first obtaining a list of all higher education institutes in the states and then looking up their provost's or chief academic officer's email. The survey was distributed across diverse geographic locations within the United States across Arkansas, Massachusetts, New Mexico, Utah and Washington to capture a wide range of perspectives. Survey responses were collected between June and September 2023.

Data Analysis. We performed χ^2 tests for each response against each of institution size, geographic location, and governance model (public or private). We also ran Pearson correlation tests for relation between need for policy, sentiment about AI tools, autonomy preference against administrators' experience length and student population. These tests were not significant.

4 Results

We received total of 102 complete survey responses - 81 High Schools (HS) and 21 Higher Education (HE) institutions. We had 21 responses from Arkansas (15 HS, 6 HE), 16 from Massachusetts (13 HS,3 HE), 23 from New Mexico (19 HS, 4 HE), 23 from Utah (18 HS, 5 HE) and 19 from Washington (16 HS, 3 HE).

4.1 RQ1: What Is the Current Landscape of Policies Related to Generative AI in Educational Settings and What Do These Policies Cover?

Existence of Current Policies. A majority of higher education respondents indicated either ongoing efforts to formulate generative AI-related policies or the existence of established policies. Specifically, over 80% of higher education institutions reported active policy development, 5% already have a policy, and 15% have no plans to enact one. In contrast, only 50% of high schools are in the process of policy formulation, while approximately 45% neither have a policy nor plans to develop one. Figure 1a depicts these data. A statistically significant difference in policy status between high school and college was observed

$\chi^2(2, N = 102) = 0.7.44, p = .0.024$ indicating that high schools are less inclined to work on policies than higher educational institutions.

Having a very small sample size in each category doesn't allow us to analyze and understand differences between the categories, but we can still understand a lot with the holistic review of the data. When asked if they need to have an AI-related policy, the prevailing sentiment among administrators was a critical need for these policies. Figure 1b shows the response on necessity of such policies. It can be seen that the necessity of AI related policy is almost universally agreed upon.

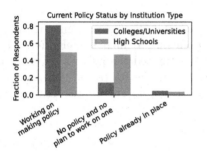

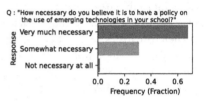

(a) Current policy status by inst type   (b) Necessity of AI policies

Fig. 1. Administrators' responses on policy status and necessity

Adequacy of Current Policies. Administrators who reported the existence or development of policies were subsequently asked about what is covered on their AI policies and their adequacy. The majority expressed that current or in-progress policies inadequately address the integration of emerging technologies. Even for many policies currently in development, administrators think these policies are not adequate. Over 45% of respondent strongly disagreed and 30% somewhat disagreed that current or in-progress policy are adequate while only about 3% respondents agreed and 11% somewhat agreed. Notably, only a small minority of these policies (less than 20%) specifically mention LLMs like Chat-GPT or Bard or image models like DALL-E.

We also asked administrators what their current or in-progress policies covered. Existing policies most commonly address issues like plagiarism, while elements like bias mitigation and algorithmic transparency are less frequently covered. 'Plagiarism' emerged as the most frequently cited motivation (37.6%) for policy development or revision. This was followed by 'Ensuring Privacy' (27.4%). Figure 2a indicates areas covered by current or in-progress policies. This indicates a perceived gap between existing governance mechanisms and the requirements for ethical and effective technology integration. Statistical tests revealed no significant associations between policy aspects and institution type, size, or location.

4.2 RQ2: What Are the Perceived Needs for Future Policy Formulation in Relation to Generative AI, and What Recommendations Can Be Made for an Effective Ethical Framework?

Our second goal of this study was to understand the key elements that educational administrators believe should be included in a policy framework for the ethical use of emerging technologies like ChatGPT in education as well as their overall sentiment on the policy and gather any additional insight and recommendation from the administrators.

Quantitatively, the focus was on the areas that respondents believe policies should primarily target and the kinds of support or resources they consider would be helpful for their institutions. Question "In which areas should policies for the use of emerging technologies in education primarily focus?" allowed multiple selections as well as free form text entry to capture administrators' focus area for policy making. Figure 2b shows the policy focus area identified by school administers. The majority of respondents highlighted 'Ethical Considerations' and 'Stopping Plagiarism' as the top two areas, with over 80% of responses, followed by ensuring students' safety and compliance with regulations.

We also asked the administrators about the support resources that would help them make or update generative AI related policies. Model guidelines from successful school or district was the most commonly resources deemed useful (81%), followed by professional development (77%) and staff training and legal/ethical consultations(55%). Collaboration and consultation with tech companies (35%) and the need for funding and resources (28%) were also identified.

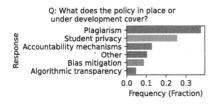
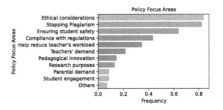

(a) Components included in existing or in-development policies (multiple selection allowed)

(b) Focus area for policy identified by the administrators (Multiple selection)

Fig. 2. Administrators' response in policy focus area and resources needed

The responses indicate a diverse perspective on who should be responsible and involved for formulating the policies governing the use of emerging technologies like ChatGPT in education. School administrators are seen as the most responsible entities, followed by teachers and students, along with school board and parent-teacher association. As for the autonomy and decision making given to schools and teachers, the respondents widely varied. For schools, the responses ranged from 'none' to 'all,' while the responses for teachers' decision-making ranged from 'none' to 'most.' Interestingly, none of the administrators responded that teacher should have all the decision making power (total autonomy). Overall, the data suggests a preference for a collaborative approach to policy formulation and implementation that includes various stakeholders at different levels of governance.

Additionally, we asked a couple of questions to understand the overall sentiment about AI tool as well as sentiment about existing detection tools. Most of the administrators are either indifferent or positive, and very few are not in favor of the technology. When asked about the use of existing tools that claim to detect AI-generated content, about half of the respondents were in favor of using such tools to narrow down, but not as a final arbiter of truth. The remaining respondents are almost evenly split between banning such tools and using such AI-detection tools.

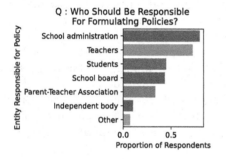 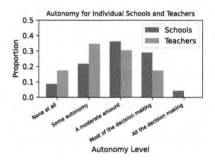

(a) Responsible entity for policy-making (Multiple selection allowed)

(b) Autonomy (decision making) for individual school and individual teachers

Fig. 3. Administrators' responses on responsible entity and autonomy

5   Conclusions and Discussion

This study aimed to address two primary research questions (RQs) regarding the policy landscape for Gen AI based tools like ChatGPT in education. RQ1 explored the current state of policies and their coverage, revealing a significant push, especially in higher education, to develop guidelines. Yet, these policies often fall short of addressing the unique challenges of technologies like LLMs. We observed that high schools are less inclined to work on policies than higher educational institutions. The necessity of policy development was universally recognized among administrators, driven by ethical considerations and student safety, though areas like algorithmic transparency and bias mitigation were less emphasized, indicating gaps in existing frameworks.

RQ2 investigated the perceived needs for future policy formulation and proposed recommendations for an policy framework. A preference for a collaborative, multi-stakeholder approach was evident, with the recognition that policies must be iterative and adaptable to keep pace with technological advances.

The findings indicate an active acknowledgment of AI and LLM's potential in education, alongside a nascent governance stage for their ethical and practical integration. Notably, the disparity in policy development between higher education and high schools-where about 40% lack any policy efforts-points to potential resource or awareness discrepancies. This study underscores the critical gaps in policy adequacy and the necessity for policies to evolve alongside educational technologies. It emphasizes the importance of multi-stakeholder dialogues for creating governance mechanisms that are robust yet flexible enough to accommodate rapid technological changes.

The study concludes that the ethical and responsible integration of AI in education demands the continuous evolution of policies, practices, and attitudes. The findings of this study suggest for strategic, ethical, and collaborative governance, highlighting the imperative for developing comprehensive, adaptable policies to navigate the advancing landscape of AI technologies in educational settings.

5.1   Future Work

This study has laid important groundwork in understanding the state and direction of policies related to AI and LLMs in educational settings. However, several avenues for future research remain. The disparity in policy development between higher education and high schools warrants a more granular investigation. Future studies could focus on identifying the barriers and facilitators that influence policy-making at these disparate educational levels, possibly extending the research to include primary schools. The evolving nature technology itself calls for longitudinal studies that can track changes in administrative attitudes, policy adequacy, and implementation efficacy over time.

5.2   Threats to Validity

Our survey was not validated and no evaluation of reliability was made. All respondents were from institutions based in the United States. Finally, generative AI is a fast-moving technology and attitudes and policies are likely also changing quickly. This work represents a snapshot of policies and attitudes in mid-2023.

References

1. Adams, C., Pente, P., Lemermeyer, G., Rockwell, G.: Ethical principles for artificial intelligence in k-12 education. Comput. Educ. Artif. Intell. **4**, 100131 (2023)
2. AI, H.: High-level expert group on artificial intelligence (2019)

3. Akgun, S., Greenhow, C.: Artificial intelligence in education: addressing ethical challenges in k-12 settings. AI Ethics 1–10 (2021). https://doi.org/10.1007/s43681-021-00096-7

4. Baidoo-Anu, D., Ansah, L.O.: Education in the era of generative artificial intelligence (AI): Understanding the potential benefits of chatgpt in promoting teaching and learning. J. AI **7**(1), 52–62 (2023)

5. Chan, C.K.Y.: A comprehensive AI policy education framework for university teaching and learning. Int. J. Educ. Technol. High. Educ. **20**(1), 1–25 (2023)

6. Chiu, T.K.: The impact of generative AI (GenAI) on practices, policies and research direction in education: a case of chatgpt and midjourney. Interact. Learn. Environ. 1–17 (2023)

7. Dwivedi, Y.K., et al.: "so what if chatgpt wrote it?" multidisciplinary perspectives on opportunities, challenges and implications of generative conversational AI for research, practice and policy. Int. J. Inf. Manage. **71**, 102642 (2023)

8. Filgueiras, F.: Artificial intelligence and education governance. Educ. Citizsh. Soc. Justice 17461979231160674 (2023)

9. Garshi, A., Jakobsen, M.W., Nyborg-Christensen, J., Ostnes, D., Ovchinnikova, M.: Smart technology in the classroom: a systematic review. prospects for algorithmic accountability. arXiv preprint arXiv:2007.06374 (2020)

10. Ghimire, A., Edwards, J.: Coding with AI: how are tools like chatgpt being used by students in foundational programming courses. In: International Conference on Artificial Intelligence in Education. Springer, Cham (2024)

11. Ghimire, A., Edwards, J.: Generative AI adoption in the classroom: a contextual exploration using the technology acceptance model (tam) and the innovation diffusion theory (IDT). In: 2024 Intermountain Engineering, Technology and Computing (IETC), IEEE (2024)

12. Ghimire, A., Prather, J., Edwards, J.: Generative AI in education: a study of educators' awareness, sentiments, and influencing factors. arXiv preprint arXiv:2403.15586 (2024)

13. Gillani, N., Eynon, R., Chiabaut, C., Finkel, K.: Unpacking the "black box" of AI in education. Educ. Technol. Soc. **26**(1), 99–111 (2023)

14. Halaweh, M.: Chatgpt in education: strategies for responsible implementation (2023)

15. Holmes, W., et al.: Ethics of AI in education: towards a community-wide framework. Int. J. Artif. Intell. Educ. 1–23 (2021). https://doi.org/10.1007/s40593-021-00239-1

16. Kooli, C.: Chatbots in education and research: a critical examination of ethical implications and solutions. Sustainability **15**(7), 5614 (2023)

17. Li, S., Gu, X.: A risk framework for human-centered artificial intelligence in education. Educ. Technol. Soc. **26**(1), 187–202 (2023)

18. Memarian, B., Doleck, T.: Fairness, accountability, transparency, and ethics (fate) in artificial intelligence (AI), and higher education: a systematic review. Comput. Educ. Artif. Intell. **5**, 100152 (2023)

19. Miao, F., Holmes, W., Huang, R., Zhang, H., et al.: AI and Education: a Guidance for Policymakers. UNESCO Publishing, Paris (2021)

20. Nguyen, A., Ngo, H.N., Hong, Y., Dang, B., Nguyen, B.P.T.: Ethical principles for artificial intelligence in education. Educ. Inf. Technol. **28**(4), 4221–4241 (2023)

21. for School Networking, C.: CoSN Issues Guidance on AI in the classroom, CoSN — cosn.org (2023). Accessed 06 Oct 2023

22. Sullivan, M., Kelly, A., McLaughlan, P.: Chatgpt in higher education: considerations for academic integrity and student learning (2023)
23. Uunona, G.N., Goosen, L.: Leveraging ethical standards in artificial intelligence technologies: a guideline for responsible teaching and learning applications. In: Handbook of Research on Instructional Technologies in Health Education and Allied Disciplines, pp. 310–330. IGI Global (2023)

Task Synthesis for Elementary Visual Programming in XLogoOnline Environment

Chao Wen[1(✉)], Ahana Ghosh[1], Jacqueline Staub[2], and Adish Singla[1]

[1] Max Planck Institute for Software Systems, Saarbrücken, Germany
{chaowen,gahana,adishs}@mpi-sws.org
[2] University of Trier, Trier, Germany
staub@uni-trier.de

Abstract. In recent years, the XLogoOnline programming platform has gained popularity among novice learners. It integrates the Logo programming language with visual programming, providing a visual interface for learning computing concepts. However, XLogoOnline offers only a limited set of tasks, which are inadequate for learners to master the computing concepts that require sufficient practice. To address this, we introduce XLOGOSYN, a novel technique for synthesizing high-quality tasks for varying difficulty levels. Given a reference task, XLOGOSYN can generate practice tasks at varying difficulty levels that cater to the varied needs and abilities of different learners. XLOGOSYN achieves this by combining symbolic execution and constraint satisfaction techniques. Our expert study demonstrates the effectiveness of XLOGOSYN. We have also deployed synthesized practice tasks into XLogoOnline, highlighting the educational benefits of these synthesized practice tasks.

Keywords: Programming Education · Task Synthesis · Block-Based Visual Programming · Logo Programming · XLogoOnline

1 Introduction

In recent years, XLogoOnline [1,2] has emerged as a new platform, which uniquely integrates the traditional Logo programming language [3] with the visual programming paradigm. XLogoOnline has been adopted in hundreds of educational courses and is utilized by tens of thousands of students every year [1,4]. XLogoOnline [1] is organized into four programming levels: *Mini*, *Midi*, *Maxi*, and *Mega*, each offering tasks tailored to specific age groups.

We focus on the *Mini* level (referred to as XLOMini), which centers around problem-solving, incorporating computing concepts like loops and basic mathematics. In XLOMini, learners are given tasks and a few code blocks, including basic commands `forward`, `back`, `left`, `right`, the state-based command `setpencolor`, and the control structure `repeat`. Each task contains a visual

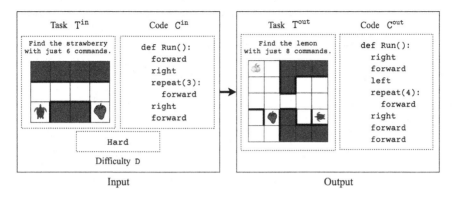

Fig. 1. Illustration of XLOGOSYN for reference task 87 from XLOMini [2]. XLO-GOSYN's input includes a reference task T^{in}, its solution code C^{in}, and the desired difficulty level D for a new practice task. The output includes a generated task T^{out} and its solution code C^{out} that satisfies the desired difficulty level w.r.t. the reference task.

grid with a turtle, descriptive text outlining the goal of the task, and code constraints. Learners must construct the code satisfying the code constraints and then execute it to direct the turtle's movement to achieve the goal.

However, XLOMini only offers a limited set of tasks. The scarcity of tasks may hinder learners from mastering computing concepts such as loops and basic mathematics, which require sufficient practice to deepen learners' understanding.

To address this, we propose XLOGOSYN, a technique for synthesizing high-quality practice tasks at a specific difficulty level. Figure 1 shows an example of the input and output of XLOGOSYN. Given a reference task, its solution code, and the desired difficulty, our technique can generate numerous varied tasks, addressing task scarcity and eliminating the need for manually crafting tasks.

Our key contributions include: (i) We develop XLOGOSYN, a technique to automatically generate high-quality tasks at a desired difficulty level; (ii) We conduct an expert study to show that XLOGOSYN can synthesize tasks with a quality close to those crafted by experts; (iii) We deploy synthesized practice tasks by XLOGOSYN on XLogoOnline and report on initial results highlighting the educational benefits.[1]

1.1 Related Work

Content Generation. Content generation has been explored in domains such as game content generation [5,6] and math problem generation [7,8]. Existing works often create a template filled with placeholders, followed by a search within this template for solutions [5–7]. To search for solutions, most works first encode the problem using a logic representation and then use Answer Set Programming

---

[1] Implementation of XLOGOSYN is publicly available at:
https://github.com/machine-teaching-group/aied2024-xlogo-tasksyn.

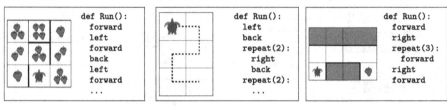

38 (Count): Collect exactly 10 strawberries.   54 (Draw): Draw the picture without "forward".   87 (Find): Find the strawberry with just 6 commands.

Fig. 2. Illustrative examples of reference tasks and their solution codes in XLOMini.

solvers [5,6] or Satisfiability Modulo Theories (SMT) solvers [9]. We employ SMT solvers; however, we further combine SMT solvers with symbolic code execution to navigate the search for solutions within the visual task grid.

Task Synthesis in Visual Programming. Recent works have studied task synthesis in visual programming domains. Most approaches utilize constraint solving techniques and symbolic code execution [10,11], with some recent works also incorporating reinforcement learning-based strategies to speed up the generation process [12]. However, these works have not considered the XLOMini domain, which is characterized by a diverse range of task types, grid elements, and state-based commands (e.g., setpencolor in XLOMini). Furthermore, existing techniques have not incorporated task difficulty into the generation process.

Large Language Models for Programming Task Synthesis. Recent works have explored large language models (LLMs) to synthesize tasks in programming domains [13,14]. Existing works have shown LLMs' potential in generating and solving tasks for text-based programming domains such as Python [15,16]. However, state-of-the-art LLMs still struggle in visual programming as they are unable to combine spatial, logical, and programming skills [12,17].

2   Preliminaries and Problem Setup

Task and Code Specifications in XLOMini. A task $T := (G, L, W)$ includes a goal G (the turtle's objective), code constraints L (constraints for solution code), and a visual grid world W (a two-dimensional grid with a turtle and elements like fruits and walls). To solve a task, a learner writes a *solution code* that meets the task's code constraints and achieves the goal when executed on the visual grid world. Figure 2 shows examples of tasks and their solution codes in XLOMini.

Levels of Task Difficulty. Given a reference task T^{in}, we define the relative difficulty D of a new practice task T^{out} as follows: (i) **Easy:** Solving T^{out} requires no additional concepts or steps beyond what is required for solving T^{in}; (ii) **Medium:** Solving T^{out} requires additional steps and understanding of concepts beyond those required to solve T^{in}; (iii) **Hard:** Solving T^{out} requires additional steps and understanding of concepts beyond those required to solve a medium task of T^{in}. In Sect. 3, we provide concrete criteria for these difficulty levels within the XLOMini domain, as part of the implementation details for XLOGOSYN.

Evaluation of Synthesized Tasks. We use a multidimensional rubric to assess the quality of a *synthesized* task-code pair $(\text{T}^{\text{out}}, \text{C}^{\text{out}})$ for a *reference* task-code pair $(\text{T}^{\text{in}}, \text{C}^{\text{in}})$ at a specified difficulty level D. The rubric consists of the following five metrics: (i) *Visual quality* evaluates the distinctiveness and aesthetic appeal of T^{out}; (ii) *Concept similarity* evaluates the alignment of concepts between the $(\text{T}^{\text{out}}, \text{C}^{\text{out}})$ and $(\text{T}^{\text{in}}, \text{C}^{\text{in}})$; (iii) *Elements utility* evaluates the usefulness of grid elements in T^{out}; (iv) *Code quality* evaluates the correctness of C^{out}; (v) *Difficulty consistency* evaluates the consistency of the difficulty of T^{out} w.r.t. the difficulty level D. Domain experts rate each metric on a three-point Likert scale: 0 for low quality, 0.5 for acceptable quality, and 1 for excellent quality. The *overall quality* of T^{out} is defined as the minimum rating across the five metrics.

Task Synthesis Objective. Given a reference task T^{in}, its corresponding solution code C^{in}, and a specified difficulty level D, our objective is to automatically synthesize a set of high-quality task-code pairs $\{(\text{T}^{\text{out}}, \text{C}^{\text{out}})\}$ (see Fig. 1).

3 Our Task Synthesis Technique: XLOGOSYN

In this section, we provide an overview of our task synthesis technique XLO-GOSYN. Figure 3a illustrates the three stages of XLOGOSYN.

Stage 1: Generation of Code, Code Constraints, and Goal. Figure 3a illustrates the first stage of XLOGOSYN. This stage first creates a template for each of C^{in}, L^{in}, and G^{in} [10,11]. The templates express the high-level structures while leaving low-level details unspecified with placeholders. Then, we fill in the placeholders with specific values using an SMT-based constraint solver [9]. For example, in Figs. 3b, 3c, and 3d, placeholder [B1] is replaced with "right" and [fruit_type] with "lemon". During instantiation, we also incorporate SMT constraints based on the input difficulty D. The difficulty D controls the difficulty of the generated outputs. At D = Easy, we maintain the original code length without extra constraints. At D = Medium, we allow code sequences up to 2 commands longer than C^{in}. At D = Hard, the code sequences need to be exactly 2 commands longer than C^{in} and we allow an extra code constraint. Moreover, G^{out} is allowed to differ from G^{in} only at D = Hard. After this stage, we obtain $(\text{C}^{\text{out}}, \text{L}^{\text{out}}, \text{G}^{\text{out}})$, aligned with the specified difficulty level D.

Stage 2: Generation of Visual Grid World. This stage synthesizes a visual grid world W^{out} using the outputs $(\text{C}^{\text{out}}, \text{L}^{\text{out}}, \text{G}^{\text{out}})$ from the previous stage. First, we create an empty grid and randomly initialize the turtle's starting location and direction (see Fig. 3a). Then, we symbolically execute code C^{out} on the empty grid using an emulator, producing a trajectory of visited grid cells $(v_1, v_2, \cdots, v_n)$, highlighted in red in Fig. 3a. Next, we use the goal G^{out} and the trajectory to formulate SMT constraints concerning the placement of various grid elements such as fruits, walls, etc. For example, if the goal is "Find the strawberry", a

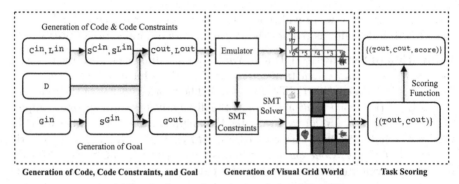

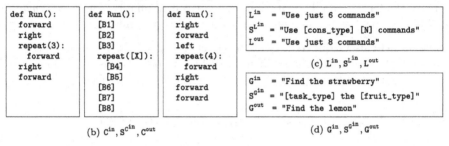

(a) Stages of our task synthesis technique XLogoSyn

def Run():	def Run():	def Run():	
forward	[B1]	right	L^{in} = "Use just 6 commands"
right	[B2]	forward	$S^{L^{in}}$ = "Use [cons_type] [N] commands"
repeat(3):	[B3]	left	L^{out} = "Use just 8 commands"
forward	repeat([X]):	repeat(4):	
right	[B4]	forward	(c) $L^{in}, S^{L^{in}}, L^{out}$
forward	[B5]	right	
	[B6]	forward	G^{in} = "Find the strawberry"
	[B7]	forward	$S^{G^{in}}$ = "[task_type] the [fruit_type]"
	[B8]		G^{out} = "Find the lemon"

(b) $C^{in}, S^{C^{in}}, C^{out}$

(d) $G^{in}, S^{G^{in}}, G^{out}$

Fig. 3. (a) illustrates the stages of XLogoSyn. (b)–(d) show examples of different components after applying these stages to Fig. 1 (Input). Specifically, (b) shows the input code, its sketch, and the output code, where B1, B2, $\cdots$, B8 $\in$ {None, left, right, ...} and X $\in$ {2, 3, ...}. (c) shows the input code constraints, its sketch, and the output code constraints, where cons_type $\in$ {AtMost, Exactly, StartBy, None} and N $\in$ {1, 2,...}. (d) shows the input goal, its sketch, and the output goal, where task_type $\in$ {Find, FindOnly, FindForbid, ...} and fruit_type $\in$ {strawberry, lemon}.

strawberry must be placed in the final grid cell v_n. An SMT solver solves these constraints to generate a visual grid world W^{out}. Finally, we merge W^{out} with the outputs from the previous stage (C^{out}, L^{out}, G^{out}), to obtain output task-code pairs (T^{out}, C^{out}), where $T^{out} = (G^{out}, L^{out}, W^{out})$.

Stage 3: Task Scoring. In the final stage, we apply a scoring function to evaluate the quality of task-code pairs {(T^{out}, C^{out})}, inspired by scoring functions considered in literature on task synthesis for visual programming [10,11].

4 Evaluation of XLogoSyn Using Expert Study

Techniques Evaluated. We compare XLogoSyn with three different techniques. Each technique accepts an input specification (T^{in}, C^{in}, D), which includes the reference task, its corresponding solution code, and the desired difficulty level respectively, and generates the output task. We consider the following baselines:

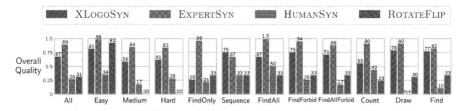

Fig. 4. The performance of our technique XLOGOSYN and three baseline techniques. On the x-axis, we present the aggregated results over all scenarios (All), followed by aggregated results based on task difficulty (Easy, Medium, and Hard) and based on 8 task types. The y-axis presents the score for *overall quality*. XLOGOSYN demonstrates performance close to EXPERTSYN, and surpasses both HUMANSYN and ROTATEFLIP.

1. EXPERTSYN involves an expert in XLOMini carefully crafting a task T^{out} and its code C^{out} based on the input specification.
2. HUMANSYN uses a collection of 1,331 user-created tasks from XLogoOnline to create tasks and solution codes [1]. During the creation of this collection, users created their tasks without specific reference tasks to guide their synthesis process. Given this collection and the input specification, this technique generates a task and its solution code as follows: the expert (the same as the one in EXPERTSYN) selects a task T^{out} that matches the input specification from the collection, considering only tasks of the same type as T^{in}. After selecting T^{out}, the expert crafts an optimal solution code C^{out} for T^{out}.
3. ROTATEFLIP generates tasks by applying rotations and flips to the input task's grid. For D = Easy, it rotates the grid 90° counterclockwise without altering C^{out}. For D = Medium, it performs a mirror flip of the grid and adjusts C^{out} to match the flipped grid. For D = Hard, it performs both rotation and mirror flip for T^{out} and adjusts C^{out} as needed.

Experimental Setup. We selected 24 reference tasks from XLOMini covering a broad range of concepts. For these reference tasks, we generated practice tasks at Easy, Medium, and Hard levels with each technique. We considered 288 scenarios (24 reference tasks × 3 difficulty levels × 4 techniques). However, for HUMANSYN, 15 scenarios of Draw type were missing from the collection of user-created tasks (5 Draw tasks × 3 difficulty levels), resulting in a final count of 273 scenarios. Two independent human evaluators, not involved in EXPERTSYN or HUMANSYN, scored each scenario based on the rubric in Sect. 2. The final score for a scenario was derived by averaging the scores provided by the two evaluators aggregated over different dimensions. During evaluation, the origin of each scenario was hidden and scenarios were presented in a randomized order.

Results. We first checked the inter-rater reliability of the two human evaluators using the quadratic-weighted Cohen's kappa score [18], achieving a near-perfect agreement of 0.84. Next, we compare XLOGOSYN's performance w.r.t. baseline techniques and report statistical significance using χ^2-test [19].

The results are shown in Fig. 4. XLOGOSYN has an overall quality score of 0.67 and is: (i) significantly lower than EXPERTSYN that has a score of 0.89 ($\chi^2 = 38.8; p < 0.01$); (ii) significantly higher than HUMANSYN that has a score of 0.26 ($\chi^2 = 75.5; p < 0.01$); and (iii) significantly higher than ROTATEFLIP that has a score of 0.31 ($\chi^2 = 125.2; p < 0.01$). All techniques show a performance decline with increasing task difficulty, indicating that generating more difficult tasks remains challenging for all, including experts. We found that HUMANSYN struggles because users create tasks without any references and tend to incorporate diverse concepts in a task. ROTATEFLIP is effective in generating Easy tasks; however, its performance drops to zero for Medium and Hard tasks, indicating that rotations and flips are not sufficient for generating tasks of higher difficulty levels.

5   Deployment on XLogoOnline and Initial Results

In this section, we present the current status of our deployment and report on initial results. We have deployed the synthesized practice tasks into the XLogoOnline platform. For each reference task on the platform, XLOGOSYN synthesized 10 tasks across three difficulty levels: 3 Easy, 4 Medium, and 3 Hard. After attempting a reference task, learners can choose to attempt the synthesized practice tasks or move to the *next reference task* by clicking the "Next" button on the platform. Preliminary statistics, based on data collected on XLogoOnline from November 2023 to March 2024, show around $13,000$ visits on the platform and over $600,000$ execution attempts to solve tasks. Out of these attempts, 87% were on reference tasks, and 13% were on practice tasks synthesized by our technique.

Next, we analyze the data to investigate the educational benefits of synthesized practice tasks. We aim to answer the following research question: *Do synthesized practice tasks enhance learners' success rates on the next reference tasks?* To this end, we analyze two groups: (i) The first group consisted of learners who failed a reference task and then moved directly to the next reference task without attempting any synthesized practice tasks; (ii) The second group consisted of learners who failed a reference task, then attempted the synthesized practice tasks of this reference task, and finally moved to the next reference task.

For comparison, we define the *success rate* of a group of learners w.r.t a task as the percentage of learners who successfully solved a task at least once. We calculate success rates for both groups across 36 pairs of consecutive reference tasks on the platform. The first group comprises $4,477$ learners, with a success rate of 49.2%. The second group includes 75 learners, with a higher success rate of 68.0%. These initial results indicate that synthesized practice tasks have the potential to enhance success rates on the next reference task.

6   Limitations and Future Work

In this section, we discuss some limitations of our current work and ideas to tackle them in the future. First, we specify the difficulty of the synthesized tasks using pre-defined rules, which may not align with learners' perception of task difficulty. In the future, it would be important to derive a more refined notion of task difficulty by analyzing learners' interactions with the platform. Second, XLogoSyn does not incorporate a learner's code during task synthesis, which limits its effectiveness in personalizing practice tasks. It would be interesting to extend our technique to generate tasks personalized to the learner's misconceptions on the platform. Third, in our current implementation, generating a single high-quality task using our technique is time-consuming as it requires synthesizing and selecting from a large pool of tasks. In future work, it would be interesting to develop learning-based strategies and explore generative AI models to accelerate the synthesis process while maintaining the high quality of the synthesized tasks.

Acknowledgments. Funded/Co-funded by the European Union (ERC, TOPS, 101039090). Views and opinions expressed are however those of the author(s) only and do not necessarily reflect those of the European Union or the European Research Council. Neither the European Union nor the granting authority can be held responsible for them.

References

1. Hromkovič, J., Serafini, G., Staub, J.: XLogoOnline: a single-page, browser-based programming environment for schools aiming at reducing cognitive load on pupils. In: Dagiene, V., Hellas, A. (eds.) ISSEP 2017. LNCS, vol. 10696, pp. 219–231. Springer, Cham (2017). https://doi.org/10.1007/978-3-319-71483-7_18
2. XLogoOnline. XLogoOnline Platform (2023). https://xlogo.inf.ethz.ch/
3. Pea, R.D.: Logo programming and problem solving (1987)
4. Staub, J.: Logo environments in the focus of time. Bull. EATCS **133** (2021). https://dblp.org/rec/journals/eatcs/Staub21.html?view=bibtex
5. Smith, A.M., Mateas, M.: Answer set programming for procedural content generation: a design space approach. IEEE Trans. Comput. Intell. AI Games **3**(3), 187–200 (2011)
6. Park, K., Mott, B., Min, W., Wiebe, E., Boyer, K.E., Lester, J.: Generating game levels to develop computer science competencies in game-based learning environments. In: Bittencourt, I.I., Cukurova, M., Muldner, K., Luckin, R., Millán, E. (eds.) AIED 2020. LNCS (LNAI), vol. 12164, pp. 240–245. Springer, Cham (2020). https://doi.org/10.1007/978-3-030-52240-7_44
7. Polozov, O., O'Rourke, E., Smith, A.M., Zettlemoyer, L., Gulwani, S., Popovic, Z.: Personalized mathematical word problem generation. In: IJCAI (2015)
8. Alvin, C., Gulwani, S., Majumdar, R., Mukhopadhyay, S.: Synthesis of geometry proof problems. In: AAAI (2014)
9. de Moura, L., Bjørner, N.: Z3: an efficient SMT solver. In: Ramakrishnan, C.R., Rehof, J. (eds.) TACAS 2008. LNCS, vol. 4963, pp. 337–340. Springer, Heidelberg (2008). https://doi.org/10.1007/978-3-540-78800-3_24
10. Ahmed, U.Z., et al.: Synthesizing tasks for block-based programming. In: NeurIPS (2020)

11. Ghosh, A., Tschiatschek, S., Devlin, S., Singla, A.: Adaptive scaffolding in block-based programming via synthesizing new tasks as pop quizzes. In: Rodrigo, M.M., Matsuda, N., Cristea, A.I., Dimitrova, V. (eds.) Artificial Intelligence in Education. AIED 2022, vol. 13355, pp. 28–40. Springer, Cham (2022). https://doi.org/10.1007/978-3-031-11644-5_3

12. Padurean, V.-A., Tzannetos, G., Singla, A.: Neural task synthesis for visual programming. Trans. Mach. Learn. Res. (2023). https://www.jmlr.org/tmlr/papers/bib/aYkYajcJDN.bib

13. OpenAI. ChatGPT (2023). https://openai.com/blog/chatgpt

14. Denny, P., et al.: Generative AI for education (GAIED): advances, opportunities, and challenges. CoRR, abs/2402.01580 (2024)

15. Sarsa, S., Denny, P., Hellas, A., Leinonen, J.: Automatic generation of programming exercises and code explanations using large language models. In: ICER (2022)

16. Phung, T., et al.: Generative AI for programming education: benchmarking Chat-GPT, GPT-4, and human tutors. In: ICER V.2 (2023)

17. Singla, A.: Evaluating ChatGPT and GPT-4 for visual programming. In: ICER V.2 (2023)

18. Cohen, J.: A coefficient of agreement for nominal scales. Educ. Psychol. Measur. **20**(1), 37–46 (1960)

19. Cochran, W.G.: The $\chi 2$ test of goodness of fit. Ann. Math. Stat. **23**(3), 315–345 (1952)

Towards Fair Detection of AI-Generated Essays in Large-Scale Writing Assessments

Yang Jiang(✉) (iD), Jiangang Hao, Michael Fauss, and Chen Li

ETS, 660 Rosedale Rd, Princeton, NJ 08540, USA
yjiang002@ets.org

Abstract. The release of ChatGPT in late 2022 triggered revolutionary advances in generative artificial intelligence (AI). With the increasing adoption of these AI tools for generating high-quality texts in response to writing assignments or assessments in educational settings, it is imperative to detect AI-generated essays to maintain academic integrity and test security. Although several automated detectors targeting AI-generated texts have been developed, an effective detector must not only exhibit high detection accuracy but also not display bias across demographic groups. In this study, we compare the effectiveness of three strategies—balancing training data, removing sensitive features, and adjusting thresholds—in mitigating the detector bias concerning native and non-native English speakers in the context of a large-scale writing assessment. Leveraging a dataset comprised of 85,567 essays, our results show that these strategies can reduce the bias in detecting AI-generated essays to varying degrees without significantly compromising detection accuracy.

Keywords: Fairness · ChatGPT · Test Security · Bias · Writing Assessment · AI

1 Introduction

Artificial Intelligence (AI) has revolutionized various aspects of life and work, including education. The advent of Large Language Models (LLMs) like ChatGPT [1] has presented both opportunities and challenges in educational settings. These models, designed to generate human-like text, have found widespread use among students and teachers. However, their potential misuse, particularly in assessments, has sparked debates and raised concerns [2]. The ability of LLMs to quickly and effortlessly produce high-quality essays poses a threat to the integrity of academic assessments, especially those conducted remotely or at home.

Previous research has demonstrated that distinguishing between AI-generated and human-written texts is difficult if not impossible for most people, including teachers, raters, and other trained experts [3]. Hence, there is an urgent need for automated detectors of AI-generated essays. Since the launch of ChatGPT, several detectors targeting ChatGPT-generated text have been developed [4, 5]. These methods have shown varying levels of detection accuracy, with some achieving over 99% [4, 5]. However, in addition to the overall accuracy, the fairness of LLM detectors also poses a significant concern.

© The Author(s), under exclusive license to Springer Nature Switzerland AG 2024
A. M. Olney et al. (Eds.): AIED 2024 Workshops, CCIS 2151, pp. 317–324, 2024.
https://doi.org/10.1007/978-3-031-64312-5_38

The fairness of AI methods has been studied extensively in various contexts; see [6], for example, for a general introduction. In an educational context, useful overviews of the types, causes, and possible mitigation strategies of algorithmic biases can be found in [7, 8]. More recently, researchers evaluated the performance of several publicly available detectors of AI-generated texts and suggested there was a detection bias that disadvantaged non-native English speakers [9], while some subsequent analysis with a larger dataset showed the bias was actually in the opposite direction, that is, disadvantaging native English speakers [5]. Obviously, detector biases must be addressed before putting them into operational use to ensure fair treatment of all test takers and learners.

The present study aims to explore ways and establish benchmarks for mitigating the bias in detecting ChatGPT-generated essays against native or non-native English speakers in the context of a large-scale writing assessment. Specifically, we targeted three possible origins of the biases: Imbalanced representations of demographic groups in the training data [10], some sensitive features that are drastically different for different demographic groups [11], and the inappropriate selection of the decision thresholds of the underlying machine learning models [12]. We evaluated the effectiveness of applying the corresponding mitigating strategies, including balancing demographic groups, removing sensitive features, and adjusting the decision thresholds, in mitigating the detection bias concerning native and non-native English speakers. We used the high-performing GPT-4 [1] to generate essays throughout this study.

2 Methods

2.1 Data

We used 85,567 human-written and AI-generated essays in response to 50 writing prompts from a large-scale standardized writing assessment in this study. We selected a total of 75,567 human-written essays submitted between 08/01/2022 and 11/30/2022 (i.e., before ChatGPT was released). Among these essays, 30.6% were submitted by test takers who identified themselves as native English speakers (NSs, $N = 23,143$) and 69.4% of the essays were submitted by those who identified themselves as non-native English speakers (NNSs). In addition, we generated 200 essays for each of the 50 writing prompts using GPT-4 [1], leading to a total of 10,000 AI-generated essays.

2.2 Detectors of AI-Generated Essays

We randomly sampled 10,000 essays (200 essays per prompt) from the pool of 75,567 human-written responses (labeled as "0") and combined them with the 10,000 GPT-4-generated essays (labeled as "1") to form a dataset of 20,000 essays for training and validating our AI essay detectors. This dataset has a balanced distribution of class ("0"s and "1"s), ensuring an equal representation of human-written and AI-generated essays. For each essay, we extracted ten NLP features to represent the essay using the e-rater® engine [13], an automated scoring engine that extracts linguistically motivated writing features, covering grammar, usage, mechanics, organization, development, word length, word choice, collocation and preposition, discourse coherence, and sentence variety. E-rater® has been in operational use for automated scoring with well-established accuracy [13] and e-rater features have shown success in detecting auto-generated essays [14].

Leveraging the e-rater features, we developed detectors for AI-generated essays using linear support vector machine (SVM) and gradient boosting machine (GBM) models, which have been shown to be effective in detecting AI-generated texts [5]. Among the 20,000 essays, 80% were used for training, and 20% were held out for testing. The models were trained and fine-tuned using 5-fold cross validation on the training set. Models with optimized hyper-parameters were tested on the 20% test set, and their performance was evaluated through standard metrics, such as classification accuracy, false positive rate (FPR), and false negative rate (FNR). The Python machine-learning framework scikit-learn was used for model training and evaluation [15]. More details about the procedures for building the detectors can be found in [5].

2.3 Detector Bias Mitigation Techniques

In this study, we explore the effectiveness of three techniques in mitigating the bias of the detector: 1) Demographic group balancing, 2) sensitive feature removal, and 3) threshold optimization. Among them, 1) and 2) were implemented in the training data construction stage prior to model training, while 3) was implemented during the model inference stage.

Balancing Demographic Groups. As described in Sect. 2.1, there were more than twice as many submissions by NNSs as those written by NSs. The data imbalance could be a potential source of bias in the predictive models [10]. To mitigate the bias accordingly, we created a balanced dataset by randomly sampling an equal number of NS (n = 100) and NNS (n = 100) essays for each writing prompt, leading to 5,000 NS and 5,000 NNS human-written essays that were combined with the AI-generated samples for model training and evaluation.

Removing Sensitive Features. Some linguistic features show significant polarity concerning the NS and NNS demographic groups, which could introduce bias to the machine learning models. We conducted Mann Whitney U tests to compare the linguistic features between the NS and NNS subgroups and identify the sensitive features that were significantly different between groups. Cohen's D was computed as a measure of effect size to infer the magnitude of difference, and linguistic features were ranked from high to low based on effect size. *Cohen's D* = 0.2, *D* = 0.5, and *D* = 0.8 were used as thresholds to determine small, medium, and large effects, respectively. Stepwise elimination was applied to explore the effects of removing the sensitive features on detection accuracy and bias.

Adjusting Thresholds. Separate decision thresholds for the NS and NNS groups that optimize the fairness metric were created using the Fairlearn Python package [16]. Specifically, all possible combinations of thresholds for the two subgroups were tested and the best combination that minimizes the false positive rate parity across groups and optimizes the accuracy was selected. More details can be found in [12, 16]. Note that Fairlearn requires the "native language" attribute from all data. In our data, while the human-written essays correspond to native and non-native English speakers, the AI-generated ones were created using the gpt-4 API and do not have the native language label. To run the module, we randomly assigned half of the AI-generated essays as NS and half as submitted by NNS, assuming that the two subgroups have a similar likelihood of cheating. However, this may not be the case, and future research involves testing the effects of this treatment on findings.

2.4 Evaluating Model Fairness

We applied the trained models with and without mitigation strategies implemented to the "independent" data set that was set aside from model training and testing to evaluate the bias. This independent data set includes 65,567 human-written essays (after excluding the 10,000 human-written essays used for model training and testing). In educational assessments, especially high-stakes ones, the costs of misclassifying a human-written essay as LLM-generated is high and can have significant consequences on individual test takers. Therefore, we conducted Chi-square tests to compare the FPR between native and non-native language subgroups. Cramer's V was computed as a measure of effect size to infer model fairness [17], with $V = 0.1, 0.3$, and 0.5 used as thresholds to determine small, medium, and large effects, respectively. In addition, False Positive Rate Parity (FPRP) assesses whether the FPR is similar across different groups by measuring the difference in FPR between subgroups and was computed as an indicator of effect size as well.

3 Results

3.1 Detector Performance

The overall performance of the detectors is presented in Table 1. In general, the linguistic features were effective in distinguishing AI-generated essays from human-written ones with an accuracy above 99% and a false positive rate below 1.2%. These results are consistent with previous findings [5].

3.2 Detector Bias

To evaluate detector bias, we applied the trained models to the independent dataset. Table 2 shows the metrics characterizing the detection bias under different conditions. For the models without any bias mitigation, there is a significantly higher FPR among the NS samples than among the NNS samples (SVM: 1.94% vs. 0.53%; GBM: 1.35% vs. 0.35%), though the effect sizes, Cramer's V and FPRP, were small or negligible (e.g., below 0.1 for Cramer's V). In the following sections, we report the effectiveness of the bias mitigation techniques.

Balancing Demographic Groups. For models trained after implementing the "balancing demographic group" mitigation procedure, FPRP decreased compared to the baseline models (0.07% lower for SVM and 0.21% lower for GBM, see Table 2). This suggested that the mitigation technique helps to reduce the bias, though not much.

Table 1. Overall model performance.

Classifier	Mitigation Technique	Accuracy	FPR	FNR
Support Vector Machine	Unmitigated	99.3%	0.86%	0.55%
	Balancing	99.1%	1.11%	0.65%
	Adjusting Threshold	99.3%	0.86%	0.50%
	Balancing + Threshold	99.1%	1.11%	0.65%
Gradient Boosting Machine	Unmitigated	99.5%	0.65%	0.45%
	Balancing	99.4%	0.76%	0.40%
	Adjusting Threshold	99.5%	0.65%	0.45%
	Balancing + Threshold	99.4%	0.76%	0.40%

Adjusting Thresholds. Table 2 also included the results for models with thresholds adjusted for subgroups. For SVM, adjusting the thresholds lowered the FPRP and Cramer's V dropped from 0.07 to 0.01. It is worth noting that the direction of difference also changed when we applied threshold optimization. For GBM, on the other hand, the postprocessing mitigation method only resulted in a 0.03% lower FPRP. If we combine the two mitigation procedures, the bias is not much different compared with using the "balancing" or "adjusting threshold" technique alone.

Table 2. Bias of detectors with and without mitigation techniques implemented.

Classifier	Mitigation technique	FPR_{NS}	FPR_{NNS}	FPRP	Cramer's V
SVM	Unmitigated	1.94%	0.53%	1.41%	0.07**
	Balancing	1.77%	0.43%	1.34%	0.07**
	Adjusting Threshold	0.98%	1.18%	−0.21%	0.01*
	Balancing + Threshold	1.19%	0.98%	0.21%	0.01*
GBM	Unmitigated	1.35%	0.35%	1.00%	0.06**
	Balancing	1.07%	0.29%	0.79%	0.05**
	Adjusting Threshold	1.31%	0.34%	0.97%	0.06**
	Balancing + Threshold	1.07%	0.29%	0.78%	0.05**

Note. $* p < .05$, $** p < .001$. FPR for NS and NNS groups, false positive rate parity (FPRP), and effect sizes (*Cramer's V*) of Chi-square tests comparing the false positive rate between native and non-native writing samples were reported,

Removing Sensitive Features. Comparisons of the linguistic features indicated that there were significant differences between NS and NNS for all features except for word length ($D < .01$, $p = .781$). This is not surprising as previous literature showed that the written products of NS and NNS have consistently different linguistic characteristics (e.g., fewer grammatical errors for NS) and quality [18]. Among the sensitive features, "Usage" showed a large effect size ($D = -0.96$, $p < .001$), three features showed medium effect sizes (Development: $D = -0.61$, $p < .001$; Mechanics: $D = -0.59$, $p < .001$; Grammar: $D = -0.55$, $p < .001$), and five showed small or close to small effect sizes (Sentence variety: $D = -0.42$, $p < .001$; Organization: $D = 0.36$, $p < .001$; Collocation and preposition: $D = -0.27$, $p < .001$; Discourse coherence: $D = -0.22$, $p < .001$; Word choice: $D = -0.17$, $p < .001$). Note these results largely align with the importance of features in predictive models where we used e-rater features to predict the native language attribute.

Figure 1 shows the overall accuracy of the detectors and the FPR within each subgroup when removing the sensitive features stepwise based on their effect size. The model accuracy dropped from above 99% (10-feature model) to 85.3% (SVM) and 83.2% (GBM) when only the non-sensitive word length feature was used in detectors, while the bias as measured by the difference between the FPRs was not lowered, suggesting that the mitigation strategy is not effective for the current application.

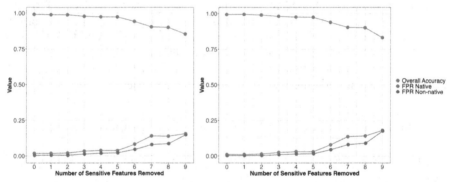

Fig. 1. Overall accuracy and FPR of SVM (left) and GBM (right) detectors with top n (n ∈ [0,9]) sensitive features removed.

4 Discussion

This paper evaluated three bias mitigation strategies for detecting AI-generated essays: Balancing demographic groups, removing sensitive features, and adjusting decision thresholds. Balancing demographic groups slightly decreased the bias, which could be due to the dataset's modest imbalance (≈1:2 ratio of native and non-native speakers), or the already small detector bias (Cramer's V ≈ 0.06) making further bias reduction difficult. Our findings largely align with prior research that balancing demographic groups alone may not ensure fair predictions [10].

Removing sensitive features did not lead to significant reductions in the bias. Moreover, removing too many sensitive features not only decreased the detection accuracy but also increased the bias. This could be due to the correlations between the sensitive features and other features in the dataset [16]. In our ongoing work, we are exploring the removal of features by taking into account the correlations among features.

Adjusting the thresholds turns out to be capable of effectively reducing the bias while maintaining the detection accuracy. However, we caution that the practice of using groupwise thresholds could be controversial in itself since it implies that the same response could be flagged or not, depending on who submitted it [19].

4.1 Implications and Directions for Future Research

Findings from this empirical study showed how different bias mitigation strategies affect the bias and accuracy for detecting AI-generated essays, which has important implications for ensuring academic integrity and test security. This study also contributes to the literature on the ethical and fair use of AI methods in education, a critical topic in the field of AIED. Our findings underscore the importance of evaluating model fairness as an integral part of selecting machine learning models.

This study has several limitations that link to the next steps. First, it is important to note that the difference in detection FPR between NSs and NNSs could be related to the differences in writing quality between the groups [5]. Previous studies showed that NSs tend to write better quality essays, which typically align with the characteristics of AI-generated essays (e.g., few grammar errors), making them less distinguishable [18]. As a next step, we will compare FPR after controlling for essay quality, i.e., the scores. Second, we examined only two machine learning algorithms, SVM and GBM, which have shown previous success in LLM detection. It will be interesting to know how these bias mitigation techniques behave for other algorithms such as random forest, logistic regression, and deep learning models, which is what we are currently exploring. In our previous study [5], we found other features, such as the lower-level e-rater features and perplexity features (characterizing how likely a sequence of words was generated by a given language model) are also effective in detecting AI-generated essays. As a next step, we will test the effectiveness of the mitigation approaches on models trained using different feature sets. In addition, other mitigation strategies such as in-process techniques like adversarial learning will be tested, and the sources of unfairness will be delved into. It is also important to point out that this study uses unedited AI-generated responses. However, in real life test takers may use only part of the AI-generated text and write the remaining by themselves, or make edits on top of the AI-generated text, all of which will make it more challenging to distinguish them from human-written text. Finally, as the writing of this paper, many powerful large language models, such as Gemini from Google and Claude 3 from Anthropic, appeared and claimed better performance than GPT-4 used in this work. So, looking at how the detectors behave with respect to these new models is also among our next steps.

References

1. OpenAI: GPT-4 technical report (2023)
2. Cotton, D.R.E., Cotton, P.A., Shipway, J.R.: Chatting and cheating: ensuring academic integrity in the era of ChatGPT. Innov. Educ. Teach. Intern. 1–12 (2023). https://doi.org/10.1080/14703297.2023.2190148

3. Clark, E., August, T., Serrano, S., Haduong, N., Gururangan, S., Smith, N.A.: All that's 'human' is not gold: evaluating human evaluation of generated text. In: Proceedings of the 59th Annual Meeting of the Association for Computational Linguistics and the 11th International Joint Conference on Natural Language Processing (Vol. 1: Long Papers), pp. 7282–7296. Association for Computational Linguistics (2021)
4. Yan, D., Fauss, M., Hao, J., Cui, W.: Detection of AI-generated essays in writing assessments. Psychol. Test Assess. Model. **65**, 125–144 (2023)
5. Jiang, Y., Hao, J., Fauss, M., Li, C.: Detecting ChatGPT-generated essays in a large-scale writing assessment: is there a bias against non-native English speakers? Comput. Educ. 105070 (2024). https://doi.org/10.1016/j.compedu.2024.105070
6. Barocas, S., Hardt, M., Narayanan, A.: Fairness and machine learning: limitations and opportunities. MIT Press (2023)
7. Baker, R.S., Hawn, A.: Algorithmic bias in education. Int. J. Artif. Intell. Educ. **32**, 1052–1092 (2022)
8. Kizilcec, R.F., Lee, H.: Algorithmic fairness in education. In: The Ethics of Artificial Intelligence in Education. pp. 174–202. Routledge, New York (2022)
9. Liang, W., Yuksekgonul, M., Mao, Y., Wu, E., Zou, J.: GPT detectors are biased against non-native English writers. Patterns **4**, 100779 (2023). https://doi.org/10.1016/j.patter.2023.100779
10. Deho, O.B., Joksimovic, S., Liu, L., Li, J., Zhan, C., Liu, J.: Assessing the fairness of course success prediction models in the face of (un)equal demographic group distribution. In: Proceedings of the Tenth ACM Conference on Learning @ Scale, pp. 48–58. ACM, New York, NY, USA (2023)
11. Wang, T., Zhao, J., Yatskar, M., Chang, K.-W., Ordonez, V.: Balanced datasets are not enough: estimating and mitigating gender bias in deep image representations. In: Proceedings of the IEEE/CVF International Conference on Computer Vision. pp. 5310–5319 (2019)
12. Hardt, M., Price, E., Srebro, N.: Equality of opportunity in supervised learning. In: 30th Conference on Neural Information Processing Systems (NIPS2016), pp. 3315–3323 (2016)
13. Attali, Y., Burstein, J.: Automated essay scoring with e-rater v 2. J. Technol. Learn. Assess. **4**, 3–30 (2006)
14. Cahill, A., Chodorow, M., Flor, M.: Developing an e-rater advisory to detect babel-generated essays. J. Writ. Anal. **2**, 203–224 (2018). https://doi.org/10.37514/JWA-J.2018.2.1.08
15. Pedregosa, F., et al.: Scikit-learn: machine learning in Python. J. Mach. Learn. Res. **12**, 2825–2830 (2011)
16. Bird, S., et al.: Fairlearn: a toolkit for assessing and improving fairness in AI (2020)
17. Fritz, C.O., Morris, P.E., Richler, J.J.: Effect size estimates: Current use, calculations, and interpretation. J. Exp. Psychol. Gen. **141**, 2–18 (2012). https://doi.org/10.1037/a0024338
18. Flor, M., Futagi, Y., Lopez, M., Mulholland, M.: Patterns of misspellings in L2 and L1 English: a view from the ETS Spelling Corpus 1. In: Learner Corpus Research: LCR2013 Conference Proceedings (2015)
19. Long, R.: Fairness in machine learning: against false positive rate equality as a measure of fairness. J. Moral. Philos. **19**, 49–78 (2021). https://doi.org/10.1163/17455243-20213439

Doctoral Consortium

Designing and Evaluating Generative AI-Based Voice-Interaction Agents for Improving L2 Learners' Oral Communication Competence

Liqun He$^{(\boxtimes)}$ ⓘ, Manolis Mavrikis ⓘ, and Mutlu Cukurova ⓘ

UCL Knowledge Lab, University College London, London, UK
{liqun.he,m.mavrikis,m.cukurova}@ucl.ac.uk

Abstract. Oral communication competence is an essential skill in 21st century society and workplaces. However, despite years of English learning, non-native learners often struggle to communicate with others in English fluently and confidently. This study aims to explore the integration of language teaching theories and AI technologies to design and evaluate a Large Language Model (LLM)-enabled voice-based agent. Specifically, the *design* integrates Communicative Language Teaching (CLT) theories to instruct generative models. The *evaluation* adopts an experimental research design, wherein we evaluate the educational effectiveness of using the voice agent among 50 Chinese tertiary students. Pre- and post-assessments will be conducted to analyse learners' oral proficiency development. To understand the mechanisms behind, a widely used coding scheme will be adapted to describe dialogue acts (DAs) at the speaking-turn level. Pattern mining algorithms will be employed to identify prominent DA patterns associated with students' improvement in oral proficiency, shedding light on effective and ineffective interactional strategies employed by AI agents for further design considerations. This research aims to contribute both locally effective solutions for enhancing L2 learners' oral communicative competence and generalisable design principles for AI applications in language teaching.

Keywords: Generative AI · Voice Agent · Computer-Assisted Language Learning · Second Language Speech

1 Problem Statement

Oral communication competence is an essential skill in 21st century society and workplaces. However, despite years of language learning, millions of English Second Language (L2) learners struggle to communicate with others in English fluently and confidently. Based on a survey conducted in University of Toledo, a large number of international students felt that their language needs were not adequately addressed, and expected more effective and relaxing ways of improving their oral (not written) communication skills by communicating with their native peers [1]. Similarly, [2] found that non-native students often need more time to find out how to say what they want to say in international classrooms. Despite accumulating substantial language knowledge, they struggle to retrieve and apply it effectively during communication.

A. M. Olney et al. (Eds.): AIED 2024 Workshops, CCIS 2151, pp. 327–333, 2024.
https://doi.org/10.1007/978-3-031-64312-5_39

Due to these challenges, it is imperative to provide English L2 learners with a more effective, less stressful, and more affordable language learning solution to enhance their communicative competence, especially in oral communication. With the rapid development in the field of Artificial Intelligence (AI) over the past decades and the growing popularity of Generative AI (GenAI), there is a potential to leverage AI to offer innovative solutions. Specifically, considering that communication in the target language is the most (at least one of the most) effective ways to enhance learners' oral communicative proficiency [3], and given GenAI's potential in achieving this by acting as the 'enablers' and 'facilitators' of the communication process, this study aims to explore:

> how to **design** voice-based AI agents for developing L2 learners' oral communicative competence, and to **evaluate** their impacts and mechanisms in real-world settings.

To attain its aims, the research will adopt the Design-Based Research (DBR) approach, which focuses on developing real-world solutions iteratively through designing, implementing, evaluating, and redesigning [4, 5]. After developing the initial solution (see Sect. 3 for details), the research will focus on real-world users' interaction with the system from an early stage for empirical measurement and iterative design [6]. The entire process will be guided by CLT theories (discussed in Sect. 2).

The following research questions will guide this exploration: first, examining the real-world impacts (RQ1) and underlying mechanisms (RQ2); then, summarising design principles based on empirical evidence (RQ3). Formally:

- **RQ1:** To what extent does interacting with the designed GenAI-based voice agent develop L2 learners' oral communicative competence?
- **RQ2:** What interactional strategies employed by the voice agent facilitate or hinder the development of L2 learners' communicative competence?
- **RQ3:** How can the design be optimised for more effectively enhancing L2 learners' communicative competence?

2   Theoretical Framework

The research is based on CLT theory, which posits that the most effective way to improve communicative competence is through communication in the target language [3]. Specifically, this section will outline three key CLT principles that underpin this research.

(**Principle 1**). The role of teachers using CLT is shifting from 'lecturers' to communication 'enablers' [2] and 'shapers' [7]. This principle is supported by the following pedagogical considerations: (1) Firstly, in pragmatic, authentic, and functional communicative situations, learners '*have to* use the language productively and receptively' [8], which activates their passive knowledge. (2) Secondly, through teachers' encouragement and support (such as scaffolding), learners may feel more comfortable and confident in retrieving their already-learned language from their 'store' [2].

(**Principle 2**). The communication activities should be real and meaningful. They should include three essential elements: 'Information gap', 'Freedom of choice', and 'Feedback for evaluation'. Firstly, participants should experience an information gap, where one person has information (such as facts, opinions, ideas, instructions) that the other does not; without this gap, communication becomes unnecessary [2]. Secondly, speakers should have the freedom to choose both what they say and how they say it, rather than being restricted by tightly controlled exercises [3]. Thirdly, feedback from others should be given to help speakers evaluate whether the information exchange was successful, i.e., whether they were understood [3], which closes the loop in real communicative situations.

(**Principle 3**). Pedagogical goals often prioritise fluency over accuracy. In this research, 'fluency' is defined as the ability to 'maintain comprehensible and ongoing communication despite limitations in his or her communicative competence' [9]. As CLT centres on effective 'information exchanging' [2] and 'meaning negotiating' [3], the primary pedagogical objectives of CLT activities are to assist learners in achieving greater fluency and confidence [2]. Ultimately, this enhances their communicative competence.

3   Proposed Solution

The proposed solution (Fig. 1) aims to enhance L2 learners' oral proficiency through voice communication with AI agents (i.e., virtual interlocutors) enabled by large language models (LLMs). In this process, L2 learners use their voice to interact with the application. The microphone array perceives their voice, sending it to the automated speech recognition (ASR) model for conversion into text. The converted text is then forwarded to instructed LLMs (text generation models) to produce a text-format human-like response. The generated response is then transformed into natural-sounding spoken text using the Text to Speech (TTS) model. The TTS audio is then streamed back to the students through the speakers of their devices, simulating a conversation with a virtual native interlocutor. This immersive experience requires learners to actively use the language both receptively and productively [8]. In this way, learners' passive knowledge is continuously retrieved and activated from their minds in a comfortable and natural manner, contributing to their oral proficiency enhancement [2].

To encourage and maintain conversations, the design of this Intelligent Learning Environment (ILE) combines CLT theories with AI technologies. The generative model will be prompted by: 'Your role is to help non-native L2 learners improve their oral language skills by discussing {topic} in a casual manner'. To increase user engagement and willingness in conversations with the chatbot, additional instructions were developed based on suggestions proposed by [10] and [11]. These instructions include: "1. Keeping responses brief, like everyday small talk; 2. Showing empathy when necessary; 3. Being friendly, interesting, and humorous; 4. Prompting further discussion with one follow-up question." This aims to enhance the model's ability to initiate and sustain meaningful conversations, aligning with CLT theories.

The learning environments will be prototyped using *Streamlit*, a Python-based tool for developing shareable web applications (https://streamlit.io). The language model employed in this study is the 'gpt-3.5-turbo' model, with three rounds of preceding user-system exchange as contextual input for generating responses.

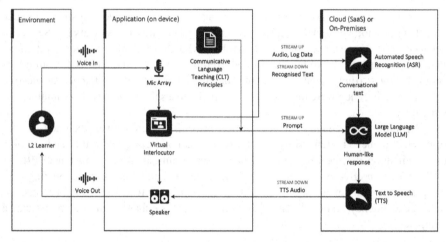

Fig. 1. Architecture of the Voice-Interaction Agent

4 Methodology

The prototype will be evaluated in real-world settings using an experimental research design. Figure 2 sequentially outlines the research activities that participants are expected to engage in. Specifically, during recruitment, researchers will introduce the research programme (step 1), and participants will opt-in voluntarily by signing a consent form (step 2). In total 50 Chinese tertiary students will be recruited with at least CEFR B1 proficiency, ensuring they can engage in voice interactions with the system. After eligibility assessment, participants will be randomly assigned to either the experimental or control group equally.

On the first day, all participants will complete a pre-test to evaluate their oral proficiency (step 3). The assessment will consist of two independent speaking tasks from the Test of English as a Foreign Language (TOEFL) Practice Online (TPO) tests. These tasks require learners to choose a preferred viewpoint from two statements and provide reasons, details, and examples to support their chosen perspective, which is a crucial indicator of their oral communicative proficiency [12]. Two experienced Chinese English teachers will assess students' audio recordings from four aspects: fluency, accuracy, complexity, and pronunciation.

After that, participants will engage in five 40-min practice sessions from Day 1 to Day 5 (step 5). In the experimental group, learners will practice their oral skills by talking with the designed voice agent in English. The control group will receive equivalent amounts (same frequency, same time) of language training in vocabulary and listening, without using the voice agent for oral practice. To control for additional language exposure, the study will recruit participants from similar educational settings, and monitor and document English language use and exposure outside of the intervention period.

After practice sessions, participants will take a post-test to measure their development of oral communication skills (step 6). The test will use the same questions to minimise test-retest effects. Participants will also be invited to a post-event survey to provide feedback on their experience using the system.

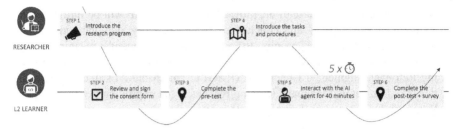

Fig. 2. Research Activities

The research questions will be addressed as follows:

1. First, to evaluate the impact of the intervention on L2 learners' development of oral communicative competence *(RQ1)*, the pre- and post-assessment scores of the experimental and control groups will be compared. This analysis will assess the overall enhancement as well as four sub-aspects: fluency, accuracy, complexity, and pronunciation. The Paired-Samples T-test will be applied in SPSS to evaluate the significance level, and Cohen's D will be employed to measure its effect size.
2. To understand the mechanism leading to effective and ineffective oral proficiency improvement *(RQ2)*, participants' pre- and post-test score changes will be taken as the ground truth. Students whose improvement exceeds the average score will be classified into the High Progress (HP) group, while those with less than average improvement will be classified into the Low Progress (LP) group. Learner-LM interactions will be audio-recorded and transcribed. Using a widely-used DA coding scheme proposed by [7], dialogic features will be annotated at the speaking-turn level. Pattern mining algorithms will then identify prevalent DA patterns associated with the HP and LP groups, revealing effective and ineffective interactional strategies. The frequency differences in DA patterns among these groups will be compared to develop additional prompts aimed at better instructing generative models for meaningful communication with non-native L2 learners.
3. A post-session survey will be conducted to gain participant feedback on using the system. The analysis of this survey, along with previous findings, will be used to optimise the design and summarise the design principles *(RQ3)*.

5   Current Progress and Next Step

- **Ethical Considerations.** The project will adhere to the Ethical Guidelines for Educational Research outlined by [13], comply with institutional and local government requirements. It was registered with UCL Data Protection Office (DPO).
- **Practical feasibility.** Regarding the intervention development, existing generation models could be easily implemented through Python Application Programming Interfaces (APIs) for responding to learners' utterances in text-based formats. ASR and TTS models could achieve voice interaction. As for analysis, the team has developed two types of automated DA annotation algorithms based on generative AI and Bidirectional Encoder Representations from Transformers (BERT) separately.

- **Next step.** In the next phase of the project, the designed ILE will be further refined. Building on the findings and conclusions from this study, our plans include: (1) developing evidence-based prompts to boost learner engagement in meaningful voice communication; (2) testing and implementing alternative LLMs; (3) incorporating Retrieval Augmented Generation (RAG) to enrich the AI's dialogue capabilities based on curated language learning resources; (4) improving personalisation to better suit learner oral proficiency levels.

6 Expected Contributions

This research aims to contribute both locally effective solutions for enhancing L2 learners' oral communicative competence and generalisable design principles for AI applications in language teaching. Specifically:

- **Theoretical advancements.** This interdisciplinary study aims to provide insights into educational effectiveness, interactional dynamics, cognitive mechanisms for developing L2 oral proficiency in an AI-enhanced learning environment. It will merge technology and language educational insights to establish evidence-based design principles for the EdTech industry and practice, influencing future LLM-enabled learning environment design and Human–computer interaction (HCI) design.
- **Language learning.** Starting with a pilot study in China and expanding global, our initiative aims to positively impact millions of students, facilitating the personalised and effective enhancement of learners' L2 oral competence.

References

1. Sherry, M., Thomas, P., Chui, W.H.: International students: a vulnerable student population. High. Educ. **60**, 33–46 (2010). https://doi.org/10.1007/s10734-009-9284-z
2. Scrivener, J.: Learning teaching: the essential guide to english language teaching, 3. edn. Hueber, Ismaning (2011)
3. Larsen-Freeman, D., Anderson, M.: Techniques and principles in language teaching, 3rd edn. Oxford University Press, Oxford, New York (2011)
4. DBRC: Design-based research: an emerging paradigm for educational inquiry. Educ. Res. **32**, 5–8 (2003). https://doi.org/10.3102/0013189X032001005
5. Scott, E.E., Wenderoth, M.P., Doherty, J.H.: Design-based research: a methodology to extend and enrich biology education research. CBE—Life Sci. Educ. **19**(3), es11 (2020). https://doi.org/10.1187/cbe.19-11-0245
6. Gould, J.D., Lewis, C.: Designing for usability: key principles and what designers think 28 (1985)
7. Walsh, S.: Classroom discourse and teacher development. Edinburgh University Press, Edinburgh (2013)
8. Brown, H.D.: Principles of language learning and teaching: a course in second language acquisition, 6th edn. Pearson Education, White Plains, NY (2014)
9. Richards, J.C.: Communicative language teaching today. Cambridge University Press, Cambridge, New York, Melbourne, Madrid, Cape Town, Singapore, São Paulo (2006)
10. Smith, E.M., Hsu, O., Qian, R., et al.: Human Evaluation of Conversations is an Open Problem: comparing the sensitivity of various methods for evaluating dialogue agents (2022)

11. Lee, M., Srivastava, M., Hardy, A., et al.: Evaluating human-language model interaction (2024)
12. Bridgeman, B., Powers, D., Stone, E., Mollaun, P.: TOEFL iBT speaking test scores as indicators of oral communicative language proficiency. Lang. Test. **29**, 91–108 (2012). https://doi.org/10.1177/0265532211411078
13. British Educational Research Association [BERA]. Ethical Guidelines for Educational Research, 4th edn. London (2018)

Advancing High School Dropout Predictions Using Machine Learning

Anika Alam$^{(\boxtimes)}$ ⓘ and A. Brooks Bowden ⓘ

University of Pennsylvania, Philadelphia, PA 19104, USA

anikaa@upenn.edu

Abstract. The importance of high school completion for jobs and postsecondary opportunities is well-documented. Schools, districts, and states are increasingly concerned about improving outcomes for vulnerable student populations. Combined with U.S. federal laws where high school graduation rate is a core performance indicator, states face pressure to actively monitor and assess high school completion. This study employs machine learning algorithms to preemptively identify students at-risk of dropping out of high school. We leverage North Carolina statewide administrative data to build an early warning prediction model that identifies students at-risk of dropping out and groups them based on similar patterns and characteristics. This study provides guidance and informed knowledge about how districts and states can capitalize "big data" to identify vulnerable students preemptively.

Keywords: Prediction models · machine learning · early warning systems

1 Introduction

There are economic and social consequences of dropping out of high school. Compared to high school graduates, adults without a high school diploma earn substantially less in the labor market, experience poorer health, are more likely to engage in criminal behavior, are more likely to require public assistance, and are less likely to vote compared to high school graduates [3, 12]. Workers whose highest education level was high school completion typically earn $26,000 more per year than those who did not complete high school [24]. Federal accountability laws such as the 2015 Every Student Succeeds Act (ESSA) emphasize high school graduation rate as a core academic performance indicator. These laws pressure school systems to actively monitor, assess, and improve high school completion.

North Carolina's high school graduation rate has gradually improved over the years to 86% in 2018. Despite these gains, subgroups such as English Language Learner (ELL) students and students with disabilities continue to have lower graduation rates (68% and 69%, respectively). Moreover, 1 in 3 teenagers who leave high school eventually complete an equivalency of a high school diploma [24]. Considering pecuniary and non-pecuniary returns of having a high school degree, on-time high school graduation merits serious attention.

© The Author(s), under exclusive license to Springer Nature Switzerland AG 2024
A. M. Olney et al. (Eds.): AIED 2024 Workshops, CCIS 2151, pp. 334–341, 2024.
https://doi.org/10.1007/978-3-031-64312-5_40

Identifying vulnerable students proactively is an important policy issue. The earlier schools can identify at-risk students, the earlier they can provide targeted support and interventions. This, in turn, would ultimately improve student engagement. When it comes to predicting high school dropout, there is no"one size fits all" where all at-risk students share exact characteristics. Inquiry into students who are stratified by characteristics (i.e. academic performance, discipline records, demographics, etc.) can provide tailored predictions for student subgroups.

The rising value of a high school diploma for postsecondary opportunities, combined with federal education laws, place pressure for schools and school systems to track and monitor student learning. Many school systems and states have developed early warning systems (EWS) or data systems designed to identify students who are at risk of missing educational milestones such as graduating high school [2, 6]. This study advances the current state of knowledge in the field by (1) providing distinct predictions for vulnerable student subgroups such as students with disabilities; and (2) examining middle school engagement to identify risk factors associated with early high school exit.

2 Research Questions

This project investigates the following research questions to predict early exit from high school:

1: How does the prediction accuracy of machine learning algorithms to predict high school dropout compare to that of traditional models (i.e., logistic regression)?
2: What are the most salient predictors of dropping out of high school for all students?
3: What are the most salient predictors of dropping out for student subgroups, such as students with disabilities?

3 Analytic Sample

The analytic sample of interest are first-time sixth-grade public school students during the 2011–2012 school year. We follow standard machine learning practices and cross-validate models by using sixth-grade students in Fall 2010 as a training sample and sixth-grade students in Fall 2011 as a testing sample [16, 18]. We limit the sample to students in districts that follow a compulsory schooling age of 16 and those with complete graduation or exit records. We retain students who have some (or all) attendance and test scores in middle grades and impute missing data with a student's unique middle school median. 90% of students in this cohort graduate by the end of 13th grade, compared to 10% who exit early.

4 Research Design

4.1 Methods

We compare the prediction accuracy of models that employ the following algorithms: logistic regression, lasso regression, ridge regression, random forests, and support vector machine (SVM) [6, 19, 20, 27, 28].

Once we identify the optimal, or highest predictive algorithm, we extract salient predictors using algorithm-specific tools. If the chosen algorithm is random forest, then we follow standard practices and employ entropy and the Gini index, or a "node purity" measure that identifies the most "important" variables. For SVM, we use a radical classification plot. LASSO and ridge regressions follow simpler extraction methods where predictors can be identified through regression coefficients [16, 18].

To prevent the models from being biased towards the majority class, we address class imbalance in two ways. First, we examine precision-recall curves that focus on the tradeoff between false positives and false negatives [11, 16]. Second, we apply Synthetic Minority Oversampling Technique (SMOTE), a feature that generates synthetic minority class observations based on the k-nearest neighbor for the minority class [10, 13]. Each model will have two versions: one that includes synthetic observations (SMOTE) to balance the skewed ratio of students who exit early to students who complete high school, and one without synthetic observations.

We apply standard metrics of optimizing accuracy with parameters and hyperparameters that are relevant to each algorithm: the optimal lambda (λ) and number of folds for lasso and ridge regression; cost penalties, various kernel, and gamma for SVM; number of trees, tree depth, and nodes at each split for random forest [18].

4.2 Outcome of Interest

The outcome of interest is if a student exits from high school prior to completion. This binary outcome takes a value of "1" if a student exits prior to graduating at the end of 13th grade, and "0" otherwise.

In binary classification models the predicted probabilities fall in the range between 0 to 1. We impose a decision threshold on converting a predicted probability into a class label (i.e., "exiting early or "not exiting early") [16, 18]. We impose the optimal probability threshold for each model using receiver operator characteristic (ROC) curve analysis for models with SMOTE observations and with precision-recall curves for models without SMOTE observations [7, 11, 23, 25].

4.3 Model Features

We follow recent literature recommendations and define student engagement with a base set of predictors: attendance, behavior, and course performance, also known as ABC [1, 2, 4, 30]. We examine student educational records from middle grades (grades 6 to 8) that capture math and reading proficiency, attendance, disciplinary infractions, and demographics as historical data to predict the probability that a student will exit high school early. We extract 8th grade demographics and create attendance, behavior, and coursework indicators for each grade between grades 6 to 8 (Table 1). We intentionally exclude high school grade data because more than one-third of all 9th grade students in the analytic sample can legally drop out.

4.4 Performance Metrics

To evaluate model performance of supervised models, we examine F1-score, accuracy precision, recall, and area under curve (AUC) derived from either the PR or ROC curve. The results will focus on false negative rate metric (1 minus the recall rate), or the proportion of student dropouts who were incorrectly labeled as not dropping out, predicted to graduate. We argue that students under false negative counts are the most vulnerable subgroup.

Students under false positive counts are less of a concern for stakeholders. An overestimation of students who exit early can still be beneficial for students who marginally graduate; these students can still benefit from additional interventions and support. Moreover, higher participation can reduce the social stigma associated with receiving targeted support(s).

5 Preliminary Findings

We find that when models do not have additional observations from SMOTE, lasso regression provides the lowest false negative rate (Table 2). However, we see substantial reductions in false negative rates in most machines after imposing SMOTE. The improved accuracy in identifying the minor class comes at a penalty of a higher misclassification rate, with the optimal model being a SMOTE logistic regression.

In examining the most predictive features from the SMOTE logistic regression, we find that middle school attendance, especially 8th grade, is most predictive of exiting early for all students and students with disabilities ($p < .001$; Table 3). Conversely, we find that 8th grade reading proficiency, followed by middle school math proficiency are associated with reductions in the likelihood of exiting early ($p < .001$; Table 4).

In closing, adding statistical techniques can be promising in building prediction models. Our evidence suggests that not all ML algorithms perform better than a logistic regression. Middle school attendance as a key predictor is consistent with prior research on absenteeism [2, 14].

A limitation of this work is that the predictive accuracy may vary when the model is tested on different student populations or in different schooling years. However, building a prediction model can be an insightful step to understand student disengagement. EWS can serve as a predecessor for providing data-driven solutions such as Integrated Student Support (ISS), tiered interventions, and more. By the time of the conference, we hope to refine our results by incorporating unsupervised algorithms to compare consistency across model features.

Acknowledgments. This research reported here was funded by the Institute of Education Sciences, U.S. Department of Education, through Grant R305B200035. The opinions expressed are those of the authors and do not represent views of the U.S. Department of Education.

Disclosure of Interests. Both authors have no competing interests to declare that are relevant to the content of this article.

Appendix

Table 1. Model Features

	Gender
	Race/Ethnicity
Demographics	Limited English Proficiency (LEP) status
	Economic disadvantage*
	Disability status
	Out of school suspension
	In school suspension
Discipline	Number of short-term suspensions
	Number of long-term suspensions
Attendance	Absence rate
	Chronically absent
Test scores	Math proficiency
	Reading proficiency

Notes: Gender, race/ethnicity, economic, and disability status are based on 8th grade student records. For features in the latter 3 categories, we create an indicator per year ("_grade") and across middle school ("__middle").
*North Carolina Department of Public Instruction (DPI) defines economic disadvantage meeting one or more of the following criteria: direct certification from food assistance programs (SNAP, TANF, FDPIR); runaway, homeless, foster, Medicaid recipient, enrolled in Head Start or state-funded pre-kindergarten, or migrant status; and community eligibility provision (CEP).

Table 2. Model Performance

	All students (n = 93,461)		Students with disabilities (n = 10, 901)	
Algorithm	Test misclassification rate	False negative rate	Test misclassification rate	False negative rate
Logistic regression	4.6%	78%	9.6%	70.8%
Lasso regression	5.6%	71.8%	11.6%	61.6%
Ridge regression	3.7%	90.3%	7.3%	87.6%
Random forest	3.9%	86.5%	7.2%	82.3%
Support vector machine (SVM)	3.75%	91.6%	7.3%	86%
SMOTE logistic regression	40.9%	13.6%	34.5%	24.4%
SMOTE lasso regression	70.3%	16.9%	14.4%	69.2%
SMOTE ridge regression	24.2%	42.6%	13.4%	70.5%
SMOTE random forest	4.5%	86.5%	17%	49.6%
SMOTE SVM	4.2%	83%	12.7%	72.9%

Table 3. Features associated with early exit

Students with disabilities (n = 10, 901)		All students (n = 93,461)	
8th grade absence rate	12.35***	8th grade absence rate	15.36***
	(0.65)		(0.27)
7th grade absence rate	4.51***	7th grade absence rate	6.46***
	(0.65)		(0.28)
6th grade absence rate	2.62***	6th grade absence rate	2.28***
	(0.65)		(0.28)
White	1.43***	Economic disadvantaged	0.70***
	(0.20)		(0.28)
Being Limited English Proficient in middle	1.26***	7th grade short-term suspensions	0.54**
	(0.37)		(0.04)
Black	0. 78***	Having out-of-school suspension in 6th grade	0.88***
	(0.20)		(0.21)
Other Race	0.70***	Having in-school suspension (ISS) in middle	0.52***
	(0.19)		(0.03)
Economic disadvantaged	0.59***	White	0.29***
	(0.21)		(0.08)
Ever chronically absent in middle	0.35**	Being Limited English Proficient in middle	0.21**
	(0.11)		(0.06)
6th grade short-term suspensions	0.29**		
	(0.09)		

Note: This table only includes statistically significant features and are listed in descending order per sample.

Table 4. Features inversely associated with early exit

Students with disabilities (n = 10, 901)		All students (n = 93,461)	
Asian	-3.75**	Asian	-0.73***
	(1.33)		(0.11)
8th grade reading proficient	-0.69***	Black	-0.30***
	(0.09)		(0.08)
7th grade math proficient	-0.58***	8th grade reading proficient	-0.27***
	(0.06)		(0.09)
8th grade math proficient	-0.58***	7th grade math proficient	-0.51***
	(0.06)		(0.02)
6th grade math proficient	-0.52***	8th grade math proficient	-1.30***
	(0.06)		(0.06)
Female	-0.37***	6th grade math proficient	-0.53***
	(0.04)		(0.02)
		Female	-0.12***
			(0.016
		8th grade reading proficient	-0.61***
			(0.01)
		7th grade reading proficient	-0.19***
			(0.02)
		6th grade reading proficient	-0.35***
			(0.03)

Note: This table only includes statistically significant features and are listed in descending order per sample.

References

1. Allensworth, E.M., Clark, K.: Are Gpas an inconsistent measure of college readiness across high schools? University of Chicago Consortium on School Research, Examining Assumptions About Grades Versus Standardized Test Scores (2019)
2. Balfanz, R., Herzog, L., Mac Iver, D.J.: Preventing student disengagement and keeping students on the graduation path in urban middle-grades schools: early identification and effective interventions. Educ. Psychol. **42**(4) (2007)
3. Belfield, C.R. Levin, H.M. (eds.):The price we pay: Economic and social consequences of inadequate education. Brookings Institution Press (2007)
4. Bowers, A.J.: Grades and graduation: a longitudinal risk perspective to identify student dropouts. J. Educ. Res. **103**(3), 191–207 (2010)
5. Bowers, A.J., Sprott, R.: Why tenth graders fail to finish high school: a dropout typology latent class analysis. J. Educ. Stud. Placed Risk **17**(3), 129–148 (2012)
6. Bowers, A.J., Sprott, R.: Examining the multiple trajectories associated with dropping out of high school: a growth mixture model analysis. J. Educ. Res. **105**(3), 176–195 (2012)
7. Bowers, A.J., Zhou, X.: Receiver operating characteristic (ROC) area under the curve (AUC): a diagnostic measure for evaluating the accuracy of predictors of education outcomes. J. Educ. Stud. Placed Risk **24**(1), 20–46 (2019)
8. Burke, A.: Early Identification of high school graduation outcomes in oregon leadership network schools. Rel 2015–079. Regional Educational Laboratory Northwest (2015)
9. Butler, M.A.: Rural-urban continuum codes for metro and nonmetro counties. US Department of Agriculture, Economic Research Service, Agriculture And . . (1990)
10. Chawla, N.V., Bowyer, K.W., Hall, L.O., Kegelmeyer, W.P.: SMOTE: synthetic minority over-sampling technique. J. Artific. Intell. Res. **16**, 321–357 (2002)
11. Cook, J., Ramadas, V.: When to consult precision-recall curves. Stand. Genomic Sci. **20**(1), 131–148 (2020)

12. Dee, T.S.: Are there civic returns to education? J. Public Econ. **88**(9–10), 1697–1720 (2004)
13. Fernández, A., Garcia, S., Herrera, F., Chawla, N.V.: SMOTE for learning from imbalanced data: progress and challenges, marking the 15-year anniversary. J. Artific. Intell. Res. **61**, 863–905 (2018)
14. Gottfried, M.A.: Chronic absenteeism and its effects on students' academic and socioemotional outcomes. J. Educ. Stud. Placed Risk **19**(2), 53–75 (2014)
15. Jackson, C.K.: The effects of an incentive-based high-school intervention on college outcomes (No. w15722). National Bureau of Economic Research (2010)
16. James, G., Witten, D., Hastie, T., Tibshirani, R., Taylor, J.: Statistical learning. In: An Introduction to Statistical Learning: with Applications in Python, pp. 15–67. Springer International Publishing, Cham (2023)
17. Kahlenberg, R.D.: All together now: creating middle-class schools through public school choice. Rowman & Littlefield (2004)
18. Kroese, D.P., Botev, Z., Taimre, T., Vaisman, R.: Data science and machine learning: mathematical and statistical methods. CRC Press (2019)
19. Kruger, J.G.C., De Souza Britto Jr, A., Barddal, J.P.: An explainable machine learning approach for student dropout prediction. Expert Syst. Appl. **233**, 120933 (2023)
20. Lee, S., Chung, J.Y.: The machine learning-based dropout early warning system for improving the performance of dropout prediction. Appl. Sci. **9**(15), 3093 (2019)
21. Leevy, J.L., Khoshgoftaar, T.M., Bauder, R.A., Seliya, N.: A survey on addressing high-class imbalance in big data. J. Big Data **5**(1), 1–30 (2018)
22. Losen, D., Orfield, G., Balfanz, R.: Confronting the graduation rate crisis in Texas. Civil Rights Project at Harvard University (2006)
23. Nakas, C., Bantis, L., Gatsonis, C.: ROC analysis for classification and prediction in practice. CRC Press (2023)
24. NCES. Common Core of Data Public Elementary/Secondary School Universe Survey(2023)
25. Pérez Fernández, S., Martínez Camblor, P., Filzmoser, P., Corral Blanco, N.O.: nsROC: an R package for non-standard ROC curve analysis. R J. **10**(2) (2018)
26. Reardon, S.F., Bischoff, K.: Income inequality and income segregation. Am. J. Sociol. **116**(4), 1092–1153 (2011)
27. Sansone, D.: Beyond early warning indicators: high school dropout and machine learning. Oxford Bull. Econ. Stat. **81**(2), 456–485 (2019)
28. Sorensen, L.C.: "Big Data" in educational administration: an application for predicting school dropout risk. Educ. Adm. Q. **55**(3), 404–446 (2019)
29. Weissman, A.: Friend Or Foe? the role of machine learning in education policy research [Doctoral Dissertation] (2022)
30. Zaff, J.F., Donlan, A., Gunning, A., Anderson, S.E., Mcdermott, E., Sedaca, M.: Factors that promote high school graduation: a review of the literature. Educ. Psychol. Rev. **29**, 447–476 (2017)

Predicting the Structural Parts of the Plot on Elementary School Written Essays

Erverson Bruno Gomes de Sousa[(✉)] and Rafael Ferreira Leite de Mello[(✉)]

Centro de Estudos Avançados de Recife, Recife, PE, Brazil
{ebgs,rflm}@cesar.school

Abstract. Essay writing is an essential task throughout all life stages. In the final years of elementary education, classes are dedicated to narrative text production, requiring students to develop and organize their ideas and thoughts in a logical sequence. The process of correcting and identifying the structural elements of the narrative plot in texts is a routine task for educators, which generates an overload and hampers the implementation of a formative, unbiased assessment that supports the enhancement of students' writing skills. Providing automatic identification of the structural elements of the narrative plot, considering this educational stage and the Brazilian Portuguese language, is an under-explored issue in the field of natural language processing. The aim of this research is to analyze the performance of predicting the structural elements of the narrative plot in narrative text productions of Brazilian elementary school students. The experimental research includes a corpus of 3099 narrative sentences obtained and annotated through the formative assessment project of the Brazil in School program of the Ministry of Education.

Keywords: Natural Language Processing · Narrative Essays · Learning Analytics

1 Motivation for Research

Despite playing a crucial role in both personal and professional advancement, effective writing instruction faces significant challenges, such as the lack of specific and constructive feedback for students, the limitation of time dedicated to instruction, and the insufficiency of adequate resources for writing practice [8]. On the other hand, factors that have been shown to facilitate this process include collaboration among teachers, the adoption of innovative teaching strategies, and access to technological resources [1]. Therefore, research identifying the elements that foster or hinder writing instruction is of utmost relevance. Such knowledge not only enables the improvement of writing instruction approaches but also provides a deeper understanding of the reasons why writing instruction can be, at present, effective or ineffective [8].

A. M. Olney et al. (Eds.): AIED 2024 Workshops, CCIS 2151, pp. 342–347, 2024.
https://doi.org/10.1007/978-3-031-64312-5_41

Textual production requires students to organize and integrate various ideas, being widely considered as the most appropriate way to assess educational skills and competencies in writing [20]. In the Brazilian context, it is evident that the development of writing ability among students represents an ongoing challenge. This is reflected in the reality of various basic education schools, which, in many cases, have faced difficulties in effectively stimulating the writing capacity of both elementary and high school students. This difficulty has been evident in the results of large-scale assessments in recent years [15].

Essay writing is a type of textual production, which represents a common practice in educational institutions to assess students' writing skills, requiring a well-defined structure [9]. In the classroom environment, essay writing plays a crucial role as a tool to assess the outcomes of learning in writing, guide the educational process, and measure student progress [26]. Despite its widespread application in education, manual essay grading is a costly activity that demands considerable time from teachers [4,10,12]. During the process of large-scale grading, it is common for teachers to face challenges in maintaining a rigorous level in the evaluation of texts [3,8]. The associated cost, reliability, and evaluator subjectivity are, furthermore, factors that limit the efficiency of manual grading of textual productions [15].

The narrative text type is utilized in the production of various textual genres in elementary education, unlike in high school, where students develop expository-argumentative text productions [19]. The narrative text type includes elements such as narrator, plot, characters, time, and space. It fosters the student's ability to "construct knowledge of the world, of the text-producing subject themselves, and, consequently, of verbal language, which can be of great educational value" [19]. Such narrative elements are essential to characterize the text as narrative, and this requires teachers to analyze textual productions to identify them and provide useful feedback to students on what is missing and what needs to be improved, offering personalized teaching.

Personalized teaching has gained momentum in recent years, especially during the period of remote education, which facilitated the generation of data in the educational context due to the massive adoption of digital tools in school environments [6]. The data generated from student interactions with digital tools can be used to monitor, analyze, predict, intervene, recommend, and, most importantly, improve the quality of the teaching and learning process through LALearning Analytics techniques [23]. In this context, Natural Language Processing (NLP) is a subfield of artificial intelligence that uses computational techniques to analyze a language, whether spoken or written, in digital format [2]. There have been applications of NLP techniques for automatic essay scoring in the context of basic education, which assist teachers in providing more assertive, effective, and timely feedback for writing improvement, allowing the teacher to use their pedagogical time focused on devising strategies to address the most recurrent writing difficulties of students [17].

2   Research Question

The narrative plot, a central element in the teaching of textual production in the context of Brazilian basic education, is traditionally divided into three structuring parts: introduction, complication, and resolution. The "introduction" establishes the setting, introducing the characters and the context, and is essential for engaging the reader from the beginning of the narrative [5]. These structuring parts of the plot are fundamental for the development of students' narrative writing skills, allowing them to express complex and creative ideas in a clear and coherent manner. Understanding and effectively applying these structures are essential for the development of students' narrative writing skills, contributing to the enhancement of their literary competences and effective communication. Studies in education emphasize the importance of narrative in learning, highlighting how the practice of narrative writing can improve critical thinking and personal expression of students [24].

Identifying the structuring parts of a narrative plot in educational texts through Automated Essay Scoring techniques represents a significant challenge, especially when considering the complexity and variety of elements that compose a narrative [21]. In this context, multi-label classification emerges as a promising approach, capable of dealing with the multifaceted nature of narrative texts [25]. Faced with a challenging and still underexplored scenario in scientific research, this study aims to evaluate narrative essays employing Machine Learning models with a multi-label classification strategy. The research is conducted with the goal of answering the following question: How to automate the process of detecting the structuring parts of the plot through the analysis of sentences from narrative essays of elementary school students?

3   Research Methodology

This section describes the methodology that will be conducted to analyze the performance of automation in detecting the structuring parts of the narrative plot in textual productions, using different Machine Learning models applying multi-label classification. To achieve the proposed objectives, the research will have a quantitative approach, classifying itself, in terms of its nature, as experimental. This study is characterized as applied research, given that it focuses on the practical application of theoretical knowledge to enhance the automatic evaluation of plots in narrative essays [7,14].

In terms of procedures, the research adopts both bibliographic and experimental approaches. The bibliographic approach is used to build a solid foundation of theoretical knowledge by reviewing existing literature on automatic text evaluation, multi-label classification, and natural language processing. On the other hand, the experimental approach is employed to test hypotheses and validate the efficacy of the proposed multi-label classification models [13]. Together, these methods provide a comprehensive and detailed understanding of the problem under study, allowing not only to describe and explore the phenomenon but also to apply and test solutions in the field of educational assessment.

The methodology employed in this study follows these steps: (i) Annotated dataset: a procedure conducted by two annotators where disagreements were resolved by a third, more experienced annotator; Sentences: division of essays into sentences for classification of the structuring parts of the plot; Structuring parts of the plot: definition of the presence or absence of each structuring part of the plot in each sentence separately; Feature extraction: extraction of features that will be used for training the models; Model training: the process of selection, training, and testing of machine learning models on sentences from narrative essays; Analysis of results: evaluation of the models and multi-label approaches selected. All stages have been executed, and preliminary results have been obtained.

4   Initial Results and Expected Contributions

In this section, we present the preliminary results of this study, whose goal is to automate the identification of the structuring parts of the plot in narrative essays by students in the final years of Brazilian elementary education. This effort aims to assist the evaluation process of teachers, using machine learning models in the context of learning analytics. The experiments employed various Machine Learning (ML) models on a corpus of narrative texts from 6th to 9th-grade students in Brazilian schools. These models were applied using a multi-label classification strategy.

The investigation carried out in this research focused on the applicability of different Machine Learning (ML) models, emphasizing multi-label classification algorithms. One of the findings is the superior performance of the Label Powerset (LP) with XGBoost (XGB) and Linguistic Inquiry and Word Count (LP-XGB-LIWC) model, which stands out significantly in the context of analyzing the detection of the structuring parts of the plot in narrative texts. When compared to other models, it shows superior performance, as evidenced by higher scores in key metrics, such as Micro-F1 and Subset Accuracy (SA), with scores of 0.742 and 0.681, respectively, as well as a Macro-F1 of 0.509 and a Hamming Loss of 0.181. This performance indicates the model's remarkable ability to accurately classify narrative text sentences according to parts of the plot, which are critical for understanding the narrative.

Recent studies have demonstrated the importance of computational analysis of narratives [16,18,21], thereby showing that narrative productions of elementary school students can benefit from this trend, where it aids teachers in more standardized corrections with less bias due to fatigue or lack of standardization. This assistance process is not limited to objectivity in evaluation but extends to enriching the educational experience of students, providing a technological solution that can enhance the understanding of key elements of narratives [11]. Through automation, it is possible to identify patterns and structures in essays that might go unnoticed in human evaluation, offering valuable educational insights and promoting a more holistic approach to teaching writing [22].

Regarding contributions, this study aims to contribute both socially and academically. Social Contribution: will be demonstrated by its ability to implement concrete improvements for the Brazilian teaching community. This will be achieved through the development of an innovative methodology for the automated correction of narrative essays, facilitating the evaluation of texts by teachers in the final years of elementary education. Such an approach aims to decrease the teaching workload, reducing the time and costs associated with the evaluation process. Moreover, the adoption of this methodology will contribute to enhancing the quality of students' textual production, acting as an unbiased and impartial assessment tool. The practical relevance of this research will be evidenced through experiments in formative writing assessments, conducted in schools participating in the Brazil School Program of the Ministry of Education (MEC), demonstrating its positive impact and applicability. Academic Contribution: This study aims to enrich the understanding of narrative essay assessment, especially through the implementation of multi-label classification models. The mentioned methodology promotes a more effective identification of the distinctive characteristics present in each structuring segment of the plot in students' textual productions. Such an approach allows for a more detailed and comprehensive evaluation of the plot as a whole. Thus, the experimental research contributes to the corpus of studies dedicated to this theme, expanding the scope of knowledge in the field of automated essay scoring and providing valuable insights for future educational applications.

References

1. Ahumada, S., Bañales, G., Graham, S., Torres, M.L.: Facilitators and barriers to writing instruction in Chile: teachers' preparation and knowledge about teaching writing. Read. Writ. **36**, 1867–1899 (2023)
2. Alhawiti, K.M.: Natural language processing and its use in education. Int. J. Adv. Comput. Sci. Appl. **5**(12) (2014)
3. de Ávila, R.L., Soares, J.M.: Uso de técnicas de pré-processamento textual e algoritmos de comparação como suporte à correção de questões dissertativas: experimentos, análises e contribuições. In: Brazilian Symposium on Computers in Education (Simpósio Brasileiro de Informática na Educação-SBIE), vol. 24, p. 727 (2013)
4. Correnti, R., Matsumura, L.C., Wang, E., Litman, D., Rahimi, Z., Kisa, Z.: Automated scoring of students' use of text evidence in writing. Read. Res. Q. **55**(3), 493–520 (2020)
5. Elliott, J.: Using narrative in social research: qualitative and quantitative approaches. In: Using Narrative in Social Research, pp. 1–232 (2005)
6. Gaftandzhieva, S., Docheva, M., Doneva, R.: A comprehensive approach to learning analytics in Bulgarian school education. Educ. Inf. Technol. **26**(1), 145–163 (2021)
7. Gil, A.C.: Como elaborar projetos de pesquisa. Atlas, São Paulo (2002)
8. Graham, S.: Changing how writing is taught. Rev. Res. Educ. **43**(1), 277–303 (2019)
9. Graham, S., Perin, D.: Writing next-effective strategies to improve writing of adolescents in middle and high schools (2007)
10. Hussein, M.A., Hassan, H., Nassef, M.: Automated language essay scoring systems: a literature review. PeerJ Comput. Sci. **5**, e208 (2019)

11. Jones, S., Fox, C., Gillam, S., Gillam, R.B.: An exploration of automated narrative analysis via machine learning. PLoS ONE **14**(10), e0224634 (2019)
12. Ke, Z., Ng, V.: Automated essay scoring: a survey of the state of the art. In: IJCAI, vol. 19, pp. 6300–6308 (2019)
13. Kerlinger, F.N., Lee, H.B.: Foundations of Behavioral Research. Harcourt College Publishers (2000)
14. Marconi, M.A., Lakatos, E.M.: Fundamentos de metodologia científica, 5th edn. Atlas, São Paulo (2003)
15. Passero, G., Ferreira, R., Dazzi, R.L.S.: Off-topic essay detection: a comparative study on the Portuguese language. Revista Brasileira de Informática na Educação **27**(03), 177–190 (2019). https://doi.org/10.5753/rbie.2019.27.03.177
16. Piper, A.: Computational narrative understanding: a big picture analysis. In: Proceedings of the Big Picture Workshop, pp. 28–39 (2023)
17. Ramesh, D., Sanampudi, S.K.: An automated essay scoring systems: a systematic literature review. Artif. Intell. Rev. **55**(3), 2495–2527 (2022)
18. Ranade, P., Dey, S., Joshi, A., Finin, T.: Computational understanding of narratives: a survey. IEEE Access **10**, 101575–101594 (2022)
19. Rezende, N.L., de Souza, M.C.: Do ensino escolar da escrita de textos narrativos. Linha D'Água **31**(2), 143–158 (2018)
20. Rodrigues, F., Oliveira, P.: A system for formative assessment and monitoring of students' progress. Comput. Educ. **76**, 30–41 (2014)
21. Santana, B., Campos, R., Amorim, E., Jorge, A., Silvano, P., Nunes, S.: A survey on narrative extraction from textual data. Artif. Intell. Rev. **56**(8), 8393–8435 (2023)
22. da Silva Filho, M.W., et al.: Automated formal register scoring of student narrative essays written in Portuguese. In: Anais do II Workshop de Aplicações Práticas de Learning Analytics em Instituições de Ensino no Brasil, pp. 1–11. SBC (2023)
23. de Sousa, E.B., Alexandre, B., Ferreira Mello, R., Pontual Falcão, T., Vesin, B., Gašević, D.: Applications of learning analytics in high schools: a systematic literature review. Front. Artif. Intell. **4**, 737891 (2021)
24. Toolan, M.J.: Narrative: A Critical Linguistic Introduction. Routledge (2013)
25. Unal, F.Z., Guzel, M.S., Bostanci, E., Acici, K., Asuroglu, T.: Multilabel genre prediction using deep-learning frameworks. Appl. Sci. **13**(15), 8665 (2023)
26. Zupanc, K., Bosnić, Z.: Automated essay evaluation with semantic analysis. Knowl.-Based Syst. **120**, 118–132 (2017)

Promoting Visual Health and Inclusive Education in Metaverse Learning Environments

Jiaqi Xu[1]([✉]), Qian Liu[2], and Xuesong Zhai[1]([✉])

[1] College of Education, Zhejiang University, Hangzhou, Zhejiang 310058, China
jiaqi.xu@zju.edu.cn
[2] Department of Education, Practice and Society,
University College London, London WC1H 0AL, UK

Abstract. With the rapid advancement of digital technology, online learning has become increasingly popular. However, extended screen time has raised concerns about students' visual health, notably Computer Vision Syndrome (CVS). How to promote visual health and respond to diverse learning needs are rarely addressed in the design of most learning environments. To fill in the gap, it is essential to consider and integrate design features and methods that focus on visual interaction, resource presentation, and inclusive, equitable education. This research aims to explore the effect of media design of metaverse learning environments incorporating visual health on learners. The research encompasses analyzing online learners' cognitive behavior and emotional responses using eye movement and EEG data, establishing visual health metrics within the online learning context, and creating online learning materials informed by media visual training. Through this research, insights into mitigating computer vision syndrome in metaverse learning environments will be uncovered, while also expanding the cognitive theory of multimedia learning within the realm of visual health. Ultimately, this study will contribute to the development of visual health intervention strategies for large-scale open courses.

Keywords: Visual Health · Online Learning Environments · Metaverse · Visual Training · Inclusive Education

1 Introduction

Online learning has been become an indispensable supplementary form of education. However, longtime online learning has impaired students' visual health, which results in inefficient learning performance and many physical issues (Wei and Chou 2020; Liu et al. 2021). Exploring the factors of influencing online learners' computer vision syndrome should consider multidisciplinary perspectives including ophthalmology, educational technology, brain science, and information science. Visual science research has shown that suitable media representation can be utilised to train the eyes, working the eye's tissues and muscles while enhancing vision. However, there is a shortage of visual intervention strategies in large-scale online learning scenarios, and the effect of online learning resources incorporating visual health media design still needs to be determined.

A. M. Olney et al. (Eds.): AIED 2024 Workshops, CCIS 2151, pp. 348–354, 2024.
https://doi.org/10.1007/978-3-031-64312-5_42

Building on this, the emergence of metaverse learning environments, where digital and physical realities converge, presents a unique opportunity to address these challenges. The metaverse, a collective virtual shared space that is created by the convergence of virtually enhanced physical reality and physically persistent virtual reality, is predicted to be the next frontier in education (Dahan et al. 2022). This immersive, interactive space offers new possibilities for promoting visual health and inclusive education. For instance, the metaverse's flexible, customizable settings allow for the adjustment of visual elements such as brightness, contrast, and text size to cater to individual learners' visual needs. Furthermore, the metaverse's inherently interactive nature can encourage learners to engage in eye exercises and breaks, potentially mitigating the effects of computer vision syndrome. Nonetheless, the application of metaverse learning environments in promoting visual health is still in its infancy, and its potential and effectiveness need to be comprehensively explored. Additionally, the metaverse also holds potential for fostering inclusive education by providing equitable learning opportunities for all learners, including those with visual impairments. However, how to design and implement inclusive metaverse learning environments remains an open question.

2 Areas of focus and Solutions

2.1 Incorporating Visual Training in Metaverse Learning Platforms for Visual Health

This area of focus centers on leveraging ophthalmology-based visual training methods to promote the visual health of online learners. Key aspects of this focus include:

Firstly, there's the identification of visual health challenges faced by online learners. Conduct a comprehensive analysis of the visual health challenges faced by online learners, with a specific emphasis on the prevalence and impact of Computer Vision Syndrome (CVS). This involves surveying and assessing the ocular health issues experienced by learners in the online learning environment. Secondly, there's the integration of scientifically validated visual training techniques into online learning platforms. This entails developing and implementing frameworks that include exercises and activities aimed at improving eye muscle coordination, reducing eye strain, and enhancing visual acuity. Collaboration with ophthalmologists and vision experts ensures the effectiveness and safety of these interventions. Thirdly, a user-centered design approach is adopted to create online learning platforms that prioritize visual health. This involves optimizing media representation, layout, color schemes, font sizes, and screen brightness to mitigate eye fatigue and discomfort. Finally, there's an emphasis on evaluating the effectiveness of integrated visual training techniques. Through systematic evaluation employing both quantitative and qualitative research methods, this aspect of the project seeks to measure improvements in visual health, reduction in CVS symptoms, and overall enhancement of the online learning experience.

2.2 Investigating Learning Processes and Experience in Metaverse Environments

In this project, we employ various cutting-edge technologies to enhance the online learning experience:

Artificial Intelligence (AI) is utilized to analyze learner behavior, engagement, and interactions with the online platform. AI-driven algorithms monitor and record learner activities, offering valuable insights into study patterns and preferences. Eye tracking technology plays a crucial role in gathering data on learners' eye movements and gaze patterns. This information provides insights into visual attention and fatigue, contributing to our understanding of the correlation between eye health and cognitive processes. Using EEG to assess the participant's consciousness level during the online learning activity. Furthermore, neuroscientific investigation focuses on studying the neurocognitive processes of learners during online learning. This involves examining neural correlates such as cognitive load, attention, and memory retention concerning different types of online learning content and delivery methods. The combination of eye tracking and EEG can explain the neural basis of learners during complex tasks (Bonhage et al. 2015).

3 Research Foundation

3.1 The Influence of Media Design on Learning Performance in Online Learning

Meyer's cognitive theory of multimedia learning and Bloom's cognitive stratification theory describe the rules of multimedia learning, the basic principles of media design and the promotion mechanism of learning efficiency in the online learning environment. For example, (1) multimedia principle. When there are multiple media simultaneously, learners save the corresponding representation forms in short-term memory simultaneously, increasing the connection between various representations (Wang et al. 2020); (2) The redundancy avoidance principle. When media use the exact channel representation simultaneously, it will cause cognitive load, so the same channel should avoid multi-information representation (Kirschner et al. 2018). From the standpoint of visual health, research has shown that prolonged exposure to a digital learning environment is significantly related to the development of visual fatigue (such as dry eye disease) (del Rosario Sánchez-Valerio et al. 2020). In addition, long-term online learning can cause computer vision syndrome characterised by glare sensitivity and slow focus change in severe cases (Gangamma & Poonam 2010).

3.2 Visual Training Promotes Learners' Visual Health.

The visual system encodes graphical and spatial information in multimedia learning and is a crucial information processing channel, according to the hypothesis of the cognitive theory of multimedia learning (Mayer 2001). Numerous studies have shown that the visual dysfunctions, such as the inaccuracy of the accommodative and vergence functions of the eye (Morad et al. 2002), impact learning performance significantly (Shin et al. 2009). Learners' eye following ability cannot keep up with the speed of the brain's analysis and processing of things, thus affecting their concentration and learning ability.

Visual training is an adjustment method for visual dysfunctions. On the one hand, visual training can increase learners' eye movement, focus, fixation ability and cooperation ability for online learning. On the other hand, visual training alleviates visual fatigue and enhances learners' visual processing ability (Eugene 2005). Oculomotor nerve, accommodation and binocular visual skills are significantly and positively corre-lated with learners' reading ability (Goldstand, Koslow, & Parush 2005). Additionally, learners' visual information processing abilities are influenced by their visual function, and these information processing skills can result in effective learning outcomes (Alvarez Peregrina et al. 2020). Online learning resources can realize the function of visual train-ing through media design. For example, flash, blur and pop-up notifications are designed to stimulate the blink, keep the eye moist and reduce the visual fatigue caused by dry eyes (Crnovrsanin et al. 2014). Enlarging an element in the picture to attract attention and blurring the text in the video to adjust the pupil are also effective approaches.

4 Research Progress

4.1 The Index Construction and Online Learning Resources Develop

Employing the combined analysis of online learners' eye tracking and EEG data, this research plans to construct cognitive behaviour and visual health indexes for online learning. In the online learning environment, learners face the dual pressure of specific learning tasks and visual stimuli, and their cognitive behaviour driven by learning tasks is more complex. Therefore, this study will draw insights from ophthalmology, build a visual health index system under a specific online learning environment based on the combined analysis of eye movements and EEG. The index system will include physiological eye movement data (e.g. pupil size, blink times, etc.) that reflect visual health and eye–machine interaction data (e.g. saccade path and gaze times) that reflect cognitive behaviour.

At present, the research on index construction is still in progress, and we have preliminarily constructed the basic index framework, as shown in Table 1. This part is only an incomplete index system, which needs to be supplemented and improved subsequently.

Table 1. The basic index framework

Eye Data Type	Classification	Indexes
Eye and machine interaction	Engaged	Fixation duration
		Fixation
	Information processing	Saccade count
		Scan path
Physiological eye movement	Optic nerve fatigue	Eyelid covering
		Pupil size
	Ciliary muscle regulation	Pupil size
	Dry eye	Blink response
		Moist eye film
	Left and right eye coordination	Squint angle

4.2 Platform Development

We've crafted an expansive online learning platform immersed within the metaverse domain. This platform boasts a plethora of functionalities, ranging from virtual gatherings and campus expeditions to collaborative group dialogues and interactive scientific experiments, all curated to furnish learners with deeply engaging educational encounters (refer to Fig. 1).

Fig. 1. Metaverse online learning platform

4.3 Experiments Design

The Impact of Visual Training Integrated Learning Resources Design in Online Learning Environments on Learners

This research proposes methods to analyse learners' cognitive behaviour and visual health with eye movements and EEG. So far, this research has completed an experiment involving 60 subjects which analyses their cognitive behaviour during online reading. The experimental scenario is shown in the Fig. 2. This research will verify the effect of the varied media design of online reading resources on learners' cognition and visual health. The experiment process is as follows: firstly, we will explore the effects of different visual stimuli on learners and then construct the eye movement and EEG data sets of online learners under four media design (blur, flicker, simulated blink and redness). Sixty college students participated in this experiment who read online learning resources (texts) containing different visual stimuli.

In addition, we will explore the effects of different types of knowledge presentation and media designs in animation videos. Specifically, we will develop animation videos incorporating two types of knowledge: declarative knowledge and procedural knowledge. These animation videos will be designed with four or more media features, including blur, flicker, simulated blink, redness, and enlarged drawing, among others. The incorporation of these diverse media designs allows us to investigate their impact on learners' engagement, comprehension, and retention within the metaverse learning environment. Through this experiment, we seek to gain insights into the optimal combination of visual training techniques and video resource design elements to enhance the learning experience in metaverse settings.

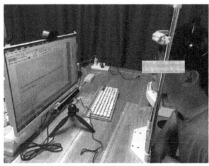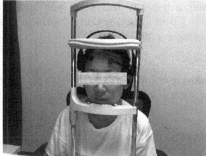

Fig. 2. Eye movement and EEG data collection experimental scenarios

The Effect of Game Design Incorporating Visual Training in Metaverse Environments on Amblyopia Learners

We focus on investigating the impact of game design integrated with visual training techniques within metaverse environments specifically tailored for amblyopia learners. Through the development of specialized educational games, our experiment aims to provide amblyopia learners with interactive experiences designed to enhance visual perception, stimulate neural pathways, and improve overall visual function. These games will incorporate a variety of visual stimuli and exercises, targeting specific areas of visual deficit commonly associated with amblyopia.

By integrating evidence-based visual training principles into game design, we seek to create engaging and effective learning experiences that address the unique needs of amblyopia learners. Through rigorous assessment methodologies, including pre- and post-intervention evaluations, subjective user feedback, and objective measures of visual performance, we aim to quantify the impact of game-based visual training interventions on amblyopia rehabilitation within metaverse environments.

5   The Anticipated Contributions and Impacts of This Research

Firstly, this research is expected to contribute a new understanding of how metaverse learning environments can be designed and utilized to promote visual health. This would involve the application of computer science principles to create adaptive, personalized visual settings and interventions within the metaverse. This contribution stands at the intersection of computer science and ophthalmology, extending the scope of AIED to include considerations of physical health in the design of educational technologies.

Secondly, this research aims to elucidate the ways in which metaverse learning environments can foster inclusive education. This involves the integration of learning science theories with the development of AI technologies to create metaverse learning experiences that are accessible and beneficial to all learners, including those with visual impairments. This contribution aligns with the learning sciences' emphasis on diversity and inclusion, while also leveraging the capabilities of AI to achieve these goals. The impact of this research could lead to the development of more health-conscious and inclusive AI educational technologies. This could improve the learning experiences and outcomes for a wide range of learners, while also promoting their physical wellbeing.

Acknowledgments. This study was funded by China Scholarship Council (grant number 202306320196), Zhejiang University Doctoral Student Academic Rising Star Cultivation Program (grant number 2022008).

Disclosure of Interests. The authors have no competing interests to declare that are relevant to the content of this article.

References

Alvarez-Peregrina, C., Sánchez-Tena, M.Á., Andreu-Vázquez, C., Villa-Collar, C.: Visual health and academic performance in school-aged children. Int. J. Environ. Res. Public Health **17**(7), 2346 (2020)

Bonhage, C.E., Mueller, J.L., Friederici, A.D., Fiebach, C.J.: Combined eye tracking and fMRI reveals neural basis of linguistic predictions during sentence comprehension. Cortex **68**, 33–47 (2015)

Dahan, N.A., Al-Razgan, M., Al-Laith, A., Alsoufi, M.A., Al-Asaly, M.S., Alfakih, T.: Metaverse framework: a case study on E-learning environment (ELEM). Electronics **11**(10), 1616 (2022)

del Rosario Sánchez-Valerio, M., Mohamed-Noriega, K., Zamora-Ginez, I., Duarte, B.G.B., Vallejo-Ruiz, V.: Dry eye disease association with computer exposure time among subjects with computer vision syndrome. Clin. Ophthalmol. **14**, 4311–4317 (2020)

Crnovrsanin, T., Wang, Y., Ma, K.L.: Stimulating a blink: reduction of eye fatigue with visual stimulus. In: Proceedings of the SIGCHI Conference on Human Factors in Computing Systems, pp. 2055–2064 (2014)

Helveston, E.M.: Visual training: current status in ophthalmology. Am. J. Ophthalmol. **140**(5), 903–910 (2005)

Gangamma, M.P., Poonam, M.R.: A clinical study on "Computer vision syndrome" and its management with Triphala eye drops and Saptamrita Lauha. Ayu **31**(2), 236 (2010)

Goldstand, S., Koslowe, K.C., Parush, S.: Vision, visual-information processing, and academic performance among seventh-grade schoolchildren: a more significant relationship than we thought? Am. J. Occup. Ther. **59**(4), 377–389 (2005)

Kirschner, P.A., Sweller, J., Kirschner, F., Zambrano, J.: From cognitive load theory to collaborative cognitive load theory. Int. J. Comput.-Support. Collab. Learn. **13**(2), 213–233 (2018)

Liu, B., Xing, W., Zeng, Y., Wu, Y.: Quantifying the influence of achievement emotions for student learning in MOOCs. J. Educ. Comput. Res. **59**(3), 429–452 (2021)

Mayer, R.E.: Multimedia learning. University Press, Cambridge (2001)

Morad, Y., Lederman, R., Avni, I., Atzmon, D., Azoulay, E., Segal, O.: Correlation between reading skills and different measurements of convergence amplitude. Curr. Eye Res. **25**, 117–121 (2002)

Shin, H.S., Park, S.C., Park, C.M.: Relationship between accommodative and vergence dysfunctions and academic achievement for primary school children. Ophthalmic Physiol. Opt. **29**(6), 615–624 (2009)

Wang, X., Han, M., Gao, Z., Wang, Z.: Research on the mechanism and optimisation strategy of visual and auditory emotional design in instructional videos. J. Dist. Educ. **8**(06), 50–61 (2020)

Wei, H.C., Chou, C.: Online learning performance and satisfaction: do perceptions and readiness matter? Distance Educ. **41**(1), 48–69 (2020)

Natural Language Processing for a Personalised Educational Experience in Virtual Reality

Nuha Alghamdi[1,2]([⊠]) and Alexandra I. Cristea[1]([⊠])

[1] Durham University, Durham, UK
{nuha.s.alghamdi,alexandra.i.cristea}@durham.ac.uk
[2] University of Jeddah, Jeddah, Kingdom of Saudi Arabia
nalghamdi@uj.edu.sa

Abstract. Virtual Reality (VR) is a technology that creates a simulated immersive environment, allowing users to be more engaged and interactive. The user can interact with a VR environment using head-mounted displays, hand controllers, and, in some cases, speech. VR has been widely used in various industries and areas, one of which is education, to build simulated and interactive experiences. However, little prior research has explored the integration of speech input in VR educational environments. Moreover, there is currently a lack of understanding of how speech and verbal/textual interaction can support personalisation in VR in general, and in the educational domain, in particular. Thus, this research targets to fill this gap. Our long-term goal is to *incorporate speech and text-based interaction in the VR learning environment, to support smooth and natural personalisation of the learning interaction.* Personalisation here is used in the classical AI in Education sense, of adapting the learning system to a learner, e.g. to their level or needs. As a first step, we have started exploring and comparing different speech recognition models that support VR applications. Further, we will personalise the user experience, by utilising the text generated from the speech input and applying NLP and adaptation techniques to it. Furthermore, we will investigate the impact of this kind of personalisation on learner engagement and outcomes.

Keywords: Natural Language Processing · Virtual Reality · Speech Recognition · Voice Recognition

1 Introduction

Virtual reality (VR) and natural language processing (NLP) are two distinct areas of research. VR focuses on creating immersive experiences, while NLP deals with processing and understanding human language [4,7].

One of the subfields of NLP is speech recognition, which identifies human spoken language and converts it into appropriate text, by processing speech analog signals and recognising speech patterns [12]. Integrating speech recognition technology into a virtual reality system not only broadens the scope of speech recognition in navigating virtual scenes, but also enhances the efficiency of user interaction with the virtual environment. Zhongxiang, Z. *et al.* showed in their work that speech commands can precisely direct the users' movements, ensuring they navigate through the scene without encountering obstacles [19].

Educators are primarily focused on improving learner engagement and educational outcomes, and researchers in the field of educational technology continually seek more effective teaching tools. VR is recognised as one such tool, with widespread agreement that it has the potential to enhance performance and deepen conceptual understanding in specific tasks [9]. The learning environment in VR has been considered by some researchers as 'more soothing', also allowing for the learning pace to be customised [18]. VR also proves advantageous in alleviating anxiety, where students can feel 'hidden' behind their avatars [2]. Moreover, VR can enhance student engagement, and elevate the occurrence of stimulating learning events for more effective learning [1]. Furthermore, statistics have shown that, aside from the gaming sector, VR is most useful in education and healthcare (agreed by 41% of the respondents [3]).

In this research, the focus is on *developing NLP techniques for text resulting from interactions in VR*, including the challenge of dealing with *free-flowing speech and learner-AI dialogue inside an educational VR experience* - and *deploying text and speech to personalise the user experience*. We will also study how this personalised experience can affect the learning engagement and outcomes.

1.1   Research Questions

Within the scope of this research, we aim to explore the following questions:

- **RQ1:** *How can we customise the learner's experience in VR based on their speech and language-based interaction?*; with emerging sub-questions e.g.:
  - **RQ1.1:** What is the best existing VR-based speech recognition model?
  - **RQ1.2:** How do current speech recognition systems need improve for appropriate use in personalised educational VR?
  - **RQ1.3:** How does speech recognition work when learners interact?
- **RQ2:** *How does customising an educational VR environment by integrating speech and NLP impact on user engagement and learning outcomes?*

2   Related Work

While there has been much research on speech recognition and Natural Language Processing (NLP), few researchers have taken speech recognition in Virtual Reality (VR) environments into consideration. A recent study [6] has shown that the speech interface in VR is significantly better than the other two interaction interfaces (2D and 3D) in multiple categories, including speed, ease of learning, and uncomplicated handling. Speech can be used for various purposes inside the VR environment. It can be used for education and training [15,17], for communication with other users [16], or for interaction with the app itself [10,19]. Here, we want to go beyond these applications, to use speech to personalise the user experience, by developing an adaptive learning system in the VR environment.

Combining VR technologies with adaptive learning systems offers optimal conditions for personalising the learning experience, leading to an enhanced quality of the educational process in terms of efficiency and effectiveness [11,14].

This work aims to bring together all these areas of research (VR, speech, NLP, education) towards novel ways of learning.

3 Methodology

To answer our research questions above, we first design a VR architecture (Sect. 3.1). Based on this architecture, we design our experiments, to evaluate the proposed solution (Sect. 3.2) and the answers to each research question (Sect. 1.1).

3.1 Architectural Design

The proposed architecture has three main parts: the *VR interface*, the *NLP module*, and the *interaction module*, as explained below.

VR Interface: The first step to answer RQ1 is to build a VR interface for the learner interaction. We use one of the well-known game engines, *Unity®*. We chose *Unity®* for several reasons; it is free for non-commercial use, and it allows the development of an interactive VR environment supporting multiple Head-Mounted Displays (HMDs). Additionally, *Unity®* Technology offers educational courses and resources for developers on their *Unity Learn*[1] platform. Moreover, they provide the *Asset Store*[2], where Unity users can download already modeled assets free, or at a very low price.

HMD is a VR headset that will be worn by the user to enter the VR experiment. We will choose *Meta Quest* (previously known as Oculus Quest) after finding that Oculus Rift (the older version) is the most popular HMD that was used in VR educational applications [5]. Oculus Rift was designed to work with a PC, but Meta Quest is a newer standalone version that runs on the Android platform. Meta Quest cameras allow the user to choose either to support 6 degrees of freedom (6DOF), which means the headset can track both the direction the user is looking (the orientation) and position, or just track the orientation (3DOF) [13].

---

[1] https://learn.unity.com/.
[2] https://assetstore.unity.com/.

NLP Module: The NLP module is needed to process all speech input and output (including creation). It will take speech input from the user, convert it into written text using a suitable voice SDK, store it in a database, then apply NLP techniques on the stored data to generate recommendations and personalise the user experience. NLP techniques could include classification and sentiment analysis.

One of the existing software development Kits (SDKs) that support speech input for VR using Unity is Voice SDK, which is powered by *Wit.ai*[3,4]. Wit.ai provides a high-quality voice experience and English language support with low latency, automatic speech recognition, and real-time transcription.

When incorporating NLP packages for VR, we will take into account the particular needs of the project, the NLP tasks we intend to perform, and the suitability of the packages with VR development engine.

Interaction/Adaptation Module: This module will focus on the interaction between learner, VR environment, and NLP module. It will handle the changes in the VR environment based on the learner's speech input and the NLP module outputs. It will define learner-AI and learner-learner interaction intelligence.

3.2   Evaluation Tool

For the VR-based NLP models, we will evaluate them using accuracy, precision, recall, and F1 score measures. For the VR experience, we will select an assessment tool based on the nature of the subject area and the learning tasks. Then, we will measure engagement (with e.g. a Web-based Learning Tools (WBLT) Evaluation Scale Questionnaire [8]) and learning outcomes using knowledge-based measures.

4   Implementation

4.1   Current Work

To answer RQ1.1, we built an initial virtual room using *Unity*®, to test *Wit.ai*, the Voice SDK (Fig. 1). This SDK allows us to understand what a learner says to the application. We tested two features so far, *voice commands* and *dictation*.

For *voice commands*, we pre-defined a set of specific intents that a learner will be able to ask the application to do. The set contains simple, generic, topic-agnostic commands for a learning environment, such as setting a timer with a specific time, modifying the timer, and deleting it. Then, we prepare a reaction for each specific command. For instance, a learner can set their alarm to go to class or to perform a learning task. The learner might accomplish this reaction via direct speech, and the system is able to make sense of it even if rephrased in various ways, for example: *'Set a timer for 3 min.'* or *'Create a timer that*

---

[3] https://developer.oculus.com/documentation/unity/voice-sdk-overview/.
[4] https://github.com/wit-ai/wit-unity.

*lasts for 3 min'...*etc. This would result, e.g., in a countdown timer appearing in front of the learner. On the training side, *Wit.ai* is trained by examples (or **utterances**). The more utterances provided, the better the AI will understand; however, not all utterances need to be provided. In addition, we may need to specify more information (**entities**) to be captured from each utterance, to help the AI understand more and respond accurately (see Fig. 2).

The *dictation feature* enables the application to efficiently transcribe speech to text in real time. We created a panel inside our VR room. The panel has an 'activate' button, to be pressed by a learner before starting to talk. Then, the dictation feature will convert their speech into text, which we will present in front of them in the same panel. This text can be stored in a database for further processing and personalisation. The dictation feature can work in the background, without an activation button or panel, but we showed it to the user in our experiment (see Fig. 3).

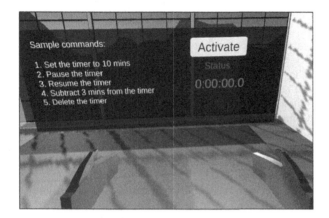

Fig. 1. Initial VR Learning Room with Pre-Defined Commands Using *Wit.ai*

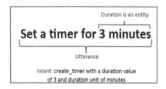

Fig. 2. Utterance, entity, and intent

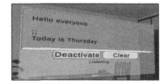

Fig. 3. Initial VR Learning Room with Dictation Using *Wit.ai*

5   Conclusion

Incorporating VR applications in the learning process is a promising subject that needs more efforts from researchers, with many parameters to explore that could enhance the learning experience and provide a personalised environment. Speech provides an interesting and convenient form of information-sharing that can be used to adapt the learning system to the user needs.

References

1. Allcoat, D., von Mühlenen, A.: Learning in virtual reality: effects on performance, emotion and engagement. Res. Learn. Technol. **26** (2018)
2. Dickey, M.D.: Three-dimensional virtual worlds and distance learning. Br. J. Edu. Technol. **36**(3), 439–451 (2005)
3. Gilbert, N.: 74 virtual reality statistics you must know in 2024 (2023). https://financesonline.com/virtual-reality-statistics/
4. Goldberg, Y.: Neural Network Methods for Natural Language Processing. Springer, Cham (2022)
5. Hamilton, D., et al.: Immersive virtual reality as a pedagogical tool in education: a systematic literature review of quantitative learning outcomes and experimental design. J. Comput. Educ. **8**(1), 1–32 (2021)
6. Hepperle, D., Weiß, Y., Siess, A., Wölfel, M.: 2D, 3D or speech? A case study on which user interface is preferable for what kind of object interaction in immersive virtual reality. Comput. Graph. **82**, 321–331 (2019)
7. Jerald, J.: The VR Book: Human-Centered Design for Virtual Reality. Morgan & Claypool (2015)
8. Kay, R.: Evaluating learning, design, and engagement in web-based learning tools (WBLTs). Comput. Hum. Behav. **27**(5), 1849–1856 (2011)
9. Lee, E.A.-L., Wong, K.W.: A review of using virtual reality for learning. In: Pan, Z., Cheok, A.D., Müller, W., El Rhalibi, A. (eds.) Transactions on Edutainment I. LNCS, vol. 5080, pp. 231–241. Springer, Heidelberg (2008). https://doi.org/10.1007/978-3-540-69744-2_18
10. Luo, J.: Research on speech recognition system based on virtual reality technology. In: 4th International Conference on ICEKIM, pp. 1547–1554. Atlantis Press (2023)
11. Marienko, V.M., Nosenko, Y.H., Shyshkina, M.P.: Personalization of learning using adaptive technologies and augmented reality. In: Burov, O., Kiv, A. (eds.) 3rd International Workshop on AR in Education. CEUR, vol. 2731, pp. 341–356 (2020)
12. Meng, J., Zhang, J., Zhao, H.: Overview of the speech recognition technology. In: 4th International Conference on Computational and Information Sciences, pp. 199–202. IEEE (2012)
13. Meta: Meta quest headset tracking (2023). https://www.meta.com/help/quest/articles/headsets-and-accessories/using-your-headset/turn-off-tracking/
14. Osadchyi, V., et al.: Conceptual model of learning based on the combined capabilities of augmented and virtual reality technologies with adaptive learning systems. In: CEUR, vol. 2731, pp. 328–340 (2020)
15. Valls-Ratés, Í., Niebuhr, O., Prieto, P.: Unguided virtual-reality training can enhance the oral presentation skills of high-school students. Front. Commun. **7**, 196 (2022). https://doi.org/10.3389/fcomm.2022.910952
16. Yan, Y., et al.: ConeSpeech: exploring directional speech interaction for multi-person remote communication in VR. IEEE Trans. Vis. Comput. Graph. **29**(5), 2647–2657 (2023). https://doi.org/10.1109/TVCG.2023.3247085
17. Yuan, J., et al.: Research on project-based learning of foreign trade English in speech recognition virtual reality environment. Soft Comput. 1–12 (2023)

18. Zhou, Y.: VR technology in English teaching from the perspective of knowledge visualization. IEEE Access (2020)
19. Zhu, Z., et al.: Application of speech recognition technology to virtual reality system. In: International Industrial Informatics and Computer Engineering Conference, pp. 103–107. Atlantis Press (2015)

From "Giving a Fish" to "Teaching to Fish": Enhancing ITS Inner Loops with Large Language Models

Yang Pian[1], Muyun Li[1], Yu Lu[1,2(✉)], and Penghe Chen[1,2]

[1] Faculty of Education, Beijing Normal University, Beijing, China
luyu@bnu.edu.cn
[2] Advanced Innovation Center for Future Education, Faculty of Education, Beijing Normal University, Beijing, China

Abstract. In this work, we aim to enhance the Intelligent Tutoring Systems (ITS) inner loop by incorporating Large Language Models (LLMs), shifting the educational approach from merely providing solution (giving a fish) to fostering deeper understanding and self-sufficiency (teaching to fish) in learning. Specifically, we propose a framework that utilizes LLMs to generate meaningful scaffoldings and facilitate insightful interactions within ITS tasks, guided by established learning science theories. Preliminary results from the model and educational experiments suggest that our LLM-generated scaffoldings could improve learning outcomes. From our current progress, we have gained valuable insights for further improvement. We hope this study will contribute to the development of smarter and more human-centered ITS solutions in the future.

Keywords: Inner loop · Intelligent tutoring system · Scaffolding · LLMs

1 Background and Related Work

Intelligent Tutoring Systems (ITS) are computer-based systems designed to mimic the interaction between a tutor and a learner [10]. By integrating artificial intelligence techniques and cognitive science theories, ITS analyzes learners' responses and progress to deliver tailored feedback, guidance, and educational content, aiming to optimize learning efficiency and outcomes across various disciplines. To achieve this goal, a key component in ITS is *the inner loop*, which serves as a real-time support mechanism within a learning task. The supports include scaffoldings such as minimal feedback (correct or incorrect), next-step hints, and solution reviews, so as to enhance knowledge mastery during tasks [1].

However, scaffoldings in today's ITS inner loops confront two main issues. Firstly, the creation of these scaffoldings often relies either on time-consuming manual efforts by teachers, or on rule-based or simplistic algorithms, leading to a lack of flexibility [11]. Secondly, most existing systems tend to provide

A. M. Olney et al. (Eds.): AIED 2024 Workshops, CCIS 2151, pp. 362–368, 2024.
https://doi.org/10.1007/978-3-031-64312-5_44

direct answers on each task step, thus failing to equip learners with independent problem-solving and critical thinking skills [12]. Given these challenges, it's crucial to re-imagine the ITS inner loops in a way that aligns with the "teaching to fish" philosophy, paving the way for constructing a learning environment that encourages autonomy and deep thinking.

In recent years, Large Language Models (LLMs), such as the GPT series, have shown remarkable potential across various domains due to their robust capabilities of language understanding, generation, and task execution [3]. In the educational context, it has been reported that LLMs can be effectively applied to automated feedback systems [4] and provide learners with interactive, immersive online learning experience [2], which indicates the educational potential of LLMs in supporting learning and facilitating natural interactions. This also presents opportunities for ITS to enhance the design of the inner loops.

Despite the above advancements, directly integrating LLMs into ITS inner loops might pose challenges, especially for disciplines requiring logical reasoning like mathematics and physics. On one hand, hallucination issues of LLMs [14] can lead to misleading content. On the other hand, LLMs' outputs are sometimes repetitive and superficial, which might not be suitable to encourage deep engagement and deep thinking [7]. These problems could compromise the educational value of ITS. Therefore, careful consideration and strategic approaches are necessary to mitigate these risks and fully realize the benefits of LLMs integration into ITS.

In this study, we aim to integrate LLMs into ITS inner loops to foster deeper learner engagement and critical thinking by automatically generating scaffoldings and enabling dynamic interaction. Our research focuses on two primary questions:

- **RQ1:** How can we effectively use LLMs to automatically generate multi-step scaffoldings of learning tasks for the ITS inner loops?
- **RQ2:** How can we integrate LLM-generated scaffoldings into ITS inner loops and design interactive mechanisms to facilitate meaningful engagement?

To address these two questions, we have proposed a framework which will be introduced in the next section. Then We will provide an update progress of our work, including preliminary results from model and educational experiments. Finally, we will discuss the anticipated outcomes and future directions.

2 Proposed Solution and Methodology

We proposed a solution framework to enhance ITS inner loops with LLMs, illustrated in Fig. 1. It encompasses two main modules: (1) the Scaffolding Generating Module for automatically creating meaningful task scaffoldings, which addresses RQ1, and (2) the Interaction Monitoring Module for providing engaging interactions and heuristic feedback, which addresses RQ2. Under the guidance of learning science theories, both modules leverage LLMs to better comprehend system-learner interactions and generate the necessary learning support for learners.

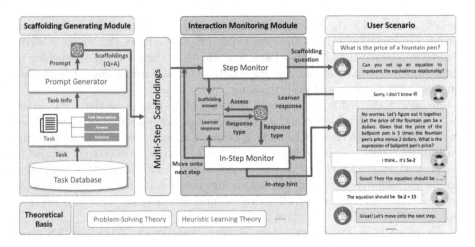

Fig. 1. System architecture of our proposed solution.

2.1   Theoretical Basis

Our solution mainly leverages two classic learning science theories to guide the integration of LLMs with the ITS inner loop, namely Problem-Solving Theory and Heuristic Learning Theory.

Problem-Solving Theory: Problem-Solving Theory provides a structured framework that empowers learners to complete learning tasks with expert-like analytical skills. In this study, we adopt a well-known four-step framework [9] to guide LLMs in generating meaningful learning scaffoldings, which include representation, planning, execution, and evaluation. This framework could develop meta-cognition for ITS learners within the inner loop.

Heuristic Learning Theory: Heuristic Learning Theory advocates for guiding learners through exploration and comprehension rather than providing direct answers to them [6]. Within the ITS inner loop, it's essential to identify learners' needs through their interactions with the scaffoldings and provide heuristic guidance, creating a more effective learning process. Hence, our solution incorporates the principles of heuristic learning into the interaction design.

2.2   The Scaffolding Generating Module

The Scaffolding Generating Module automates the generation of multi-step scaffoldings for various learning tasks. Its core component is **the Prompt Generator**. Specifically, the Prompt Generator takes the learning task information and the 4-steps problem-solving framework as input to create LLMs-friendly prompts. Adopting the problem-solving framework, rather than solely relying on the Chain-of-Thought (CoT) strategy [13], facilitates the generation of universal problem-solving steps and enhances learners' metacognitive skills. The generated

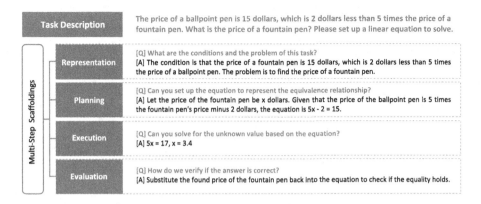

Fig. 2. An example of the generated 4-steps scaffoldings.

prompts, incorporating elements like role-playing, task outlines, examples, and output limitations, aim to precisely convey the scaffolding generation requirement to the LLMs.

The workflow of the Scaffolding Generating Module is as follows: it first retrieves learning tasks (e.g., exercise problems) from the ITS task bank, typically including their text description, answers, and explanations. The Prompt Generator then transform the gathered data, together with the description of our 4-steps framework, into customized prompts, which are then fed into the LLMs. The final output of this module includes 4-steps scaffoldings, each consisting of a scaffolding question and its answer. Figure 2 showcases an example of the generated 4-steps scaffoldings.

2.3   The Interaction Monitoring Module

The Interaction Monitoring Module coordinates learning tasks and learner interactions within the ITS inner loop. Its two key components, namely the Step Monitor and the In-Step Monitor, work together to promote ITS users' understanding and engagement. Specifically, **the Step Monitor** monitors the interaction process within the Inner Loop and presents multi-step scaffolding questions to learners based on the learning task, guiding them to systematically complete the tasks. During each scaffolding interaction, **the In-Step Monitor** evaluates learners' responses to each scaffolding question, providing heuristic feedback based on their responses to encourage in-depth thinking rather than direct answer-seeking.

The workflow is as follows: when a new learning task begins, the Step Monitor first retrieves the corresponding multi-step scaffolding from the scaffolding bank and presents the first scaffolding question on the ITS interface, awaiting the learner's response. Once receiving the learner's answer, the In-Step Monitor evaluates it against the standard answer, categorizing the response as *correct, incorrect, help-seeking,* or *irrelevant*. Each response type triggers a specific

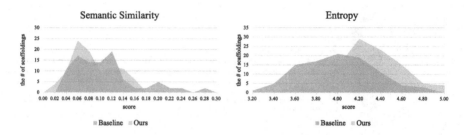

Fig. 3. Comparative analysis results of scaffolding quality: our solution vs. GPT 3.5.

heuristic strategy. (1) For correct responses, the In-Step Monitor informs the Step Monitor, which subsequently introduces the next scaffolding question to the learner. (2) For incorrect, help-seeking, or irrelevant responses, the In-Step Monitor administers heuristic prompts to foster deeper thinking. If the learner persistently fails after multiple interaction cycles (exceeding the pre-set threshold), the In-Step Monitor would explain the correct answer to the learner, and then inform the Step Monitor to proceed to the next scaffolding question.

Throughout this process, LLMs are employed to analyze the consistency between learners' responses and standard scaffolding answers in real-time, and categorize them so as to provide proper heuristic feedback. This function is achieved through prompt engineering, where the LLMs is tailored with role-specific adjustments, enabling it to adopt distinct guiding strategies.

3 Current Progress and Preliminary Results

At present, we have developed the Scaffolding Generating Module, and obtained preliminary findings from the model and educational experiments. These results, together with feedback from participants in the educational experiment, will inform future improvements to the Interaction Monitoring Module as well as the whole framework. The preliminary experimental results are presented below.

3.1 Results from the Model Experiment

To assess the quality of scaffoldings generated by our proposed Scaffolding Generating Module, we conducted model experiments using the most recent GPT 3.5 series model (gpt-3.5-turbo-0125). Specifically, we collected 96 middle school-level mathematics tasks from a real-world ITS [8] as our evaluation dataset. For comparison, we generated scaffoldings using our solution and directly using GPT 3.5 (serving as the baseline). We evaluated the results using BERT-based semantic similarity for relevance and information entropy [5] for richness. Results are presented in Fig. 3.

As illustrated in Fig. 3, firstly, the semantic similarity scores are positive, indicating that our generated scaffoldings maintained semantic consistency with the task statements. Most similarity scores range from 0.00 to 0.15, indicating that the generated content is not just a paraphrase of the task, but has additional revealing information. Secondly, the information entropy of our solution ($Mean = 4.2$) is higher than the baseline ($Mean = 3.9$), indicating that our solution offers richer content and provides more diverse inspiration for learners.

3.2 Results from the Educational Experiment

To evaluate the educational effectiveness of our generated scaffoldings, We conducted a quasi-experiment with 28 middle school students during a math learning session on *Linear Equations*. Within a 15-min task, the experimental system delivered scaffoldings in dialogue form, and students responded in kind. Paired-t test on pre-($Mean = 0.23$ out of 1) and post-knowledge ($Mean = 0.68$ out of 1) assessments on Linear Equations ($t = -.45$, $p < 0.001$) indicate that our generated scaffoldings could facilitate knowledge acquisition. Notedly, in the post-knowledge assessment, students displayed well-structured problem-solving processes, demonstrating that the generated scaffoldings effectively embody the "teaching to fish" principle. Additionally, results on technology acceptance ($Mean = 4.17$, $SD = 1.19$) and learning satisfaction ($Mean = 4.20$, $SD = 1.08$) also indicate the effectiveness of our solution. After the experiment, we conducted interviews with 6 students to obtain further insights for improving our design.

4 Future Work and Expected Contribution

Based on our current progress, several major challenges should be addressed in the future. Firstly, although a quasi-experiment has been conducted, it is necessary to expand the sample size and conduct comparative experiments in the future. Secondly, it is necessary to further enrich the model evaluation metrics, e.g., explore the correlation between scaffolding steps. Thirdly, cross-disciplinary validation is required to verify the generality and reliability of the proposed methods.

Our study also holds certain potential. Firstly, based on the interviews in our quasi-experiment, learners expect the scaffoldings to be more in line with their current level of knowledge mastery. In the future, we plan to introduce personalized design in the Interaction Monitoring Module, in order to cater to learners with different ability levels. Furthermore, the key modules constructed in this study, such as scaffolding generation, and heuristic interaction, possess distinct educational functionalities and can be utilized either independently or in an integrated manner. This flexibility enables the construction of AI Agents tailored for ITS scenarios, thereby facilitating the development of more intelligent and human-centered ITS solutions.

Acknowledgement. This work was supported by National Key Research and Development Program of China (2022YFC3303600).

References

1. Azevedo, R., Hadwin, A.F.: Scaffolding self-regulated learning and metacognition-implications for the design of computer-based scaffolds. Instr. Sci. **33**(5/6), 367–379 (2005)
2. Bicknell, K.: GPT-4 completes Duolingo's language learning nest (2023). https://analyticsindiamag.com/gpt-4-completes-duolingos-language-learning-nest/
3. Brown, T., et al.: Language models are few-shot learners. In: Advances in Neural Information Processing Systems, vol. 33, pp. 1877–1901 (2020)
4. Dai, W., et al.: Can large language models provide feedback to students? A case study on ChatGPT. In: 2023 IEEE International Conference on Advanced Learning Technologies (ICALT), pp. 323–325. IEEE (2023)
5. Gray, R.M.: Entropy and Information Theory. Springer, New York (2011). https://doi.org/10.1007/978-1-4419-7970-4
6. Hjeij, M., Vilks, A.: A brief history of heuristics: how did research on heuristics evolve? Humanit. Soc. Sci. Commun. **10**(1), 1–15 (2023)
7. Kasneci, E., et al.: ChatGPT for good? On opportunities and challenges of large language models for education. Learn. Individ. Differ. **103**, 102274 (2023)
8. Lu, Y., Pian, Y., Chen, P., Meng, Q., Cao, Y.: RadarMath: an intelligent tutoring system for math education. In: Proceedings of the AAAI Conference on Artificial Intelligence, vol. 35, pp. 16087–16090 (2021)
9. Polya, G.: How to Solve It: A New Aspect of Mathematical Method, vol. 85. Princeton University Press (2004)
10. Shute, V.J., Psotka, J.: Intelligent tutoring systems: past, present, and future. Armstrong Laboratory, Air Force Materiel Command (1994)
11. Tenório, T., Isotani, S., Bittencourt, I.I.: Authoring inner loops of intelligent tutoring systems using collective intelligence. In: Rodrigo, M.M., Matsuda, N., Cristea, A.I., Dimitrova, V. (eds.) AIED 2022. LNCS, vol. 13356, pp. 400–404. Springer, Cham (2022). https://doi.org/10.1007/978-3-031-11647-6_79
12. VanLehn, K.: The behavior of tutoring systems. Int. J. Artif. Intell. Educ. **16**(3), 227–265 (2006)
13. Wei, J., et al.: Chain-of-thought prompting elicits reasoning in large language models. In: Advances in Neural Information Processing Systems, vol. 35, pp. 24824–24837 (2022)
14. Zhang, Y., et al.: Siren's song in the AI ocean: a survey on hallucination in large language models. arXiv preprint arXiv:2309.01219 (2023)

The Effectiveness of AI Generated, On-Demand Assistance Within Online Learning Platforms

Aaron Haim[✉][iD], Eamon Worden[✉][iD], and Neil T. Heffernan[✉][iD]

Worcester Polytechnic Institute, Worcester, MA 01609, USA
{ahaim,elworden,nth}@wpi.edu

Abstract. Since GPT-4's release it has shown novel abilities in a variety of domains. This paper explores the use of LLM-generated explanations as on-demand assistance for problems within the ASSISTments platform. In particular, we are studying whether GPT-generated explanations are better than nothing on problems that have no supports and whether GPT-generated explanations are as good as or better than teacher-authored explanations. This study contributes to existing literature since as of yet, there are no studies on the scale of ASSISTments evaluating the effectiveness of GPT support in education. Should GPT explanations prove effective then we plan to continue developing and evaluating explanations, hints, and other supports with GPT within ASSISTments.

Keywords: AI Generated Assistance · On-Demand Assistance · Online Learning Platform

1 Introduction

Over the last few years, large language models, or LLMs, have become increasingly mainstream. LLMs such as GPT4 from OpenAI[1], Llama from Meta[2], Bard from Google[3], and many others have been integrated into several online learning platforms, such as Khan Academy [6]. Using the generated content, the platforms can provide more accessible and nuanced information that can help their intended audience.

Within education research, LLMs were used for numerous experiments such as simulating students [7] and generating assistance for a problem [9]. These studies have shown that LLM-generated content generally improves outcomes given some level of filtering for incorrect responses and appropriate prompt engineering. When compared to teacher-generated content, LLM generated content was generally less effective [19]. However, most of these applications were distributed

---

[1] https://openai.com/gpt-4.
[2] https://ai.meta.com/llama/.
[3] https://bard.google.com/.

A. M. Olney et al. (Eds.): AIED 2024 Workshops, CCIS 2151, pp. 369–374, 2024.
https://doi.org/10.1007/978-3-031-64312-5_45

with a small number of participants; few within online learning platforms have been deployed or evaluated at scale.

The goal of this work is to expand on these prior works by running an experiment with LLM-generated on-demand assistance, in the form of explanations, within an online learning platform at scale. First, we will compare the LLM-generated explanations to when there is no content available. There have been previous studies on the effectiveness of on-demand assistance [10,12], so this experiment will be a replication with the authors being an LLM. Then, we will compare the LLM-generated explanations to those written by experts already within the online learning platform. Unlike previous studies, this study is looking for where there is no significant negative correlation between student learning and LLM-generated explanations. If so, then LLM generated explanations could be used in at-scale environments more broadly given some level of preprocessing, with the confidence it will not be significantly worse than teacher generated content.

Specifically, this work aims to answer the following questions (henceforth referred to as Research Questions, or RQ):

1. Do LLM-generated explanations improve student learning compared to when no assistance is available?
2. Do LLM-generated explanations not decrease student learning compared to expert-generated explanations already within the online learning platform?

2   Background

2.1   The ASSISTments Platform and On-Demand Assistance

ASSISTments[4] is an online learning platform that provides immediate feedback to students and detailed analytics for instructors to better inform classroom instruction [2,4]. Similar to other online learning platforms, instructors can assign sets of problems from open educational resources, the majority of which are K-12 mathematics. When a piece of assistance has been written for a problem by the assigning instructor or approved for general consumption, students can request the assistance at any point while completing the current problem. Assistance within the ASSISTments platform comes in numerous forms such as hints guiding the student towards the answer [4,10,13], worked examples to similar problems [8], examples with errors [1,13], feedback on common wrong answers [3,14,15], and worked solutions to the current problem [10,17,18].

2.2   AI Generation and Prompt Engineering

Previous attempts to generate explanations using large language models found a 50% error rate, but utilized the formerly state-of-the-art model GPT3 [11]. Since then, superior model and prompting techniques have been able to achieve up to

---

[4] https://assistments.org.

a 90% accuracy rate [5]. However, as of yet these have not been deployed on a scale as large as ASSISTments, nor compared to teacher-authored explanations, both of which we will investigate in this study.

In this paper, we utilize a chain of thought [16] prompting technique to improve GPT 4's ability to solve and explain math questions. Requesting a step-by-step explanation GPT4 provides both a readable explanation and a more accurate explanation.

3 Methodology

Below are the prompts we used for generating explanations:

Prompt 1: Generating Explanations for Close Answered Questions

"Write a step by step explanation for how to solve this {problem_type} problem: {problem}. This is the correct answer: {correct_answer}. Here are all the answers: {all_answers}. Use language and ideas appropriate for {grade}. Use the active voice. Be concise. Write a step by step explanation. Do not repeat the question. Format the output with HTML. Avoid long paragraphs. Write each step as its own paragraph."

Prompt 2: Generating Explanations for Open Answered Questions

"Write a step by step explanation for how to solve this problem: {problem}. This is the correct answer: {correct_answer}. Use language and ideas appropriate for {grade}. Use the active voice. Be concise. Write a step by step explanation. Do not repeat the question. Format the output with HTML. Avoid long paragraphs. Write each step as its own paragraph."

We provide one explanation generated by GPT as it would appear for students using ASSISTments:

Step 1: First, we need to find the difference between the actual weight of the egg and the weight it's supposed to be. In this case, the egg is supposed to weigh 2.5 oz, but it actually weighs 2.4 oz. So, subtract 2.4 oz from 2.5 oz. The difference is 0.1 oz.

Step 2: Next, we need to divide this difference by the weight the egg is supposed to be. This will give us the error as a decimal. So, divide 0.1 oz by 2.5 oz.

Step 3: The result from step 2 is a decimal. To convert this decimal into a percentage, we need to multiply it by 100.

Step 4: The result from step 3 is the percent error. This is the answer to the problem. In this case, the percent error is 4.

Below is the prompt used to check explanations

Prompt 3: Checking Explanation

"The following is a step by step explanation for how to solve this {problem_type} problem: {problem}. Here is the explanation: {explanation}. Only write an explanation for part {position}. This is the correct answer: {correct_answer}. Here are all the answers: {all_answers}. The explanation should use language and ideas appropriate for {grade}. Score the explanation in 'correctness' as a 0 if the explanation is incorrect and a 1 if the explanation is correct and makes sense and score the explanation in 'appropriateness' as a 0 if it uses language or ideas inappropriate for the grade and a 1 if it uses language or ideas appropriate for the grade. Output the answer as a JSON."

3.1   Experiment 1: AI Generated Assistance Vs No Assistance

We selected 145 problems from the Illustrative Math and EngageNY curriculums which were easily interpretable by ChatGPT (no images or other information), and did not have an existing teacher explanation in the ASSISTments platform. We removed HTML tags from the problem bodies as so that they would not affect GPT4's performance on math questions. We used prompt 1 to write an explanation for all 'multiple choice', 'select all that apply' and 'rank the options' questions in our 145 problems. For fill-in-the-blank questions, we used prompt 2. We then used prompt 3 to identify the potentially incorrect or inappropriate prompts so we could manually remove them.

We manually reviewed every explanation that GPT4 determined was either incorrect or not grade-appropriate. We then checked 10 random explanations and found each of them to be correct. We ended up with 130 explanations which were deployed into ASSISTments for experiment 1.

Students are randomized on the problem level into either the treatment group where the student can request a GPT-generated explanation, or the control group where no explanation will be available. We plan to use the following linear mixed-effects model to analyze the next problem correctness:

$$next\_problem\_correctness\_binary \sim$$
$$control\_treatment\_assignment+$$
$$prior\_5pr\_avg\_correctness+ \qquad (1)$$
$$(1|problem) + (1|class)$$

3.2   Experiment 2: AI Generated Assistance vs Expert Created Assistance

We selected 277 problems from the Illustrative and EngageNY curriculums which were easily interpretable by ChatGPT (no images or other information), and did have an existing teacher explanation in ASSISTments. We removed HTML tags from the problem bodies. We utilized the same process and prompts 1–3 to generate explanations for these 277 problems. We determined that 233 were good enough to be deployed into ASSISTments. Each assignment is randomized into either the treatment group, where the student can request a GPT-generated explanation, or the control group, where the student can request a teacher-generated assistance. As the randomization is on the assignment level, the outcome will measure the average correctness of all problems after the problem the student views the first assistance on, as shown in Eq. 2.

$$
\begin{aligned}
average\_correctness\_after\_first\_assistance \sim \\
control\_treatment\_assignment+ \\
prior\_5pr\_avg\_correctness+ \\
(1|problem) + (1|class)
\end{aligned}
\tag{2}
$$

The study was anonymously preregistered on OSF.[5] It is currently anonymous until the embargo period has passed.

Acknowledgements. We would like to thank NSF (2118725, 2118904, 1950683, 1917808, 1931523, 1940236, 1917713, 1903304, 1822830, 1759229, 1724889, 1636782, & 1535428), IES (R305N210049, R305D210031, R305A170137, R305A170243, R305A180401, R305A120125, & R305R220012), GAANN (P200A120238, P200A180088, P200A150306, & P200A150306), EIR (U411B190024 S411B210024, & S411B220024), ONR (N00014-18-1-2768), NIH (via a SBIR R44GM146483), Schmidt Futures, BMGF, CZI, Arnold, Hewlett and a $180,000 anonymous donation. None of the opinions expressed here are those of the funders.

References

1. Adams, D.M., et al.: Using erroneous examples to improve mathematics learning with a web-based tutoring system. Comput. Hum. Behav. **36**, 401–411 (2014)
2. Feng, M., Heffernan, N.T.: Informing teachers live about student learning: reporting in the assistment system. Technol. Instr. Cogn. Learn. **3**(1/2), 63 (2006)
3. Gurung, A., et al.: How common are common wrong answers? Crowdsourcing remediation at scale. In: Proceedings of the Tenth ACM Conference on Learning @ Scale, L@S 2023, pp. 70–80. Association for Computing Machinery, New York (2023). https://doi.org/10.1145/3573051.3593390
4. Heffernan, N.T., Heffernan, C.L.: The ASSISTments ecosystem: building a platform that brings scientists and teachers together for minimally invasive research on human learning and teaching. Int. J. Artif. Intell. Educ. **24**(4), 470–497 (2014)
5. Imani, S., Du, L., Shrivastava, H.: MathPrompter: mathematical reasoning using large language models (2023)
6. Khanmigo (2023). https://khanmigo.ai/
7. Markel, J.M., Opferman, S.G., Landay, J.A., Piech, C.: GPTeach: interactive ta training with GPT-based students. In: Proceedings of the Tenth ACM Conference on Learning @ Scale, L@S 2023, pp. 226–236. Association for Computing Machinery, New York (2023). https://doi.org/10.1145/3573051.3593393

---

[5] https://osf.io/b7p6v/?view_only=03b19d094a9440a0ae5df4177907a1d1.

8. McLaren, B.M., van Gog, T., Ganoe, C., Karabinos, M., Yaron, D.: The efficiency of worked examples compared to erroneous examples, tutored problem solving, and problem solving in computer-based learning environments. Comput. Hum. Behav. **55**, 87–99 (2016)
9. Pardos, Z.A., Bhandari, S.: Learning gain differences between ChatGPT and human tutor generated algebra hints (2023)
10. Patikorn, T., Heffernan, N.T.: Effectiveness of crowd-sourcing on-demand assistance from teachers in online learning platforms. In: Proceedings of the Seventh ACM Conference on Learning @ Scale, L@S 2020, pp. 115–124. Association for Computing Machinery, New York (2020). https://doi.org/10.1145/3386527.3405912
11. Prihar, E., et al.: Comparing different approaches to generating mathematics explanations using large language models. In: Wang, N., Rebolledo-Mendez, G., Dimitrova, V., Matsuda, N., Santos, O.C. (eds.) AIED 2023. CCIS, vol. 1831, pp. 290–295. Springer, Cham (2023). https://doi.org/10.1007/978-3-031-36336-8_45
12. Prihar, E., Patikorn, T., Botelho, A., Sales, A., Heffernan, N.: Toward personalizing students' education with crowdsourced tutoring. In: Proceedings of the Eighth ACM Conference on Learning @ Scale, L@S 2021, pp. 37–45. Association for Computing Machinery, New York (2021). https://doi.org/10.1145/3430895.3460130
13. Razzaq, L.M., Heffernan, N.T.: To tutor or not to tutor: that is the question. In: Dimitrova, V. (ed.) AIED, pp. 457–464. IOS Press (2009)
14. Schnepper, L.C., McCoy, L.P.: Analysis of misconceptions in high school mathematics. Netw. Online J. Teach. Res. **15**(1), 625–625 (2013). https://newprairiepress.org/networks/vol15/iss1/7/
15. VanLehn, K., Siler, S., Murray, C., Yamauchi, T., Baggett, W.B.: Why do only some events cause learning during human tutoring? Cogn. Instr. **21**(3), 209–249 (2003). https://doi.org/10.1207/S1532690XCI2103_01
16. Wei, J., et al.: Chain-of-thought prompting elicits reasoning in large language models. In: Koyejo, S., Mohamed, S., Agarwal, A., Belgrave, D., Cho, K., Oh, A. (eds.) Advances in Neural Information Processing Systems, vol. 35, pp. 24824–24837. Curran Associates, Inc. (2022). https://proceedings.neurips.cc/paper_files/paper/2022/file/9d5609613524ecf4f15af0f7b31abca4-Paper-Conference.pdf
17. Whitehill, J., Seltzer, M.: A crowdsourcing approach to collecting tutorial videos–toward personalized learning-at-scale. In: Proceedings of the Fourth ACM Conference on Learning@ Scale, pp. 157–160 (2017)
18. Williams, J.J., et al.: AXIS: generating explanations at scale with learnersourcing and machine learning. In: Proceedings of the Third ACM Conference on Learning@ Scale, pp. 379–388 (2016)
19. Xiao, C., Xu, S.X., Zhang, K., Wang, Y., Xia, L.: Evaluating reading comprehension exercises generated by LLMs: a showcase of ChatGPT in education applications. In: Proceedings of the 18th Workshop on Innovative Use of NLP for Building Educational Applications (BEA 2023), pp. 610–625 (2023)

Culturally Relevant Artificial Intelligence Education with Music for Secondary School Students

Nora Patricia Hernández López[(✉)] and Xiao Hu[(✉)]

The University of Hong Kong, Pok Fu Lam, Hong Kong S.A.R., China
noraphl@connect.hku.hk, xiaoxhu@hku.hk

Abstract. Artificial Intelligence (AI) has increasingly gained attention in recent years, and with it, the need to involve youth in responsible uses the technology. I propose a method that aims to equip secondary school students with the necessary knowledge and skills to identify, describe, interact, and create with AI responsibly. Particularly, my research aims to engage students in culturally relevant learning integrating music as a cultural signifier. Using a Constructionist approach, I propose a method in which students will learn about AI through making personally meaningful musical artefacts. By situating the study in two different sociocultural contexts, this study aims to characterise the extent to which different local cultures influence learning outcomes (cognitive, affective, and behavioural) when learning about AI. This study aims to provide a systematic method for collecting and analysing data from students' interactions with an AI-enabled music creation platform, thus enabling a holistic understanding of students' learning processes.

Keywords: AI Education · Multimodal Learning Analytics · Culturally relevant education · Music

1 Background and Motivation

In recent years, educational research in AI education has led to evidence-based guidelines, recommendations, and design considerations about what students in K-12 should learn about AI [12]. Nonetheless, while recent developments in AI curricula for K-12 aim to provide a comprehensive understanding of technical and ethical aspects of AI, little do we know about the influence of students' cultural background and personal interests and identities over their learning outcomes, including cognitive, affective, and behavioural.

Culture is defined as the set of practices within a group that are different to other groups, and that gives identity and belongingness to members of said group [9]. Culture is embedded in many everyday practices, for example, language, rituals, and artefacts. In my research, I am interested in the use of music as a maker of culture. Music helps in establishing a collective identity well differentiated from others through style and genre, particularly among the

A. M. Olney et al. (Eds.): AIED 2024 Workshops, CCIS 2151, pp. 375–381, 2024.
https://doi.org/10.1007/978-3-031-64312-5_46

youth. In this way, music provides the sense of authenticity and belonging to a particular identity of the local culture and makes possible the expression of a plurality of identities through construction and consumption of different musical genres [3]. Taking advantage of this, music has been widely used in computing education research to study its potential for providing meaningful and authentic learning, especially among K-12 students, e.g., [7]. Specifically for AI education, Fiebrink [8] implemented an AI curriculum and creation tool for artists and musicians to learn about AI by creating musical artefacts using Machine Learning (ML) algorithms. Nonetheless, the potential of music for leveraging AI education in the K-12 context is yet to be explored.

The use of music as part of the cultural framing of this study calls for a culturally relevant educational practice. Framed under social constructivist theory, the work of Au [1] and Ladson-Billings [11] has been instrumental in the field of culturally relevant education (CRE). CRE places learning at the intersection of culturally relevant practices and individual knowledge construction, and acknowledges that social interactions between students, their peers, teachers, and the environment shape knowledge and understanding. In this way, sociocultural perspectives in learning place culture as central piece in knowledge construction, highlighting the locality of such processes [13]. While these approaches have been studied in classrooms where a variety of marginalised identities often collide with, or are oppressed by, the dominant culture [14], I intend to use the sociocultural framework to understand how local culture, which encompasses heritage culture and at the same time is highly dynamic and community oriented (ibid), can influence learning. In this study, local students in a fairly homogeneous learning environment will enact musical practices, unique to them but similar to their local peers. This study is situated in two different sociocultural contexts, namely Mexico and Hong Kong. Thus, by understanding how students build from within their own local culture to experiment with sound and music for learning about AI, we can have a better understanding of students' learning outcomes (cognitive, affective, and behavioural), allowing an exploration of any possible differences between the two regions, and whether these differences can be attributed to a culturally relevant curriculum.

Through a purposeful curation of the curriculum to include music that is familiar, or not, for students as a way to promote culturally relevant learning, this research aims to answer two research questions:

For secondary school students, (1) what are the best practices for designing and implementing a culturally relevant AI curriculum?, and (2) to what extent using music and a culturally relevant AI curriculum influences learning outcomes?

2   Learning Analytics in Computing Education, the Arts, and AI

The use of data from integrated development environments (IDEs) where students learn how to code has been instrumental in understanding students' behaviours and learning processes. In this sense, learning analytics (LA) offer great advantages for analysing these type of data for understanding and enhancing learning. For instance, [2] implemented implement LA methodologies to understand individual learning progressions, and identify when a student is more like to exhibit learning of one skill. In their model, students' interactions are first characterised by the probability of learning a specific skill at one moment in time, conditional of previous responses. Using a similar understanding of the current state of student knowledge at a given point in time as evidence of learning, [6] study coding behaviours of novice programmers in learning programming methodologies in Java, while [5] developed their own coding platform where students program soccer playing robots to assess tinkering practices as demonstration of increased programming proficiency. In both studies, the assessment of learning is done using metrics of code editions, and not directly evaluating the quality of the outcome. While this approaches can still be found in more recent literature in the same domain, other metrics to evaluate goodness of the solution are included as well, such as syntax or compilation errors, on top of keystroke level data.

Specifically in the field of artistic computing education, implementations of learning analytics techniques are scarce. [16] evaluated the programming patterns of undergraduate students in an introductory computer science course. They collected code-editing logs and error messages to compare the coding patterns of students who participated in a STEAM-based pedagogy course to their peers who received traditional instruction, for the assignment of creating visualisations and sound. Similarly, [17] analysed middle school students programming behaviours while using a block-based programming platform to create music. They collected data from editing events and execution. Results of both studies show that tinkering behaviours are encouraged by the design of the learning activities, i.e., when students need to meet certain requirements, the number of editing events increases compared with practices when students are left to free exploration. These results put the emphasis in purposefully designed activities that guide students through learning.

A similar approach was taken by [15] in an introductory programming course for a multicultural classroom. Students, who belonged to two different Chinese ethnic groups, received training in programming during four weeks, and their performance in solving programming tasks was recorded to analyse how motivation, social and self-expectancy influence learning performance. Data from programming, i.e., syntax errors and time spent on task, reveal differences in the influence of the factors under analysis over programming performance and learning effectiveness between students from different ethnic groups. Thus, the authors recommend implementing culturally relevant programming education to address students differences, specially in motivation [15].

More recently, only a few studies have investigated the behaviours of students while learning about AI. For example, [10] video-recorded students while participating in project-based, cooperative AI learning, and coded their activities into 14 behaviours, whose codes emerged from the observations. Their results show an iterative process for refining the model, where students engage in collecting data and training the model several times to improve the desired result. While this is indicative of students' engagement in the learning activities, these behaviours did not influence performance in knowledge tests, i.e., high and low achievers behaved in the same way. However, when analysing the cooperative behaviours of students, simulating the results before testing and executing the model was significantly correlated with positive collaboration and decreased discord [10]. Although these results offer an innovative methodology for studying learners' behaviours during the learning activities, to the best of our knowledge, there is no technological tool for learning AI that can systematically collect system logs to examine students' behavioural engagement in-depth with a robust methodology.

3   Method

3.1   Study Design

Participants are secondary school students between 12 and 15 years old, who reside in the two contexts under study. This study will take a quasi-experimental approach in a between-subject design. Since the purpose of the study is to identify the effect of culturally relevant AI curriculum over learning outcomes, music will be defining factor of cultural relevancy for the learning activities. Culturally relevant music in this study refers to traditional music practices and popular music familiar to the youth culture in each of the two specific regions under study. Conversely, culturally irrelevant music refers to the music that is not popular in a specific context. For example, the rhythms and beats characteristic of reggaeton are widely popular in current Mexican pop music and are potentially more relevant in that context. Similarly, Cantopop melodies and lyrics are instantly recognisable for students in Hong Kong and will be almost unintelligible for students in Mexico. Culturally (ir)relevant curricula will be tested in the two regions along with a control curriculum, thus creating three different conditions in each: AI curriculum that uses culturally relevant music, AI curriculum that uses culturally irrelevant music, and AI curriculum without music or any other culturally relevant aspect.

Following a Constructionist approach, in which knowledge and meaning are built through making, a web-based integrated development environment (IDE) that supports music creation with machine learning (ML) models will facilitate the learning activities. With this platform, students will collect data from mouse movements or web-camara to create musical outputs that dynamically respond to these inputs by implementing ML models in real time. The platform is intended to be accessible to students without prior experience with AI or music, or even coding.

The study consists of four main stages: co-design, pre-assessment of cultural relevance, pilot study, and main study. All stages are accompanied by experts in either teaching information and communication technologies (ICT) related subjects in K-12 level and music in each context. The co-design stage focuses on curriculum design and tool development. In the pre-assessment stage, students are asked to rate the relevance of different sets of music, including current pop music in their own context and in other contexts, and traditional music in their own and in other contexts. These ratings serve as a baseline for analysing relevance, while recognising that this stage has limitations, such potential personal preference bias. The pilot study will serve to test the curriculum and AI-music creation tool, and to validate the instruments used for assessment. Finally, the main study will provide the relevant data to answer the research questions.

3.2 Data Collection and Analysis

To answer the research questions, data from different sources will be collected. The learning outcomes will be evaluated via content knowledge test, attitude questionnaire, semi-structured interviews, artefacts, video-recordings of lessons, and system logs. All data will be handled with considerations in transparency, privacy, and confidentiality, and complying with ethical considerations for research purposes. Ethical approval for the collection of the comprehensive data set has been granted by the Human Research Ethics Committee of the University. The analysis of data will be guided by Constructionist theory, which emphasises personal knowledge construction, integrating data from multiple sources. I aim to understand student learning as they create and tinker personally meaningful musical artefacts and advance a Constructionist perspective for multimodal LA [4]. The analyses of the data integrate findings from qualitative and quantitative data to allow for a comprehensive evaluation of students' learning outcomes. Specific techniques will include descriptive and inferential statistics (e.g., MANOVA), thematic content analysis, and temporal analysis of system logs (e.g., sequential pattern mining, Bayesian knowledge tracing), emphasising the understanding of the processes that occur during learning.

4 Current State of the Research and Expected Contributions

I am currently finalising the codesign of the curriculum for all conditions in the study, collaborating with one ICT teacher and one musician/music teacher in Mexico, and one sound artist and educator in Hong Kong. Concurrently, the web-based platform is being constructed with a team of developers. Data collection will take place during Spring and Fall 2024 in Mexico and Hong Kong, respectively, starting with pre-assessment of cultural relevance, followed by pilot study. Thus, the opportunity for discussing research design and pilot results during the conference has the potential to illuminate some challenges and opportunities for the Fall round of data collection and later analyses.

Outcomes of this research aim to illuminate opportunities and challenges for teaching AI in a given cultural context, and whether students' learning outcomes (cognitive, affective, and behavioural) are influenced by culturally relevant curriculum. Identifying and understanding these differences, if any, can contribute to more effective and culturally responsive AI education practices. Moreover, the systematic collection, analysis, and evaluation of data from students' practices and interactions with the AI-enabled music creation platform will provide rich data from students' learning process, thus facilitating insightful findings that allow us to identify practices that improve AI learning. This study proposes to develop a practice of multimodal LA that looks at learning using quantitative and qualitative data that serves both the analysis of artistic computing and AI education.

Disclosure of Interests. The authors have no competing interests to declare that are relevant to the content of this article.

References

1. Au, K.H.: Social constructivism and the school literacy learning of students of diverse backgrounds. J. Literacy Res. **30**(2), 297–319 (1998). https://doi.org/10.1080/10862969809548000
2. Baker, R.S., Hershkovitz, A., Rossi, L.M., Goldstein, A.B., Gowda, S.M.: Predicting robust learning with the visual form of the moment-by-moment learning curve. J. Learn. Sci. **22**(4), 639–666 (2013). https://doi.org/10.1080/10508406.2013.836653
3. Bennett, A.: Popular Music and Youth Culture: Music, Identity and Place. Macmillan Press LTD. Springer (2000)
4. Berland, M., Baker, R.S., Blikstein, P.: Educational data mining and learning analytics: applications to constructionist research. Technol. Knowl. Learn. **19**(1), 205–220 (2014). https://doi.org/10.1007/s10758-014-9223-7
5. Berland, M., Martin, T., Benton, T., Petrick Smith, C., Davis, D.: Using learning analytics to understand the learning pathways of novice programmers. J. Learn. Sci. **22**(4), 564–599 (2013). https://doi.org/10.1080/10508406.2013.836655
6. Blikstein, P., Worsley, M., Piech, C., Sahami, M., Cooper, S., Koller, D.: Programming pluralism: using learning analytics to detect patterns in the learning of computer programming. J. Learn. Sci. **23**(4), 561–599 (2014). https://doi.org/10.1080/10508406.2014.954750
7. Engelman, S., Magerko, B., McKlin, T., Miller, M., Edwards, D., Freeman, J.: Creativity in authentic STEAM education with EarSketch. In: Proceedings of the 2017 ACM SIGCSE Technical Symposium on Computer Science Education, SIGCSE 2017, pp. 183–188. Association for Computing Machinery, New York (2017). https://doi.org/10.1145/3017680.3017763
8. Fiebrink, R.: Machine learning education for artists, musicians, and other creative practitioners. ACM Trans. Comput. Educ. **19**(4), 31:1–31:32 (2019). https://doi.org/10.1145/3294008
9. Hofstede, G.: Culture's Consequences: Comparing Values, Behaviors, Institutions and Organizations Across Nations, 2nd edn. Sage, Thousand Oaks (2001)
10. Hsu, T.C., Abelson, H., Lao, N., Tseng, Y.H., Lin, Y.T.: Behavioral-pattern exploration and development of an instructional tool for young children to learn AI. Comput. Educ. Artif. Intell. **2**, 100012 (2021). https://doi.org/10.1016/j.caeai.2021.100012. https://www.sciencedirect.com/science/article/pii/S2666920X21000060

11. Ladson-Billings, G.: Toward a theory of culturally relevant pedagogy. Am. Educ. Res. J. **32**(3), 465–491 (1995). https://doi.org/10.3102/00028312032003465

12. Long, D., Magerko, B.: What is AI literacy? Competencies and design considerations. In: Proceedings of the 2020 CHI Conference on Human Factors in Computing Systems, pp. 1–16. Association for Computing Machinery, New York (2020). https://dl-acm-org.eproxy.lib.hku.hk/doi/10.1145/3313831.3376727

13. Nasir, N.S., Hand, V.M.: Exploring sociocultural perspectives on race, culture, and learning. Rev. Educ. Res. **76**(4), 449–475 (2006). https://doi.org/10.3102/00346543076004449. ERIC Number: EJ759792

14. Paris, D., Alim, H.S.: What are we seeking to sustain through culturally sustaining pedagogy? A loving critique forward. Harv. Educ. Rev. **84**(1), 85–100 (2014). https://doi.org/10.17763/haer.84.1.982l873k2ht16m77

15. Tsai, C.W., Ma, Y.W., Chang, Y.C., Lai, Y.H.: Integrating multiculturalism into artificial intelligence-assisted programming lessons: examining inter-ethnicity differences in learning expectancy, motivation, and effectiveness. Front. Psychol. **13** (2022). https://www.frontiersin.org/articles/10.3389/fpsyg.2022.868698

16. Yee-King, M., Grierson, M., d'Inverno, M.: Evidencing the value of inquiry based, constructionist, learning for student coders. Int. J. Eng. Pedagogy **7**(3), 109–129 (2017). https://doi.org/10.3991/ijep.v7i3.7385. https://doaj.org/article/5c491393bad94b478922744705a0c57e

17. Zhang, Y., Krug, D.L., Mouza, C., Shepherd, D.C., Pollock, L.: A case study of middle schoolers' use of computational thinking concepts and practices during coded music composition. In: Proceedings of the 27th ACM Conference on on Innovation and Technology in Computer Science Education, ITiCSE 2022, vol. 1, pp. 33–39. Association for Computing Machinery, New York (2022). https://doi.org/10.1145/3502718.3524757

Virtual Reality (VR) in Safety Education: A Case Study of Mining Engineering

Haoqian Chang[1,2](✉)(iD), Ziqi Pan[1](iD), and Alexandra I. Cristea[1](✉)(iD)

[1] Department of Computer Science, Durham University, Durham DH1 3LE, UK
{haoqian.chang,ziqi.pan2,alexandra.i.cristea}@durham.ac.uk
[2] School of Economics and Management, Anhui University of Science and
Technology, Huainan 232001, China

Abstract. Safety education and training are vital in the mining industry. However, traditional training relies on passive modalities, such as lectures, videos and brochures. These suffer from sever limitations - poor reproducibility, inefficient resource utilisation, and a lack of interactive feedback in one-size-fits-all training scenarios. Addressing these challenges, we introduce a *novel hybrid approach combining virtual reality (VR) and electroencephalography (EEG) for training in the use of underground self-contained self-rescuers (SCSR)*. The VR component provides an interactive and immersive training experience, to facilitate a higher level of engagement compared to traditional methods. Initial EEG testing showed that VR training could elevate trainees' brain activity, which may result in higher ratings and satisfaction. Beyond EEG we use also after-scenario questionnaire (ASQ) and system usability scale (SUS).

Keywords: VR · Electroencephalography · Safety training · Mining

1 Introduction

Virtual reality (VR) is revolutionising mining industry operations, particularly in enhancing safety training methods [7]. Simulating the operation of equipment, such as *self-contained self-rescuers* (SCSR) [11] and emergency countermeasures through VR, are promising. However, determining the *metrics for measuring VR's effectiveness* [9] in safety training and elucidating its advantages over traditional methods requires further study. Our focus includes constructing a virtual SCSR device and simulating mine disasters, complemented by the use of electroencephalography (EEG), to monitor participants' brain activity, thereby offering an innovative approach to safety training. Via the *analysis of EEG signal fluctuations by deep learning algorithms* [5] (e.g. deep convolutional neural networks - CNN); time series methods (e.g. recurrent neural networks - RNN), self-attention mechanisms, alongside *evaluations from the after-scenario questionnaire* (ASQ) and *system usability scale* (SUS), we aim to provide a comprehensive evaluation of VR's effectiveness in enhancing safety training outcomes.

A. M. Olney et al. (Eds.): AIED 2024 Workshops, CCIS 2151, pp. 382–387, 2024.
https://doi.org/10.1007/978-3-031-64312-5_47

Underground mines are characterised by harsh environments and a high propensity for accidents [15]. Most of the air in a mine has to come from the surface ventilation. By pumping air from the inlet to the outlet of the mine, it will supply oxygen to the miners and take away the exhaust gases from the longwall face[1]. While accidents, such as gas explosion, fires, ventilation failures have to be stopped, we focus here on accidents leading to the longwall face filling with unbreathable exhaust gas, which requires miners to urgently evacuate.

Self-contained self-rescuers (SCSR) are emergency respiratory devices consisting of a.o. an oxygen tank, air bladder, air conduit, mouthpiece, harness. Introduced since 1947 [4], their function is to provide miners with oxygen for a period of time, in the case of an emergency; all miners working underground are required to carry an SCSR kit. However, after more than half a century of its use, miners' proficiency in SCSR is still insufficient. On the 24th September 2023 a major fire accident [1] occurred in a coal mine in Guizhou Province, China, resulting in 16 deaths. The cause of the accident was a fire in the belt of the conveyor (a type of long-distance coal transport equipment), which quickly filled the longwall face with toxic and hazardous gases. Following this, the National Mine Safety Administration of China conducted a random sampling of workers at several coal mines, to check SCSR mastery, and found that up to 75% of the miners could not use it correctly or failed to wear it at the right time [2].

This shows that traditional methods may fail to effectively explain hazardous events to trainees, beside not providing timely feedback on errors. By simulating emergency scenarios, VR training can target the gap in miners' proficiency with SCSR, offering hands-on practice in a controlled, risk-free environment.

2 Related Work

Applications of VR in the mining industry mostly focus on longwall face simulations [14], dust modeling and control [13], and virtual entity modulation of excavation equipment [3]. As artificial intelligence plays an increasingly important role in education, researchers are looking for ways to use VR to improve the quality of safety training for miners. Related work in the field [10] presents several hypotheses and develops a set of necessary skills and knowledge for emergency operations. E.g., [12] implemented VR safety training to the New South Wales Mine Rescue Team in Australia and conducted a post-training questionnaire with positive results, trainees perceiving the benefits as far outweighing the costs, and an overall tendency to recommend VR training to other colleagues.

Despite the growing popularity of VR in safety training and mining operations simulation, there is still limited research on the psychology and brain activity of the miners. The integration of VR with EEG technology provides an effective method for monitoring and analysing the brain activity of trainees during safety training. This allows research on the impact of the environment on the mood of people [8], assessing participants' levels of anxiety and concentration

---

[1] Longwall mining face is a form of underground coal mining where a long wall of coal is mined in a single slice.

[6]. The latter are particularly suited for mining safety training, due to mining's hazardous nature. Assessing the brain activity and psychology of miners under different training scenarios, evaluating the brain activity and the psychology of trainees under different training conditions can also improve the VR model [16].

3   Methodology

The initial study will include systematic data collection on the components of self-contained self-rescuers (SCSR), the donning process, and the requirements related to SCSR in mines in different countries, with specific reference to the types of data such as technical specifications, operating protocols, regulatory standards, etc. This data will guide the training feedback for the virtual entity. To the best of our knowledge, there are currently no well-established virtual models for mining SCSR and corresponding VR safety training methods.

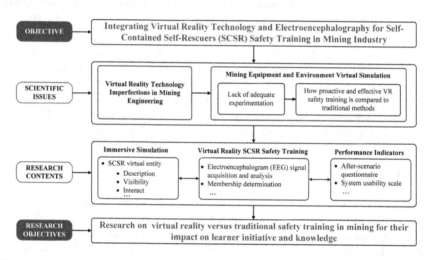

Fig. 1. Enhancing Mining Safety Training through Virtual Reality and EEG Analysis

The research framework, illustrated in Fig. 1, encompasses the research objectives, questions, and content, which collectively guide our experimental approach. The experimental platform will utilise 3D modeling software such as Blender and 3D Max for model construction, along with Unity for developing animation effects, to be tested on Meta Quest 2. EEG data analysis will be conducted using Python libraries, including PyTorch for machine learning algorithms, and Pandas and NumPy for data manipulation and numerical analysis.

The pilot will compare traditional safety training (manuals, safety lectures, and videos) with VR training. Six people will be invited to test the experiment and will be required to wear an EEG throughout. The final experiment will be upgraded based on the performance of these 6 trainees, with a minimum of 20 people. The study will apply for review and approval from the Durham University Ethics Committee. All participants will be fully informed about the purpose, methods, potential risks, and possible benefits of the study before participating. Written informed consent will be obtained from all before any data collection. The study strictly adheres to the Declaration of Helsinki and/or other relevant ethical guidelines to ensure participant privacy and data security.

Research Question in Progress

1. What is the comparative effectiveness of virtual reality (VR) in enhancing safety training methods against traditional training approaches within the mining industry?
2. In what ways can a virtual self-contained self-rescuer (SCSR) device be constructed and integrated into VR simulations to elevate miners' proficiency and their response efficacy in the event of mine disasters?
3. Considering the pronounced gap in miners' proficiency with SCSR devices, which innovative training methodologies can be formulated to bolster their skillset and preparedness for confronting emergency scenarios?

Pre-experiment. All subjects will be people who have never been exposed to SCSR equipment, to test the effectiveness of the training. Trainees will complete three training components (manual, lecture, video) in a randomised order, and each time they complete a stage, they will fill out a questionnaire related to the SCSR procedure. This process is designed to closely simulate real-life training scenarios. The reason for selecting subjects who had never been exposed to SCSR was to simulate the learning process of acquiring unfamiliar knowledge. EEG signals and questionnaire responses are collected at all stages.

Data analysis. The collected EEG signals will be processed as time-series data, which will be sliced and denoised through data pre-processing. A Deep Convolutional Neural Network (CNN) will be built to classify the data. We will also use time series algorithms to analyse the signal strength of the EEG, such as fluctuations, trends, etc. Machine learning methods such as Support Vector Machines (SVM) a.o. comparator baselines will evaluate the effectiveness of data classification. The questionnaire will evaluate trainees' understanding of SCSR.

Feedback. Meaningful feedback based on trainees' responses will be used, to enhance the VR simulations. Specific metrics for comfort and engagement will be established, to assess the effectiveness of the virtual environment. Trainees' feedback will directly inform the iterative development process, aiming to improve both the realism and educational value of the simulations. Following development, the device and its software will be rigorously evaluated for validity and reliability. This evaluation is expected to contribute positively to the subsequent VR modeling of coal mining longwall faces and provide insights for future accident simulation exercises.

4   Expected Contribution

We propose a *virtual reality (VR) and electroencephalography (EEG) based safety training process for coal mining self-contained self-rescuers (SCSR)* to help miners get a better grasp of the equipment, and assessing the benefits and effectiveness of VR-enhanced safety training, specifically within the context of the mining industry. This is a significantly unexplored area for the AIED community, and its research can make major difference in coal mining. We further aim to contribute with the design of the SCSR virtual training environment and development of a performance evaluation system for VR versus traditional safety training, including parametrising ease of use, effectiveness and proactivity. Moving forward, we aim to enhance the VR safety training module to include the operation of mining equipment, such as coal mining machines, and to simulate scenarios involving mining accidents and emergency evacuation strategies. This initiative is believed to have significant long-term implications for enhancing the integration of AI technologies in industrial engineering practices.

References

1. Guizhou "9-24" major fire accident (2023). https://www.chinamine-safety.gov.cn/xw/mkaqjcxw/202309/t20230928_464600.shtml
2. National mine safety administration of china investigation on the use of underground self-rescuers (2023). https://www.chinamine-safety.gov.cn/fw/kp/202311/t20231118_469090.shtml
3. Dong, M., Xie, J., Li, J., Du, W., Cui, T., Huo, P.: A virtual planning method for spatial pose and performance fusion advancement of mining and transportation equipment in complex geological environment. Min. Metall. Explor. **40**(1), 231–251 (2023)
4. Forbes, J.J.V., Grove, G.W.: Protection Against Mine Gases, vol. 35. US Government Printing Office (1954)
5. Gao, Z., et al.: Complex networks and deep learning for EEG signal analysis. Cogn. Neurodyn. **15**, 369–388 (2021)
6. Guedj, C., et al.: Self-regulation of attention in children in a virtual classroom environment: a feasibility study. Bioengineering **10**(12), 1352 (2023)
7. Gürer, S., Surer, E., Erkayaoğlu, M.: Mining-virtual: a comprehensive virtual reality-based serious game for occupational health and safety training in underground mines. Saf. Sci. **166**, 106226 (2023)
8. Hou, K., Liu, X., Kong, Z., Wang, H., Lu, M., Hu, S.: Impacts of corridor design: an investigation on occupant perception of corridor forms in elderly facilities. Front. Archit. Res. **12**(6), 1047–1064 (2023)
9. Isleyen, E., Duzgun, H.S.: Use of virtual reality in underground roof fall hazard assessment and risk mitigation. Int. J. Min. Sci. Technol. **29**(4), 603–607 (2019)
10. Obukhov, A.D., Krasnyanskiy, M.N., Dedov, D.L., Nazarova, A.O.: The study of virtual reality influence on the process of professional training of miners. Virtual Reality **27**(2), 735–759 (2023)
11. Onifade, M.: Towards an emergency preparedness for self-rescue from underground coal mines. Process Saf. Environ. Prot. **149**, 946–957 (2021)

12. Pedram, S., Ogie, R., Palmisano, S., Farrelly, M., Perez, P.: Cost-benefit analysis of virtual reality-based training for emergency rescue workers: a socio-technical systems approach. Virtual Reality **25**(4), 1071–1086 (2021)
13. Ren, T., et al.: Development and evaluation of an immersive VR-CFD-based tool for dust exposure mitigation in underground tunnelling operations. Tunn. Undergr. Space Technol. **143**, 105496 (2024)
14. Tao, C., Jiacheng, X., Wang, X., Xin, Z., Suhua, L., Mengyao, D.: Constructing a high-precision virtual scene of mining equipment and coal seam roof and floor using actual mining data. Min. Technol. **131**(1), 12–24 (2022)
15. Zhang, J., Fu, J., Hao, H., Fu, G., Nie, F., Zhang, W.: Root causes of coal mine accidents: characteristics of safety culture deficiencies based on accident statistics. Process Saf. Environ. Prot. **136**, 78–91 (2020)
16. Zhang, Y., Zhang, Y., Jiang, Z., Xu, M., Qing, K.: The effect of EEG and fNIRS in the digital assessment and digital therapy of Alzheimer's disease: a systematic review. Front. Neurosci. **17**, 1269359 (2023)

Prediction of Essay Cohesion in Portuguese Based on Item Response Theory in Machine Learning

Bruno Alexandre Barreiros Rosa[1]([⊠]), Hilário Oliveira[2]([⊠]),
and Rafael Ferreira Mello[1,3]([⊠])

[1] C.E.S.A.R School, Recife, PE, Brazil
babr@cesar.school
[2] Federal Institute of Espírito Santo (IFES), Serra, ES, Brazil
hilario.oliveira@ifes.edu.br
[3] Federal Rural University of Pernambuco (UFRPE), Recife, PE, Brazil
rafael.mello@ufrpe.br

Abstract. The essay is considered a useful mechanism for evaluating learning outcomes in writing. Essay correction is a manual task that presents difficulties related to time, cost, reliability, and the subjectivity of the examiner. Cohesion is a fundamental aspect of the text, as it helps to establish a meaningful relationship between its different parts. The automated scoring of cohesion in essays presents a challenge in the field of artificial intelligence in education. This is primarily due to the fact that machine learning algorithms, commonly employed for text evaluation, often overlook the unique characteristics of individual instances within the analyzed corpus. To address this issue, item response theory can be adapted to the machine learning context. This adaptation involves characterizing aspects such as ability, difficulty, discrimination, and guessing in the utilized models. This research aims to analyze the performance of cohesion score prediction in Brazilian basic education essays, using item response theory to estimate the scores generated by machine learning models. The research extracted 325 linguistic features and treated it as a regression problem. Initial results indicate that the proposed approach has the potential to outperform conventional models. The research presents a promising avenue for a more precise evaluation of cohesion in educational essays.

Keywords: Automated essay scoring · textual cohesion · natural language processing · item response theory

1 Introduction

Writing texts is a fundamental skill for an individual to succeed in academic, corporate, or even social life [6]. The production of descriptive, dissertative, injunctive, or narrative essays requires the correct use of linguistic mechanisms

© The Author(s), under exclusive license to Springer Nature Switzerland AG 2024
A. M. Olney et al. (Eds.): AIED 2024 Workshops, CCIS 2151, pp. 388–394, 2024.
https://doi.org/10.1007/978-3-031-64312-5_48

essential for the development of writing [16]. This rigor becomes even more crucial when we consider that the evaluation of an essay is a subjective process that involves several criteria [6]. Textual cohesion is one of the most important criteria in this assessment [7–9,13,14]. Cohesion is an indispensable aspect for providing grammatical links and connections between text elements such as words, sentences, and phrases [7]. A cohesive text presents interconnected ideas, which allows the reader to follow the writer's reasoning fluidly [8]. Cohesion is achieved through the appropriate use of linguistic mechanisms necessary for textual construction [7]. In a text devoid of cohesion, the fundamental ideas may be present, but the lack of connections compromises the clarity and effectiveness of communication, making it difficult for the reader to understand and interpret [8]. Therefore, it is expected that improving cohesion in essays will benefit the overall quality of other aspects of writing [1,8].

The use of Automated Essay Scoring (AES) methods brings several advantages, such as reducing the time and costs involved in the correction process, together with minimizing possible biases and human errors [15]. However, when it comes to assessing cohesion in essays, the methods still have limitations. For example, they often fail to capture the reference and sequence of semantic relationships, as well as the logical interconnection between the different parts of the text [1]. In addition, identifying elements of cohesion often requires a contextual understanding that the methods cannot fully replicate [13]. There is also the challenge of correctly interpreting cohesive elements that can have multiple meanings based on the context in which they are used [8]. This makes the task of automatically scoring cohesion in essays an open problem [1,9,13,14].

In English, promising approaches to cohesion have been proposed [1]. In Portuguese, it is still a challenge to treat cohesive elements in essays in an automatic way [9,13,14]. One of the possibilities for dealing with these limitations is to create a committee of estimators that can be selected to perform the scoring in specific contexts. This approach is reinforced by the recurring practice of using techniques to evaluate the performance of Machine Learning (ML) algorithms in order to select the most suitable ones for the various challenges inherent in their application. These evaluation techniques seek to understand the advantages and limitations of the algorithms. Recent studies have taken a different approach to evaluation, in which algorithm performance is evaluated according to the instance level using Item Response Theory (IRT) [12,17].

In this respect, IRT and ML are integrated in a complementary way. The IRT serves as an essential statistical tool. It is used to measure the probability of a respondent answering a specific item correctly based on their latent ability [2]. In this integration, the instances of a dataset in ML can be equated to the items of a test in IRT, while the ML algorithms act as the respondents whose abilities are being assessed. Recent studies have linked this integration and produced promising results [12,17]. However, given a challenging and recently explored context in the scientific field, namely the evaluation of cohesion in essays using ML and IRT models combined in regression problems, this research aims to answer the research question below: *How to predict the cohesion score in essays to support teacher correction using item response theory to estimate the scores generated by machine learning models?*

To answer this research question, the general and specific objectives were established, as described below:

General Objective

- Analyze the performance in predicting the cohesion score in essays to support teacher correction, using item response theory to estimate the scores generated by machine learning models.

Specific Objectives

- Relate the works that discuss automated essay scoring and item response theory;
- Extract characteristics from essays to run machine learning models for scoring cohesion levels in a real corpus of Portuguese-language essays;
- Propose an algorithm that uses different machine learning models in regression problems to evaluate the level of cohesion in Brazilian basic education essays using item response theory;
- Evaluate the proposed model in relation to traditional machine learning approaches.

2   Background

Recent research has examined the relationship between computationally extracted writing features and human evaluations of cohesion in Portuguese writing. For example, [14] conducted an investigation using 151 features that encompass aspects such as the use of connectors, lexical diversity, readability, and similarity between adjacent sentences, along with several features extracted from the Coh-Metrix tool. The authors compared various regression algorithms to estimate the cohesion score using the Essay-BR database [10]. Following the same research line, the study developed in [13] explored regression algorithms through an attribute-based approach and the BERT language model to estimate scores related to textual cohesion in Portuguese and English. Additionally, explainability methods were employed to provide interpretations of the decisions made by the models for the estimated scores.

The integration between IRT and ML aims to measure the ability of ML models and the difficulty of learning from datasets, showing promising results in recent work [12,17]. An example is the work proposed by [12] that compares the use of IRT across different regressors and provides a theoretical analysis of its parameters and abilities in ML models. This paper models absolute error as a function based on IRT, following a Gamma distribution (Γ), to deal with unlimited positive responses. The research was limited to analyzing the parameters of the proposed Γ-IRT model applied to answers to open-ended Statistics exam questions, without exploring other writing assessment contexts or extensions of IRT approaches.

In the research [17], the authors proposed a method that applies IRT to evaluate the characteristics of the scores assigned by the AES models and integrates these scores to generate an estimated final score. The study demonstrated that the proposed method, using the IRT Generalized Many-Facet Rasch Model (GMFRM), achieved a higher average precision (QWK 0.7562) compared to individual AES models (QWK 0.7209) and conventional integration methods (QWK 0.7395). This result demonstrates that integrating multiple AES models using IRT can significantly improve performance, evidencing its potential. However, the authors limited themselves to testing a restricted collection of AES and IRT models without cross-validation analysis and only applied the method to one English corpus.

Thus, the use of IRT to analyze and adjust ML models enables more research that can improve the precision and reliability of automated cohesion evaluations in essays. The study by [12] compares the use of IRT in different regressors and provides a theoretical analysis of their parameters and abilities in ML models. [17] focused on integrating the results of AES models using IRT. However, in contrast to previous proposals, this research uses IRT to adjust the output prediction of ML models to produce a new approach for predicting cohesion scores.

3   Method

This study intends to treat the analysis of textual cohesion of essays written in Portuguese as a regression ML problem. Additionally, it plans to integrate IRT to improve score prediction according to the ability of each algorithm. To achieve the proposed objectives, this experimental research will be developed using mixed methods through the following stages:

1. **Data Description:** This stage aims to describe the database to be processed. This research intends to use a Portuguese dataset that includes university-level essays from the Essay-BR extended corpus collected by [10]. This corpus has essays written following the same style as the textual production test of the National High School Exam (ENEM) and includes scores for five competencies used to assess writing, including cohesion [13,14]. There are 6,579 essays, divided into 151 topics, collected from December 2015 to August 2021. These essays were written by high school students, respecting the stipulated limit of a minimum of 8 and a maximum of 30 lines. At least two experts evaluated the essays, and the final score for each competency is the arithmetic mean. In this research, the focus will be on Competency IV, which refers exclusively to textual cohesion in essays;

2. **Feature Extraction:** In this step, the aim is to convert texts into feature vectors for ML models processing while preserving the text's original meaning. These features have previously been used to analyze the rhetorical structure of essays [5], analyze online discussions [11] and evaluate cohesion [13,14].

The intention is to extract measures of linguistic features that include tools such as NILC-Metrix, Coh-Metrix, LIWC, and connective elements. These approaches help to identify, for example, semantic cohesion, referential cohesion, syntactic complexity, lexical diversity, textual simplicity, and readability. In total, 325 measures of linguistic characteristics will be considered to develop the cohesion automated scoring model;

3. **Machine Learning Model Processing:** In this stage, the purpose is to select, train, validate, and test different regression algorithms to estimate essay cohesion scores. For this prediction task, we selected algorithms based on the literature [4,13] that use different approaches, including statistical, decision trees, classical neural networks, Bayesian models, and ensembles. The ML algorithms we intend to adopt in the experiment are: *Bayesian Ridge, CatBoost Regressor, Decision Tree Regressor, Extra Trees Regressor, LGBM Regressor, Linear Regression, MLP Regressor, Random Forest, SVR,* and *XGB Regressor*. We also plan to use a conventional ensemble of ML model integrations, such as *Stacking* and *Voting* regressors;

4. **Item Response Theory Processing:** In this step, the aim is to predict the score for the level of cohesion assessed in the essays. The IRT calculates the parameters of ability, difficulty, and discrimination of the ML models from the training validation data. Thus, we plan to employ the BIRT-GD library proposed in [3] to calculate the error expectation per model and cohesion level. Subsequently, the ML models will be classified based on cohesion level, so the ML models with the lowest error expectation obtained by IRT are chosen. Finally, the aim is to adjust the output cohesion score by identifying the model instance and cohesion level with the best ability;

5. **Evaluating Results:** In this stage, we will use the evaluation process recommended in the literature [13,14] to compare the results of the regression models and the IRT approach. To ensure the consistency and integrity of the analysis results, we will adopt a 10-fold stratified cross-validation in combination with the following measures: linear Kappa coefficient, Quadratic Weight Kappa (QWK), and Accuracy. It is important to note that these metrics are traditionally used in classification problems. However, in this context, they also apply to regression problems because the values predicted by the regressors are categorized into predefined score ranges.

4 Concluding Remarks

This research aims to improve the evaluation of cohesion in essays to support the correction of Brazilian basic education teachers. Research advances include the development of a model using IRT parameters in order to improve the prediction of ML models for cohesion scores. The study has a detailed timetable that guides the development and deadlines until the thesis defense, scheduled for December 2024. Proposed experiments include: Firstly, we plan to apply the methodology to dissertative-argumentative essays written by Brazilian high school students; Secondly, we intend to incorporate regression models based on Bidirectional

Encoder Transformer (BERT) representations; Finally, we plan to apply the methodology to narrative essays written by Brazilian elementary school students. It is important to mention that this study does not intend to analyze the practical application of the proposed method. This would include developing a learning analysis tool and evaluating instructor and student satisfaction based on the results of our method. We intend to develop this tool in a future line of research.

References

1. Crossley, S.A., Kyle, K., Dascalu, M.: The tool for the automatic analysis of cohesion 2.0: integrating semantic similarity and text overlap. Behav. Res. Methods **51**(1), 14–27 (2019)
2. Embretson, S.E., Reise, S.P.: Item Response Theory. Psychology Press (2013)
3. Ferreira-Junior, M., Reinaldo, J.T., Neto, E.A.L., Prudencio, R.B., et al.: β^4-IRT: a new β^3-IRT with enhanced discrimination estimation. arXiv preprint (2023)
4. Ferreira-Mello, R., André, M., Pinheiro, A., Costa, E., Romero, C.: Text mining in education. Wiley Interdisc. Rev. Data Min. Knowl. Discov. **9**(6), e1332 (2019)
5. Ferreira Mello, R., Fiorentino, G., Oliveira, H., Miranda, P., Rakovic, M., Gasevic, D.: Towards automated content analysis of rhetorical structure of written essays using sequential content-independent features in Portuguese. In: LAK22: 12th International Learning Analytics and Knowledge Conference, pp. 404–414 (2022)
6. Graham, S.: Changing how writing is taught. Rev. Res. Educ. **43**(1), 277–303 (2019)
7. Halliday, M.A., Hasan, R.: Cohesion in English. Longman (1976)
8. Koch, I.G.V.: A Coesão Textual, vol. 22. São Paulo Contexto (2010)
9. Lima, F., Haendchen Filho, A., Prado, H., Ferneda, E.: Automatic evaluation of textual cohesion in essays. In: 19th International Conference on Computational Linguistics and Intelligent Text Processing (2018)
10. Marinho, J., Anchiêta, R., Moura, R.: Essay-BR: a Brazilian corpus to automatic essay scoring task. J. Inf. Data Manag. **13**(1) (2022)
11. Mello, R.F., Fiorentino, G., Miranda, P., Oliveira, H., Raković, M., Gašević, D.: Towards automatic content analysis of rhetorical structure in Brazilian college entrance essays. In: Roll, I., McNamara, D., Sosnovsky, S., Luckin, R., Dimitrova, V. (eds.) AIED 2021. LNCS (LNAI), vol. 12749, pp. 162–167. Springer, Cham (2021). https://doi.org/10.1007/978-3-030-78270-2_29
12. Moraes, J.V., Reinaldo, J.T., Ferreira-Junior, M., Silva Filho, T., Prudêncio, R.B.: Evaluating regression algorithms at the instance level using item response theory. Knowl.-Based Syst. **240**, 108076 (2022)
13. Oliveira, H., et al.: Towards explainable prediction of essay cohesion in Portuguese and English. In: LAK23: 13th International Learning Analytics and Knowledge Conference, pp. 509–519 (2023)
14. Oliveira, H., et al.: Estimando coesão textual em redações no contexto do enem utilizando modelos de aprendizado de máquina. In: Anais do XXXIII Simpósio Brasileiro de Informática na Educação, pp. 883–894. SBC (2022)
15. Ramesh, D., Sanampudi, S.K.: An automated essay scoring systems: a systematic literature review. Artif. Intell. Rev. **55**(3), 2495–2527 (2022)

16. Travaglia, L.C.: Tipologia textual e ensino de língua. Domínios de Lingu@gem **12**(3), 1336–1400 (2018)
17. Uto, M., Aomi, I., Tsutsumi, E., Ueno, M.: Integration of prediction scores from various automated essay scoring models using item response theory. IEEE Trans. Learn. Technol. **16**(6), 983–1000 (2023)

Effects of Generative Artificial Intelligence on Instructional Design Outcomes and the Mediating Role of Pre-service Teachers' Prior Knowledge of Different Types of Instructional Design Tasks

Kristina Krushinskaia$^{(\boxtimes)}$, Jan Elen, and Annelies Raes

Centre for Instructional Psychology and Technology, KU Leuven, Leuven, Belgium
kristina.krushinskaia@kuleuven.be

Abstract. Generative artificial intelligence (GenAI) is increasingly becoming a tool used in many spheres, including education. A key question is whether GenAI can support teachers in one of their major roles: designing learning environments. This research aims to bridge the gap in understanding how different types of GenAI use impact instructional design outcomes. To achieve this, the study design includes three experimental conditions: 1) with GenAI, 2) with a GenAI bot pre-trained as a co-instructional designer, 3) automated by GenAI and one control condition (no use of GenAI). Participants, who are pre-service teachers, are tasked with designing a lesson blueprint and developing a lesson plan. The quality of instructional design outcomes and any common patterns across conditions are evaluated. Moreover, the impact of teachers' interactions with GenAI on the instructional design outcomes is analyzed. Additionally, the study aims to reveal a relationship between participants' prior familiarity with instructional design tasks and the way they design instruction with and without different uses of GenAI. This research study is crucial for exploring GenAI's potential to become a valuable partner for teachers in co-designing instruction and providing students with better learning opportunities.

Keywords: Generative AI · GenAI · Artificial intelligence · Instructional design · Lesson planning · Design for learning

1 Generative AI and Instructional Design

Since its public launch, generative artificial intelligence (GenAI) has shown explosive growth across different spheres. ChatGPT, perhaps the most famous GenAI tool, outperformed students on assignments [3], while a survey by GitHub reported that 92% of US-based developers use GenAI in their daily work [16]. As a result, many speculations have arisen, particularly regarding GenAI's potential to revolutionize existing practices. According to Hodges and Kirschner [14], strong claims about the revolutionary effect often accompany the expansion of new technologies. However, as noted by researchers,

GenAI stands out significantly from other technological tools due to its capacity to "generate original work that is virtually indistinguishable from that of human authors" [14]. Therefore, the impact of GenAI on many spheres, including education, might indeed be transformative.

The study by Montenegro-Rueda et al. [24] identified emerging research areas in education that require attention in light of the rise of GenAI. According to the study, the first area is to explore the redefinition of the role of teachers [24]. A recent review revealed four groups of opportunities brought by AI in general to teachers, in-cluding the ability to provide personalized learning, to make data-driven decisions, to outsource administrative load to AI, and to apply new forms of assessment. These opportunities also apply to GenAI, which has made their realization more technologi-cally feasible. As an example, in an empirical study [17], teachers successfully per-sonalized learning based on students' prior knowledge using GenAI. They provided learning experiences for three knowledge levels (basic, intermediate, and advanced) and were able to adjust learning materials not only before but also during the session . Another example can be studies that aim to automate grading using GenAI to reduce educators' administrative burden, such as the study [27].

Designing learning environments and lesson planning are important tasks of teachers [4, 11, 20], and some research has explored how GenAI can support teachers in their role as instructional designers. For instance, one study investigated the poten-tial of ChatGPT in creating lesson plans [4]. The findings revealed that ChatGPT can serve as a valuable tool for initiating the development of lesson plans by providing a general structure, ideas and resources [4], what might be particularly advantageous for novice teachers [8]. Given that teachers must critically evaluate the responses provid-ed by ChatGPT [4, 14], researchers assumed that the use of ChatGPT might also en-hance teachers' critical thinking skills in lesson planning [4]. Research focusing on the special features of GenAI-designed instructional outcomes and the evaluation of their overall quality is scarce. One study discovered that ChatGPT can successfully complete graduate-level instructional design assignments [25]. However, it only briefly assessed the quality of these ChatGPT-generated outcomes, noting that they met the standards to pass an assignment.

While there is a consensus that ChatGPT can be used to design instruction and create lesson plans [4, 5, 12, 19], it remains an open question what this collaboration with GenAI would entail. A default ChatGPT does not rely on a single instructional design model; instead, it uses a mixture of them. Therefore, its support in instructional design cannot be considered systematic. Moreover, a default ChatGPT possesses a rather "eager-to-please" communicational style [13] and may readily change its response if the user indicates that it was wrong. These features might make collaboration with Chat-GPT problematic. Additionally, when using a non-pre-trained ChatGPT, teachers have no choice but to rely on their prior knowledge of designing instruction to critically judge the ChatGPT's output [4, 14]. This implies that teachers' knowledge of instructional design must be sufficient for successful collaboration with ChatGPT. However, studies indicate that teachers' instructional design proficiency is often not robust [15, 18]. There-fore, co-creating lessons with a GPT bot, pre-trained on an instructional design model and designed for collaborative partnership, could support teachers where necessary to achieve better instructional design outcomes. For this study, a GPT bot (GPT-4) has been developed following the AI chatbot design framework conceptualized by Sonderegger and Seufert [26]. The bot was trained based on the Dick and Carey instructional design model [7] to support the teacher's lesson design process and potentially partially compensate for knowledge gaps. Nevertheless, in this case, the need for critical assessment of GPT's responses would still remain. Lastly, given that AI is often expected to automate processes, including the provision of personalized learning [21], claims regarding GenAI's potential to automate lesson design are anticipated to emerge. While studies, such as [1], argue that AI-supported "co-teaching" can be a game-changer, more research is needed to identify the optimal ways for integrating GenAI to co-design instruction.

This study aims to address the gap in understanding how GenAI affects instruc-tional design outcomes and how teachers' prior knowledge influences this process. According to Ausubel [2], prior knowledge plays a crucial role in how we engage in our activities. When learners lack the cognitive schemas, or 'schemata' in Glaser's terminology [10], necessary for assimilating new information, they may experience cognitive overload, leading to ineffective learning [17]. The significance of this re-search study lies in exploring the newly emergent field of GenAI-assisted teaching, potentially enhancing teachers' ability to co-design instruction with GenAI and offer better learning opportunities for students.

2 Experimental Design

In this research, first, we plan to investigate how different types of GenAI use may impact instructional design outcomes, namely lesson blueprints and lesson plans. To achieve this, the study design includes three experimental conditions: 1) with GenAI, 2) with GenAI bot pre-trained as a co-instructional designer, and 3) automated by GenAI; along with one control condition, which involves no use of GenAI. Addition-ally, we intend to analyze teacher interactions with GenAI and reveal the connection between these interactions and instructional design outcomes.

Second, we aim to replicate the experiment for different types of instructional de-sign tasks: designing a lesson with learning personalization and designing a lesson using the 4C/ID instructional design model [22]. This is done to explore the relation-ship between participants' prior familiarity with various instructional design tasks and their implementation of these tasks with and without GenAI. The level of participants' prior familiarity with the instructional design task is assessed before the experiment.

Thus, the study design includes two independent variables: different types of GenAI use and the level of instructional design task familiarity. The dependent variables are instructional design outcomes and teacher interactions with GenAI.

The study's proposal has been approved by the KU Leuven's Ethics Committee. Between December 2023 and January 2024, a series of pilot studies was conducted to test the experiments under various conditions. The study has been scaled up starting from March 2024, aiming to include at least 30 participants for each condition, with the exception of experimental condition 3, which involves no human participants.

3 Research Methodology

3.1 Participants and Recruitment

Participants are pre-service teachers from Flemish university colleges, currently in their 3rd year of "Teacher education" programs, and Flemish universities, pursuing their Master's degree in Teaching. No exclusion criteria regarding the subject taught were applied in the study. The choice of pre-service teachers as participants is based on the premise that they are future educators who have already undergone sufficient training and completed several internships. Simultaneously, they may be more psychologically prepared for designing lessons with GenAI, as their everyday practice of preparing for lessons has not yet become "automatic", which allows for greater flexibility.

Participant recruitment is organized through a research study website created specifically for this study: https://research-study-generative-ai.tilda.ws/.

3.2 Experimental Setup and Data Collection

Participants need to design a two-hour lesson about any historical figure of their choice, focusing on their main accomplishment or a phenomenon associated with them. The task is intentionally made to be open, giving teachers the space to make design decisions and formulate their own learning objectives. To help future teachers concentrate on the design phase, they are provided with a script containing nine questions related to design phase of instruction, such as "What is the overarching aim or purpose of this instruction?" and "What specific outcomes should your students achieve?".

In experimental condition 1, participants first engage in a short training on how to interact with ChatGPT and are provided with recommendations on effective prompt engineering. Following this, participants are tasked with designing a lesson blueprint and lesson plan with GenAI. In experimental condition 2, where a GPT co-instructional design bot is applied, participants are briefed on the GPT bot's capabilities and the effective ways to interact with it, along with training on prompt engineering. In experimental condition 3, a similar GPT based on the Dick and Carey instructional design model [7] is employed; however, it is created to produce instructional design outcomes automatically based solely on the task description, implying that the learning objectives and all the instructional design components are generated by GenAI without the input of a teacher. In the control condition, participants are given the same task in a digital format, but they do not have access to GenAI. All the data collected, including lesson blueprints and lesson plans, screen recordings, and links to chats with ChatGPT and GPT bots, have a qualitative nature.

3.3 Data Analysis

To determine which instructional design outcomes exhibit higher quality, they will be analyzed first by instructional design experts using the pairwise comparison technique [6]. This means that experts will compare two items at a time to holistically determine which one is better, based on the provided criteria, relying on Merrill's first principles of instruction [23] for the tasks "design instruction" and "design instruction with learn-ing personalization" and on the 4C/ID guidelines [22] for the task "design instruction using 4C/ID". For this purpose, the Comproved tool[1] for comparative judg-ment will be employed.

The qualitative analysis of chatbot dialogues will be based on the framework pro-posed by Følstad and Taylor [9]. However, since this framework was initially designed for customer experience in chatbot interactions, some modifications might be applied.

4 Expected Contributions

The contribution of this research to the AIED community lies in the exploration of the newly emerging field of teaching with GenAI. From the Learning Sciences perspec-tive, this research is needed because it aims to bridge the gap in understanding how GenAI can be applied to support teachers in instructional design to provide students with better learning opportunities. From the Computer Science perspective, this re-search is note-worthy for its attempt to use the latest GenAI technology in a meaning-ful and effective way. The research is experimental, intended to test theory-based hypotheses. It is based on prior knowledge theory [2, 10], the Dick and Carey [7] and 4C/ID instructional design models [22], Merrill's principles of instruction [23], and the framework for designing chatbots in education [26].

References

1. Alasadi, E.A., Baiz, C.R.: Generative AI in education and research: opportunities, concerns, and solutions. J. Chem. Educ. **100**(8), 2965–2971 (2023). https://doi.org/10.1021/acs.jchemed.3c00323
2. Ausubel, D.P.: Educational psychology: a cognitive view. Holt, Rinehart and Winston, New York, NY (1968)
3. Baglivo, F., et al.: Exploring the possible use of AI Chatbots in public health education: feasibility study. JMIR Med. Educ. 9, e51421 (2023). https://doi.org/10.2196/51421
4. van den Berg, G., du Plessis, E.: ChatGPT and generative AI: possibilities for its contribution to lesson planning, critical thinking and openness in teacher education. Educ. Sci. (Basel) **13**, 10 (2023). https://doi.org/10.3390/educsci13100998
5. Bolick, A.D., da Silva, R.L.: Exploring artificial intelligence tools and their potential impact to instructional design workflows and organizational systems. TechTrends (2023). https://doi.org/10.1007/s11528-023-00894-2
6. Coertjens, L., et al.: Teksten beoordelen met criterialijsten of via paarsgewijze vergelijking: een afweging van betrouwbaarheid en tijdsinvestering (2017)
7. Dick, W., Carey L., Carey J.: The systematic design of instruction. 8th edn. Pearson Higher Education Inc. (2015)
8. Farrokhnia, M., et al.: A SWOT analysis of ChatGPT: implications for educational practice and research. Innov. Educ. Teach. Int. (2023). https://doi.org/10.1080/14703297.2023.2195846
9. Følstad, A., Taylor, C.: Investigating the user experience of customer service chatbot interac-tion: a framework for qualitative analysis of chatbot dialogues. Qual User Exp. **6**, 1 (2021). https://doi.org/10.1007/s41233-021-00046-5

---

[1] https://comproved.com/en/.

10. Glaser, R.: Education and thinking: the role of knowledge. Am. Psychol. **39**(2), 93–104 (1984)
11. Goodyear, P.: Teaching as design. HERDSA review of higher education, vol. 2. 2, (2015)
12. Grassini, S.: Shaping the future of education: exploring the potential and consequences of AI and ChatGPT in educational settings (2023). https://doi.org/10.3390/educsci13070692
13. Guha, A., et al.: Generative AI and marketing education: what the future holds. J. Mark. Educ. (2023). https://doi.org/10.1177/02734753231215436
14. Hodges, C.B., Kirschner, P.A.: Innovation of instructional design and assessment in the age of generative. Artif. Intell. (2023). https://doi.org/10.1007/s11528-023-00926-x
15. Huizinga, T., et al.: Teacher involvement in curriculum design: need for support to enhance teachers' design expertise. J. Curric. Stud. **46**(1), 33–57 (2014). https://doi.org/10.1080/002 20272.2013.834077
16. Shani, I.: Survey reveals AI's impact on the developer experience
17. Jauhiainen, J.S., Guerra, A.G.: Generative AI and ChatGPT in school children's education: evidence from a school lesson. Sustainability (Switzerland). **15**, 18 (2023). https://doi.org/10. 3390/su151814025
18. Kerr, S.T.: How teachers design their materials: Introduction: implications for instructional design. Teacher Thinking and Instructional Design (1981)
19. Kim, M., Adlof, L.: Adapting to the future: ChatGPT as a means for supporting constructivist learning environments. TechTrends (2023). https://doi.org/10.1007/s11528-023-00899-x
20. Kirschner, P.A.: Do we need teachers as designers of technology enhanced learning? Instr. Sci. **43**(2), 309–322 (2015). https://doi.org/10.1007/s11251-015-9346-9
21. Krushinskaia, K., Elen J., Raes A.: The impact of artificial intelligence on the teacher's role: a systematic literature review. Under review
22. van Merrienboer, J.J.G.: Training complex cognitive skills: a four-component instructional design model for technical training. Educ. Technol. Publ. (1997)
23. Merrill, D.M.: First Principles of Instruction. Educ. Technol. Res. Dev. (2002)
24. Montenegro-Rueda, M. et al.: Impact of the implementation of ChatGPT in education: a systematic review (2023). https://doi.org/10.3390/computers12080153
25. Parsons, B., Curry, J.H.: Can ChatGPT pass graduate-level instructional design assignments? potential implications of artificial intelligence in education and a call to action. TechTrends (2023). https://doi.org/10.1007/s11528-023-00912-3
26. Sonderegger, S., Seufert, S.: Chatbot-mediated learning: conceptual framework for the design of Chatbot use cases in education. In: International Conference on Computer Supported Education, CSEDU – Proceedings, pp. 207–215. Science and Technology Publications, Lda (2022). https://doi.org/10.5220/0010999200003182
27. Tobler, S.: Smart grading: A generative AI-based tool for knowledge-grounded answer evaluation in educational assessments. MethodsX. **12** (2024). https://doi.org/10.1016/j.mex.2023. 102531

How Could Be Used Student Comments for Delivering Feedback to Instructors in Higher Education?

Gabriel Astudillo[1]($\boxtimes$) (ID), Isabel Hilliger[2] (ID), and Jorge Baier[1] (ID)

[1] Departamento de Ciencia de la Computación, Pontificia Universidad Católica de Chile, Santiago, Chile
gastudillo@uc.cl
[2] Pontificia Universidad Católica de Chile, Santiago, Chile

Abstract. In higher education, open-text comments from Student Evaluations of Teaching (SET) provide valuable insights into instructional strategies. However, processing these comments can be challenging, leading to limited feedback for instructors. This research aims to develop Natural Language Processing (NLP) strategies to transform student comments into actionable feedback. Two research questions guide this study: 1) How can NLP methods diagnose the effectiveness or mismatch of instruction in higher education? and 2) How can these diagnoses inform personalized recommendations for contextually relevant teaching practices? Using cosine similarity between vector representations of student comments and literature-based statements it is diagnosed the presence of effective teaching practices. This diagnosis will inform personalized feedback recommendations.

Preliminary work has used Exploratory Factor Analysis was used to analyze latent dimensions in the comment-statement similarity matrix and results suggest that correlations are linked to pedagogically relevant latent variables. This methodology seems to be a valid strategy for diagnosing the effectiveness or mismatch of teaching practices in higher education. Future research directions include exploring text data representations from different theoretical perspectives on education and investigating the impact and implementation of teaching practices suggested by language models compared to those recommended by human agents.

Keywords: Natural Language Processing · Student comments · Effective instruction · Higher Education · recommender systems

1 Introduction to the Research Problem

In higher education, open-text comments from Student Evaluations of Teaching (SET) have proven to be key to providing specific feedback on instructional strategies. According to McDonald et al. [1], these comments allow students to express, in their own words, the aspects that were most relevant to their learning experience in a particular course.

However, Hujala et al. [2] argue that given the high effort required for processing comments, in most institutions instructors receive this raw data, or at best, they are given

© The Author(s), under exclusive license to Springer Nature Switzerland AG 2024
A. M. Olney et al. (Eds.): AIED 2024 Workshops, CCIS 2151, pp. 401–408, 2024.
https://doi.org/10.1007/978-3-031-64312-5_50

feedback based on the identification of topics [2]. Since merely identifying topics is not sufficiently informative, some works, such as that of Raaf et al. [3] have attempted to complement topics with sentiment analysis to facilitate inference about how different dimensions of teaching and course design are evaluated. However, given that instructors in higher education are recruited from their own fields of research, they do not necessarily have pedagogical [4, 5] from which to construct educationally relevant interpretations of the results of analysis.

Since university instructors do not necessarily have pedagogical training, it is very important to deliver quality feedback to them in ways that they have the ability to interpret. To address this need as a research problem, the overall goal of this project is to develop strategies based on Natural Language Processing (NLP) models to transform teaching survey comments into quality feedback. This is operationalized in three specific objectives: 1) to develop an NLP-based comment analysis strategy to deliver specific and actionable feedback to higher education instructors; 2) to develop a recommender system-based strategy for generating suggestions on contextually relevant effective teaching practices; and 3) to evaluate the perceived usefulness and relevance of the suggestions delivered by the model to higher education instructors and experts.

2 Theoretical Framing and Proposed Solutions

Although there is not necessarily a consensus in the literature on what constitutes "good" university teaching, several researchers have examined in depth both its theoretical-epistemological foundations and the diversity of pedagogical practices [6, 7]. Moreover, these previous studies have led to the conclusion that, in higher education, a pedagogical practice is valued to the extent that it enables students to achieve the learning objectives for which it was designed [6, 8]. Consequently, the impact on the achievement of the expected learning objectives is used as an empirical criterion for comparison between different teaching practices implemented in university classrooms.

This is why in this research, we rely on the works of Smith and Baik [6] and Schneider and Preckel [8] to synthesize the instructional practices with robust evidence of having the largest positive effect sizes on student learning. On the one hand, Schneider and Preckel [8] reviewed 38 meta-analyses published between 1980 and 2014, estimating the effects of 105 variables on learning achievement in higher education. On the other hand, Smith and Baik [6], analyzed 78 publications, ranking the quality of the evidence provided in terms of its methodological validity and generalization of its conclusions.

In this research, we will use the instructional practices systematized in those studies as a basis to compare instructors' ability to solve teaching-learning problems in a real university context. By teaching-learning problems, we mean instructors' ability to identify instructional problems and solve them autonomously, and to identify problems that not only persist over time but whose resolution requires support. Or, to use the concepts coined by Vygotsky [9], the zone of real development and the zone of proximal development of instructors.

Previous work on feedback about teaching in higher education [10, 11] has emphasized that, in addition to identifying what instructors are doing well and what they are not doing in their instructional practices, it is crucial that it can provide suggestions on possible solutions to address the problems that instructors are not yet able to solve autonomously, or in other words, propose modifications in teaching-learning practices.

Taking into account previous efforts to analyze SET [2, 3, 12, 13], this paper focuses on diagnosing the presence of effective teaching practices in student comments and providing recommendations to instructors to improve their teaching. In this way, the following research questions are addressed:

RQ 1. How can a methodological strategy based on NLP diagnose the effectiveness or misalignments of instruction in higher education?

RQ 2. How to use this diagnosis to suggest instructors' effective modifications in their practices that are pedagogically and contextually relevant, by means of a recommender system?

2.1 A Strategy to Diagnose Teaching Practices

In order to address Research Question 1, NLP methods will be used to detect the presence of effective teaching practices or mismatches with respect to them in student comments. Specifically, phrases will be constructed to represent effective teaching practices identified in the reviewed literature [6, 8], using similar wording and phrases to what students might use.

Then, a BERT model [14] will be used to obtain vector representations of the student comments and of the statements. Subsequently, cosine similarity will be calculated for each comment-statement pair to detect the presence of the investigated instructional practices. By cosine similarity we will refer to "a measure of similarity between two vectors of an inner product space that measures the cosine of the angle between them [15, p. 14] on a scale from 0, when the vectors have no shared information, to 1, when the vectors have exactly the same information. The output is a matrix where each row represents a comment, each column a formulated statement, and the cells contain cosine similarity.

In this matrix it is possible to identify which of the expected teaching practices are most and least present in a specific course imparted by a specific instructor. And with this, a diagnosis of the teaching problems is obtained by identifying those instructors who are capable of solving these problems effectively and those who are not.

2.2 Some Ideas for Designing a Strategy for Delivering Quality Feedback

Generative language models have 'revolutionized' recommender systems [16]. Based on the new opportunities created by this technological progress, not only teaching staff could be provided with quality feedback about their effective teaching practices and unresolved challenges, but also with alternative improvement actions that instructors may not envision autonomously.

In principle, the problem of raising contextually relevant modifications in pedagogical practices can be understood as a recommendation problem: delivering personalized suggestions given some representation of previous behavior. The particularity with respect to other recommendation problems is that, for example, recommendations of consumption items -such as streaming music and video, or items to purchase- do not require evidence-based criteria or learning theories. In other words, the problem is how to make the recommendations delivered by a generative language model consistent with an effective evidence-based teaching perspective.

Work that has used language models for recommendation problems [17–19], what they have done is a process in which first a traditional recommender system selects a list of candidate items, which is given to a language model along with previous user preferences at a prompt, and then the language model makes the final selection. This approach is known as In Context Learning [20, 21] and takes advantage of the reasoning capabilities of Large Language Models to make inferences about new information.

Following these concepts, as a solution to the problem of recommending effective teaching practices, an In Context Learning strategy via prompting is proposed as a solution. To do so, the steps are: take the results of the similarity matrix between comments and statements, filter n most detected effective teaching practices, n most detected teaching misalignments, and n least detected effective practices. This is concatenated into a template prompt along with additional course information such as learning objectives. Finally, the model is instructed to select recommendations among the n least detected teaching practices. This allows the language model to make the final recommendation considering the diagnosis and the objectives of the specific course.

3   Preliminary Work

So far, progress has been made regarding research question 1, having tested the NLP based strategy in a synthetic dataset of 9,921 comments. The comments came from teaching evaluation surveys conducted at the end of 2022, aiming to evaluate teaching practices in the 636 engineering course sections offered at a Latin American university premises in a Spanish-speaking country. These course sections were imparted by different engineering departments at both undergraduate and graduate level, including computer science, industrial engineering, electrical engineering, mechanical engineering, structural engineering, among others. The surveys included two open-ended questions: "Comment on aspects you consider positive in the course or the professor's work" and "Comment on aspects you think should be improved in the course or the professor's work". In order to ensure privacy of teaching staff and students, comments were rephrased by using ChatGPT.

Since texts were in Spanish, to obtain vector representations of both comments and statements, the bert-base-spanish-wwm-uncased model [22] was used, which is a BERT-type model trained in Spanish Subsequently, cosine similarity was calculated for each comment-statement pair to detect the presence of teaching practices.

Table 1. Examples of student comments, statements and cosine similarity

Student comment	Statement	Cos. Sim
It is a difficult course, but it is very clear that the professor knows a lot and is passionate about it	It is evident that he/she really enjoys teaching the course	0.82
The professor always said to ask questions	Always encouraged us to ask questions when we didn't understand	0.83
The course is well structured and well organized	The course is very well organized	0.84
The course lectures are on interesting topics. They helped me to get an idea of the major and motivated me to learn more on my own	The course provided me with things I wanted to learn	0.82
the professor is very close to his students	The instructor is very approachable with students	0.82

Finally, it is important to provide evidence of the extent to which similarity between statements and comments can be explained by latent pedagogical variables, and not simply by linguistic or semantic coincidences in the vector representation provided by the BERT-type model. To meet this, Exploratory Factor Analysis, was applied to the similarity matrix between comments and statements. Maximum Likelihood (ML) was used to extract the factors, and oblimin rotation, assuming correlated latent factors.

An exploratory factor analysis yielded a factorial solution that retained 17 out of the 20 proposed statements, from which three latent dimensions emerged, collectively explaining 74% of the variance in the original variables.

Table 2 presents the loadings of the statements on the emergent factors. ML1 groups statements that could be interpreted as related to meaningful learning, in addition to aspects reasonably correlated to it, such as "It is evident that he really enjoys teaching the course" (.51) and "Good instructor-student relationship" (.35). Statements with high factor loadings in ML2 appear to account for the teacher-student relationship in two different senses: aspects related to the mediation of learning—such as clarity of explanations and asking good questions in class—and elements linked to interpersonal relationships. Finally, ML3 is less straightforward, but an interpretation could suggest that promoting student questions, providing feedback, and making classes interesting are related to content comprehension.

Beyond the set of statements used in these tests, the results of the exploratory factor analysis suggest that the correlations in the similarity matrix do not seem to be explained by general linguistic coincidences, but seem to be explained by pedagogically relevant latent variables, linked to theoretical constructs. Thus, the exploratory factor analysis provides evidence in favor of the methodology of estimating the cosine similarity between the vector representations of literature-based statements and student comments, could be a valid methodological strategy to diagnose the effectiveness or mismatch of teaching practices in higher education (Table 1).

Table 2. Factor loadings matrix[a].

Statements[b]	ML1	ML2	ML3	H2[c]	Com[d]
Contributed to the development of the project	0.85			0.59	1.17
We could apply what we learned	0.83			0.71	1.00
The course provided me with things I wanted to learn	0.83			0.80	1,11
I understood clearly what the course was about	0.82			0.73	1.08
I could connect what we saw in the course to the real world	0.78			0.74	1.17
The course was demanding, but a lot was learned	0.75			0.79	1.21
The course is very well organized	0.72			0.72	1.10
The instructor made us reflect deeply	0.53	0.44		0.66	2.14
It is evident that he/she really enjoys teaching the course	0.51	0.50		0.77	2.00
He/She/They asked us good questions in class		0.94		0.97	1.04
The professor is very approachable with students		0.81		0.70	1.02
Always available to answer the questions and doubts we have		0.78		0.56	1.05
The explanations were very clear		0.69		0.70	1.18
Always encouraged us to ask questions when we didn't understand		0.54	0,54	0.72	2.00
Good instructor-student relationship	0.35	0.38		0.55	2.49
The classes were very interesting			0,94	0.94	1.02
The feedback was very useful			0,62	0.56	1.30

[a] SS loadings: ML1 = 6.23, ML2 = 4.51, ML3 = 1.82; Proportion Var: ML1 = .37, ML2 = .27, ML3 = .74

[b] Phrases constructed to detect effective teaching practices in student-similar language

[c] Communality: proportion of variance in an observed variable that is accounted for by the common factors in the model

[d] Item complexity: how much constructs reflect each item

4 Expected Research Contributions

This research will develop natural language processing strategies for utilizing comments from student teaching evaluation surveys to deliver quality feedback to instructors in higher education, targeting crucial aspects that affect learning.

A first contribution, currently more advanced, is the strategy to diagnose teaching based on student comments. This represents an advance for the Artificial Intelligence in Education community in that it allows specific and actionable inferences about teaching practices in critical aspects that would not necessarily be detected by state-of-the-art methods [2, 3, 12, 13], and can facilitate intervention by instructors themselves, as well as by decision makers or support units. This strategy can be easily replicated in different institutions. But it is also flexible enough to test different sets of statements, including aspects of teaching that have not been considered by the studies on which our set was based.

A second expected contribution is the development of a teaching practices recommender system. This will allow the efficient suggestion of effective teaching practices relevant to the context of the course, its learning objectives, and the diagnosis of key aspects of teaching.

On the other hand, new lines of research are also open that can use strategies derived from the one presented here to produce text data representations oriented from different theoretical perspectives on education. For example, sets of ad-hoc statements can be constructed for different research questions and their effects on quantitative aspects of teaching surveys can be modeled in an explainable way: satisfaction or recommendation with the course, perception of learning, among others. Likewise, lines of research oriented towards the intervention of teaching in higher education are also opened. What differences in impact on teaching can there be between the recommendations delivered by language models compared to those delivered by human agents? What factors facilitate or hinder the implementation of teaching practices suggested by a language model?

References

1. McDonald, J., Moskal, A.C.M., Goodchild, A., Stein, S., Terry, S.: Advancing text-analysis to tap into the student voice: a proof-of-concept study. Assess. Eval. High. Educ. **45**(1), 154–164 (2020). https://doi.org/10.1080/02602938.2019.1614524
2. Hujala, M., Knutas, A., Hynninen, T., Arminen, H.: Improving the quality of teaching by utilising written student feedback: a streamlined process. Comput. Educ. **157**, 103965 (2020). https://doi.org/10.1016/j.compedu.2020.103965
3. Raaf, S.A., Knoos, J., Dalipi, F., Kastrati, Z.: Investigating learning experience of MOOCs learners using topic modeling and sentiment analysis. In: 2021 19th International Conference on Information Technology Based Higher Education and Training (ITHET), pp. 01–07. IEEE (2021). https://doi.org/10.1109/ITHET50392.2021.9759714
4. Benassi, V.A., Buskist, W.: Preparing the new professoriate to teach. In: Effective College and University Teaching: Strategies and Tactics for the New Professoriate, pp. 1–8. SAGE (2012)
5. Groccia, J.E., Buskist, W.: Need for evidence-based teaching. New Directions for Teaching and Learning, vol. 128 (2011)
6. Smith, C.D., Baik, C.: High-impact teaching practices in higher education: a best evidence review. Stud. High. Educ. **46**(8), 1696–1713 (2021). https://doi.org/10.1080/03075079.2019.1698539
7. Ambrose, S.A., Bridges, M.W., DiPietro, M., Lovett, M.C., Norman, M.K.: How learning works: seven research-based principles for smart teaching. Wiley (2010)
8. Schneider, M., Preckel, F.: Variables associated with achievement in higher education: a systematic review of meta-analyses. Psychol. Bull. **143**(6), 565–600 (2017). https://doi.org/10.1037/bul0000098.supp
9. Vygotsky, L.S.: Mind in society: development of higher psychological processes. Harvard University Press, United States of America (1978)
10. Jeffs, C., Nelson, N., Grant, K.A., Nowell, L., Paris, B., Viceer, N.: Feedback for teaching development: moving from a fixed to growth mindset. Prof. Dev. Educ. **49**(5), 842–855 (2023). https://doi.org/10.1080/19415257.2021.1876149
11. Gormally, C., Evans, M., Brickman, P.: Feedback about teaching in higher ed: neglected opportunities to promote change. CBE—Life Sci. Educ. **13**(2), 187–199 (2014). https://doi.org/10.1187/cbe.13-12-0235

12. Marshall, P.: Contribution of open-ended questions in student evaluation of teaching. High. Educ. Res. Dev. **41**(6), 1992–2005 (2022). https://doi.org/10.1080/07294360.2021.1967887
13. Kastrati, Z., Kurti, A., Dalipi, F., Ferati, M.: Leveraging topic modeling to investigate learning experience and engagement of MOOC completers, pp. 54–64 (2023). https://doi.org/10.1007/978-3-031-41226-4_6
14. Devlin, J., Chang, M.-W., Lee, K., Toutanova, K.: BERT: pre-training of deep bidirectional transformers for language understanding (2018)
15. Gomaa, W.H., Fahmy, A.A.: A survey of text similarity approaches. Int. J. Comput. Appl. **68**(13), 13–18 (2013). https://doi.org/10.5120/11638-7118
16. Fan, W., et al.: Recommender systems in the era of large language models (LLMs) (2023)
17. Gao, Y., Sheng, T., Xiang, Y., Xiong, Y., Wang, H., Zhang, J.: Chat-REC: towards interactive and explainable LLMs-augmented recommender system (2023)
18. Liu, P., Zhang, L., Gulla, J.A.: Pre-train, prompt and recommendation: a comprehensive survey of language modelling paradigm adaptations in recommender systems (2023)
19. Zhang, J., Xie, R., Hou, Y., Zhao, W.X., Lin, L., Wen, J.-R.: Recommendation as instruction following: a large language model empowered recommendation approach (2023)
20. Rubin, O., Herzig, J., Berant, J.: Learning to retrieve prompts for in-context learning (2021)
21. Kim, H.J., Cho, H., Kim, J., Kim, T., Yoo, K.M., Lee, S.: Self-generated in-context learning: leveraging auto-regressive language models as a demonstration generator (2022)
22. Cañete, J., Chaperon, G., Fuentes, R., Ho, J.-H., Kang, H., Pérez, J.: Spanish pre-trained BERT model and evaluation data (2023)

An Augmented Intelligence Framework as an Enabler of Digital Transformation and Innovation in the Pedagogical Evaluation of the Brazilian National Textbook Program (PNLD)

Luciane Silva[1,2](✉) ⓘ, André Araújo[2](✉) ⓘ, and Rafael Araújo[1,2](✉) ⓘ

[1] Federal University of Uberlândia, Uberlândia, Brazil
lucianefatsilva@gmail.com
[2] Center for Excellence in Social Technologies (NEES), Federal University of Alagoas, Maceió, Brazil
{luciane.silva,andre.araujo,rafael.araujo}@nees.ufal.br

Abstract. This project investigates the integration of Artificial Intelligence (AI) into the Brazilian National Textbook Program (PNLD), aiming to expand human capacities and improve pedagogical assessment. The focus is to specify, develop, and evaluate a technological framework that uses Augmented Intelligence and Natural Language Processing (NLP) to automate repetitive tasks, reduce errors, and optimize decision-making through efficiently used data. The research proposes not only to enhance the interaction between humans and data but also to contribute to the improvement of public educational policies in Brazil, through prototypes that demonstrate the potential of AI to transform pedagogical assessment, promoting continuous development in the educational field.

Keywords: Augmented Intelligence · Digital Transformation in Government · NLP · PNLD

1 Introduction

Education is a fundamental pillar for the social and intellectual development of societies, providing equal opportunities for learning and personal growth. Technological advancements have significantly influenced the transformation of educational systems worldwide, a trend exemplified by the Brazilian National Textbook Program (PNLD). With a history spanning over 85 years, the PNLD has been instrumental in supplying quality educational resources to public education across Brazil. This government initiative not only facilitates access to educational materials across various regions but also promotes equal educational opportunities by overcoming logistical and socio-economic challenges, thus enhancing education quality nationwide [2,5,7].

© The Author(s), under exclusive license to Springer Nature Switzerland AG 2024
A. M. Olney et al. (Eds.): AIED 2024 Workshops, CCIS 2151, pp. 409–417, 2024.
https://doi.org/10.1007/978-3-031-64312-5_51

Within the context of the PNLD, the pedagogical evaluation stage, which is the main focus of this doctoral research project, faces considerable challenges. The magnitude of evaluating educational materials in a country as vast as Brazil, marked by significant diversity in cultural, regional aspects, and specific needs of each public call, makes this process particularly complex. Besides the large scale, it is crucial to consider cultural and regional nuances that directly impact the effectiveness of the evaluation. This reality imposes a considerable human effort and the mobilization of significant resources to ensure comprehensive and accurate assessment, highlighting the importance of innovative approaches, such as the application of smarter processes focused on optimization [2].

With an eye on expanding the capabilities of an educational policy that has existed for so many years and driven by Digital Information and Communication Technologies (DICTs), it is also possible to make an important reference to digital transformation and its significant impact on society, including at the governmental level. This concept refers to the comprehensive integration of digital technologies to enhance the efficiency, innovation, and quality of operations, notably since the last decades of the 20th century [1,4].

Digital transformation in public policies represents a significant shift in how governments create, design, and implement their strategies [6,8]. Digital technologies provide a powerful set of tools to improve the efficiency and responsiveness of public policies. Automation and digitization simplify administrative processes, reducing bureaucratic inefficiencies while simultaneously increasing the overall agility of public institutions. Furthermore, digital platforms facilitate ongoing communication between government agencies and citizens, promoting greater transparency and inclusion in policy-making. Digital transformation in the Brazilian government has played a crucial role in reshaping Brazil's public policies, especially in the educational context exemplified by the PNLD, the subject of study in this doctoral work.

Moreover, the collaboration between academia, the public and private sectors within the PNLD represents an inclusive approach that seeks to elevate educational policies to standards of excellence. In this context, the proposal to incorporate technologies such as augmented intelligence, artificial intelligence (AI), and data analysis emerges as a unique opportunity to optimize the implementation and evaluation of policies. Augmented intelligence, in particular, stands out as a strategic tool for digital transformation, recognizing the effectiveness of collaboration between human skills and artificial capabilities [3,9].

This doctoral research project aims to investigate and analyze the use of artificial intelligence resources to enhance human capacity in the context of the PNLD's pedagogical evaluation and drive digital transformation in this setting. The research seeks to contribute to understanding the implications of augmented intelligence in public educational policies, highlighting the challenges, opportunities, and potential benefits of this integration. Given the complexity inherent in the pedagogical evaluation, which involves various book categories and spans multiple domains, the use of AI, particularly Natural Language Processing (NLP) in Portuguese, emerges as a promising approach to support the evaluation team in preparing opinions. The difficulty of conducting pedagogical evaluations, due to the wide diversity of book categories and domains, is evidenced, and AI, through NLP, arises as a valuable tool in this context.

Therefore, the project aims to enhance the pedagogical evaluation processes of the PNLD through the use of metrics and some deeper analyses such as time and effort, with the help of tools and technologies like data analysis, artificial intelligence, augmented intelligence, and NLP. These metrics will quantify the similarity between reports at different stages of review, the consistency between evaluators, and the efficiency in time and effort of the evaluations. The adoption of these indicators has the potential to make the evaluation faster, more accurate, and aligned with the educational needs of Brazil, significantly contributing to the objectivity, precision, and suitability of the educational materials to the national pedagogical context.

2 The Research Questions and Methodological Approach

This section presents the research questions and methodological approach under-pinning this doctoral thesis.

2.1 Research Objectives

The central objective of this doctoral thesis is to specify, develop, and evaluate a technological framework based on augmented intelligence, aimed at expanding human capacity and enhancing the evaluation process of the PNLD, contributing to the pedagogical assessment process. This framework will seek to integrate advanced technologies to automate repetitive tasks, reduce errors, and improve decision-making through efficient data use.

The specific objectives include:

- Analyze the current scenario to understand the current pedagogical evaluation process of the PNLD, identifying its main nuances and challenges.
- Establish evaluation metrics to define and assess the effectiveness and efficiency of the current process.
- Conduct in-depth data analyses to collect and analyze detailed information about the pedagogical evaluation process, including evaluators' opinions, evaluation effort, and similarities between opinions, to better understand the internal dynamics and efficiency of the process.
- Investigate the application of Artificial Intelligence technologies and explore the integration of Augmented Intelligence and NLP technologies in the context of digital transformation in Brazilian public educational policies, focusing on improving the interaction between humans and data.
- Develop improvement proposals through prototypes that use augmented intelligence and Natural Language Processing (NLP) to optimize the pedagogical evaluation process, automating repetitive tasks and improving the accuracy and speed of evaluations.

These objectives were formulated to ensure that the research contributes significantly to scientific and practical knowledge, providing results for the continuous improvement of public educational policies in Brazil. The next section will detail the proposed hypotheses to achieve these objectives, followed by a discussion on the study's impact expectations.

2.2 Methodology

This doctoral project investigates the application of technology and artificial intelligence (AI) in Brazil's National Textbook Program (PNLD) with the goal of enhancing the pedagogical evaluation process. The study focuses on specifying, developing, and evaluating a technological framework that incorporates augmented intelligence and Natural Language Processing (NLP) to optimize tasks, reduce errors, and improve decision-making. Furthermore, it seeks to understand how these technologies can meet the contemporary demands of the Brazilian educational system, facilitating the integration of technologies in education and meeting the evolving expectations of educators.

To understand the current scenario of the system and consider relevant adjustments for users, detailed questionnaires were employed that investigate the evaluators' experience with the PNLD's digital evaluation environment. These questionnaires, along with strategic workshops involving all stakeholders at the end of each evaluation cycle, provided valuable insights. These activities not only highlighted how artificial intelligence can be effectively applied but also created an efficient channel for identifying and correcting deficiencies, guiding adaptive improvements in the evaluation process.

The adopted methodology comprises several structural stages, briefly described as follows:

- Immersion in the context of the PNLD through interviews with stakeholders, real-time monitoring of the evaluation process, and review of official documents to gain a deep understanding of the environment and the parties involved.
- Literature review and theorization to build a theoretical foundation on augmented intelligence and Natural Language Processing (NLP), with a systematic review of the literature focused on applications in educational and governmental assessments.
- Diagnosis of the current process focusing on analyzing the efficiency and effectiveness of the current pedagogical evaluation process using flow diagrams and existing data.
- Development and testing of prototypes aiming to implement augmented intelligence and NLP solutions to optimize the evaluative process, with pilot tests and validation based on user feedback.
- Empirical investigation of the application of NLP for text analysis and decision support systems, aiming to improve decision-making in the evaluation process.

– Data analysis and formulation of recommendations using statistical techniques to evaluate the results and formulate recommendations for the effective implementation of technologies.
– Thesis writing and defense with the compilation of the results.

3 Preliminary Research Results and Current Challenges

This section outlines preliminary results and ongoing challenges within the pedagogical evaluation phase of the PNLD. As the diversity of educational materials grows and the workload increases, the PNLD confronts significant challenges that risk diminishing the quality and efficiency of evaluations.

3.1 Current Challenges

The PNLD confronts significant challenges in its pedagogical evaluation phase due to the diversification of materials and increased workload, which threaten the quality and efficiency of evaluations. In response, this project explores the use of Augmented Intelligence and Natural Language Processing (NLP) to enhance this process. It hypothesizes that advanced systems can notably decrease the time and effort required for evaluations, improve assessment quality, facilitate error diagnosis, and provide better feedback to publishers. Additionally, the project will investigate how the adoption of technologies such as consistent metrics and decision-support systems can standardize evaluations, increase their reliability and precision, and reduce discrepancies among evaluators.

Preliminary analyses of data from recent pedagogical evaluations reveal that nearly 30% (83,909) of specific failures correspond to spelling and grammatical errors, as shown in Fig. 1. These types of errors demand significant time from evaluators for identification and correction, involving publishers in the revision process and reviewers in validating the corrections. The need for multiple interventions not only prolongs the time required to complete the evaluations but also increases the associated costs. This finding emphasizes the potential to implement technological solutions that can automate the detection and correction recommendation of these errors, thereby increasing the efficiency of the pedagogical evaluation process and potentially reducing operational costs.

In addition to information about grammatical errors, the analysis of data from the PNLD database on public calls has revealed important insights into the pedagogical evaluation process. There are significant variations in the average number of flaws per work, indicating heterogeneity in the quality of the materials evaluated. The specific characteristics of each public call directly impact the type and number of flaws identified, which lead to feedback for publishers and the evolution of public calls. Moreover, a notable increase in the number of evaluators and improvements in the self-registration of reviewers highlight a positive response to digital transformation in the PNLD, increasing professional involvement in different regions.

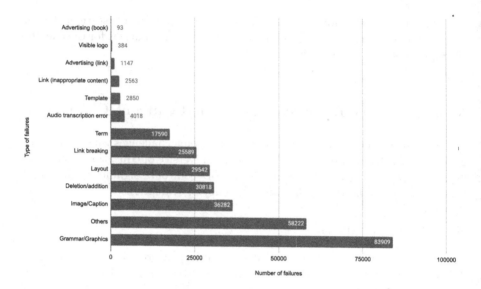

Fig. 1. Types of failures

As can be seen in Fig. 2, there was a significant increase in the evaluator base from September to January. Continuing with the data analysis, particularly regarding the geographical distribution of evaluators, a substantial growth in the registered user base was observed over time. In September 2023, there were 5,682 registered evaluator teachers, as shown on the left side of Fig. 2. However, following updates to the sign-up module and a promotional campaign by the MEC, the number of reviewers in the database has significantly increased, reaching 14,571, as illustrated on the right side of the same figure.

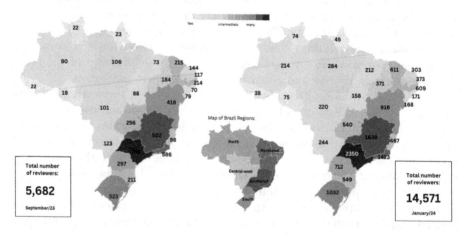

Fig. 2. Distribution of reviewers per state in Brazil before (left) and after (right) the redesign of the sign-up module.

This considerable expansion in the evaluator base can primarily be attributed to the introduction of the new self-registration module and the promotional and outreach efforts by government agencies. This redesign of the registration functionality not only represents a practical advancement but also reflects ongoing technological evolution. It contributes to the modernization of the software platform, introducing technological innovations and driving digital transformation in the evaluation process. Moreover, it is clear from the analysis of Fig. 2 that there remains a significant disparity in the distribution of evaluators across Brazilian regions, with the Southeast, Northeast, and South regions having higher concentrations compared to the Central-West and North. The Southeast region alone accounts for 40.35% of the evaluators (5,879 out of the total of 14,571), indicating the need for a more balanced geographical distribution to ensure a comprehensive and contextualized evaluation that reflects the diverse cultural and educational realities present across the country.

Regarding limitations and areas needing attention, several issues stand out that could influence the evaluation outcomes. The "Others" category in pinpoint flaws requires a more detailed description due to its significant volume, and individual interpretation by evaluators introduces subjectivity into the results. Variations in evaluator density across regions suggest opportunities to optimize distribution for equitable representation nationwide. Moreover, the adaptation of Natural Language Processing for Brazilian Portuguese may face challenges in addressing linguistic variations and cultural nuances, potentially affecting the accuracy of AI analyses. Finally, decision-making in educational contexts involves complex values and ethical considerations, emphasizing the importance of continuous human involvement to complement AI systems, thereby ensuring that technological integration supports a more thorough and efficient evaluation process.

3.2   Expected Contributions

This project primarily aims to explore the use of Augmented Intelligence and Natural Language Processing (NLP) in the context of the pedagogical evaluation of the PNLD. The integration of these technologies has the potential to promote digital transformation and bring innovation to the pedagogical assessment process, making it more efficient, accurate, and equitable. The expected results of this study are anticipated to impact both the theoretical and practical fields of pedagogical evaluation and the management of public educational policies.

Firstly, a considerable improvement in operational efficiency is expected, characterized by reduced time and effort needed to evaluate educational materials, along with a decrease in manual workload. This will be possible through the automation of tasks and simplification of processes, especially in the correction of grammatical and spelling errors, using Natural Language Processing (NLP). Moreover, the quality of the evaluated educational materials will be enhanced by reducing content errors, and the feedback to publishers will be more constructive and efficient, allowing for a more agile and precise revision cycle.

Decisions during the evaluation process will be data-driven, with the development and use of analytical dashboards and decision support systems that leverage historical and current data to provide actionable insights. This not only improves the accuracy of the decisions made by evaluators and technical committees but also increases the ability to respond quickly if a relevant issue arises during the process. Strategies will be implemented to form evaluation teams that reflect the geographical and cultural diversity of Brazil, promoting equity and inclusion and ensuring that all regions and cultural groups are adequately represented.

Beyond the practical benefits, the project will contribute to the academic literature with studies on the application of innovative technologies in education and public policy management. There is a proposal for a replicable model of using augmented intelligence in pedagogical evaluation, which can be adapted to other educational and governmental contexts. This model will not only demonstrate how the integration of augmented intelligence and NLP can transform governmental practices, making them more efficient, transparent, and accessible, but also establish a more sustainable pedagogical evaluation process that uses resources more effectively, reduces redundancies and operational costs.

References

1. Albertin, A.L., de Moura Albertin, R.M.: Transformação digital: gerando valor para o novo futuro. GV-EXECUTIVO **20**(1), 26–29 (2021)
2. Batista, A.A.G.: Recomendações para uma política pública de livros didáticos. Ministério da Educação, Secretaria de Educação Fundamental (2001)
3. Choi, Y., Gil-Garcia, R., Aranay, O., Burke, B., Werthmuller, D.: Using artificial intelligence techniques for evidence-based decision making in government: random forest and deep neural network classification for predicting harmful algal blooms in new york state. In: DG. O2021: The 22nd Annual International Conference on Digital Government Research, pp. 27–37 (2021)
4. Edelmann, N., Albrecht, V.: Designing public participation in the digital age: Lessons learned from using the policy cycle in an austrian case study. In: Proceedings of the 24th Annual International Conference on Digital Government Research, pp. 300–308 (2023)
5. FNDE: Encontro do pnld aborda a importância do programa para o aprendizado dos estudantes (2022), https://www.gov.br/fnde/pt-br/assuntos/noticias/encontro-do-pnld-aborda-a-importancia-do-programa-para-o-aprendizado-dos-estudantes, acessado em: Janeiro de 2024
6. Hoekstra, M., Van Veenstra, A.F., Bharosa, N.: Success factors and barriers of govtech ecosystems: a case study of GovTech ecosystems in The Netherlands and Lithuania. In: Proceedings of the 24th Annual International Conference on Digital Government Research, pp. 280–288 (2023)
7. MEC, M.d.E.: Pnld. Governo Brasileiro (2023). http://portal.mec.gov.br/component/content/article?id=12391:pnld

8. Mergel, I., Edelmann, N., Haug, N.: Defining digital transformation: results from expert interviews. Gov. Inf. Q. **36**(4), 101385 (2019)
9. Young, M., Himmelreich, J., Honcharov, D., Soundarajan, S.: The right tool for the job? assessing the use of artificial intelligence for identifying administrative errors. In: DG. O2021: The 22nd Annual International Conference on Digital Government Research, pp. 15–26 (2021)

Reducing University Students' Exam Anxiety via Mindfulness-Based Cognitive Therapy in VR with Real-Time EEG Neurofeedback

Ziqi Pan$^{(\boxtimes)}$ ⓘ, Alexandra I. Cristea$^{(\boxtimes)}$ ⓘ, and Frederick W. B. Li$^{(\boxtimes)}$ ⓘ

Department of Computer Science, Durham University, Durham, UK
{ziqi.pan2,alexandra.i.cristea,frederick.li}@durham.ac.uk

Abstract. This research aims to develop and evaluate a novel approach to reduce university students' exam anxiety and teach them how to better manage it using a personalised, emotion-informed Mindfulness-Based Cognitive Therapy (MBCT) method, delivered within a controlled and immersive Virtual Reality (VR) environment. By integrating real-time Electroencephalography (EEG) data for anxiety recognition through a Brain-Computer Interface (BCI), the approach will attempt to dynamically adapt interventions, based on *individual anxiety levels*.

Keywords: Exam Anxiety · Mental Health Education · Mindfulness-Based Cognitive Therapy (MBCT) · Brain-Computer Interface (BCI) · Personalisation

1 State of the Art and Resulting Problem to Address

As one of the most prevalent concerns amongst students, *exam anxiety* is characterised by intense worry and apprehension before or during examinations, and impacts students' academic performance and well-being [12]. Traditionally, Cognitive Behavioural Therapy (CBT) has been the mainstay treatment for exam anxiety. For instance, Grassi et al. (2011) [4] applied CBT with relaxation exercises and multimedia audio-video content, to reduce exam anxiety in university students. Ugwuanyi et al. (2020) [13] investigated the effectiveness of CBT combined with music interventions in reducing physics test anxiety in secondary school students. Similarly, Safari et al. (2022) [8] found that the effectiveness of CBT with game interventions can reduce exam anxiety in Indian students.

Despite its documented effectiveness in treating exam anxiety, CBT possesses certain inherent limitations. To start with, CBT primarily focuses on identifying and challenging negative thought patterns. This can be difficult for some individuals, specifically those with high anxiety, who might find the process of analysing and restructuring thoughts itself anxiety-provoking. Additionally, CBT provides limited focus on emotional experience, thus not fully addressing the emotional aspects of anxiety, such as difficulty regulating emotions [6].

A. M. Olney et al. (Eds.): AIED 2024 Workshops, CCIS 2151, pp. 418–423, 2024.
https://doi.org/10.1007/978-3-031-64312-5_52

Building on CBT, Mindfulness-Based Cognitive Therapy (MBCT) proposed a better method for treating exam anxiety, providing increased self-awareness and acceptance along with improved emotional regulation [14]. Compared with CBT, MBCT adds mindfulness practices to cultivate present-moment awareness and acceptance of thoughts and emotions. However, while effective, the clinical setting in traditional MBCT sessions may exacerbate anxiety and discomfort for some individuals, as it requires them to disclose their conditions in front of others, which may hinder open communication and progress [1]. In addition, access to traditional face-to-face MBCT can be constrained by both geographical limitations in therapist availability and individual time flexibility [7].

Seeking to overcome the limitations of traditional MBCT, the emergence of virtual reality (VR) is revolutionising the therapeutic landscape in MBCT treatment. A growing body of evidence highlights the potential of VR to address a diverse range of mental health concerns with increased engagement, personalised interventions, and enhanced therapeutic outcomes. Compared to traditional methods, VR offers greater flexibility, control, and accessibility, particularly for individuals facing obstacles, such as limited mobility or social stigma [2].

However, existing VR-based MBCT approaches for anxiety often lack real-time personalised adjustments, relying on predetermined scenarios and interventions that may not effectively address individual fluctuations in emotional states [5]. This gap highlights the potential of incorporating real-time emotion recognition using Electroencephalography (EEG) data into VR-based MBCT for anxiety. EEG is one of the main data types used in Brain Computer Interfaces (BCIs), which measures electrical activity in the brain in a non-invasive manner, and offers a promising avenue for *objective emotional assessment*. During anxious states, specific brain regions, such as amygdala, exhibit characteristic changes in activity in various frequency bands [11].

2 Research Design

2.1 Research Aim

While extensive research has studied EEG in education [9] and VR in education [3] separately, *combining EEG and VR for real-time experiences in educational settings* remains under-explored. By utilising a Brain-Computer Interface (BCI) to interpret EEG signals in real-time, we aim to develop a personalised, emotion-based MBCT approach delivered within VR environments, to reduce university students' exam anxiety and teach them how to better manage their anxiety. This novel intervention will dynamically adapt its therapeutic elements based on individual anxiety levels, ensuring targeted and immediate support during moments of heightened emotional distress.

By *harnessing the combined power of VR and BCI-driven emotion recognition and interpretation, we hope to pave the way for a more personalised and effective approach to adaptively managing exam anxiety for students.*

Based on the above research aim, we formulate the umbrella research question as:

How can we mitigate exam anxiety among university students in a more effective way?

Our hypothesis is that combined application of BCI and VR technology, deeply rooted in psychological approaches, such as MBCT, will demonstrate the greatest efficacy in mitigating exam anxiety. Thus, sub-research questions are:

1. *How can an effective MBCT session be designed in VR for students?*
2. *How can an exam-related anxiety recognition model be designed based on EEG data and incorporated into the MBCT session in VR?*
3. *To what extent can an emotion-based MBCT approach in VR ease individual student exam anxiety?*

2.2   Methods

To address these research questions, the first objective of this research is to develop a personalised MBCT program that provides an immersive VR experience and dynamically adapts to students' real-time anxiety level. This level will be determined by the EEG data derived from a BCI device. By first training on each participant's unique anxiety profile, the personalised MBCT program aims to accurately recognise their in-session anxiety level and adjust therapeutic interventions, accordingly. Secondly, this research aims to examine the effectiveness of such a program in mitigating exam anxiety in university students.

The Underlining Mechanism for Anxiety Recognition and Interpretation. Given the current state-of-the-art (SoA) performance of Convolutional Neural Network (CNN) in EEG-based emotion recognition, this study will employ in a first instance a CNN architecture for automatic feature learning from preprocessed EEG data. However, further investigation into fine-tuning existing CNN and other SoA architectures and hyperparameters will be pursued, to optimise network performance for this specific task. All human-studies will be undergoing rigorous ethical checks via our regular university channels. This includes all participants having the right to withdraw from the study at any point.

This study will first build individual anxiety profile based on EEG data. University students who self-report experiencing exam anxiety will be guided through *a passive anxiety elicitation procedure*. The passive anxiety elicitation procedure will involve prompting participants to imagine scenarios associated with exam anxiety, such as feeling overwhelmed by the perceived difficulty of an exam, being unable to formulate answers, and simultaneously observing classmates appearing relaxed and progressing confidently. EEG data from the BCI equipment will be concurrently recorded during such procedure, labelling anxious moments versus non-anxious moments for all participants. After the anxiety elicitation experiments, collected EEG data will be cleaned and pre-processed, to remove noise and artifacts. A CNN model will then automatically learn relevant patterns and relationships in this data, followed by optimising its internal weights and biases.

Next, when the same participants take part in the actual MBCT sessions, their EEG data will be processed in real-time, using the same feature extraction methods (e.g., time-frequency analysis) employed during training. The extracted features from the real-time data will be fed into the trained CNN model. The model will predict the anxiety levels for each data segment, providing a continuous stream of real-time anxiety state estimates. Based on the real-time predictions of anxiety level, the MBCT system will dynamically adjust its interventions. This may include but is not limited to adding a relaxation exercise, when the anxiety level is too high, delivering adjusted instructions under different anxiety levels, to enhance the treatment effect. Additionally, the system will include initial calibration and continuous training data acquisition.

Evaluation Procedure. To evaluate the proposed approach, there will be *technical evaluations*, as well as, as mentioned, *experiments with real students*. As noted, the latter will undergo ethical review and approval from the university before commencement. After small-scale think-aloud methodology with small scale experimentation with a limited number of students (≈ 5, to explore any usability issues[1]), larger scale (> 200) experimentation involving *mixed design* will be performed, as follows. First, all participants will undergo an experiment to build their *unique anxiety profile*, then all participants will be randomly assigned to one of three groups (VR group, VR+BCI group, and control group) in the second experiment (see Fig. 1). The *control group* will receive a standard MBCT sessions, without any BCI or VR integration. Through the session, participants will learn to recognise unhelpful thought patterns and gently let them go without judgment. This practice, following standard MBCT, fosters present-moment awareness and helps individuals develop a more flexible and accepting relationship with their inner experience and build personalised coping strategies for effective problem-solving. Unlike the control group, the *VR group* will receive MBCT, however in a VR environment. Additional features include choosing the scene base on their preference (e.g., forest, seaside) and therapeutic interactions with the elements. Finally, the *VR + BCI Group* will experience MBCT delivered within a preferred VR environment, while also receiving real-time anxiety level feedback, based on their EEG data collected through a BCI system. The dependent variable will be the anxiety level, measured by the widely validated State-Trait Anxiety Inventory (STAI) by Spielberger et al. (1971) [10], before and after second experiment.

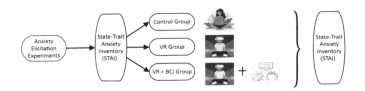

Fig. 1. Experimental Procedure

---

[1] https://www.nngroup.com/articles/why-you-only-need-to-test-with-5-users/.

3   Progress to Date

The initial phase of this project has focused on laying the groundwork for a robust and effective MBCT intervention utilising VR and BCI technology. Extensive literature review has delved into existing research on MBCT, VR in education and psychotherapy, and EEG/BCI applications in anxiety detection and biofeedback. Progress has also been made in designing the content and structure of the MBCT sessions, focusing on three perspectives: acceptance and non-judgment, present moment awareness, and increased self-awareness and regulation. Some initial work has been done on incorporating relevant therapeutic elements and optimising the flow of the intervention in the VR environment (see Fig. 2 for an example environment: beach during sunset).

Simultaneously, investigations in maximising the functionality of BCI equipment have been initiated. Several types of BCI equipment (e.g., open BCI - the complete Ultracortex, Emotiv - Epoc x, Emotiv - Flex) have been evaluated by factors like accuracy, portability, ease of use, and the optimal equipment has been selected (Emotiv - Flex 32-channel wireless saline EEG head cap system) as it can be fitted with the VR headset and performs well in delivering real-time neurofeedback. Additionally, initial work has begun on building the anxiety-detection program, leveraging machine learning algorithms to process BCI data (see Fig. 3 for example data) and translate it into meaningful cues of anxiety level.

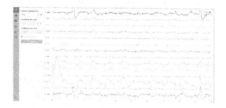

Fig. 2. An Example VR Environment   Fig. 3. Sample EEG data

4   Expected Contributions and Impact

While traditional research in education with VR or EEG often focuses solely on academic skills and performance, this research *prioritises student wellbeing*, by recognising the often-neglected realm of mental health and emotional regulation within the learning environment and addressing their crucial role in student success. This focus on holistic student development sets this research apart, potentially revolutionising personalised learning and mental health support for students. In the education domain, our research holds *important promise for equipping students with effective coping strategies for exam anxiety, potentially leading to immediate reductions in anxiety symptoms and improved academic performance.* In addition, our study contributes to the burgeoning field of EEG-VR integration, by pioneering the use of this technology to create real-time, dynamic experiences. This addresses the scarcity of research in this area and paves the way for: (1) Informing the development of future AI-powered personalised learning interventions for emotional well-being in educational settings. (2) Deepening our understanding of VR's potential to enhance MBCT efficacy and personalise interventions based on real-time physiological data. (3) Helping students struggling with anxiety or facing communication difficulties due to conditions like autism or aphasia, whose emotions often play a crucial role but can be challenging to express verbally or publicly.

References

1. Aliwi, I., et al.: The role of immersive virtual reality and augmented reality in medical communication: a scoping review. J. Patient Experience **10**, 23743735231171560 (2023)
2. Carl, E., et al.: Virtual reality exposure therapy for anxiety and related disorders: a meta-analysis of randomized controlled trials. J. Anxiety Disord. **61**, 27–36 (2019)
3. Freina, L., Ott, M.: A literature review on immersive virtual reality in education: state of the art and perspectives. In: The International Scientific Conference Elearning and Software for Education, vol. 1, pp. 10–1007 (2015)
4. Grassi, A., et al.: New technologies to manage exam anxiety. Annu. Rev. Cyberther. Telemed. **2011**, 57–62 (2011)
5. Greenberg, L.S.: Emotion-focused therapy. Clin. Psychol. Psychother. Int. J. Theory Pract. **11**(1), 3–16 (2004)
6. Lovell, K., Richards, D.: Multiple access points and levels of entry (MAPLE): ensuring choice, accessibility and equity for CBT services. Behav. Cogn. Psychother. **28**(4), 379–391 (2000)
7. Olatunji, B.O., et al.: Efficacy of cognitive behavioral therapy for anxiety disorders: a review of meta-analytic findings. Psychiatr. Clin. **33**(3), 557–577 (2010)
8. Safari, R., et al.: The effectiveness of cognitive-behavioral play therapy on exam anxiety and academic vitality. J. Adolesc. Youth Psychol. Stud. (JAYPS) **3**(1), 244–252 (2022)
9. Saleh, S., et al.: A review of electroencephalography (EEG) application in education. Int. J. Early Child. Spec. Educ. (INT-JECSE) (2022)
10. Spielberger, C.D., et al.: The state-trait anxiety inventory. Revista Interamericana de Psicologia/Interamerican J. Psychol. **5**(3& 4) (1971)
11. Teplan, M., et al.: Fundamentals of EEG measurement. Meas. Sci. Rev. **2**(2), 1–11 (2002)
12. Trifoni, A., Shahini, M.: How does exam anxiety affect the performance of university students. Mediterr. J. Soc. Sci. **2**(2), 93–100 (2011)
13. Ugwuanyi, C.S., et al.: Effect of cognitive-behavioral therapy with music therapy in reducing physics test anxiety among students as measured by generalized test anxiety scale. Medicine **99**(17) (2020)
14. Zandi, H., et al.: The effectiveness of mindfulness training on coping with stress, exam anxiety, and happiness to promote health. J. Educ. Health Promot. **10**(1) (2021)

Workshops and Tutorials

AfricAIED 2024: 2nd Workshop on Artificial Intelligence in Education in Africa

George Boateng[1,2(✉)] and Victor Kumbol[1,3(✉)]

[1] Kwame AI Inc., Claymont, USA
{jojo,victor}@kwame.ai
[2] ETH Zurich, Zürich, Switzerland
[3] Charité - Universitätsmedizin Berlin, Berlin, Germany

Abstract. Recent AI advancements offer transformative potential for global education, yet their application often overlooks Africa's unique educational landscape. AfricAIED 2024 will address this gap, spotlighting efforts to develop AI in Education (AIED) systems tailored to Africa's needs. Building on the success of the inaugural workshop, AfricAIED 2024 will feature an online AI Hackathon focused on democratizing preparation for Ghana's National Science & Maths Quiz (NSMQ). Participants will create open-source AI tools leveraging resources from the Brilla AI project to level the academic playing field and enhance science and math education across Africa. The workshop will showcase top competitors' solutions, invite discussions on AIED opportunities and challenges in Africa, and highlight the latest advancements in AI education integration. AfricAIED 2024 aims to foster collaboration and innovation, amplifying African voices in the AIED community and driving positive change in African education through AI.

Keywords: AI in Education · Science Education · NLP · Speech Processing

1 Introduction

Recent advances in AI systems such as BERT [7]and GPT-4 [1], have demonstrated their capacity to transform education globally. Prominent figures in the EdTech industry, including Duolingo [8], Quizlet [11], Chegg [6], and KhanAcademy [9], have actively incorporated these AI advancements into their platforms to enrich learning experiences. Nevertheless, the deployment and evaluation of these AI systems primarily focus on Western educational contexts, overlooking the distinct requirements and obstacles encountered by students in Africa. Notably, the introduction of GPT-4 in March 2023 featured various academic exams as benchmarks, none of which were from Africa [1]. This omission underscores the tendency to marginalize Africa in the utilization of state-of-the-art AI innovations, disregarding the diverse educational environments and needs

A. M. Olney et al. (Eds.): AIED 2024 Workshops, CCIS 2151, pp. 427–431, 2024.
https://doi.org/10.1007/978-3-031-64312-5_53

of African students. Consequently, the potential of AI to address educational disparities and hurdles prevalent in Africa remains largely unexplored, emphasizing the pressing need for greater inclusivity and customized solutions tailored to the specific challenges faced by students across the continent. AfricAIED - workshop on AI in Education in Africa - is the genesis of a movement to make pronounced the building and advancement of AIED systems that work well in the African context.

2   Content and Themes

Building upon a successful inaugural edition of this workshop last year - AfricAIED 2023 [2] - this year's workshop - AfricAIED 2024[1] - aims to crowdsource and highlight efforts to build and deploy AIED systems in Africa as well as discuss potential opportunities and challenges. This workshop will be centered around an online AI Hackathon that builds upon our open-source project, Brilla AI [4,5] which is building an AI Contestant to address our proposed grand challenge in education - **NSMQ AI Grand Challenge** - *"Build an AI to compete in Ghana's National Science & Maths Quiz Ghana (NSMQ) competition and win - performing better than the best contestants in all rounds and stages of the competition"* [3]. The NSMQ is an annual live science and mathematics competition for senior secondary school students in Ghana in which 3 teams of 2 students compete by answering questions across biology, chemistry, physics, and math in 5 rounds over 5 progressive stages until a winning team is crowned for that year [10]. The NSMQ is an exciting live quiz competition with interesting technical challenges across speech-to-text, text-to-speech, question-answering, and human-computer interaction. An AI that conquers this grand challenge could have real-world impact on education such as enabling millions of students across Africa to have one-on-one learning support from this AI.

3   AI Hackathon

3.1   Motivation

There is a lot of inequity in preparations for the NSMQ where resources such as study materials, great teachers, etc., are available only to the big-name high schools that then almost always make it to the semis and finals, and win. This problem is quite representative of the inequity in Ghana's educational system. The goal of this challenge is to crowdsource AI-powered tools to democratize preparation for the NSMQ and give all schools a fairer chance at winning the NSMQ. More broadly, the tools could be extended to improve science and math learning across Africa.

---

[1] https://www.africaied.org.

3.2   Challenge

The AI hackathon challenge will be as follows: "Build an open-source, AI-powered tool that enables students to prepare for the NSMQ using any of or a combination of the following open-source outputs from the Brilla AI project: (1) dataset consisting of quiz questions and open-source textbooks (2) code, (3) models for speech-to-text (STT) - transcribes speech with a Ghanaian accent, text-to-speech (TTS) - speaks with a Ghanaian accent (we have a model for the Quiz Mistress' voice), and question-answering (QA). You are allowed to additionally use open-source code, models, and datasets. You are NOT allowed to use commercial models such as GPT-3.5." Some potential ideas include an audio-based assessment system where a question is asked in the Quiz Mistress' voice, accepts audio answers and says whether it's correct or wrong along with an explanation, an audio-based question and answering system where you ask a question and get an answer and explanations, a multiplayer game of the quiz for assessment that tracks metrics on how quickly questions were answered, number of wrong answers, etc.

3.3   Timeline and Participation

It will run from May 15th to June 15th, 2024. Participants will be asked to make a submission by releasing their outputs with the Apache 2.0 license (so it is compatible with our open-source project) via GitHub and share (1) the link to the repo, (2) a 5-min video explaining and demoing their system, and (3) a technical report (max 5 pages) describing the architecture of their technical system, and various design decisions made. The Brilla AI team will review the submissions and select the top 3 based on innovativeness and potential impact. They will receive cash prizes and be invited to present and demo their solution at AfricAIED 2024.

4   Relevance and Importance to the AIED Community

With the underrepresentation of AIED work from Africa, this workshop will highlight such work. It will increase participation from African scientists and engineers in the 2024 AIED conference. Also, the style of this workshop centered around a hackathon focused on AIED in Africa is likely to have a good number of contributions as there will be a lower barrier to participation compared to paper submissions as an example.

5   Format and Activities

AfricAIED 2024 will be run as a half-day workshop on Monday, 8th July 2024 in a hybrid format with the in-person component in Accra, Ghana. It will bring together educators, researchers, entrepreneurs, policymakers, and AI experts to discuss and collaborate on innovative ideas, best practices, and future developments in leveraging AI to enhance learning experiences and outcomes in Africa. The workshop will consist of technical presentations by the top 3 competitors on their winning approaches, invited talks, and panel discussions on opportunities for AIED, and challenges in developing and deploying AIED in Africa. The workshop will also feature presentations and demos from team leads of the Brilla AI team on the 2024 version of Brilla AI. The target audience is students from Africa, African researchers in the area of education, and researchers with an interest in AIED research in Africa. We expect to have a maximum of 50 participants in person and 50 participants online.

6   Previous Editions of the Workshop: AfricAIED 2023

We successfully ran the inaugural event of AfricAIED - 1st workshop on AI in Education in Africa - at Google Research Ghana with sponsorship from Google and ETH for Development. We had about 35 people in person and 40 online [2]. Present was also JoyNews, a National TV news broadcaster in Ghana. We had an incredible lineup of 16 speakers spanning educators, education researchers, education entrepreneurs, and AI experts who gave talks on the ways AI is already being used in education in Africa, a primer on AI, and Brilla AI, an AI being built to win Ghana's NSMQ. There was also a lively panel session that discussed the opportunities and challenges of leveraging AI to improve education in Africa. One of the key highlights was the demo of Brilla AI, the first version of our AI Contestant for the 2023 NSMQ that transcribed speech with a Ghanaian accent, generated an answer to a scientific riddle, and said that answer with a Ghanaian accent.

References

1. Achiam, J., et al.: Gpt-4 technical report. arXiv preprint arXiv:2303.08774 (2023)
2. Africaied 2023 recap. https://www.africaied.org/africaied-2023/recap
3. Boateng, G., Kumbol, V., Kaufmann, E.E.: Can an AI win Ghana's national science and maths quiz? an AI grand challenge for education. arXiv preprint arXiv:2301.13089 (2023)
4. Boateng, G., et al.: Towards an AI to win Ghana's national science and maths quiz. In: Deep Learning Indaba 2023 (2023)
5. Boateng, G., et al.: Brilla AI: Ai contestant for the national science and maths quiz. arXiv preprint arXiv:2403.01699 (2024)
6. Chegg announces cheggmate, the new ai companion, built with GPT-4 (2023). https://www.businesswire.com/news/home/20230417005324/en/Chegg-announces-CheggMate-the-new-AI-companion-built-with-GPT-4
7. Devlin, J., Chang, M.W., Lee, K., Toutanova, K.: BERT: pre-training of deep bidirectional transformers for language understanding. In: Burstein, J., Doran, C., Solorio, T. (eds.) Proceedings of the 2019 onference of the North American Chapter of the Association for Computational Linguistics: Human Language Technologies, Volume 1 (Long and Short Papers), pp. 4171–4186. Association for Computational Linguistics, Minneapolis, Minnesota (2019). https://doi.org/10.18653/v1/N19-1423, https://aclanthology.org/N19-1423

8. Introducing duolingo max, a learning experience powered by GPT-4 (2023). https://blog.duolingo.com/duolingo-max/
9. Harnessing GPT-4 so that all students benefit. a nonprofit approach for equal access (2023). https://blog.khanacademy.org/harnessing-ai-so-that-all-students-benefit-a-nonprofit-approach-for-equal-access/
10. National science and maths quiz. https://nsmq.com.gh/
11. Introducing q-chat, the world's first AI tutor built with openai's chatgpt (2023). https://quizlet.com/blog/meet-q-chat

The Learning Observer: A Prototype System for the Integration of Learning Data

Piotr Mitros[1]([✉]) [iD], Paul Deane[1] [iD], Collin Lynch[2] [iD], and Brad Erickson[1]

[1] Educational Testing Service, Princeton, NJ 08541, USA
{pmitros,pdeane,berickson}@ets.org
[2] North Carolina State University, Raleigh, NC 27695, USA
cflynch@ncsu.edu

Abstract. In this tutorial workshop, we will introduce *The Learning Observer*. This is a project whose goal is to set up an open, transparently-governed consortium to manage student data in student interest and public interest. The open-source platform we are developing under this umbrella is at a work-in-progress stage, and is ready for community feedback. It is a modular platform designed to enable (1) the integration of diverse learning data, (2) the use of sophisticated machine learning techniques over that data, and (3) the presentation of real-time teacher dashboards. We are further extending it to support open-science and SEER-aligned [1] research methodology, as well as family rights. This workshop will be split into two closely related [2] components, one focused on policy and one on technology. Participants may come to either or both.

Keywords: learning process data · policy, privacy and diversity · writing data · systems and architectures · psychometrics and measurement · stealth assessment · measuring complex constructs

1 Introduction

This interactive workshop will consist of two related but independent sections:

1. In the first section, we aim to explore how such systems should be designed from an organizational, legal, cultural, and ethical perspective in order to guarantee that they operate in the public interest.
2. In the second section, we will present our work-in-progress open source platform, we aim to explore how it, and systems like it, should be structured to embody the concepts of the first section, as well as to be flexible enough and simple enough to help to support participants' research.

The Learning Observer is a working prototype system, currently collecting keystroke-level writing process data from a handful of schools. These kinds of data is cross-cutting across subjects, and while currently collected at a middle school level, across grades. It can surface a broad range of constructs, including non-cognitive constructs, such as time-on-task, creativity, and collaboration.

It has also been used to generate dashboards from several mathematics activities for studies with individual students and small groups of students. The modular design allows very rapid development of new realtime teacher dashboards from learning process data. The next step is to collect feedback from a broader community in order to understand how it might be useful for your projects. *Ergo*, this workshop.

An ideal workshop would have between 12–30 participants, as this gives enough participants to form small groups, and enough time for a public presentation/readout from each group, using standard classroom practice. However, there is significant flexibility here.

2   Policy Segment

In the policy segment, the goal is to understand how we should, as a society, manage student data. We've seen many educational projects fail to make impact due to unaligned incentives. There is a fascinating set of questions here:

- How do structure an organization which has sufficient checks-and-balances — both legal and technological — to be guaranteed to operate in the student interest forever?
- How do we set this up so that researchers, ed-techs, and schools wish to participate?
- How do we make this platform supportive of the broad diversity of concepts about what equity, fairness, or integrity are, or values around student privacy? Ethical uses of student data are quite different in China, Jordan, and Australia (or even among families in the same school).
- How do we make sure we finally align incentives to support instruction aligned to the diverse goals of students from different cultures, with different interests, and at different academic levels? For all the talk of DEI, differentiation, and cultural responsiveness, most schools are forced to align to standardized curricula which don't account for that diversity.

The format will depend on number of registered participants, but we tentatively plan to use a jigsaw format, where we will break into small groups, each focused on one area, such as:

- Legal family rights and transparency frameworks (e.g. FERPA [4], PPRA [5], open meetings, public records [6], COPPA [7], GDPR [8], etc.
- Psychometric standards [9,10]
- Evolving AI safety/fairness/ethics
- Diversity, equity, and inclusion (DEI) and culturally-responsive instruction
- Open governance/checks-and-balances
- Org design/incentive structures
- Open science

- Open source/free software
- Cross-cultural issues
- Security, and issues such as reidentification
- Evolving risk landscape and potential misuses, compromises, and abuses of student data

If the number of registered participants is amenable to a jigsaw format, and as these are individually complex areas, we will provide each group with basic background material on these issues to discuss. Once this segment is finished, we will regroup into mixed groups in order to draft concrete proposals. Finally, we will have a read-out.

Our approach to this workshop is informed by the significant changes to the landscape, especially as our ability to gather and make inferences from education grows, and as new priorities (such as DEI) enter the mainstream. We hope that a small group, structured collaborative formats can help solicit a broad set of ideas about how to keep such a system and organization honest and aligned to public interest in the long term, and integrate and synthesis those with ideas with the long history of similar ideas in domains such as open source, open governance, open science, family rights, and psychometrics.

In particular, increasingly important is the trend of increasing potential risks from student data, from a perfect storm of developments: Significant progress on data reidentification, increasing number of compromised machines, increasingly complex data collected, and increasing techniques for making sensitive inferences from that data.

2.1 Optional: The Learning Observer Approach

If time allows, following the above activity, we may provide background about how the *Learning Observer* project is currently planning to approach these issues. Specifically, the genesis of the *Learning Observer* project was an exploration of why effective educational initiatives fail to broadly disseminate and evidence-based practices aren't adopted at scale. This exploration converged on issues around governance, transparency, and incentive structures. Specifically, while we know many better methods of teaching-and-learning, as we've organized ourselves, market and organizational forces prevent their adoption. We would like to address this in multiple ways:

- By providing parents with increased transparency, we hope to make them more effective student advocates and citizens.
- By providing teachers with improved visibility into their classrooms, we hope to enable them to orchestrate richer classroom processes.
- By providing open-science research platforms, we hope to improve the integrity of education research.
- By providing administrators with measures of more complex constructs, we hope to incentive implementation of richer classroom instruction.

– By measuring diverse constructs, supporting children from diverse cultures, who might have different prior knowledge, interests, and goals, and so be disadvantaged by a homogenous curriculum.

By doing this with maximal checks-and-balances, we hope to have stakeholder buy-in and to address many issues where student data is currently being mismanaged or misused.

We have thought about these issues for many years [11–14], and in particular, focused on the design of mechanisms such as:

– Transparent governance, including open meetings and responding to public records requests.
– Legally-binding privacy policy, focused on family rights, and explicitly providing auditing and enforcement mechanisms, so families can understand what data is collected, how it was processed, ask to have it removed, or ask to have it corrected.
– Privacy-preserving research access to data under reasonable and non-discriminatory terms
– Oversight from a broad range of organizations (e.g. on the board, as well as through the use of a genuine consortium model)
– 501(c)3 corporate registration, with stronger limits on profit and on conflict-of-interest in bylaws than required by the IRS
– Full audit trails for how data was used (a portion of which requires open-source) to support both privacy-focused audits and open-science.
– An app store model to analytics algorithms, dashboards, etc. which would allow a broad set of researchers, ed-techs, and other organizations to develop means to analyze data, while still maintaining the data in-house.

We would welcome feedback.

2.2   Optional: Participant Projects

If participants have worked on related problems, and let us know in advance, we will allot short time segments for them to present their work as well.

3   Technology Segment

In the technology segment, our goal will be to understand how to design data-processing architectures which meet the needs of the requirements from the policy segment. We will present what we have designed and built for feedback, and depending on participant background and interest, take the time to understand what participants have built, and to look for connections.

3.1   Architectures for Data Integration

In this segment, we will give an overview of the open-source components we have already designed or implemented, in part for understanding how to better align them to the requirements of participant research.

1. A modular framework for processing of learning process data. This is based on a reducer model, which means an unsophisticated developer (e.g. an undergrad) ought to be able to write a function, and we ought to be able to scale it our horizontally as far as necessary. For a simple example, see the time-on-task function in our writing analysis suite.
2. A query language for compiling the outputs of said reducers. This is based on SQL as much as reasonable, but with significant differences since we are querying very different types of data.
3. A system for rapidly developing teaching-facing dashboards. This is designed to allow non-sophisticated users to rapidly develop simple visualizations using the friendly plotly library [15], and more sophisticated users to develop sophisticated ones in react [16]. The two play nicely together too!
4. A suite of NLP algorithms for extracting features from student texts. This is being used in our system for a dashboard where teachers can highlight e.g. supporting arguments, or citations.
5. An architecture (not yet implemented) for managing student data in ways which comply with family rights and open science. This is complex, as we need to balance the right to be forgotten with having robust laboratory notebooks and archiving research methods (if a students' data is removed, we cannot exactly replicate studies). However, by storing past analyses, we should be able to, for example, automatically rerun before/after analyses, to see how a removal request changes research results.
6. Mid-implementation, a system designed to allow the use of all of these, on live student data, using Jupyter notebooks, with smooth pathways from prototyping to live deployment. Critically, we capture all analyses run. Why? Lab notebooks used to do this for open science. Plus, families have a right to know how their data was used.
7. Mid-implementation, a we-once-thought-novel architecture which allows us to build educational activities which, by design, capture complete process data about what students did, to where we are guaranteed to be able to reconstruct state at any point in time. This is a significant step ahead of current standards, such as xAPI/Tincan, IMS Caliper, and Open edX events.
8. In use, we have an event streaming library featuring pretty robust queuing, metadata, etc., as well as the ability to capture a broad diversity of browser events (such as keystroke logging), and closely integrate with react/redux.

All of this is in various stages ranging from architectural design to production-ready code. Our goal is to collect feedback to make sure the system develops in ways helpful for supporting the diverse projects at AI-ED. In terms of activities, I think these might best be done in a tutorial format - installing the system, having participants build a few dashboards, and letting us know what they think, as well as what would be needed to make this useful for their own work. An ideal outcome would be research collaborations.

3.2 Demo: Writing Observer

To ground the above in a concrete example, we will present the most mature Learning Observer component, the Writing Observer. This subsystem provides real-time dashboard for teachers to manage writing processes in Google Docs. These can, for example:

– At-a-glance, show a classrooms' worth of writing, highlighting specific text features through classical NLP algorithms
– At-a-glance, allow teachers to run all texts through a large language model.
– Allow teachers to interactively navigate a visualization of common student grammar and writing errors

Depending on participant interest, we may walk through the iterative, design-based approach for developing these dashboards.

3.3 Optional: Participant Demos

If participants have worked on related problems, and let us know in advance, we may allot short time segments for them to present their work as well.

3.4 Optional: Design Mini-Workshop

Depending on participant interest and time, we may run a short design workshop on creating learning analytics dashboard mockups, focused on the research of the participants. We have run many similar design workshops for both teacher and researcher audiences, and in most cases, these focus on a set of steps for

– Working in small groups to identify changes they would like to see in class-rooms, likely aligned to their research
– Coming up with information people need to support those changes (incentive structures, classroom orchestration, etc.)
– Coming up with source of information to enable those (keeping in mind the rich data being generated and the possibilities of processing that with AI.
– Coming up with dashboards to present that information
– Classroom readout.

Acknowledgment. This work is funded under Institute of Education Sciences award R305A210297, and initial work on this project was supported Schmidt Futures, as well as by the Educational Testing Service. This workshop presents a work-in-progress, and has not been reviewed or endorsed by ETS, IES, or Schmidt Futures, and does not represent their views.

References

1. Institute of Education Sciences. (n.d.). Standards for Excellence in Education Research. https://ies.ed.gov/seer/index.asp
2. Lessig, L.: Code and Other Laws of Cyberspace. Basic Books (1999)
3. Bueno de Mesquita, B., Smith, A.: The Dictator's Handbook: Why Bad Behavior is Almost Always Good Politics. PublicAffairs (2011)
4. Family Educational Rights and Privacy Act, 20 U.S.C. §1232g (1974)
5. Protection of Pupil Rights Amendment, 20 U.S.C. §1232h (1978)
6. Freedom of Information Act, 5 U.S.C. §552 (1966)
7. Children's Online Privacy Protection Act, 15 U.S.C. §§91 (1998)
8. General Data Protection Regulation, Regulation (EU) 2016/679. European Parliament and Council of the European Union (2016)
9. American Psychological Association: National Council on Measurement in Education: Standards for Educational and Psychological Testing. American Psychological Association, Washington, DC (2014)
10. ETS Standards for Quality and Fairness. Educational Testing Service (2013)
11. "Implications of Privacy Concerns for Using Student Data for Research" National Academy of Education Workshop, Big Data: Balancing Research Needs and Student Privacy (Panel Discussion) (2016)
12. Brend, Littlejohn, Kern, Mitros, Shacklock, Blakemore. "Big data for monitoring educational systems." Luxembourg: Publications Office of the European Union (2017)
13. "Big Data in Education: Balancing the Benefits of Educational Research and Student Privacy: Learning Process Data in Education." National Academy of Education (final report)
14. Mitros, N., Foltz, S., Katz, G.: Why Assess: The Role of Assessment in Learning, Science, and Society. Chapter Design Recommendations for Intelligent Tutoring Systems: Volume 5 - Assessment (2017)
15. Plotly Dash Documentation. https://dash.plotly.com/
16. React Documentation. https://react.dev/

Ethical AI and Education: The Need for International Regulation to Foster Human Rights, Democracy and the Rule of Law

Christian M. Stracke[1]([✉]) [iD], Irene-Angelica Chounta[2] [iD], Vania Dimitrova[3] [iD], Beth Havinga[4] [iD], and Wayne Homes[5] [iD]

[1] University of Bonn, Bonn, Germany
stracke@uni-bonn.de
[2] University of Duisburg-Essen, Duisburg, Germany
[3] University of Leeds, Leeds, UK
[4] European EdTech Alliance, Edinburgh, UK
[5] University College London, London, UK

Abstract. The need for AI regulation is true in particular in the fields of education. Education is a special case and different from other sectors. Therefore, it reuqires careful considerations: How can we develop and implement legal and organisational requirements for ethical AI introduction and usage? How can we safeguard and maybe even strengthen human rights, democracy, digital equity and rules of law through AI-supported education and learning opportunities? And how can we involve all educational levels (micro, meso, macro) and stakeholders to guarantee a trustworthy and ethical AI usage? These open and urgent questions will be discussed during a workshop that aims to identify and document next steps towards establishing ethical and legal requirements for Artificial Intelligence and Education.

Keywords: ethics · Artificial Intelligence · human rights · democracy · rule of law

1 Introduction

It is obvious and demanded from latest publications, studies and events that Artificial Intelligence (AI) requires legal and organizational regulations [10, 13, 15]. That is reflected in the research of the Artificial Intelligence and Education (AI&ED) community on the needs of ethics in AI and Education [3, 4], Fairness, Accountability, Transparency, and Ethics in AI&ED [9], educational stakeholders' perceptions of AI [1] and Scrutable AI [6]. A similar picture is depicted by the public announcements and actions of international organisations such as the EU (AI Act[1]), UNESCO (International Forum on Artificial Intelligence and Education 2022[2], and the Council of Europe (CoE) leading to their first guidelines for policy makers [7, 8].

---

[1] https://artificialintelligenceact.eu/
[2] https://www.unesco.org/en/articles/international-forum-artificial-intelligence-and-education-2022.

A. M. Olney et al. (Eds.): AIED 2024 Workshops, CCIS 2151, pp. 439–445, 2024.
https://doi.org/10.1007/978-3-031-64312-5_55

The need for AI regulation is true in particular in the fields of education. Education is a special case and different from other sectors: It does not deliver a product like all producing sectors but it offers a promise for future learning processes that cannot be predicted and pre-assessed [14, 16]. Furthermore, education is a human right and learners are often in a particular situation (mandatory school education).

Therefore, the Council of Europe (CoE) took action as it is the international organisation founded in 1949 to uphold human rights, democracy and the rule of law in Europe. Currently, it has 46 member states (19 more than the European Union), with a population of approximately 675 million.

In 2021, the CoE established an Expert Group to investigate, and propose a legal instrument for, the application (AIED) and teaching (AI Literacy) of AI in education. The project, entitled, "Artificial Intelligence and Education" (AI&ED), is working towards developing an actionable set of recommendations for Member States, helping ensure that the application and teaching of AI in education is for the common good.

To that end, the CoE AI&ED Expert Group have researched and written an initial report ("AI and Education. A critical view through the lens of human rights, democracy and the rule of law.") and undertaken a survey of Member States, both of which have been designed to inform the project and to stimulate further critical debate [2]. Currently the AI&ED Expert Group appointed by the CoE is working on the analysis of the survey and developing the legal instrument as international convention [11, 12].

All workshop organizers are members of the CoE AI&ED Expert Group and contributed to the two CoE Expert Conferences "Artificial Intelligence and Education" held by the CoE in Strasbourg on 18th / 19th of October 2022 and on 21st / 22nd of November 2023.

To establish the context, in advance of the workshop, the CoE Expert Group's initial report and results will be shared with the workshop participants [2, 13]. We envision that our discussions as planned for the workshop are necessary to set solid foundations regarding the ethical development, deployment, and use of AI systems in education respecting human rights and principles, ensuring fair and ethical use and promoting human agency [10, 15].

2   Objectives, Relevance, and Importance

2.1   Workshop Objectives

This workshop is the continuation of the very successful workshop held at the AIED 2023 Conference in Tokyo, Japan (see: http://aied2023.learning-innovations.eu) [13]. That allows ongoing updates and awareness raising on the important topic of ethical AI&ED as well as direct communication and discussions for contributions to the current Council of Europe (CoE)'s development of an international regulation as binding law in Europe and design of comprehensive AI literacy tools.

The workshop intends to achieve the following objectives:

- To explore potential guidelines for trustworthy and ethical AI usage and to critically assess their structure and usage scenarios

- To develop ideas around personal future ramifications (including trustworthiness and ethical implications) and AI usage in education, and to collate, present and discuss these
- To explore in group work and in the plenary how legal regulations and guidelines for trustworthy and ethical AI usage can be developed and designed, in particular related to legal and organizational requirements
- To identify methods for dealing with the quick changing developments in AI, providing a legal framework that is flexible but also appropriately effective

Therefore, these guiding aspects will be addressed:

- How can we develop and implement legal and organizational requirements for ethical AI introduction and usage?
- How can we keep and maybe even strengthen human rights, democracy, digital equity and rules of law through AI-supported education and learning opportunities?
- And how can we involve all educational levels (micro, meso, macro) and stakeholders to guarantee a trustworthy and ethical AI usage?

Main focus will be on the legal and organizational requirements to achieve a regulation in the fields of AI and Education (AI&ED). Basis will be the work and activities of the Council of Europe (CoE) and its appointed AI&ED Expert Group and in particular its initial report. All these open and urgent issues will be discussed and next steps in relation to these requirements and goals will be defined.

2.2 Workshop Relevance and Importance to the AIED Community

To prepare for the Council of Europe Council of Ministers meeting in 2024, when the proposed text of the legal instrument on AI in Education will be debated and (hopefully) agreed, and when a final report from the CoE Expert Group will be published, the CoE Expert Group is currently eliciting expert input from key stakeholder groups: students, teachers, commercial organisations, NGOs, and academia.

The members of the AIED community are key stakeholders, such that it is critical that their views are properly addressed by the Council of Europe's in their work (particularly the proposed legal instrument) on AI and education.

The workshop is also closely aligned with the AIED 2024 conference theme - "AI in Education for a World in Transition". It will provide an arena for the AIED community to familiarise themselves with the CoE initiative and to engage in follow on activities. Therefore, it will present the results from CoE AI&ED Expert Conference "Artificial Intelligence and Education" (held on 21st/22nd of November 2023 in Strasbourg) and introduce in the initial report of the CoE Expert Group: *Artificial Intelligence and Education. A critical view through the lens of human rights, democracy and the rule of law* (2022) [2].

The workshop will also facilitate the forming of a wider community of practice within the AIED community to consider the wider societal implication of AIED.

3   Workshop Format

This half-day workshop will be conducted in a highly interactive manner. It consists of two sections that are interrelated and complement each other. In the first section, the workshop will follow traditional collaboration in small working groups on four questions. In the second section, the workshop will follow the World Café method offering up to 10 prepared questions that can be selected. The whole workshop will be hybrid with small working groups that are on-site as well as online and always followed by plenary sessions. Thus, the groups inform each other on the discussions and results.

3.1   Program

In the first section, the interactive Workshop will begin with an introduction to the workshop and a summary of the key messages from the CoE work on ethical AI including the initial report by the AI&ED Expert Group appointed by CoE (15 min).

The core part of the first section will centre on four key questions related the CoE work and results developed by the AI&ED Expert Group. These four questions are:

- Q1. What issues does the report misunderstand or misrepresent, and how should this be corrected?
- Q2. What issues does the report ignore, miss or forget, and what other information should be included?
- Q3. What regulations, if any, should be put in place to protect human rights, democracy and the rule of law whenever AI is applied (AIED) or taught (AI Literacy) in educational contexts?
- Q4. What case studies can the participants suggest to enhance the public understanding of the issues raised in the report?

Each question will be introduced by a 2-min presentation. Participants will then work in small groups, to respond to the issues introduced, before reporting back their views and ideas to the workshop.

The first section will conclude with a discussant summarising the workshop participants' input (15 min) followed by a wrap up (5 min).

In the second section, the interactive Workshop follows the World Café method and consists of four parts:

In the first short part (5 min), we will briefly present the World Café method and the topic "Legal and organizational requirements for ethical AI introduction and usage".

In the second core part, we will discuss different and pre-prepared themes and aspects in groups of ideally 4 to 10 participants.

We will prepare 10 themes before the conference and will decide depending on the number of participants how many group tables we will offer during the workshop. The participants will work on one selected group table focusing the presented open question of this table and documenting the discussion. Afterwards, they will change the group table and work on another open question. In total, three rounds of group table work will be facilitated (each round of 25 min).

The procedure will be identical on all group tables in each round:

1. The theme will be read that always consists of an open question and clarified.
2. Afterwards, the contributions (= answers on the open question of the theme) will be collected as post-it at a flipchart next to the group table.
3. Finally, all collected contributions will be sorted in open discussion and be clustered according defined categories or criteria so that they can be presented. That can be done independently or using the categories or criteria from the other rounds before.

In the third part (35 min), the results from the group tables will be presented in the plenary and critically discussed and reflected. These can then be provided as summaries for publication after the event.

Table 1. The agenda of the AIED 2024 Workshop "Ethical AI and Education: The need for international regulation to foster human rights, democracy and the rule of law"

Sessions	Agenda Item	Duration (minutes)
1. Start	Welcome to the workshop	5'
	Introduction to the CoE work on AIED	10'
	Discussion of four questions (Q1 to Q4)	40'
	Summarising participants' input and wrap up	20'
2. World Café	Introduction to the World Café	5'
	First round of group discussions	25'
	First round of group discussions	25'
	First round of group discussions	25'
3. Plenary	Results from the group tables	35'
4. Closing	Collecting participants' input for future CoE work	15'
	Envisioning future steps and directions	5'

In the fourth part, we will collect all issues that the workshop participants wish to discuss and bring to the attention of the CoE Expert Group (15 min). Finally, we will define next steps and activities and conclude with a wrap up of the whole Workshop (5 min). The proposed agenda is presented in Table 1.

3.2 Target Audience

We aim to reach out and involve all AIED community members regardless their expertise and experience. The interactive workshop is dedicated to interested beginners without any pre-knowledge as well as to experts. To promote inclusiveness, we chose a hybrid participation format (on-site and online) that we envision to foster discussions with all AIED community members worldwide.

3.3 Expected Outcomes

We envision that this workshop will results in three main outcomes. First, the presentation, discussion and collection of legal and organizational criteria, dimensions and requirements that are needed for the development and design of strategies and guidelines defining and regulating trustworthy and ethical AI and its introduction and usage (in education and beyond). Second, the workshop discussions will directly inform the Council of Europe's Expert Group on AI&ED's work towards the final report and the potential for a legal instrument to be debated by Council Ministers in 2024. Third, we intended to collaborate on a common publication based on the workshop discussions and results.

References

1. Chounta, I.-A., Bardone, E., Pedaste, M., Raudsep, A.: Exploring teachers' perceptions of Artificial Intelligence as a tool to support their practice in Estonian K-12 education. IJAIED (2021). https://doi.org/10.1007/s40593-021-00243-5
2. Holmes, W., Persson, J., Chounta, I.-A., Wasson, B., Dimitrova, V.: Artificial intelligence and education: a critical view through the lens of human rights, democracy and the rule of law. Council of Europe (2022). https://rm.coe.int/artificial-intelligence-and-education-a-critical-view-through-the-lens/1680a886bd
3. Holmes, W., et al.: Ethics of AI in education: towards a community-wide framework. Int. J. Artific. Intell. Educ. 1–23 (2021)
4. Holmes, W., Porayska-Pomsta, K. (eds.): The ethics of AI in education. practices, challenges, and debates. Routledge, New York (2023)
5. Wayne, H., Tuomi, I.: State of the art and practice in AI in education. Eur. J. Educ. Res. Develop. Polic. n/a (n/a) (2022). https://doi.org/10.1111/ejed.12533
6. Kay, J., Kummerfeld, B., Conati, C., Porayska-Pomsta, K., Holstein, K.: Scrutable AIED. In: Handbook of Artificial Intelligence in Education, pp. 101–125. Edward Elgar Publishing (2023)
7. Leslie, D., Burr, C., Aitken, M., Cowls, J., Katell, M., Briggs, M.: Artificial intelligence, human rights, democracy, and the rule of law: a primer, Council of Europe (2021). https://www.turing.ac.uk/news/publications/ai-human-rights-democracy-and-rule-law-primer-prepared-council-europe
8. Miao, F., Holmes, W., Huang, R., Zhang, H.: AI and education: A guidance for policymakers. UNESCO Publishing (2021)
9. Woolf, B.: Introduction to IJAIED special issue, FATE in AIED. Int. J. Artific. Intell. Educ. **32**, 501–503 (2022)
10. Bozkurt A, et al.: Speculative futures on ChatGPT and generative Artificial Intelligence (AI): a collective reflection from the educational landscape. Asian J. Distan. Educ. **18**(1), 53–130 (2023). https://www.asianjde.com/ojs/index.php/AsianJDE/article/view/709/394
11. Council of Europe: Draft Framework Convention on Artificial Intelligence, Human Rights, Democracy and the Rule of Law (2023a). https://rm.coe.int/cai-2023-28-draft-framework-convention/1680ade043
12. Council of Europe: Regulating Artificial Intelligence in education (2023b). https://rm.coe.int/regulating-artificial-intelligence-in-education-26th-session-council-o/1680ac9b7c
13. Holmes, W., et al.: AI and education. a view through the lens of human rights, democracy and the rule of law. legal and organizational requirements. In: Artificial Intelligence in Education. Communications in Computer and Information Science, vol. 1831, pp. 79–84 (2023). https://doi.org/10.1007/978-3-031-36336-8_12
14. Stracke, C.M.: Quality frameworks and learning design for open education. Int. Rev. Res. Open Distrib. Learn. **20**(2), 180–203 (2019). https://doi.org/10.19173/irrodl.v20i2.4213

15. Stracke, C.M., Chounta, I.-A., Holmes, W., Tlili, A., Bozkurt, A.: A standardised PRISMA-based protocol for systematic reviews of the scientific literature on Artificial Intelligence and education (AI&ED). J. Appl. Learn. Teach. **6**(2), 64–70 (2023). https://doi.org/10.37074/jalt.2023.6.2.38
16. Stracke, C.M., et al.: Impact of COVID-19 on formal education: an international review on practices and potentials of Open Education at a distance. Int. Rev. Res. Open Distrib. Learn. **23**(4), 1–18 (2022).. https://doi.org/10.19173/irrodl.v23i4.6120

Promoting Open Science in Artificial Intelligence: An Interactive Tutorial on Licensing, Data, and Containers

Aaron Haim[1]([⊠])(iD), Stephen Hutt[2]([⊠])(iD), Stacy T. Shaw[1]([⊠])(iD), and Neil T. Heffernan[1]([⊠])(iD)

[1] Worcester Polytechnic Institute, Worcester, MA 01609, USA
{ahaim,sshaw,nth}@wpi.edu
[2] University of Denver, Denver, CO 80210, USA
stephen.hutt@du.edu

Abstract. Across the past decade, the open science movement has increased its momentum, making research more openly available and reproducible across different environments. In parallel, artificial intelligence (AI), especially within education, has produced effective models to better predict student outcomes, generate content, and provide a greater number of observable features for teachers. However, there is a discernible gap between the understanding and application of open science practices in artificial intelligence. In this tutorial, we will expand the knowledge base towards open data and open analysis. First, we will introduce the complexities of intellectual property and licensing within open science. Next, we will provide insights into data sharing methods that preserve the privacy of participants. Finally, we will conclude with an interactive demonstration on sharing research materials reproducibly. We will tailor the content towards the needs and goals of the participants, enabling researchers with the necessary resources and knowledge to implement these concepts effectively and responsibly.

Keywords: Open Science · Reproducibility · Licensing

1 Introduction

Open science and robust reproducibility practices are becoming increasingly adopted within numerous scientific disciplines. Within subfields of educational technology, however, the adoption and review of these practices are sparsely implemented, typically due to a lack of time or incentive to do so [1,14]. Some subfields of education technology have introduced open science practices (special education [4], gamification [6], education research [13]); however, others have seen little to no adoption. Authors have numerous concerns and minimal experience in what can be made publicly available, such as datasets and analysis code [7]. Additionally, research made publicly available is not typically reproducible without additional effort to fix unnoticed issues [8]. A lack of discussion can lead to repetitive communication, irrecoverable processes, or a reproducibility crisis within a field of study [2]. As such, there is a need for accessible resources, providing an understanding of open science practices, how they can be used, and how to mitigate potential issues that may arise at a later date.

Following the success of previous tutorials at the *13th International Conference on Learning Analytics* [10], *24th International Conference on Artificial Intelligence in Education* [12], *15th and 16th International Conference on Educational Data Mining* [9,15], and *10th ACM Conference on Learning @ Scale* [11] along with an accepted tutorial to be presented at the *14th International Conference on Learning Analytics*[1], this tutorial aims to expand the knowledge base of participants towards two concepts of open science: open data and open analysis. First, we will discuss the issues of intellectual property and licensing within openly available work. Next, we will provide an overview of data sharing methods that preserve participant privacy and align with data collection agreements, through the use of data enclaves and anonymized datasets. Finally, we will conclude with an interactive example of how to share materials in a reproducible way using the current best practices. Throughout the tutorial, we will adapt to the needs and goals of participants, addressing concerns and providing resources tailored to them.

2   Background

Open Science is a transformative movement that advocates for the democratization of scientific knowledge. At its core, Open Science seeks to make scientific research, data, and dissemination accessible to all, breaking down the barriers of paywalls, proprietary databases, and closed-access publications. It is built on the principles of transparency, collaboration, and shared knowledge. The goals of Open Science are to advance the pace of discovery but also foster a more inclusive, equitable, and accountable scientific community.

As with many things, the translation from ideals and principles into real-world implementation comes with considerable challenges. For example, open access publication typically comes with a higher cost for the researcher (in turn damaging goals of equity and accessibility). Similarly, in education research, data sharing often poses challenges. Data are typically collected in partnership with educators, administrators, and students, who authorize the collection of data for a specific study/set of research questions, and often actively prohibit the distribution of data to third parties. Data can be deidentified, but given how intrinsically personal educational data can be, this task can be labor-intensive. Worse, some of the easier forms of deidentification (such as removing all forum post data prior to sharing[2]) lead to data no longer being usable for a wide range of research and development goals.

Sharing data on a by-request basis (e.g., Wolins, 1962 [17]) and carefully crafting data agreements has long been a potential solution, but it is often inef-

---

[1] https://doi.org/10.17605/osf.io/kja8r.
[2] https://edx.readthedocs.io/projects/devdata/en/latest/using/package.html.

fective. For example, (Wicherts, Borsboom, Kats, & Molenaar, 2006) [16] contacted owners of 249 datasets, only receiving a response from 25.7%., a response rate similar to that noted in (Wolins, 1962) [17] following requesting data from 37 APA articles (though many years earlier and prior to email). Within education technology, (Haim et al., 2023) [8] contacted the authors of 594 papers, only receiving a response from 37, or 6.2%, of which only 19 responded that their dataset is public or could be requested. Some of the reasons cited were a lack of rights necessary to release the dataset, personally identifiable information was present, or that the dataset itself was part of an ongoing study. The task of sharing data requires a time investment from researchers, typically with no incentive. Moreover, the process can be stalled by changes in email addresses or institutions.

Work has been done to address Open Science principles specifically in education research, through Open Education Science [18], a subfield of Open Science [5]. This movement seeks to address problems of transparency and access, specifically in education research, addressing issues of publication bias, lack of access to original published research, and the failure to replicate. The practices proposed by Open Education Science fall into four categories, each related to a phase in the process of educational research: 1) open design, 2) open data, 3) open analysis, and 4) open publication.

Of most relevance to the current tutorial are Open Data and Open Analysis. **Open Data** is all about ensuring research data and materials are freely available on public platforms, aiding in replication, assessment, and close examination. However, there can be challenges, especially with educational data. There might be initial agreements that prevent the sharing of data or issues related to personal identifiable information (PII) which restrict what can be made public. **Open Analysis**, on the other hand, emphasizes that analytical methods should be reproducible. This is commonly achieved by sharing the code used for analyses. Such code is typically shared on platforms like GitHub or preregistration websites. But there is a catch: the code is often of limited value without the associated data. Simply put, without Open Data, achieving Open Analysis can be tough. Moreover, there are challenges like "code rot" and "dependency hell" [3], where changing libraries can render older code unrunnable.

3   Tutorial Goals

The tutorial focuses on introducing some common open science practices and their usage within education technology along with some interactive examples on how to apply the concepts in research. The target audience is researchers, as the practices offer structure and robustness. Based on past tutorials, we anticipate 5–10 participants and will design an interactive session tailored to their experiences and questions. This approach will allow us to present a responsive tutorial and foster additional community around open science topics.

3.1 Prior to the Conference

Prior to the conference, we will be compiling and organizing all relevant materials and resources. These will be published on a dedicated website, ensuring participants have easy access both during and after the tutorial. In addition, we will request all registrants to complete a pre-survey (using the participant registration list following the author registration/early registration deadlines). This survey aims to gather insights about participants' prior experience with the topics and their specific expectations from the tutorial. This data will be instrumental in allowing us to customize the tutorial, ensuring it meets the individual needs of participants and fostering an engaging and interactive session.

3.2 During the Conference

Our tutorial session will be an interactive and responsive session split into three sections. These sections are outlined below:

1. We will begin the tutorial by discussing how Intellectual Property (IP) intersects with the Open Science Framework. We'll tackle any questions or concerns from attendees about this topic. Our focus will be on code licensing, guided by the principles from Creative Commons. We will discuss why licensing code is important, strategies to safeguard a researcher's intellectual property, and provide guidelines for both Tech Transfer and University IP protection.

2. In the next segment of our tutorial, we discuss Open Data relative to the needs of participants. We anticipate opening this section by again addressing participant concerns to frame our future discussion. This will include identifying personal, moral, institutional, or legal concerns regarding open data. Participants will be introduced to the concept of Data Enclaves. This will cover understanding the primary objectives of sharing data (including identifying the goals of the individual research team, the relationship between Data Enclaves and GDPR/Privacy legislation, and real-world examples of accessing information via these enclaves. Furthermore, we will provide valuable resources on establishing and efficiently using Data Enclaves.
We will also discuss how researchers can share data sets after they have been anonymized, ensuring the identity of participants remains confidential. We will also provide insight into the creation and sharing of synthetic datasets that mimic real datasets without using actual data.
We will close this segment of the tutorial with a general discussion, weighing the advantages and drawbacks of each of the aforementioned approaches. This will not only deepen participant understanding but also help them draw parallels to their own research objectives and needs. Throughout this section, we will emphasize that there is not a "one size fits all" solution and that researchers should make choices based on individual goals and requirements.

3. Finally, we will provide instruction towards sharing materials in a reproducible manner, including best practices on storage, documentation, and privacy. This will be demonstrated with an interactive example using development containers via Visual Studio Code[3] and Docker[4]. The specific example used will depend on information gathered from the participants in the survey prior to the conference.

3.3 Following the Conference

After the conference, all additional resources created for the tutorial will be uploaded to the project's homepage for preservation. As this tutorial wants to repeat and expand upon open science and reproducibility at prior tutorials across conferences, an additional project will be created on the OSF website containing components pointing to all previous conferences and resources. A post-survey will be available at the end and after the tutorial to obtain feedback about the presentation for future use. An aggregate of the response will also be made public on the project's homepage. A community group on Discord will be created to collect, communicate, and discuss open science and reproducibility following the tutorial.

References

1. Armeni, K., et al.: Towards wide-scale adoption of open science practices: the role of open science communities. Sci. Public Policy **48**(5), 605–611 (2021). https://doi.org/10.1093/scipol/scab039
2. Baker, M.: 1,500 scientists lift the lid on reproducibility. Nature **533**(7604), 452–454 (2016). https://doi.org/10.1038/533452a
3. Boettiger, C.: An introduction to docker for reproducible research. SIGOPS Oper. Syst. Rev. **49**(1), 71-79 (2015).https://doi.org/10.1145/2723872.2723882
4. Cook, B.G., Collins, L.W., Cook, S.C., Cook, L.: A replication by any other name: a systematic review of replicative intervention studies. Remedial Spec. Educ. **37**(4), 223–234 (2016). https://doi.org/10.1177/0741932516637198
5. Fecher, B., Friesike, S.: Open science: one term, five schools of thought. In: Bartling, S., Friesike, S. (eds.) Opening Science, pp. 17–47. Springer, Cham (2014). https://doi.org/10.1007/978-3-319-00026-8_2
6. García-Holgado, A., et al.: Promoting open education through gamification in higher education: the opengame project. In: Eighth International Conference on Technological Ecosystems for Enhancing Multiculturality, pp. 399–404. TEEM'20, Association for Computing Machinery, New York, NY, USA (2021). https://doi.org/10.1145/3434780.3436688
7. Haim, A., Baxter, C., Gyurcsan, R., Shaw, S.T., Heffernan, N.T.: How to open science: analyzing the open science statement compliance of the learning @ scale conference. In: Proceedings of the Tenth ACM Conference on Learning @ Scale, pp. 174–182. L@S '23, Association for Computing Machinery, New York, NY, USA (2023). https://doi.org/10.1145/3573051.3596166

---

[3] https://code.visualstudio.com/.
[4] https://www.docker.com/.

8. Haim, A., Gyurcsan, R., Baxter, C., Shaw, S.T., Heffernan, N.T.: How to open science: debugging reproducibility within the educational data mining conference. In: Proceedings of the 16th International Conference on Educational Data Mining, pp. 114–124. International Educational Data Mining Society (2023). https://doi.org/10.5281/zenodo.8115651

9. Haim, A., Shaw, S., Heffernan, N.: How to open science: promoting principles and reproducibility practices within the educational data mining community. In: Feng, M., Kåser, T., Talukdar, P. (eds.) Proceedings of the 16th International Conference on Educational Data Mining. pp. 582–584. International Educational Data Mining Society, Bengaluru, India (2023). https://doi.org/10.5281/zenodo.8115776

10. Haim, A., Shaw, S.T., Heffernan, Neil T, I.: How to open science: promoting principles and reproducibility practices within the learning analytics community (2023). https://doi.org/10.17605/OSF.IO/KYXBA, osf.io/kyxba

11. Haim, A., Shaw, S.T., Heffernan, N.T.: How to open science: promoting principles and reproducibility practices within the learning @ scale community. In: Proceedings of the Tenth ACM Conference on Learning @ Scale, pp. 248–250. L@S 2023, Association for Computing Machinery, New York, NY, USA (2023). https://doi.org/10.1145/3573051.3593398

12. Haim, A., Shaw, S.T., Heffernan, N.T.: How to open science: promoting principles and reproducibility practices within the artificial intelligence in education community. In: Wang, N., Rebolledo-Mendez, G., Dimitrova, V., Matsuda, N., Santos, O.C. (eds.) AIED 2023. CCIS, vol. 1831, pp. 74–78. Springer, Cham (2023). https://doi.org/10.1007/978-3-031-36336-8_11

13. Makel, M.C., Smith, K.N., McBee, M.T., Peters, S.J., Miller, E.M.: A path to greater credibility: large-scale collaborative education research. AERA Open 5(4), 2332858419891963 (2019). https://doi.org/10.1177/2332858419891963

14. Nosek, B.: Making the most of the unconference (2022). https://osf.io/9k6pd

15. Shaw, S., Sales, A.: Using the open science framework to promote open science in education research. In: Proceedings of the 15th International Conference on Educational Data Mining, pp. 853–853. International Educational Data Mining Society (2022). https://doi.org/10.5281/zenodo 6852996

16. Wicherts, J.M., Borsboom, D., Kats, J., Molenaar, D.: The poor availability of psychological research data for reanalysis. Am. Psychol. 61(7), 726 (2006). 10.1037/0003-066X.61.7.726

17. Wolins, L.: Responsibility for raw data. Am. Psychol. 17(9), 657–658 (1962). https://doi.org/10.1037/h0038819

18. van der Zee, T., Reich, J.: Open education science. AERA Open 4(3), 2332858418787466 (2018). https://doi.org/10.1177/2332858418787466

Adaptive Lifelong Learning (ALL)

Alireza Gharahighehi[1,2]([✉]) [iD], Rani Van Schoors[1,3] [iD], Paraskevi Topali[4] [iD],
and Jeroen Ooge[5] [iD]

[1] KU Leuven, Itec, imec research group, Kortrijk, Belgium
{alireza.gharahighehi,rani.vanschoors}@kuleuven.be
[2] Department of Public Health and Primary Care, KU Leuven Campus Kulak,
Kortrijk, Belgium
[3] KU Leuven, Centre for Instructional Psychology and Technology, Leuven, Belgium
[4] NOLAI — National Education Lab AI, Behavioural Science Institute,
Radboud University, Nijmegen, The Netherlands
evi.topali@ru.nl
[5] Department of Information and Computing Sciences, Utrecht University, Utrecht,
The Netherlands
j.ooge@uu.nl

Abstract. This half-day workshop emphasizes the role of adaptive lifelong learning in the dynamic landscape of Artificial Intelligence in Education (AIED). The pervasive influence of digitization mandates continuous learning, responding to the challenges of automation and skill gaps.
While learners and trainees are pivotal in implementing AI in lifelong
learning education, they face challenges such as information overload and
the necessity to exhibit high degrees of flexibility to adapt to the rapid
changes and continuous evolution of AI tools. By highlighting the virtues
of adaptive lifelong learning, the workshop examines the impact of AI on
a diverse spectrum of learners. Adaptive AI tools emerge as a source of
support for bridging learning gaps, particularly in the post-pandemic
landscape, offering essential support to learners through streamlined
automation. Furthermore, the workshop emphasizes the critical necessity for adaptation in the evolving AIED domain, moving beyond its
roots in computer science to encompass a broader educational perspective. With a specific focus on multi-criteria adaptive lifelong learning,
the workshop advocates for collaboration among learners, educators, policymakers, researchers, and EdTech companies. The ultimate goal is to
facilitate the development of relevant and evidence-based adaptive AI
tools that significantly enhance lifelong learning.

1 Relevance and Importance

To illustrate the relevance of adaptive lifelong learning, we discuss three questions
in the following paragraphs.

1.1 Why Lifelong Learning?

Driven by digitization and datafication, innovative technologies are increasingly integrated into educational research and policies [16,20,27]. Within this

context, Artificial Intelligence in Education (AIED) is a timeless issue for educational practitioners and researchers worldwide [3,34] because the field of AI is characterized by rapid progress and technological pushes towards more automation. This results in skill gaps [33] and forces individuals and organizations to continually update their knowledge and capabilities to remain relevant in an increasingly AI-driven world. In education specifically, learners and trainees face significant challenges related to demonstrating high flexibility and keeping up with the rapid changes and continuous development of new AI tools [44]. As a result, and further accelerated by the COVID-19 pandemic, millions of lifelong learners have turned to online learning communities. In sum, the upsurge of AI encompasses an increasing need for lifelong learning and training.

1.2   Why Adaptive Lifelong Learning?

There is a growing interest in AI systems that facilitate adaptation because of its many presumed benefits [8,23,25]. For example, studies often indicate a positive impact on cognitive and non-cognitive learning outcomes [17,42,43]. Furthermore, many praise adaptive learning for considering the heterogeneity of learners and trainees and their needs by appropriately personalizing exercises, scaffolds, and assessments [16]. Moving away from the traditional 'one-size-fits-all' learning and training approach, it is believed that adaptive AI tools can remediate learning gaps, especially in the post-pandemic context [4,20,28,38,42].

To personalize learning with adaptive learning systems, learner models play a key role [35]. These dynamic models represent learners' evolving knowledge and understanding [7] and can be based on various cognitive, pragmatic, or data-driven approaches. Furthermore, the rise of ubiquitous contextual information in adaptive learning has inspired ubiquitous user modeling, which refers to tracking and analyzing user behavior across different systems that share user models [15]. While promising, ubiquitous user modeling comes with several challenges, including addressing the disparities in user modeling techniques, domains, and contexts; efficiently and effectively initializing user models for on-demand services, and continuously updating user models as needed [21]. Aligned with the call for more transparent and explainable AI systems [1,2,26], researchers have argued for transparent learner models. For example, Bull and Kay [6,7] proposed Open Learner Models through which learners can better understand and maintain their learner models. Furthermore, Kay and Kummerfeld [19] discussed the importance of scrutable user modeling and personalization for addressing challenges such as privacy, the invisibility of personalization, errors in user models, wasted user models, and the controllability of user models.

Compared to formal learning contexts, adaptation is even more essential in lifelong learning because learners are afforded more choices and flexibility, requiring learners to adjust to the evolving educational landscape. PHelpS (Peer Help System) [14] and i-Help [5] are examples of adaptive lifelong learning systems designed to assist learners in completing their tasks. These systems offer support and guidance, facilitate connections with peer helpers when needed, and act as mediators for task-related communication. Adaptive AI-supported learning technologies such as intelligent dashboards are moreover seen as valuable means to support teachers and trainers and make teaching and training more efficient [4,20,28,42–44]. Considering these promising prospects, advocates predict that AI will continue to change the educational ecosystem [16,20].

1.3 Why Adaptation Based on Multiple Criteria?

AIED was initially rooted in computer science but later merged with educational sciences. As a distinctive research domain, the focus now encompasses diverse educational and training settings, reflecting broader perspectives on AI in formal and lifelong learning environments. With the increasing interest in AI for education, many new research initiatives emerge, including critical reflections on traditional educational objectives; exploring ethical, technical, and implementation challenges; and developing new and more efficient algorithms [22,37].

This workshop focuses on multi-criteria adaptive lifelong learning, i.e., the broad domain of AIED technology, ranging from descriptive, predictive, and prescriptive learning analytics to language processing, speech, and image recognition/processing [24,25]. We also look at the multi-dimensionality of learning contexts, i.e., adaptation on both individual and group levels. Moreover, our approach transcends traditional learning environments such as classrooms, facilitating learning anywhere and anytime; for example, at home. We focus on research that involves multiple actors as it becomes apparent that establishing a robust partnership among learners/trainees, teachers/trainers, policymakers, researchers, and EdTech companies is essential when integrating AI into education and training. Such collaborations hold significant value: by sharing knowledge, co-designing interventions, and engaging in ongoing dialogue, more evidence-based AI tools can be developed that are applicable and valorized in real-world educational settings.

2   Team

2.1   Program Organizers

Alireza Gharahighehi is a senior researcher specializing in AI within itec, imec research group at KU Leuven. His primary focus lies in personalization and recommendation systems for educational applications such as MOOCs [13]. He is currently leading a use-case in the field of education within the Flanders AI Research Program. Throughout his PhD in computer science, he explored various aspects of personalization, including diversity [10], fairness [12], sparsity [9], and multi-stakeholder settings [11].

Rani Van Schoors is a postdoctoral researcher in itec (research group imec) and CIP&T at KU Leuven. During her PhD she worked on the topic of digital personalized learning (DPL) [39] and specifically examined the role of the teacher during the implementation of DPL tools [40,41]. Now, her focus lies on the potential of artificial intelligence in education [18] and the potential of virtual reality in a medical training context.

Paraskevi Topali holds a BS degree in primary education, a MS degree in information and communications technology, and a PhD in transdisciplinary research in education. Currently, she is a postdoctoral researcher at the National Educational Lab AI (NOLAI) from Radboud University. Her main research interests lie in technology-enhanced learning, including human-centered AI/LA, personalized feedback and scaffolding, learning design, MOOCs, and K-12.

Jeroen Ooge is an assistant professor at Utrecht University, passionate about multidisciplinary science and lifelong learning. He holds two MS degrees (fundamental mathematics and applied informatics) and a PhD in computer science focused on human-centered explainable AI. His research mainly focuses on how tailored visualizations can support explaining [32] and controlling [31,36] AI-based adaptive learning systems, and how motivational techniques can be personalized in learning contexts [29,30].

2.2   Program Committee

In forming the program committee, we considered including scholars from the AIED community, as well as experts in AI and educational sciences:

- *Ifeoma Adaji*, assistant professor, computer science, The University of British Columbia
- *Bita Akram*, assistant professor, computer science, North Carolina State University
- *Judy Kay*, professor, computer science, University of Sydney
- *Mirko Marras*, assistant professor, artificial intelligence, University of Cagliari
- *Sameh Metwaly*, postdoctoral researcher, psychology and educational sciences, itec, imec research group
- *Felipe Kenji Nakano*, postdoctoral researcher, artificial intelligence, KU Leuven
- *Wim van den Noortgate*, full professor, psychology and educational sciences, KU Leuven
- *Chen Sun*, research associate, Education and Development, The University of Manchester
- *Anaïs Tack*, postdoctoral researcher, language technology, KU Leuven and UCLouvain
- *Celine Vens*, professor, artificial intelligence, KU Leuven Campus Kulak
- *Diego Zapata-Rivera*, Distinguished Presidential Appointee, artificial intelligence in education, Educational Testing Service at Princeton NJ

3 Organization

We will announce a call for short and long papers about lifelong learning. Submissions must be anonymous, as we adhere to the double-masked review process. Depending on the number of accepted papers, we will determine whether short papers should be presented in oral presentations or a poster session. The tentative timeline for announcing the call for papers is as follows:

- Paper submission deadline: May 15th, 2024
- Notification of acceptance: June 2nd, 2024
- Camera-ready version deadline: June 9th, 2024

Accepted papers will be included in the workshop proceedings and published on the workshop website and shall be submitted to CEUR-WS.org for online publication. Additionally, we will invite the authors of selected papers to extend their submissions by at least 30% for potential inclusion in a special issue of a journal. All accepted papers will be presented and discussed interactively during the workshop, and a keynote speaker will present their research on adaptive lifelong learning.

4 Content and Themes

This workshop aims to unite a community of AIED researchers interested in various aspects of lifelong learning, particularly in adaptation and personalization. We invite contributions covering all topics related to adaptive lifelong learning, with a specific focus on, but not limited to, the following list:

- Tailoring lifelong learning to various factors, including knowledge, skills, motivation, engagement, and learning objectives
- Extending the range of adaptation criteria beyond mere relevance, incorporating factors such as fairness, diversity, bias, and pedagogical aspects
- Group-aware and context-aware adaptations in lifelong learning
- Lifelong learning adaptation involving multiple stakeholders, including learners, trainers, and EdTech companies
- Adaptive learning in Massive Open Online Courses (MOOCs)
- Quantifying learners' engagement and dropout risk
- Adaptive or personalized nudging strategies within lifelong learning
- Multi-modal adaptive lifelong learning
- Adaptive educational games for enhancing lifelong learning
- Adaptive computer-assisted language learning
- Adaptive communication with learners (e.g., feedback, dashboard, etc.)
- Chatbots for lifelong learning
- Adaptive simulations in workstations
- Explainable adaptations in lifelong learning
- Generating adaptive learning trajectories

5   Expected Target Audience

The workshop is designed for researchers, PhD researchers, and master's students who are actively engaged in lifelong learning, adaptive learning, learning in MOOCs, language learning, recommendation systems, educational data mining, and learning analytics. It is also relevant for employees at EdTech companies interested in technologies for lifelong learning. Overall, we anticipate 15–25 participants.

Acknowledgments. This workshop is supported by Itec, imec research group at KU Leuven. The authors also acknowledge support from the Flemish Government (AI Research Program).

References

1. Adadi, A., Berrada, M.: Peeking inside the black-box: a survey on explainable artificial intelligence (XAI). IEEE Access **6**, 52138–52160 (2018). https://doi.org/10.1109/ACCESS.2018.2870052
2. Barredo Arrieta, A., et al.: Explainable artificial intelligence (XAI): concepts, taxonomies, opportunities and challenges toward responsible AI. Inf. Fusion **58**, 82–115 (2020). https://doi.org/10.1016/j.inffus.2019.12.012
3. du Boulay, B., Mitrovic, A., Yacef, K.: Handbook of Artificial Intelligence in Education. Edward Elgar Publishing, Cheltenham (2023)
4. Breines, M.R., Gallagher, M.: A return to teacherbot: rethinking the development of educational technology at the university of Edinburgh. Teach. High. Educ. **28**(3), 517–531 (2023)
5. Bull, S., Greer, J., McCalla, G., Kettel, L., Bowes, J.: User modelling in i-help: what, why, when and how. In: Bauer, M., Gmytrasiewicz, P.J., Vassileva, J. (eds.) UM 2001. LNCS (LNAI), vol. 2109, pp. 117–126. Springer, Heidelberg (2001). https://doi.org/10.1007/3-540-44566-8_12
6. Bull, S., Kay, J.: Student models that invite the learner in: the smili:() open learner modelling framework. Int. J. Artif. Intell. Educ. **17**(2), 89–120 (2007)
7. Bull, S., Kay, J.: SMILI: a framework for interfaces to learning data in open learner models, learning analytics and related fields. Int. J. Artif. Intell. Educ. **26**, 293–331 (2016)
8. Cardona, M.A., Rodríguez, R.J., Ishmael, K., et al.: Artificial intelligence and the future of teaching and learning: Insights and recommendations (2023)
9. Gharahighehi, A., Pliakos, K., Vens, C.: Addressing the cold-start problem in collaborative filtering through positive-unlabeled learning and multi-target prediction. IEEE Access **10**, 117189–117198 (2022)
10. Gharahighehi, A., Vens, C.: Personalizing diversity versus accuracy in session-based recommender systems. SN Comput. Sci. **2**(1), 39 (2021)
11. Gharahighehi, A., et al.: Multi-stakeholder news recommendation using hypergraph learning. In: Koprinska, I., et al. (eds.) ECML PKDD 2020. CCIS, vol. 1323, pp. 531–535. Springer, Cham (2020). https://doi.org/10.1007/978-3-030-65965-3_36
12. Gharahighehi, A., Vens, C., Pliakos, K.: Fair multi-stakeholder news recommender system with hypergraph ranking. Inf. Process. Manag. **58**(5), 102663 (2021)

13. Gharahighehi, A., Venturini, M., Ghinis, A., Cornillie, F., Vens, C.: Extending Bayesian personalized ranking with survival analysis for mooc recommendation. In: Adjunct Proceedings of the 31st ACM Conference on User Modeling, Adaptation and Personalization, pp. 56–59 (2023)
14. Greer, J.E., Mccalla, G., Collins, J.A., Kumar, V.S., Meagher, P., Vassileva, J.: Supporting peer help and collaboration in distributed workplace environments. Int. J. Artif. Intell. Educ. 9, 159–177 (1998)
15. Heckmann, D.: Ubiquitous User Modeling, vol. 297. IOS Press (2006)
16. Holmes, W., Bialik, M., Fadel, C.: Artificial intelligence in education: promises and implications for teaching and learning (2019)
17. Holmes, W., Porayska-Pomsta, K.: The Ethics of Artificial Intelligence in Education: Practices, Challenges, and Debates. Taylor & Francis (2022)
18. Itec: Learning, teaching & training in the era of artificial intelligence: Challenges and opportunities for evidence-based educational research. Acco (2024)
19. Kay, J., Kummerfeld, B.: Creating personalized systems that people can scrutinize and control: drivers, principles and experience. ACM Trans. Interact. Intell. Syst. (TiiS) 2(4), 1–42 (2013)
20. Knox, J.: AI and Education in China: Imagining the Future, Excavating the Past. Taylor & Francis (2023)
21. Kuflik, T., Kay, J., Kummerfeld, B.: Challenges and solutions of ubiquitous user modeling. Ubiquit. Display Environ. 7–30 (2012)
22. Luan, H., et al.: Challenges and future directions of big data and artificial intelligence in education. Front. Psychol. 11, 580820 (2020)
23. Maslej, N., et al.: Artificial intelligence index report 2023. arXiv preprint arXiv:2310.03715 (2023)
24. Miao, F., Shiohira, K.: K-12 AI curricula. a mapping of government-endorsed AI curricula (2022)
25. Miao, F., Holmes, W., Huang, R., Zhang, H., et al.: AI and Education: A Guidance for Policymakers. UNESCO Publishing (2021)
26. Miller, T.: Explanation in artificial intelligence: insights from the social sciences. Artif. Intell. 267, 1–38 (2019). https://doi.org/10.1016/j.artint.2018.07.007
27. Nowotny, H.: IN AI WE TRUST: Power Illusion and Control of Predictive Algorithms. Wiley, Hoboken (2021)
28. OECD.: OECD Digital Education Outlook 2021 Pushing the Frontiers with Artificial Intelligence, Blockchain and Robots. OECD Publishing (2021)
29. Ooge, J., De Braekeleer, J., Verbert, K.: Nudging Adolescents Towards Recommended Maths Exercises With Gameful Rewards. In: Artificial Intelligence in Education. Springer Nature Switzerland, Cham (2024)
30. Ooge, J., De Croon, R., Verbert, K., Vanden Abeele, V.: Tailoring gamification for adolescents: a validation study of big five and Hexad in Dutch. In: Proceedings of the Annual Symposium on Computer-Human Interaction in Play, pp. 206–218. ACM, Virtual Event Canada (2020). https://doi.org/10.1145/3410404.3414267
31. Ooge, J., Dereu, L., Verbert, K.: Steering recommendations and visualising its impact: effects on adolescents' trust in e-learning platforms. In: Proceedings of the 28th International Conference on Intelligent User Interfaces, pp. 156–170. IUI 2023, Association for Computing Machinery, New York, NY, USA (2023). https://doi.org/10.1145/3581641.3584046
32. Ooge, J., Kato, S., Verbert, K.: Explaining recommendations in e-learning: effects on adolescents' trust. In: 27th International Conference on Intelligent User Interfaces, pp. 93–105. IUI 2022, Association for Computing Machinery, New York, NY, USA (2022). https://doi.org/10.1145/3490099.3511140

33. Outlook, O.S.: Skills for a resilient green and digital transition (2023)
34. Salomon, G.: Technology and pedagogy: why don't we see the promised revolution? Educ. Technol. **42**(2), 71–75 (2002)
35. Self, J.: The defining characteristics of intelligent tutoring systems research: ITSs care, precisely. Int. J. Artif. Intell. Educ. **10**, 350–364 (1998)
36. Szymanski, M., Ooge, J., De Croon, R., Vanden Abeele, V., Verbert, K.: Feedback, control, or explanations? supporting teachers with steerable distractor-generating AI. In: Proceedings of the 14th Learning Analytics and Knowledge Conference, pp. 690–700. LAK 2024, Association for Computing Machinery, New York, NY, USA (2024). https://doi.org/10.1145/3636555.3636933
37. Tuomi, I.: Beyond mastery: toward a broader understanding of AI in education. Int. J. Artif. Intell. Educ. 1–12 (2023)
38. UNICEF: Policy guidance on AI for children (2021)
39. Van Schoors, R., Elen, J., Raes, A., Depaepe, F.: An overview of 25 years of research on digital personalised learning in primary and secondary education: a systematic review of conceptual and methodological trends. Br. J. Edu. Technol. **52**(5), 1798–1822 (2021)
40. Van Schoors, R., Elen, J., Raes, A., Depaepe, F.: Tinkering the teacher-technology nexus: the case of teacher-and technology-driven personalisation. Educ. Sci. **13**(4), 349 (2023)
41. Van Schoors, R., Elen, J., Raes, A., Vanbecelaere, S., Depaepe, F.: The charm or chasm of digital personalized learning in education: teachers' reported use, perceptions and expectations. TechTrends **67**(2), 315–330 (2023)
42. Zhai, X., et al.: A review of artificial intelligence (AI) in education from 2010 to 2020. Complexity **2021**, 1–18 (2021)
43. Zhang, K., Aslan, A.B.: Ai technologies for education: recent research & future directions. Comput. Educ. Artif. Intell. **2**, 100025 (2021)
44. Zimmerman, M.: Teaching AI: exploring new frontiers for learning. Int. Soc. Technol. Educ. (2018)

7th Edition of the International Workshop on Culturally-Aware Tutoring Systems (CATS)

Emmanuel G. Blanchard[1](✉) 🆔, Isabela Gasparini[2] 🆔,
and Maria Mercedes T. Rodrigo[3] 🆔

[1] Le Mans Université, Le Mans, France
emmanuel.blanchard@univ-lemans.fr
[2] Santa Catarina State University, Joinville, Brazil
isabela.gasparini@udesc.br
[3] Ateneo de Manila University, Quezon City, the Philippines
mrodrigo@ateneo.edu

1 Content and Theme

Research has shown that culture has an impact on educational practices [5, 9] as well as on many learning-related domains such as cognition, motivation, and emotions [4, 10, 11, 16], which naturally attracted the attention of the AIED community.

Although the most influential intelligent educational technologies have been created by and for the Developed World [1, 3, 12], an increasing number of these systems are designed with different contexts in mind [7, 9], and more and more intercultural evaluations are reported [13, 15]. Recent publications also discuss large-scale efforts to use AI to reach students in under-resourced, and marginalized contexts with the hope of improving life outcomes of their inhabitants [6, 8].

Therefore, the 2024 edition of the CATS workshop aims to engage interested researchers in a conversation on how to take culture and context into account in the design and operations of educational systems, and discuss issues such as:

- What features of culture are important to consider in the design process of AIED systems?
- Can educational technologies designed and developed in a specific cultural context transfer to other parts of the World and remain effective?
- How to embed culturally-adaptive mechanisms into intelligent educational technologies?

2 Importance for the AIED Community

This increased sensitivity of the AIED community to cultural awareness and related issues such as inclusiveness and diversity is evidenced by at least three efforts of the AIED society:

- The AIED conference now includes a statement of diversity and inclusion[1].

---

[1] See https://aied2024.cesar.school/diversity-and-inclusion/diversity-and-inclusion.

- The AIED conference now calls for papers for the WideAIED track[2]. This track solicits papers from underrepresented countries in order to broaden participation in the community and to share more global perspectives of AI use in a greater variety of educational contexts.
- A recent AIED Society program on Diversity, Equity, Inclusion, and Accessibility (AIED DEIA) seeks to link scholars from underrepresented communities with mentors who can help them develop their research and prepare it for presentation in AIED.

Furthermore, several past [14] and present[3] special issues of the *International Journal of Artificial Intelligence in Education* have emerged from, or conform to the aim of the CATS workshop.

3 Past Editions

The CATS workshop has been held 6 times to date, in conjunction with ITS and AIED conferences: ITS2008, AIED2009, ITS2010, AIED2013, ITS2014, and AIED2015. It has grown in popularity since its first edition, a trend reflected by an increasing number of submissions, a steady participation of researchers, and the enthusiasm we faced when assembling our program committee for the 2024 edition. This shows that many researchers in our community clearly see the relevance of this workshop and the importance of having an event to discuss culture and AIED in a transversal way. For many colleagues, it is indeed becoming increasingly crucial to understand and integrate diverse cultural perspectives to ensure that developments, technologies and methods arising from AIED research are inclusive, equitable and globally relevant.

4 Organizers

The 7th CATS workshop is organized by Emmanuel G. Blanchard, Isabela Gasparini, and Maria Mercedes T. Rodrigo, who contributed to create the workshop and organize its 6 previous editions. You will find below paragraphs summarizing their professional biographies. The three co-organizers are similarly involved in the preparation of the workshop. Their information is therefore presented in alphabetical order.

Emmanuel G. Blanchard is an Associate Professor at *Le Mans Université*, France, and a member of its research laboratory in computer science. He is also an Adjunct Professor in the *Department of Surgery of McGill University*, Canada. He holds a PhD in computer science from the *Université de Montréal*, and has completed a postdoctoral fellowship in the *Department of Educational and Counselling Psychology* of *McGill University*. In the past, he was a visiting researcher at *Osaka University* where he focused his efforts on formal ontology engineering, a Senior Lecturer in Software Engineering at *Polytechnic of Namibia*, and an Assistant Professor at the *Department of Architecture*,

---

[2] See https://aied2024.cesar.school/call-for-papers/wideaied.
[3] Call for papers: Special Issue on AIED in the Global South. https://link.springer.com/journal/40593/updates/25187332.

Design and Medialogy of *Aalborg University, Denmark*. From 2014 to 2023, he worked in the startup ecosystem of Montreal, Canada. His research interests include, but are not limited to, human factors (cognition, motivation, affect, culture) and technology, expert-centered design, data engineering, formal ontology engineering, virtual simulations for training and education, and human-computer interaction at large. He has co-created the series of international workshops on *Culturally-Aware Tutoring Systems*, was an invited participant of the *Dagstuhl seminar* on *Computational Models of Cultural Behaviors for Human-agent Interactions*[4], and was the principal editor of the *Handbook of Research on Culturally-Aware Information* [2], one of the first publications on the links between culture and technology.

Isabela Gasparini received her Ph.D. from the *Federal University of Rio Grande do Sul (UFRGS - Brazil)* with a sandwich period at *TELECOM Sud Paris* (France). She is an Associate Professor at the *Department of Computer Science at the Santa Catarina State University (UDESC)*, where she is involved in Human-Computer Interaction and Technology-Enhanced Learning fields, with a special interest in adaptive e-learning systems, gamification, learning analytics, recommender systems, infoviz, context-awareness, and cultural issues. She has a Productivity Scholarship in Technological Development and Innovative Extension from *CNPq (National Council for Scientific and Technological Development - Brazil)* and is currently a Member of the *Brazilian Computer Society Council* (2021–2025). She was the editor-in-chief of the *Brazilian Journal of Computers in Education* (2019–2021) and coordinator of the *Special Human-Computer Interaction Committee* of *SBC (Brazilian Computer Society)* (2019–2021). She was a member of the special informatics in education committee of SBC (2019–2021). She is currently part of the *Digital Girls Program* steering committee, a nonprofit initiative by the Brazilian Computing Society. More details in her CV: http://lattes.cnpq.br/3262681213088048.

Maria Mercedes T. Rodrigo is a professor at the *Department of Information Systems and Computer Science, Ateneo de Manila*, the Philippines. Her research interests include learning analytics, artificial intelligence in education, technology in education, and educational games. She is the head of the *Ateneo Laboratory for the Learning Sciences (ALLS)*. Since 2007, she has received approximately US$800,000 in research grants from government and non-government organizations. In collaboration with colleagues and students, Dr. Rodrigo has published on computers and related technologies in education, human-computer interaction, computer science education and others. Her current research projects include the establishment of a virtual, augmented, and mixed reality laboratory at the *Ateneo de Manila University*, analyses of learning management systems logs collected during the pandemic, deployment of Minecraft to increase STEM interest among Filipino learners, and an eye tracking study on programmer reading and debugging skills. She serves on the editorial boards of several high-impact journals including *Internet and Higher Education* and *Research and Practice in Technology-Enhanced Learning*. Dr. Rodrigo is on the Executive Committee of the *Artificial Intelligence Education Society*. In 2021, Dr. Rodrigo received the Distinguished Researcher Award from the *Asia-Pacific Society for Computers in Education* (APSCE). She is President of APSCE.

---

[4] https://www.dagstuhl.de/en/seminars/seminar-calendar/seminar-details/14131.

5 Program Committee

CATS program committees have always been tailored to express cultural diversity through their members, which is a prerequisite for the workshop's credibility given its theme (culture and AIED). The CATS2024 program committee is no exception and includes researchers not only from the leading countries in the AIED community, but also academics from less present countries. Similarly, this committee is made up mainly of members of our community (senior and emerging scholars) as well as a few researchers from related disciplines (e.g., HCI). Finally, some of these PC members have been involved in previous editions of CATS. The list of program committee members is presented below:

Faisal Bin Badir, Charles Darwin University, Australia
Ryan Baker, University of Pennsylvania, USA
Geoffray Bonnin, University of Lorraine, France
Benedict du Boulay, University of Sussex, UK
Corentin Coupry, Le Mans Université, France
Victoria Eyharabide, Sorbonne Université, France
Kateryna Holubinka, Open University, The Netherlands
Ioana Jivet, Goethe University Frankfurt, Germany
Aditi Kothiyal, IIT Gandhinagar, India
Paul Libbrecht, IU International University of Applied Science, Germany
Bruce McLaren, Carnegie Mellon University, USA
Guilherme Medeiros Machado, ECE-Paris, France
Riichiro Mizoguchi, Japan Advanced Institute of Science and Technology, Japan
Phaedra Mohammed, University of the West Indies, Trinidad & Tobago
Benjamin Nye, University of Southern California, USA
Jaclyn Ocumpaugh, University of Pennsylvania, USA
Roberto Pereira, Federal University of Parana, Brazil
Genaro Rebolledo Mendez, AffectSense, Mexico
Guillermo Rodriguez, ISISTAN, Argentina
Rod Roscoe, Arizona State University, USA
Silvia Schiaffino, ISISTAN, Argentina
Shashi Kant Shankar, Amrita Vishwa Vidyapeetham, India
Julita Vassileva, University of Saskatchewan, Canada
Lung-Hsiang Wong, Nanyang Technological University, Singapore
Su Luan Wong, Universiti Putra Malaysia, Malaysia

6 Format and Activities

CATS2024 is designed according to the following objectives:

- To propose an event that presents and discuss efforts to connect culture and AIED
- To convince new members of the AIED community of the importance of considering culture in their research.
- To provide theoretical and methodological entry points to facilitate new members' first steps in this particularly complex interdisciplinary field

- To stimulate discussion and reflection around a culturally relativistic approach to AIED. i.e., many of the approaches and technologies developed by our community are not as universal as expected, including some particularly established elements. We don't know what we don't know.

To address these objectives, the workshop is a **half day event** that consists of several complementary activities.

Introduction Session. This activity will present the theme of the workshop and give time for the organizers and participants to introduce themselves. A brief overview of several conceptual approaches of interest to the development of culturally-aware tutoring systems will then be provided to the audience. It will last around 10 min, and equip participants with the theoretical grounding they need to better understand the presentations and discussions to come.

Paper Sessions/Discussion Sessions. Several paper sessions will be organized during the CATS workshop, each targeting a specific topic that will be determined following the submission/review process. Each session will consist of 3 papers and each one of them will be allocated 10 min for presentation. This format will require a concise yet comprehensive overview of each study, ensuring that all key findings and concepts are effectively communicated within a limited time frame. Our aim is to offer a hybrid workshop.

After a paper session, we will transition to a 30 min discussion session where the audience will discuss both the paper findings and the session topic altogether. During the final program preparation, we will have identified possible bridges between papers that can be highlighted during the discussion. We hope this format will encourage people to share examples and challenges from their personal cultural contexts, thus enriching the cultural awareness and the relativistic perspective of all participants.

Design Session. A special activity is planned to engage participants in a unique, culturally-focused design challenge. The audience will be divided into groups. Each group will have the choice of either discussing a way to bring cultural awareness into an existing educational technology familiar to them, or describing modules and functionalities for a hypothetical technology that would address an educational challenge in a culturally-intelligent way. Each group will then share its ideas, providing a platform for inter-group learning and stimulating a wider dialogue on culturally-aware design approaches.

Focused Discussion Session. A focused discussion session will conclude the workshop. This 30 min activity will be a collaborative way of triggering reflection on a topic of prime importance to the AIED community: the digital divide within and between societies. It will therefore address a number of questions, such as:

- How does (the lack of) cultural awareness in AIED technologies relate to the digital divide?
- How can AIED systems help overcome educational challenges in the Global South?
- What AIED technologies or initiatives are currently missing that could help reduce the negative effects induced by the digital divide?

We will frame this session in such a way as to connect the ensuing discussions with the reflections and ideas that have emerged from previous sessions.

7 Conclusion

The CATS workshop has been designed as a half day hybrid event to help participants (especially newcomers) develop a richer, and more nuanced understanding of the challenges and opportunities of taking culture into account in AIED.

With its different sessions with complementary formats and objectives, this event will help AIED researchers deepen their knowledge and understanding of existing CATS initiatives and perspectives. We hope that new ideas will be sparked by these engaging discussions, and that international collaborations will emerge from this event.

References

1. Blanchard, E.G.: Socio-cultural imbalances in AIED Research: Investigations, implications and opportunities. Int. J. Artif. Intell. Educ. **25**, 204–228 (2015)
2. Blanchard, E.G., Allard, D. (eds.): Handbook of research on culturally-aware information technology: perspectives and models. IGI Global (2011)
3. Blanchard, E., Ogan, A.: Infusing cultural awareness into intelligent tutoring systems for a globalized world. In: Nkambou, R., Mizoguchi, R., Bourdeau, J. (eds.), Advances in intelligent tutoring systems, pp. 485–505 (2010)
4. Henrich, J., Heine, S., Norenzayan, A.: Most people are not WEIRD. Nature **466**, 29 (2010)
5. Hofstede, G.: Cultural differences in teaching and learning. Int. J. Intercult. Relat. **10**, 301–320 (1986)
6. Isotani, S., Bittencourt, I., Walker, E.: Artificial intelligence and Educational Policy: bridging research and practice. In: Proceedings of AIED2023, pp. 63–68 (2023)
7. Karumbaiah, S., Ocumpaugh, J., Baker, R.S.: Context matters: Differing implications of motivation and help-seeking in educational technology. Int. J. Artif. Intell. Educ. **32**(3), 685–724 (2022)
8. McReynolds, A. A., Naderzad, S. P., Goswami, M., Mostow, J.: Toward learning at scale in developing countries: lessons from the global learning XPRIZE Field Study. In: Proceedings of LAKS2020, pp. 175–183 (2020)
9. Melis, E., Goguadze, G., Libbrecht, P., Ullrich, C.: Culturally aware mathematics education technology. Handbook of research on culturally-aware information technology: perspectives and models, pp. 543–557 (2011)
10. Mesquita, B.: Between us: how cultures create emotions. W. W. Norton & Company (2022)
11. Nisbett, R. E., Norenzayan, A.: Culture and cognition. In: H. Pashler, D. Medin (eds.) Steven's handbook of experimental psychology: memory and cognitive processes, 3rd edn., 5pp. 61–597 (2002)
12. Nye, B.D.: Intelligent Tutoring Systems by and for the Developing World: a review of trends and approaches for educational technology in a global context. Int. J. Artif. Intell. Educ. **25**, 177–203 (2015)
13. Ogan, A., Yarzebinski, E., Fernández, P., Casas, I.: Cognitive tutor use in Chile: Understanding classroom and lab culture. In: Lecture notes in computer science, pp. 318–327 (2015)
14. Ogan, A., Johnson, W.L.: Preface for the special issue on culturally aware educational technologies. Int. J. Artif. Intell. Educ. **25**, 173–176 (2015)

15. Rodrigo, M.M.T., Baker, R.S.J.D., Rossi, L.: Student off-task behaviour in computer-based learning in the Philippines: Comparison to Prior Research in the U.S.A. Teacher College Record **115**(10), 1–27 (2013)
16. Rustamovna, S.N., Vladimirovna, E.A., Lynch, M.F.: Basic needs in other cultures: using qualitative methods to study key issues in Self-Determination theory research. Psychology. J. High. School Econ. **17**, 134–144 (2020)

Hack Beyond the Code: Building a Toolbox of Human-Centred Strategies for AI Literacy

Cleo Schulten[1]([⊠])(iD), Li Yuan[2]([⊠]), Kiev Gama[3]([⊠])(iD), Wayne Holmes[4]([⊠]),
Alexander Nolte[5,6]([⊠])(iD), Tore Hoel[7]([⊠]), and Irene-Angelica Chounta[1]([⊠])(iD)

[1] University of Duisburg-Essen, Duisburg, Germany
{cleo.schulten,irene-angelica.chounta}@uni-due.de
[2] Beijing Normal University, Zhuhai, China
l.yuan@bnu.edu.cn
[3] Universidade Federal de Pernambuco, Recife, Brazil
kiev@cin.ufpe.br
[4] University College London, London, UK
wayne.holmes@ucl.ac.uk
[5] Eindhoven University of Technology, Eindhoven, The Netherlands
a.u.nolte@tue.nl
[6] Carnegie Mellon University, Pittsburgh, PA, USA
[7] Oslo Metropolitan University, Oslo, Norway
tore.hoel@gmail.com

Abstract. Artificial Intelligence (AI) Literacy is quickly becoming an essential 21st-century skill. Be it discerning AI art and pictures from human-made ones, challenging answers from chatbots, or addressing ethical consequences of AI – with AI being ever more present in daily life, it is crucial for everyone to be aware of its utility but also potential pitfalls and challenges. This workshop aims to facilitate a hackathon in which AIED conference participants and local stakeholders will come together, exchange ideas and experiences, and collaborate with the aim of building a toolbox to promote AI Literacy. Our goal for this toolbox is to contain a spread of tools – in the form of lesson plans, apps, activities, and others – to support educators in teaching AI Literacy and students in learning or practicing AI Literacy.

Keywords: Artificial Intelligence · AI Literacy · toolkit · hackathon

1 Content and Themes

Hackathons are collaborative events spanning from a couple of hours to multiple days or weeks [6,13]. Typically, these events have a general theme, which provides an umbrella for participants to define projects they would like to work on [11]. Teams can, for example, be formed based on common interests, capabilities, and preexisting friendships or be determined by event organizers [11]. Related to these themes, organizers also often have specific goals in mind when

A. M. Olney et al. (Eds.): AIED 2024 Workshops, CCIS 2151, pp. 467–472, 2024.
https://doi.org/10.1007/978-3-031-64312-5_59

creating an event, which can vary from ideation, design, and/or prototyping to (software) development [11,13]. The contexts in which hackathons take place are manifold; among others, hackathons are used in educational contexts in classrooms [7,14], as informal [1], as company events [2], and as open innovation tool in government [18], among other contexts.

Hackathons have been organized in or around conferences as community events, for example:

- the LAKathon[1] in LAK
- HPC in the City[2] connected to the Supercomputing Conference
- HackHPC@ADMI23[3] in connection with the ADMI conference aimed towards minorities in the computing field
- FacultyHack@Gateways23[4] at the Gateways conference which aims to bring together scientists from various disciplines with computer scientists [12]

Likewise, there are university hackathons bringing together students of different programs, such as at the University of British Columbia[5]. Indeed, learning is often cited as the motivation for participating or the reason for organizing hackathons [15].

This workshop will be organized as a one-day hackathon event with the theme of **promoting AI Literacy**. To that end, participants will work in groups to contribute to a toolbox for training AI Literacy.

AI literacy is defined by Long and Magerko as *"a set of competencies that enables individuals to critically evaluate AI technologies; communicate and collaborate effectively with AI; and use AI as a tool online, at home and in the workplace"* [10]. The key word here is *"critical"*. While the AI and machine learning community work hard to improve the technology, - for example, by preventing hallucination, avoiding bias, strengthening the guardrails - we experience increasing criticism from the public, especially parents, now warning against the adverse effects of too early introduction of AI tools in schools. With the rise of publicly available AI tools and chatbots, AI Literacy is becoming an invaluable skill [3]. Those who have not yet acquired that skill, may fall under the illusion that AI tools are infallible. Holmes et al. [9] pointed towards the urgent need for training and promoting AI Literacy for everyone to prepare to live with AI as a measure of safeguarding human values and fundamental ethical principles. At the same time, research shows that students currently use tools like ChatGPT as an "upgraded search engine" [16]. Additionally, Shoufan [16] explored the impact of students using generative AI (in this case Chat-GPT) to solve tasks and then evaluate the accuracy of the results. One takeaway was the importance of prior

---

[1] https://lakathon.org/.
[2] https://hackhpc.github.io/hpcinthecity23/.
[3] https://hackhpc.github.io/admi23/.
[4] https://hackhpc.github.io/facultyhack-gateways23/.
[5] https://ctlt.ubc.ca/2019/04/23/students-hack-away-at-educational-data-during-learning-analytics-canvas-api-hackathon/.

knowledge to assess the accuracy of ChatGPT's answers [16]. Singh et al. [17] found that of their 430 students, only 11% view "providing misinformation" as the biggest threat, whereas 6% chose biased responses and a total of 63% see either plagiarism, reduced thinking or over-dependence as the biggest threats [17]. This could stem from confidence in detecting misinformation or a lack of awareness. On the other hand, teachers often have a limited understanding of AI and how they can use it to support their practice, although they agree that AI can be a powerful resource [4].

But even for more knowledgeable users AI usage can have its pitfalls, thus AI literacy education is also supposed to grapple with "the grown ups"' problem of "falling asleep at the wheel" while using powerful AI [5]: one tends to trust AI too much, become lazy and forget to apply critical judgment.

2   Relevance and Importance to the AIED Community

The AIED 2024 Call for Papers[6] includes a newly-added topic of interest regarding the importance of AI Literacy: "Furthermore, we envision that the triad of AIED, AI Literacy, and Fair, and Ethical AI will play a fundamental role in this world in transition and be the drivers for shaping meaningful changes in pedagogical practice, educational policies, and regulations." Our proposed workshop explicitly addresses this topic of interest and aims to further align with the theme of AIED 2024 *"AI in Education for a World in Transition,"* thus contributing to shaping education and its needs for this World in Transition.

We envision that the workshop will provide an arena for members of the AIED community to familiarize themselves with aspects of AI Literacy, brainstorm how to promote AI Literacy to prepare learners to live with AI, and go even further to pinpointing fundamental concepts about AI Literacy and designing prototypes for training AI Literacy as tools for educators and learners. The workshop also aims to involve local stakeholders – local AI researchers and PhD students from local universities, EITA! Recife's open innovation initiative – and, thus, bring them in touch and engage them with the AIED community.

3   Format and Activities

The workshop will take place as an interactive, full-day, in-person event in the form of a hackathon. This means that we will dedicate most of the time to group work (see Table 1 for an overview). The organizers will provide guidance on applying proven creativity techniques that they have mapped [8] and that are appropriate for the dynamic hackathon environment to help the participants to use their time efficiently.

We will start the day with a welcoming session and an impulse talk on the topic of AI Literacy to engage and motivate participants, as well as lay some grounding on what the event will be about. Afterward, we will have a discussion

---

[6] https://aied2024.cesar.school/call-for-papers/general-call-for-papers.

about AI Literacy where we will also go into the ethical connotations of AI usage. This is followed by a short ideation and team formation session, supported by appropriate dynamics format. Teams will then have time to discuss and fine-tune their topic and goals before presenting them to the group. In this process we will also guide them in applying heuristics to quickly generate and choose idea.

The teams will spend most of the workshop day working on their projects. We will have one checkpoint after the lunch break where teams will share the teams progress and present their plans for the remainder of the day. This checkpoint will provide an opportunity for organizers and participants to provide feedback to teams. The organizers will be available to support teams during the whole duration of the workshop. The day will conclude with presentations of the produced artifacts by the groups.

We envision these artifacts to include lesson plans, interventions, app prototypes or mock-ups, information materials, or plans for hands-on activities. This list is, of course, non-exhaustive. The **output** of the workshop will be a **toolbox**, in the form of an online repository, accompanied by documentation that can be used as a basis for creating training materials to promote AI Literacy.

The workshop will be communicated and disseminated via a website that we will share with participants and the general public before the AIED conference and will be maintained after the workshop in order to promote output's sustainability.

After the hackathon, we will share the artifacts produced during the hackathon via GitHub[7]. Additionally, we will update the workshop's website with reports regarding the hackathon activity. Finally, we plan to organize a follow-up event for local participants with the support of the Innovation Office of Recife.

Table 1. Proposed agenda for the workshop.

Session	Agenda	Duration
Morning session	Welcoming, schedule and logistics	15 min
	"Impulse" talk	15 min
	Scenarios and discussion	45 min
	Ideation and team formation	30 min
	Team presentations	15 min
	Team working time	60 min
Afternoon session	Checkpoint	15 min
	Team working time	120 min
	Team final presentations	30 min
Closing session	Reflection and future plans	15 min

---

[7] https://github.com/.

4   Expected Target Audience and Expected Maximum Number of Participants

We welcome practitioners, researchers and students at all stages and look forward to an active exchange between participants of different backgrounds. The expected target audience is individuals participating in the conference and local participants, especially students. We expect a maximum of 20–30 participants. We would potentially request to schedule this workshop (provided that it is accepted) on a day that does not conflict with the AIED Doctoral Consortium.

References

1. Armstrong, J., Longmeier, M.M.: An informal learning program as a replicable model for student-led, industry-supported experiential learning. In: ASEE Annual Conference and Exposition, Conference Proceedings, vol. 2020-June (2020)
2. Backert, M., Jeberla, F.K., Kumar, S., Paulisch, F.: Software Engineering Learning Landscape: an experience report from Siemens Healthineers. 2022 IEEE/ACM 4th International Workshop on Software Engineering Education for the Next Generation (SEENG), pp. 43–50 (2022)
3. Casal-Otero, L., Catala, A., Fernández-Morante, C., Taboada, M., Cebreiro, B., Barro, S.: AI literacy in K-12: a systematic literature review. Int. J. STEM Educ. **10**(1), 29 (2023)
4. Chounta, I.A., Bardone, E., Raudsep, A., Pedaste, M.: Exploring teachers' perceptions of artificial intelligence as a tool to support their practice in estonian k-12 education. Int. J. Artif. Intell. Educ. **32**(3), 725–755 (2021)
5. Dell'Acqua, F., et al.: Navigating the jagged technological frontier: field experimental evidence of the effects of AI on knowledge worker productivity and quality (2023)
6. Falk Olesen, J., Halskov, K.: 10 years of research with and on hackathons. In: Proceedings of the 2020 ACM Designing Interactive Systems Conference, pp. 1073–1088. ACM, Eindhoven Netherlands (2020). https://doi.org/10.1145/3357236.3395543
7. Gama, K., Gonçalves, B.A., Alessio, P.: Hackathons in the formal learning process. In: Proceedings of the 23rd Annual ACM Conference on Innovation and Technology in Computer Science Education, pp. 248–253. ACM, Larnaca Cyprus (2018). https://doi.org/10.1145/3197091.3197138
8. Gama, K., Valença, G., Alessio, P., Formiga, R., Neves, A., Lacerda, N.: The developers' design thinking toolbox in hackathons: a study on the recurring design methods in software development marathons. Int. J. Hum. Comput. Interact. **39**(12), 2269–2291 (2023)
9. Holmes, W., Persson, J., Chounta, I.A., Wasson, B., Dimitrova, V.: Artificial intelligence and education: a critical view through the lens of human rights, democracy and the rule of law. Council of Europe (2022)
10. Long, D., Magerko, B.: What is AI literacy? Competencies and design considerations. In: Proceedings of the 2020 CHI Conference on Human Factors in Computing Systems, pp. 1–16. CHI 2020, Association for Computing Machinery, New York, NY, USA (2020). https://doi.org/10.1145/3313831.3376727

11. Medina Angarita, M.A., Nolte, A.: What do we know about hackathon outcomes and how to support them? – A systematic literature review. In: Nolte, A., Alvarez, C., Hishiyama, R., Chounta, I.-A., Rodríguez-Triana, M.J., Inoue, T. (eds.) CollabTech 2020. LNCS, vol. 12324, pp. 50–64. Springer, Cham (2020). https://doi.org/10.1007/978-3-030-58157-2_4

12. Nolte, A., Hayden, L.B., Herbsleb, J.D.: How to support newcomers in scientific hackathons - an action research study on expert mentoring. In: Proceedings of the ACM on Human-Computer Interaction, vol. 4, no. (CSCW1), pp. 1–23 (2020)

13. Pe-Than, E.P.P., Nolte, A., Filippova, A., Bird, C., Scallen, S., Herbsleb, J.D.: Designing corporate hackathons with a purpose: the future of software development. IEEE Softw. 36(1), 15–22 (2019)

14. Porras, J., et al.: Hackathons in software engineering education - lessons learned from a decade of events. In: Proceedings of the 2nd International Workshop on Software Engineering Education for Millennials, pp. 40–47. ACM, Gothenburg Sweden (2018). https://doi.org/10.1145/3194779.3194783

15. Schulten, C., Chounta, I.A.: How do we learn in and from hackathons? a systematic literature review. Educ. Inf. Technol. (2024)

16. Shoufan, A.: Exploring students' perceptions of CHATGPT: thematic analysis and follow-up survey. IEEE Access (2023). https://ieeexplore.ieee.org/abstract/document/10105236/

17. Singh, H., Tayarani-Najaran, M.H., Yaqoob, M.: Exploring computer science students' perception of ChatGPT in higher education: a descriptive and correlation study. Educ. Sci. 13(9), 924 (2023), https://www.mdpi.com/2227-7102/13/9/924, publisher: MDPI

18. Yuan, Q., Gasco-Hernandez, M.: Open innovation in the public sector: creating public value through civic hackathons. Public Manag. Rev. 23(4), 523–544 (2021)

Workshop on Automatic Evaluation of Learning and Assessment Content

Luca Benedetto[1]($\boxtimes$)(iD), Shiva Taslimipoor[1], Andrew Caines[1](iD),
Diana Galvan-Sosa[1](iD), George Dueñas[2], Anastassia Loukina[3](iD),
and Torsten Zesch[4](iD)

[1] ALTA Institute, Department of Computer Science and Technology, University of
Cambridge, Cambridge, UK
{luca.benedetto,shiva.taslimipoor,andrew.caines,diana.galvan-Sosa}@cl.cam.ac.uk
[2] National Pedagogical University, Bogota, Colombia
geduenasl@upn.edu.co
[3] Grammarly Inc, San Francisco, USA
anastassia.loukina@grammarly.com
[4] FernUniversität, Hagen, Germany
torsten.zesch@fernuni-hagen.de

Abstract. The evaluation of learning and assessment content has always been a crucial task in the educational domain, but traditional approaches based on human feedback are not always usable in modern educational settings. Indeed, the advent of machine learning models, in particular Large Language Models (LLMs), enabled to quickly and automatically generate large quantities of texts, making human evaluation unfeasible. Still, these texts are used in the educational domain – e.g., as questions, hints, or even to score and assess students – and thus the need for accurate and automated techniques for evaluation becomes pressing. This workshop aims to attract professionals from both academia and the industry, and to to offer an opportunity to discuss which are the common challenges in evaluating learning and assessment content in education.

Keywords: Automated Evaluation · Natural Language Processing · Language Models · Educational NLP

1 Content, Themes, and Relevance to the AIED Community

The evaluation of learning and assessment content has always been very important in the educational domain. Assessment content, such as questions and exams, is commonly evaluated both with traditional approaches based on students' responses – e.g., Item Response Theory [8] – and more modern approaches based on machine learning and Natural Language Processing [1–3]. On the other hand, the evaluation of learning content – such as single lectures, and whole courses and curricula – still relies heavily on experts from the educational domain. The same is true for several other components of the educational pipeline: for instance, distractors (i.e., the plausible incorrect options in multiple choice questions) are commonly evaluated with manual labelling [4,6,7], since automatic evaluation approaches proposed so far have some limitations [12]. The need to develop accurate metrics for evaluating learning and assessment content became even more pressing with the rapid growth and adoption of LLMs in the educational domain, both open (e.g., the family of Llama models [13] and Vicuna [14]) and closed (e.g., GPT-* models [11]). Indeed, previous research showed that LLMs can be used for a variety of educational tasks – from feedback generation and automated assessment to question and content generation [5,9,10] – and being able to accurately evaluate the output of LLMs in an automated manner becomes of crucial importance to ensure the effectiveness of their application, since traditional approaches based on human feedback are not easily scalable to large amounts of data.

Importantly, the evaluation needs to consider both the educational require- ments of the generated content and the biases that might emerge from the gen- eration models. For instance, the generated content must align to the learning objectives of the specific course (or exam) where it is being used, as well as to the language level suitable for the target students. Moreover, similarly to what happens when applying language models to other domains, the evaluation must assess the factual accuracy, and the EDIB (Equity, Diversity, Inclusion, & Belonging) appropriateness of the generated text.

This workshop focuses on approaches for automatically evaluating learning and assessment content. Topics of interests include but are not limited to:

- Question evaluation (e.g., in terms of the pedagogical criteria listed above: alignment to the learning objectives, factual accuracy, language level, cogni- tive validity, etc.).
- Estimation of question statistics (e.g., difficulty, discrimination, response time, etc.).
- Evaluation of distractors in Multiple Choice Questions.
- Evaluation of reading passages in reading comprehension questions.
- Evaluation of lectures and course material.
- Evaluation of learning paths (e.g., in terms of prerequisites and topics taught before a specific exam).
- Evaluation of educational recommendation systems (e.g., personalised curric- ula).
- Evaluation of hints and scaffolding questions, as well as their adaptation to different students.
- Evaluation of automatically generated feedback provided to students.
- Evaluation of techniques for automated scoring.
- Evaluation of bias in educational content and LLM outputs.

Human-in-the-loop approaches are welcome, provided that there is also an auto- mated component in the evaluation and there is a focus on the scalability of the proposed approach. Papers on generation are also very welcome, as long as there is an extensive focus on the evaluation step.

We expect this workshop to attract professionals from both industry and academia, in terms of both submissions and attendees, and to offer an opportunity to discuss which are the common challenges in evaluating learning and assessment content in education. From the submitted papers and through the discussions which will take place during the workshop, we aim at collecting some guidelines and best-practices for the evaluation of educational content, and we believe these will be a contribution very relevant for the AIED community, as they could be used as reference for future research on the evaluation (and generation) of learning and assessment content.

1.1 Length, Format, and Activities

We are envisioning a **full-day** workshop, with a focus on the paper presentations and talk(s) from the invited speaker(s). We are also planning to have a discussion session at the end of the workshop, to clarify the main takeaways from the workshop and some guidelines which can be helpful when evaluating learning and assessment content.
Proposed format:

- talks from invited speaker(s);
- paper presentations;
- discussion to summarise the key takeaways from the workshop.

In terms of **papers format**, the call for papers will be open to both short (5 pages, excluding references) and long papers (10 pages, excluding references). The proceedings volume will be submitted for publication to *CEUR Workshop Proceedings*.

2 Proposed Timeline

- *Notification of acceptance (of the workshop): March 13, 2024*
- First call for papers: between March 13, 2024 and March 21, 2024
- *Workshop website published on the conference website: March 21, 2024*
- Submission deadline: May 17, 2024
- Notification of acceptance: June 4, 2024
- *Workshop: 8 July 2024*

3 Expected Target Audience

With a focus on the automated evaluation of educational content, both manually and automatically generated, this workshop can be of interest to both industry and academia, as it will shed some light on possible approaches to automatically evaluate learning content. In this sense, it can appeal most of the AIED community that will be attending the conference.

4   Program Committee Members

Several distinguished researchers from both academia and industry agreed to serve as members of the program committee:

- Nicolas Ballier, Université Paris Cité (France)
- Paula Buttery, University of Cambridge (UK)
- Andrea Cappelli, CloudAcademy, Mendrisio (Switzerland)
- Sagnik Ray Choudhury, NBME and University of Michigan (US)
- Thomas François, UCLouvain (Belgium)
- Sergio Jimenez, Caro and Cuervo Institute (Colombia)
- Sylwia Macinska, Cambridge University Press & Assessment, Cambridge (UK)
- Daniel Montenegro, Aprendizaje Profundo SAS (Colombia)
- Russell Moore, University of Cambridge (UK)
- Gabriel Moreno, Javeriana University (Colombia)
- Andrew Mullooly, Cambridge University Press & Assessment, Cambridge (UK)
- Ulrike Padó, HFT Stuttgart (Germany)
- Radek Pelanek, Masaryk University (Czech Republic)
- Marek Rei, Imperial College London (UK)
- Saed Rezayi, NBME (US)
- Sue Sentance, Raspberry Pi Computing Education Research Centre, Cambridge, (UK)
- Roberto Turrin, CloudAcademy, Mendrisio (Switzerland)
- Yiyun Zhou, NBME (US)

References

1. AlKhuzaey, S., Grasso, F., Payne, T.R., Tamma, V.: Text-based question difficulty prediction: a systematic review of automatic approaches. Int. J. Artif. Intell. Educ. 1–53 (2023)
2. Benedetto, L.: A quantitative study of NLP approaches to question difficulty estimation, pp. 428–434 (2023)
3. Benedetto, L., Cremonesi, P., Caines, A., Buttery, P., Cappelli, A., Giussani, A., Turrin, R.: A survey on recent approaches to question difficulty estimation from text. ACM Comput. Surv. (CSUR) (2022)
4. Bitew, S.K., Deleu, J., Develder, C., Demeester, T.: Distractor generation for multiple-choice questions with predictive prompting and large language models. arXiv preprint arXiv:2307.16338 (2023)
5. Caines, A., et al.: On the application of large language models for language teaching and assessment technology (2023)
6. Chamberlain, D.J., Jeter, R.: Creating diagnostic assessments: automated distractor generation with integrity. J. Assess. High. Educ. 1(1), 30–49 (2020)
7. Ghanem, B., Fyshe, A.: Disto: Evaluating textual distractors for multi-choice questions using negative sampling based approach. arXiv preprint arXiv:2304.04881 (2023)

8. Hambleton, R.K., Swaminathan, H.: Item Response Theory: Principles and Applications. Springer, Cham (2013)

9. Jeon, J., Lee, S.: Large language models in education: a focus on the complementary relationship between human teachers and chatgpt. Educ. Inf. Technol. 1–20 (2023)

10. Kasneci, E., et al.: ChatGPT for good? on opportunities and challenges of large language models for education. Learn. Individ. Differ. **103**, 102274 (2023)

11. OpenAI: Gpt-4 technical report. ArXiv **abs/2303.08774** (2023)

12. Rodriguez-Torrealba, R., Garcia-Lopez, E., Garcia-Cabot, A.: End-to-end generation of multiple-choice questions using text-to-text transfer transformer models. Expert Syst. Appl. **208**, 118258 (2022)

13. Touvron, H., et al.: Llama 2: open foundation and fine-tuned chat models. arXiv preprint arXiv:2307.09288 (2023)

14. Zheng, L., et al.: Judging LLM-as-a-judge with MT-bench and chatbot arena. arXiv preprint arXiv:2306.05685 (2023)

LAFe: Learning Analytics Solutions to Support On-Time Feedback

Rafael Ferreira Mello[1,2](✉), Gabriel Alves[2](✉), Elaine Harada[1,3](✉),
Mar Pérez-Sanagustín[4](✉), Isabel Hilliger[5](✉), Esteban Villalobos[4],
Esther Félix[4], and Julien Broisin[4]

[1] Centro de Estudos Avançados de Recife, Recife, Brazil
`rflm@cesar.org.br`
[2] Universidade Federal Rural de Pernambuco, Recife, Brazil
`gabriel.alves@ufrpe.br`
[3] Universidade Federal do Amazonas, Manaus, Brazil
`elaine@icomp.ufam.edu.br`
[4] Université Toulouse III - Institut de Recherche en Informatique de Toulouse,
Toulouse, France
`mar.perez-sanagustin@irit.fr`
[5] Pontificia Universidad Católica de Chile, Santiago, Chile
`ihillige@uc.cl`

Abstract. Feedback given to students by instructors is essential to guide students and help them improve from their mistakes. However, in higher education, instructors feel unable to give quality and timely feedback due to work overload and lack of time. In this context, this tutorial intends to discuss possible data-based and AI solutions for supporting students and instructors in the feedback process. It will include: a panel discussion about the importance of automating the feedback process, a demo of tools for this goal, and a card sorting activity to understand important aspects of developing tools to support on-time feedback.

Keywords: Provision of feedback using artificial intelligence ·
Feedback tools · qualitative research

1 Introduction

Feedback is crucial for students and teachers in terms of clarifying expectations, monitoring the current progress of learners, and moving towards desired learning goals [7,8]. However, substantial evidence shows that higher education struggles to deliver consistent, timely, and constructive feedback to meet the needs and expectations of students [1,9,14,19]. Moreover, recent case studies conducted in Latin America identified that providing timely feedback is crucial for continuously improving the higher education curriculum [9]. The inadequacy in delivering effective feedback to students is partly due to conflicts between an increasing focus on 'massiveness', 'inclusiveness', and 'personalization' in higher education and the unmatched capacity of teaching staff to produce feedback that speaks to the needs of individual students [9].

A. M. Olney et al. (Eds.): AIED 2024 Workshops, CCIS 2151, pp. 478–485, 2024.
https://doi.org/10.1007/978-3-031-64312-5_61

Furthermore, learning has become more digital, especially after the COVID-19 pandemic, so higher education institutions have accumulated unprecedented amounts of data produced from learners' interaction with online learning environments and Learning Management Systems (LMS). Also, thinking about possible future pandemics and lockdowns is essential from now on. In this context, Artificial Intelligence in Education (AIED) has become a promising means to improve feedback delivery. So far, one of the most used techniques to support decision-making is to provide visual dashboards that transform trace data into "actionable insights". For students, prior work proposes visual dashboards to support students' meta-reflection and self-monitoring to produce behavioral changes in terms of how they self-regulate their learning both online [10, 16] and in blended learning contexts [16]. For teachers, recent studies propose tools for assisting teaching staff in providing contextualized feedback [9, 15].

This tutorial explores the application of data-based and Artificial Intelligence in Education (AIED) solutions, specifically focusing on their role in delivering timely feedback. Research, including studies by [15, 16], has shown that such feedback can significantly enhance student satisfaction and academic performance. Within this framework, we propose a practical, hands-on tutorial designed to explore various AIED tools that have been effectively implemented in our institutions. This session aims to provide participants with a deeper understanding and first-hand experience of how AIED can foster a more engaging and effective learning environment.

2 Relevance and Methodology

Several tools have been developed to assist instructors in the feedback process [12, 13, 15]. However, the majority focused on sending automatic feedback messages. Recently, research started to shift to the goal of assisting instructors in constructing quality feedback instead of sending automatic messages [1]. Yet, the accountability and quality assurance of the feedback process is still an open issue to be addressed due to the expedition and the limited capacity of staff to produce feedback that speaks to the needs of individual students [9].

The primary contribution of the tutorial lies in disseminating and popularizing innovative methodologies and tools that utilize trace-data and AI models to enhance the feedback process in educational settings. This initiative aims to bridge the gap between advanced AI technologies and their practical application in education, making these tools more accessible and understandable to educators and other stakeholders. By focusing on data and AI-enhanced feedback mechanisms, the tutorial seeks to foster a deeper understanding of how these technologies can be effectively integrated into the educational process, thereby changing how feedback is provided and received in learning environments. To summarize, this tutorial aims to i) highlight the significance of feedback in education and demonstrate how data and AI-based tools can assist instructors throughout this process; ii) outline a practical methodology to help stakeholders grasp the context and implementation of feedback tools; iii) introduce Tutoria, a feedback tool that leverages Natural Language Processing (NLP) and Large Language Models (LLMs) to streamline and enhance the feedback process, and Note-MyProgress, a plug-in for Moodle that tracks students' activity on the course and provides teachers with a dashboard for monitoring their actions and sending them personalized feedback.

3   Methodology to Understand Stakeholders

One of the main challenges in designing feedback tools, especially those that rely on the use of large amounts of data and AI models, is to provide interpretable and actionable indicators [11] for students and teachers making good decisions and providing good feedback. When defining these indicators, and according to recent literature, we should define both the information to provide (the explanation or what we want to show) and the format (the way in which we want to provide it [3]. For this purpose, we must adopt human-centered design approaches that are capable of capturing and understanding stakeholders' perspectives.

Under this approach, we designed a card sorting activity [18]. This type of activities have been extensively used in Human-Computer Interaction (HCI) as a user-centered design method. Card sorting is an eliciting technique that explores how different items could be classified. It aims to identify consistent patterns in how people think about particular items, which is especially useful at the beginning of a design project. Aiming to inform the development of AI solutions, we designed an activity to support researchers and designers when defining indicators and suitable visual representations of feedback from the perspective of its potential users [6,20]. Based on Spencer's guidelines [18], our proposed activity is structured into the following steps:

1. Deciding what you want to learn;
2. Selecting the method;
3. Choosing the cards' content;
4. Recruiting participants;
5. Running the card sorting and recording the data;
6. Analyzing outcomes.

4   Tutoria Platform

Tutoria [4] is a platform designed to assist educators in reviewing and providing written feedback on assignments, which can be imported from Google Classroom or Moodle. This platform allows instructors to easily view questions for each assignment within the Learning Management System and the corresponding student responses. Once the assignment is imported, the instructor can choose to correct it by question or student, evaluating an entire assignment for each student or all students' responses to a question, respectively [5].

In grading, instructors can highlight specific sections of a student's response and assign tags, indicating an error, a partially correct, or a correct statement.

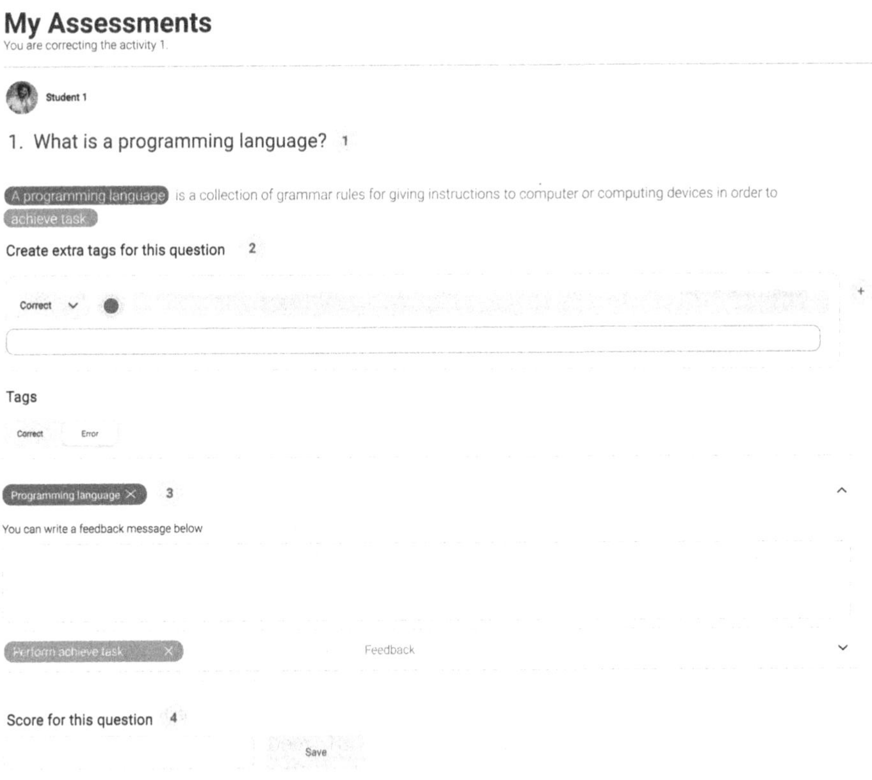

My Assessments
You are correcting the activity 1

👤 Student 1

1. What is a programming language? 1

A programming language is a collection of grammar rules for giving instructions to computer or computing devices in order to achieve task

Create extra tags for this question   2

Correct ∨   ●   +

Tags

Correct   Error

Programming language ✕   3   ∧
You can write a feedback message below

Perform achieve task   ✕   Feedback   ∨

Score for this question   4

Save

Fig. 1. Assessment of an open-ended question.

They can apply previously created tags during the correction or generate new ones. Whenever creating a new tag, the instructor must provide a description related to that highlighted excerpt of the students' response [5]. This description is a requirement as it contributes to the feedback provided to students. The instructor can also use a Generative AI feature of Tutoria to create a description. A description is thus automatically created based on the text excerpt, the type of the tag (error, partially correct, or correct), and the tone of the previous descriptions given by the instructor. It is important to emphasize that, regarding the generated description, the instructor can accept it as is, modify it, or even completely rewrite it.

Instructors are incited to use tags for correct (or partially correct) statements made by students, promoting positive feedback to recognize students' achievements, a pedagogical approach that can boost student motivation [5]. Feedback can also be given when a student leaves out a critical point or offers general comments about the response without thus being related to an excerpt of the text. In such a scenario, the instructor can provide a description for broader remarks related to the whole student's response. Figure 1 presents an overview of the grading process.

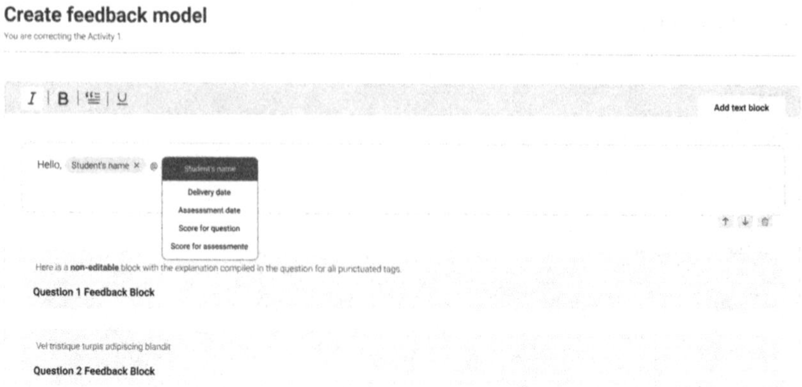

Create feedback model
You are correcting the Activity 1

I | B | ≡ | U

Add text block

Hello, Student's name × @ Student's name

Delivery date
Assessment date
Score for question
Score for assessments

↑ ↓ ⊟

Here is a **non-editable** block with the explanation compiled in the question for all punctuated tags.

Question 1 Feedback Block

Vel tristique turpis adipiscing blandit

Question 2 Feedback Block

Fig. 2. Construction of template for feedback message.

Once an instructor completes grading an assignment, the system will show a screen where all tags created for that specific assignment are presented, allowing the instructors to update tag explanations or add descriptions where needed. When grading a response from another student, Tutoria uses AI models to seek an excerpt similar to another one that is associated with a tag. The related tag is thus recommended for this new excerpt, reusing its description.

The instructor can also configure a default feedback template. Figure 2 presents the final feedback screen. This template features predefined text blocks that group tag explanations for each question, which will be automatically customized for each student based on the tags applied to their responses. On this screen, instructors can add additional text blocks, like greetings, closing remarks, and transitions, to connect feedback sections or general comments about the activity. This process is performed once, as the template will be applied to all students' feedback.

Once the template is configured, the instructor can preview the final feedback messages generated from the template and the tags used for each student. They can adjust individual messages as needed. Instructors can send the feedback to all students simultaneously or to selected individuals. This flexibility allows instructors to tailor the timing and content of feedback, providing autonomy in managing student communications.

5   NoteMyProgress Moodle Plug-in

NoteMyProgress (NMP) [17] is a Moodle plugin designed to monitor students' activities within a course and support teachers in providing timely feedback. The plugin aims to support learners' self-regulation and assist teachers in monitoring students' actions, enabling intervention with personalized feedback. The tool features various interactive visualizations for monitoring students' time management, strategic planning, and self-evaluation.

On one hand, students can interact with these visualizations to monitor their own activity and progress within the course. On the other hand, teachers can access aggregated data from all students and compare their activities. Additionally, the plugin enables personalized feedback through two mechanisms: (1) group messages via interactive graphics (Fig. 3), and (2) personal messages through dashboards displaying individual student activities. Using the first mechanism, teachers can identify which students have completed a specific activity or viewed certain content and directly email them by clicking on the graph (Fig. 4). Thus, the graphs not only filter the students engaged in specific actions but also facilitate group email communication under similar circumstances. Using the second mechanism, teachers can send personalized feedback to individual students through the individual visualization section. Since all visualizations are updated in real-time in response to student actions, NMP effectively facilitates timely feedback, serving as a communication channel between students and teachers.

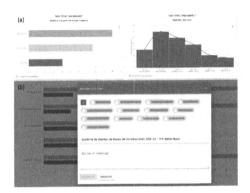

Fig. 3. Teacher Dashboard for group feedback. The teacher has dashboards for displaying group activities (a) and has the option to send messages to a group of students with similar behavior (b).

Fig. 4. Teacher Dashboard for individual feedback. The teacher can access a visualization section for each student to send individual feedback messages.

6 Final Remarks

This tutorial will demonstrate that data and AI can play a significant role in enhancing the educational feedback process. Through a hands-on session and practical demonstrations, participants will gain insights into the use of various AI-driven tools to support instructors in providing timely and meaningful feedback to students [1,2], along with examining a method to inform the design of this type of tools from a user-centered perspective. Thus, this tutorial offers educators a more straightforward path toward improving their feedback practices by focusing on AI applications such as natural language processing and large language models.

These approaches to feedback aim to contribute to the growing evidence that AI can help address some of the challenges in education, including workload management, personalized feedback, and timely student engagement, especially in large student groups. Moreover, the tutorial expects to encourage instructors to explore AI solutions focusing on both efficiency and pedagogy, ultimately creating a more supportive learning environment.

Overall, this tutorial will be a valuable resource for educators, administrators, and other stakeholders interested in leveraging AI to improve educational outcomes. Bridging the gap between technology and education fosters a community of practice where AI-driven feedback tools become an integral part of modern teaching and learning. The discussion, the tools, and the user-centered method that will be presented in this tutorial aim to inspire participants to implement AI solutions within their educational contexts, paving the way for a more effective and personalized feedback experience.

7 Funding

The authors of the workshop have an active project funded under the grant STIC-AMSUD-20222157702P that will fund the attendance of all proponents at the workshop.

References

1. Cavalcanti, A.P., et al.: Automatic feedback in online learning environments: a systematic literature review. Comput. Educ. Artif. Intell. **2**, 100027 (2021)
2. Cavalcanti, A.P., et al.: How good is my feedback? a content analysis of written feedback. In: Proceedings of the Tenth International Conference on Learning Analytics & Knowledge, pp. 428–437 (2020)
3. Donoso-Guzmán, I., Ooge, J., Parra, D., Verbert, K.: Towards a comprehensive human-centred evaluation framework for explainable AI. In: Longo, L. (ed.) xAI 2023. CCIS, vol. 1903, pp. 183–204. Springer, Cham (2023). https://doi.org/10.1007/978-3-031-44070-0_10
4. Falcão, T.P., et al.: Tutoria: a software platform to improve feedback in education. J. Interact. Syst. **14**(1), 383–393 (2023)

5. Falcao, T.P., et al.: Tutoria: supporting good practices for providing written educational feedback. In: Anais do XXXIII Simpósio Brasileiro de Informática na Educação, pp. 668–679. SBC (2022)
6. Félix, E., Oliveira, E.H.T., Ramos, I.M.M., Pérez-Sanagustín, M., Villalobos, E., Hilliger, I., Ferreira Mello, R., Broisin, J.: Designing actionable and interpretable analytics indicators towards explainable AI-based feedback. In: European Conference on Technology Enhanced Learning. Springer (2024). Submitted
7. Garcia, S., Marques, E., Mello, R.F., Gašević, D., Falcão, T.P.: Aligning expectations about the adoption of learning analytics in a Brazilian higher education institution. In: Proceedings of the Conference of Artificial Intelligence in Education, pp. 1–6 (2021)
8. Hattie, J., Timperley, H.: The power of feedback-review of educational research. American Education Research Association and SAGE, p. 86 (2011)
9. Hilliger, I., Celis, S., Perez-Sanagustin, M.: Engaged versus disengaged teaching staff: a case study of continuous curriculum improvement in higher education. High Educ. Pol. **35**(1), 81–101 (2022)
10. Jivet, I., Scheffel, M., Specht, M., Drachsler, H.: License to evaluate: preparing learning analytics dashboards for educational practice. In: Proceedings of the 8th International Conference on Learning Analytics and Knowledge, pp. 31–40 (2018)
11. Jørnø, R.L., Gynther, K.: What constitutes an 'actionable insight' in learning analytics? J. Learn. Anal. **5**(3), 198–221 (2018)
12. Krusche, S., Seitz, A.: Artemis: an automatic assessment management system for interactive learning. In: Proceedings of the 49th ACM Technical Symposium on Computer Science Education, pp. 284–289 (2018)
13. Marin, V.J., Pereira, T., Sridharan, S., Rivero, C.R.: Automated personalized feedback in introductory java programming MOOCs. In: 2017 IEEE 33rd International Conference on Data Engineering (ICDE), pp. 1259–1270. IEEE (2017)
14. Pardo, A.: A feedback model for data-rich learning experiences. Assess. Eval. High. Educ. **43**(3), 428–438 (2018)
15. Pardo, A., Jovanovic, J., Dawson, S., Gašević, D., Mirriahi, N.: Using learning analytics to scale the provision of personalised feedback. Br. J. Edu. Technol. **50**(1), 128–138 (2019)
16. Pérez-Álvarez, R.A., Maldonado-Mahauad, J., Sharma, K., Sapunar-Opazo, D., Pérez-Sanagustín, M.: Characterizing learners' engagement in MOOCs: an observational case study using the NoteMyProgress tool for supporting self-regulation. IEEE Trans. Learn. Technol. **13**(4), 676–688 (2020)
17. Pérez-Sanagustín, M., Pérez-Álvarez, R., Maldonado-Mahauad, J., Villalobos, E., Sanza, C.: Designing a moodle plugin for promoting learners' self-regulated learning in blended learning. In: Hilliger, I., Muñoz-Merino, P.J., De Laet, T., Ortega-Arranz, A., Farrell, T. (eds.) EC-TEL 2022. LNCS, vol. 13450, pp. 324–339. Springer, Cham (2022). https://doi.org/10.1007/978-3-031-16290-9_24
18. Spencer, D.: Card sorting: Designing usable categories. Rosenfeld Media (2009)
19. Tsai, Y.S., Mello, R.F., Jovanović, J., Gašević, D.: Student appreciation of data-driven feedback: A pilot study on OnTask. In: LAK21: 11th International Learning Analytics and Knowledge Conference, pp. 511–517 (2021)
20. Villalobos, E., Hilliger, I., Pérez-Sanagustín, M., González, C., Celis, S., Broisin, J.: Analyzing learners' perception of indicators in student-facing analytics: a card sorting approach. In: Viberg, O., Jivet, I., Muñoz-Merino, P., Perifanou, M., Papathoma, T. (eds.) EC-TEL 2023. LNCS, vol. 14200, pp. 430–445. Springer, Cham (2023). https://doi.org/10.1007/978-3-031-42682-7_29

Author Index

© The Editor(s) (if applicable) and The Author(s), under exclusive license
to Springer Nature Switzerland AG 2024
A. M. Olney et al. (Eds.): AIED 2024 Workshops, CCIS 2151, pp. 487–491, 2024.
https://doi.org/10.1007/978-3-031-64312-5

Printed in the United States
by Baker & Taylor Publisher Services